Electroless Nickel Plating

Foreword

I am delighted to write a foreword for this book entitled *Electroless Nickel Plating: Fundamentals to Applications*, edited by Prof. Véronique Vitry, Prof. Fabienne Delaunois, and Dr. Luiza Bonin, with contributions from many eminent researchers in this field. Electroless plating has emerged significantly since its accidental discovery in 1946 by Brenner and Riddell. During the past three decades, there has been intense research activity in this field. Bringing out a book on electroless plating now is timely and essential, as the earlier books on this topic date back to 1991. This book consists of nine chapters, with a good blend of fundamental aspects of electroless plating and its recent developments. The coverage of topics is comprehensive and it ranges from the chemistry of electroless plating, the green chemistry approaches in developing suitable formulations and the characteristics of electroless Ni-P and Ni-B coatings with special emphasis on corrosion resistance and tribological behavior. All the chapters are integrated well. The inclusion chapters on modified electroless nickel coatings, which consists of electroless Ni-P and Ni-B composite coatings, poly-alloy, multilayer, and graded coatings, is one of the highlights of this book. Presentation of a chapter on the emerging applications of electroless plating is unique and interesting. The contributors of this book have done a wonderful job in bringing out this book. I congratulate Prof. Véronique Vitry and other eminent contributors for their valiant effort in compiling a useful reference material. I foresee that this book shall be a valuable addition in university libraries worldwide and as a reference book for researchers in surface engineering.

T. S. N. Sankara Narayanan
Department of Mechanical Engineering
Ulsan National Institute of Science and Technology (UNIST)
Ulsan, Republic of Korea

Acknowledgments

This book was born as one of those crazy ideas that are usually put to the side, but this one did not accept to be forgotten and when we finally decided to go for it, none of us expected to enter in such an adventure and to bring it to fruition only a handful of months after Dr. Bonin's graduation.

There are many people who have made this possible and who we'd like to thank.

- The metallurgy lab at UMONS that has dealt with three very busy ladies during last 2 years.
- UMONS and the FNRS for the travel grant that allowed Prof. Vitry time to finalize this book.
- The University of Coventry, and mostly its Functional Materials Research Group, that has not only hosted Prof. Vitry but also provided help and moral support through the last rush. Shakiela, Andrew, and the others, thanks for all.
- The University of Gent, where Dr. Bonin continues her career, for allowing her to continue writing this book.
- All our friends who have helped us to alleviate stress during the last months of the writing process.
- Last, but not least, our families who have had to live with this book for the last year, and mostly Leonardo, who is not only Dr. Bonin's husband but also her chief copy-editor, as well as Jonas and Sonia, who have accepted to see their mother go away for weeks without complaint.

Editors

Fabienne Delaunois received her engineering degree in metallurgy and her PhD from the Engineering Faculty of Mons (FPMs) in 1993 and 2002, respectively. She's been working on electroless nickel-boron for more than 20 years and has authored approximately 20 publications in that field.

She's very active as an expert for industries in various fields such as failure analysis and corrosion. She is currently a Professor and the Department Head in the Metallurgy Laboratory at the Engineering Faculty of the University of Mons, Belgium, where she teaches principally Physical Metallurgy and Recycling.

She chaired the Chemistry and Materials Science department of the Engineering Faculty of UMONS from 2010 to 2017 and is a member of the board of two research centers in the field of materials: Materia Nova and BCRC-INISMa.

Véronique Vitry received her engineering degree in materials science from the Engineering Faculty of Mons in 2003 and first worked on optical PVD coatings for a research center. She then pursued her PhD in engineering from the University of Mons and graduated in 2010.

Her main research interest is electroless nickel plating, which she has been studying for more than 10 years, leading to more than 30 publications in international journals. She's also involved in the research on powder metallurgy, corrosion, recycling, and physical metallurgy.

She's currently a part-time associate professor and senior research and teaching associate at the Engineering Faculty of the University of Mons (UMONS), where she teaches process metallurgy and process modeling, thin films, nanotechnologies and spectroscopic methods in chemical and materials analysis.

She's the Vice president and Regional Chair for the North of France and Belgium of A3TS (the French Heat and Surface Treatment Association). She also chairs the International Relations Committee of the Failure Analysis Society (FAS), an affiliate of ASM.

She's currently an advisor to the Rector of UMONS on Internationalization at Home.

Luiza Bonin started graduate studies in materials engineering at the Federal University of Santa Catarina, in 2009. Since then, Luiza has been quite attracted to the principles of sustainable development and had the opportunity to work with new and greener synthesis processes. In 2013, she was granted with an academic mobility program and had the opportunity to perform a Master thesis at CIRIMAT (Toulouse, France) related to greener chemistry routes for the production of zirconia coatings. Immediately after her graduation in 2014, she moved to Belgium to start her PhD at the Metallurgy lab of the University of Mons, where she worked on the replacement of lead stabilizer in electroless nickel-boron deposits and the electrochemical/mechanical characterization of coatings that originated from greener baths. She has been working with new compositions of electroless nickel-boron since 2014, and she has authored 11 papers in the field of electroless nickel. Luiza received her PhD degree in 2018 and is currently working at the Ghent University.

Contributors

Shakiela Begum
Coventry University
Coventry, United Kingdom

Luiza Bonin
Ghent University
Ghent, Belgium

Juan G. Castaño
CIDEMAT, Universidad de Antioquia
Medellin, Colombia

Andrew J. Cobley
Coventry University
Coventry, United Kingdom

Esteban Correa
Universidad de Medellin
Medellin, Colombia

Suman Kalyan Das
Jadavpur University
Kolkata, India

Fabienne Delaunois
University of Mons
Mons, Belgium

Félix Echeverría
Universidad de Antioquia
Medellin, Colombia

Eva García-Lecina
Cidetec
San Sebastian, Spain

Sudagar Jothi
VIT-AP University
Amaravati, India

R. Muraliraja
Vels University
San Sebastian, Spain

Asier Salicio-Paz
Cidetec
San Sebastian, Spain

Prasanta Sahoo
Jadavpur University
Kolkata, India

A. Selvakumar
Bannari Amman Institute of Technology
Sathyamangalam, India

T. R. Tamilarasan
B.S. Abdur Rahman Crescent Institute of Science and Technology
Chennai, India

Sanjith Udayakumar
University Sains Malaysia
Penang, Malaysia

Mustafa Urgen
Department of Metallurgical and Materials Engineering
Istanbul Technical University
Istanbul, Turkey

Véronique Vitry
University of Mons
Mons, Belgium

Alejandro Alberto Zuleta Gil
Universidad Pontificia Bolivariana
Medellin, Colombia

List of Acronyms

AA	Aluminium Alloy
AFM	Atomic Force Microscope
AISI	American Iron and Steel Institute
AMS	Aerospace Materials Standard
ASTM	American Society for Testing and Materials
BF	Bamboo Fabric
CEM	Cation Exchange Membrane
CFX	Fluorinated Carbon
CMC	Critical Micelle Concentration
CMMC	Compositionally Modulated Multilayer Coatings
CNT	Carbon Nanotubes
COF	Coefficient of Friction
CSP	Crab Shell Particles
CTAB	Cetyl Trimethyl Ammonium Bromide
DMAB	Dimethyl Amine Borane
DSC	Differential Scanning Calorimetry
DTAB	Trimethyl Ammonium Bromide
E	Elastic Modulus
ED	Elecfrodeposition
EDX	Energy Dispersive X-ray (Spectroscopy)
EIS	Electrochemical Impedance Spectroscopy
ELV	End of Life Vehicle
EMF	Electromotive Force Series
EMI	Electro-Magnetic Interference
EN	Electroless Nickel
EN	European Normalization
EP	Electroless Plating
ESP	Electric Submersible Pump
FCC	Face Centered Cubic
FGM	Functionally Graded Material
GDOES	Glow Discharge Optical Emission Spectroscopy
GFRP	Glass Fiber Reinforced Plastic
H	Hardness
HEED	Hybrid Electro-Electroless Deposited Coatings
HK	Knoop Hardness Unit (Macro Hardness)
hk	Knoop Hardness Unit (Micro Hardness)
HRC	Rockwell Hardness Unit
HRTEM	High Resolution Transmission Electron Microscope
HV	Vickers Hardness Unit (Macro Hardness)
hv	Vickers Hardness Unit (Micro Hardness)
LM	UK/India Standard for Aluminium Alloys
MEMS	Micro Electro Mechanical System

MIL	USA Military Standard
MMC	Metal Matrix Composite
MTO	Metal Turnover
NACE	National Association of Corrosion Engineers
NCZ	Electroless Nickel Coated ZrO_2
NDP	Diamond Nanoparticles
NiB	Nickel-Boron
NiP	Nickel-Phosphorus
PC	Polycarbonate
PM	Powder Metallurgy
PP	Polypropylene
PS	Polystyrene
PTFE	Polytetrafluoroethylene
PVP	Polyvinylpyrrolidone
R&D	Research and Development
RA	Activated Rosin
$\mathbf{R_a}$	Average Roughness
RH	Reductant (Chapter 1)
RH	Relative Humidity (Chapter 4)
RMA	Rosin Mildly Activated
RoHS	Restriction of Hazardous Substances (UK Directive)
$\mathbf{R_p}$	Peak Roughness
$\mathbf{R_q}$	Quadratic Roughness
$\mathbf{R_t}$	Maximal Peak/Valley Distance
$\mathbf{R_v}$	Depth of Valleys
$\mathbf{R_z}$	Total Roughness
SAE	Society of Automotive Engineers
SCE	Standard Calomel Electrode
SDS	Sodium Dodecyl Sulfate
SEM	Scanning Electron Microscope
SHE	Standard Hydrogen Electrode
SLS	Sodium Lauryl Sulfate
SMC	Step-Wise Multilayer Coatings
STM	Scanning Tunneling Microscope
TEM	Transmission Electron Microscope
TWI	Taber Wear Index
UI	Ultrasonic Impulsion
WEEE	Waste Electrical and Electronic Equipment
XRD	X-ray Diffraction
YSZ	Yttria Stabilized Zirconia

Introduction

Electroless plating is a process that is at the same time very simple in its principles and very complex in its execution. It involves reduction of metallic salts (nickel for the coatings described in this book) to the metallic state by a chemical agent in an aqueous solution. The process thus does not use any external current sources, which brings it some really interesting advantages over electroplating such as the possibility to plate nonconductive materials and the absence of edge effects that allows to plate very complex-shaped parts with perfect throwing power and dimensional compliance as long as the solution can reach the surfaces to be plated. Other differences with electroplating include the absence of a metallic anode—metallic ions that are reduced are provided directly in the aqueous solution form—and the catalytic nature of the process, which is not only one of its most distinctive features but also one of its most complex ones. However, similar to electroplating, the reaction leading to the formation of the metallic deposit is electrochemical and includes charge transfer in the solution.

The versatility of electroless nickel plating, illustrated by a variety of coatings obtained with this method, from the electroless nickel-phosphorus to the more exotic nickel-boron, multi-alloyed coatings, multilayers and composites, and the specific characteristics of the process make electroless nickel coatings increasingly popular in large areas of industry. They are among the best candidates to replace hard chrome plating in most of its applications.

There are not many books on electroless nickel plating, and the field has not been the object of much scientific interest before the late 1990s, as shown in Figure I.1. Out of the 25,700 papers on the subject retrieved by Google Scholar, 75 are from the 1950s, 500 from the 1960s, 700 from the 1970s, 1000 from the 1980s, 3000 from the 1990s, more than 8000 from the 2000s, and 13,000 from the last 10 years. The amount of knowledge that has been accumulated since the 2000s is astonishing. This period coincides, according to B. Zhang, with electroless nickel's entrance in its fourth stage of development: in-depth development including deep investigation into the mechanism and theory of electroless plating and the nanoelectroless plating stage.[1] Furthermore, as said by Weill as early as the 1980s, "the understanding that has been gained is to a great extent responsible for changing plating from an art to a science."[2] The increase in knowledge about the field more than justified the publication of a new book that would include not only the theory and concepts of electroless nickel plating but also the new developments of the field and the deeper knowledge gained in recent years.

This book is our attempt to summarize the thriving and complex field of electroless nickel and to present in a concise form the present state of knowledge of both theoretical and applied aspects of electroless nickel. The theoretical part of this book includes not only a discussion of electrochemical thermodynamics of the process but also information about kinetics, nucleation, growth, and morphology of the coatings that are formed by electroless plating of nickel. In the more practical part,

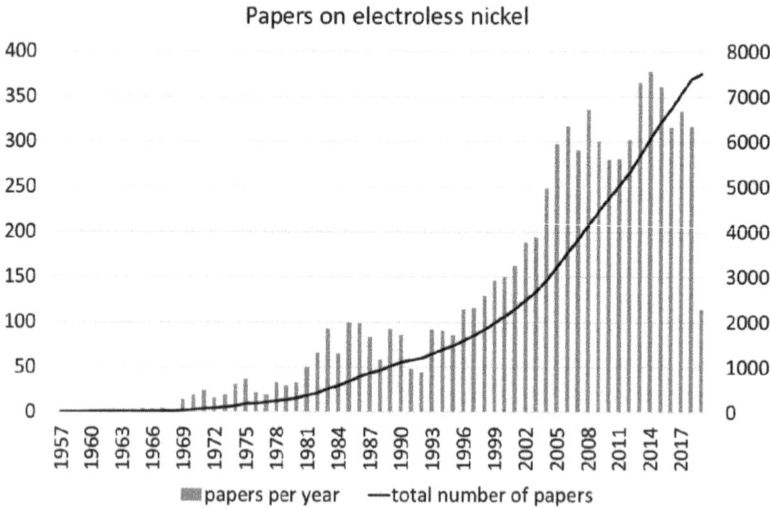

FIGURE I.1 Publications on electroless nickel as listed in Scopus on April 16, 2019.

information about the properties and applications of electroless nickel-based coatings are included, and information about recent developments in electroless plating such as the use of composites, ultrasound, multi-alloyed deposits, and multilayers, as well as when possible, meta-analysis of available data and quantitative discussion, are also included.

The first chapter addresses the thermodynamical conditions of electroless plating as well as the kinetics aspects of the process. It also discusses the catalytic nature of electroless plating and the role of various components in the electroless plating baths.

The second chapter is dedicated to the growth of electroless nickel coatings. In this chapter, some practical aspects such as substrate cleaning and activation are discussed and catalytic activity of substrate and growth modes of coatings are also discussed in depth.

Chapter 3 examines all aspects of electroless nickel-phosphorus, which is the most popular avatar of the technology and is currently used in a large array of industrial applications. These coatings are described from plating bath chemistry and operation to properties, effect of heat treatment, and applications. Recent trends in the electroless nickel-phosphorus field, including the search for more environmentally friendly plating solutions, are also discussed.

The fourth chapter, divided into two parts, is dedicated to electroless nickel-boron. The first part focuses on plating bath chemistry and properties of the coatings while the second part contains information about posttreatments and the way they affect electroless nickel-boron coatings.

The next four chapters are dedicated to modifications of the classic electroless plating process by various means. First, ultrasound assistance is described in Chapter 5. The use of ultrasound during plating is an increasingly popular method for improving coating properties and increasing plating rate. It is also very useful for the synthesis of composite coatings. Next, Chapter 6 describes polyalloyed electroless nickel, which are coatings in which at least one other element is added to the electroless nickel-phosphorus or nickel-boron coating by co-reduction of metallic salts. These polyalloyed coatings present several very interesting properties and open various new areas of application to electroless nickel coatings. The next modification, discussed in Chapter 7, is multilayer electroless nickel coatings. Multilayer coatings can be formed with a single material or by mixing materials, and they bring increased corrosion and wear resistance to electroless nickel coatings. Finally, composites electroless nickel coatings are discussed in Chapter 8. The aim of using composite coatings is usually to decrease wear of the material by carrying out one of the following methods: addition of hard particles or addition of solid lubricant particles.

The last chapter of this book is dedicated to more practical aspects of electroless nickel coatings. The first part focuses on applications and both standard applications, such as increase in wear and/or corrosion resistance of the substrate materials, and more specific applications—catalytic materials, magnetic shielding, and so on—are discussed. The last section of the book presents advice for people who wish to begin using or making electroless nickel coatings.

This book is intended as a companion for professional electrochemists, advanced undergraduate and postgraduate students, and also for electroless plating specialists with a physical, technical, or chemical education, engineers, and specialists engaged in research on new coating technologies related to metallic layers and amorphous alloys.

The methods of metal and alloy deposition without the use of external current sources are very important in modern technology, especially in the production of new materials for applications in electronics, wear and corrosion resistant materials, medical devices, battery technologies, and so on. We hope this book will be useful for the readers and will help them develop a better understanding of those specific processes.

Prof. Véronique Vitry
Coventry, April 16, 2019

REFERENCES

1. Zhang, B. Amorphous and Nano Alloys Electroless Depositions: Technology, Composition, Structure and Theory, n.d.
2. Weill, R. 1982. "Plat." *Surf. Finish.* 69 (46): 18.

1 Chemistry of Electroless Plating

Mustafa Urgen

CONTENTS

1.1 INTRODUCTION

Electroless metal deposition, also known as chemical deposition, is conducted in an electrolyte similar to metal plating, in which a chemical reducing agent is added. Thus, the metal ions are reduced not by polarizing the cathodic reaction by an external source but by the oxidation of the chemical reductant on the metal surface. In electroless deposition, a homogeneous coating thickness is achieved, independent of the shape of the metal parts to be deposited.

In this chapter, the basic principles of electroless plating will be outlined for providing the readers a scientific and technological perspective on the interdisciplinary and complex nature of electroless plating. This chapter starts with the comparison of electroless plating with other methods of deposition that use metal ions dissolved in liquids as a source of depositing metals. As the reactions involved in the process are chemical and electrochemical, thermodynamic and kinetic principles relevant to electroless plating are also summarized by giving emphasize on the hydrogen evolution reaction and its relation to the catalytic selective surfaces. Section 1.7 is devoted to mixed potential theory, which is extensively used to explain the kinetics of electroless deposition. In Section 1.8, the role of deposition parameters and bath constituents in the electroless nickel deposition processes is outlined.

1.2 CHEMISTRY OF ELECTROLESS PLATING

The history of electroless plating dates back to 1844; there are very comprehensive book chapters and reviews that cover the history of this plating method (Riedel 1991; Popov et al. 2002; Mallory and Hajdu 1991). Here, only the major milestones and a brief review on the progress of electroless plating will be given. Wurtz (1844) observed that nickel ions can be reduced by hypophosphite ions. The first bright electrodeposits were obtained in 1911 by Breteau (Breteau 1911). In 1916, Roux received the first patent on electroless Ni plating. However, these baths either decomposed spontaneously or very slow. In 1946, Brenner and Riddell (1946) described the proper conditions for obtaining electroless deposition. The main contribution of Riddell and Brenner was to understand and define the kinetic conditions to keep these thermodynamically unstable baths from spontaneous decomposition. Today, many metals and alloys can be electroless deposited relying on the principles put forth by Riddell and Brenner. In the periodic table given below (Table 1.1), the metals and metalloids that can be deposited by electroless method are outlined. Though the table indicates that many elements have the ability to be deposited on different substrates, major coatings produced by electroless method are those of Ni–P, Ni–B, Cu, and Ag, among which Cu- and Ni-based ones have the widest technological application area.

TABLE 1.1
Metals and Metalloids That Can Be Electroless-deposited

1 H 1																	2 He 4
3 Li 7	4 Be 9											**5 B 11**	6 C 12	7 N 14	8 O 16	9 F 19	10 Ne 20
11 Na 23	12 Mg 24											13 Al 27	14 Si 28	**15 P 31**	16 S 32	17 Cl 35	18 Ar 40
19 K 39	20 Ca 40	21 Sc 45	22 Ti 48	**23 V 51**	**24 Cr 52**	**25 Mn 55**	**26 Fe 56**	**27 Co 59**	**28 Ni 59**	**29 Cu 64**	**30 Zn 65**	**31 Ga 70**	**32 Ge 73**	**33 As 75**	**34 Se 79**	35 Br 80	36 Kr 84
37 Rb 85	38 Sr 88	39 Y 89	40 Zr 91	41 Nb 93	**42 Mo 96**	**43 Tc (98)**	44 Ru 101	**45 Rh 103**	**46 Pd 106**	**47 Ag 108**	**48 Cd 112**	**49 In 115**	**50 Sn 119**	**51 Sb 122**	**52 Te 128**	53 I 127	54 Xe 131
55 Cs 133	56 Ba 137	57 *La 139	72 Hf 178	73 Ta 181	**74 W 184**	**75 Re 186**	76 Os 190	77 Ir 192	**78 Pt 195**	**79 Au 197**	80 Hg 201	**81 Tl 204**	**82 Pb 207**	**83 Bi 209**	84 Po (209)	85 At (210)	86 Rn (222)
87 Fr (223)	88 Ra (226)	89 †Ac (227)	104 Rf (267)														

Source: Kanani, N., *Electroplating Basic Principles*, Elsevier, Burlington, VT, 2004.

The readers interested further in the history of electroless plating and the major contributors to the development of electroless baths may refer to the papers published by Shipley (1984) and Brenner (1984) on the history of the commercialization of electroless plating beyond the discovery of electroless nickel (Shipley 1961; Brenner 1984).

A very comprehensive study on the progress of electroless plating until 2010 can also be found in the book of Zhang (2016). After the collection of the number of papers and patents from different databases, they classified electroless deposition process stages as early stage of development (mid-1940s to 1959), slow growth period (1960–1979), rapid development period (1980–1999), and fundamental mechanistic studies and development of nanoelectroless plating (2000–present).

According to this study, during the first two periods of slow growth is a kind of incubation period. In these stages, studies on the development of the stability of baths, efforts for application of electroless plating to larger scales, development of methods for electroless deposition of other metals (especially copper), and commercial applications in small scale have occurred.

Increase in the scientific interest on the mechanism of electroless plating, determination of interesting physical and chemical properties of electroless deposits (mechanical, magnetic, chemical resistance, etc.), extension of large-scale electroless plating to different industrial fields, development of ternary and multicomponent alloys and composites, and participation of China and India in the research and developments in the plating field are discussed in the same papers as the main reasons of the rapid development stage. Figure 1.1 shows the number of papers published on electroless nickel deposition since 1940.

Today, more than thousands of electroless plating facilities around the world have extended the use of electroless plating in different industry sectors (mainly automotive, petrochemical, and electronics) and estimated 15 billion USD output value in the world market (Zhang et al. 2014), which makes this deposition method an important player in the surface treatment field. The unique properties of this deposition method

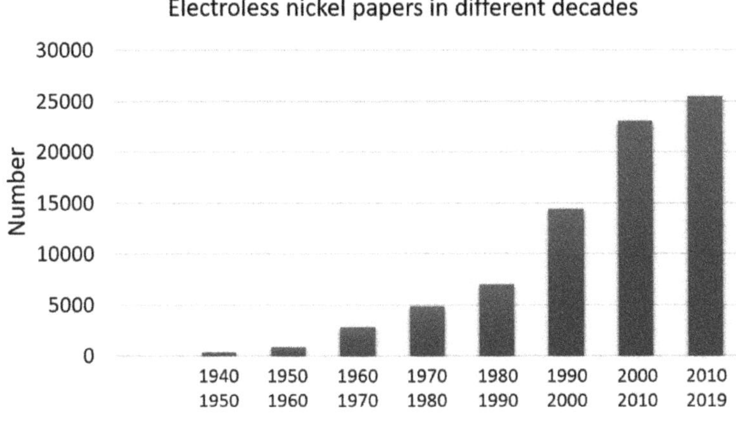

FIGURE 1.1 Number of papers and US patents published on electroless nickel deposition since 1940 (from Zhang (2014).)

such as geometry-independent homogeneous deposition on any kind of substrate; possibility of deposition in amorphous and/or nanocrystalline, composite, and nanocomposite state; ability to be deposited in alloy form; possibility of deposition on conductive and nonconductive powders, and fibers; and deposition on selected areas are some of the reasons for continuing expansion of R&D activities and industrial use of electroless nickel (EN) plating.

1.3 COMPARISON OF ELECTROLESS PLATING WITH OTHER ELECTROLYTIC METAL DEPOSITION PROCESSES

Electroless metal plating is an autocatalytic process that proceeds in accordance with the electrochemical kinetic principles. This property differentiates this method from other metal deposition processes that use metal ions as source of plating. Namely, in electroplating, we supply the electrons required for the reduction of metal ions using an external power supply, or in galvanic displacement/immersion metal plating, the electrons required for the reduction of the metal ions are supplied through corrosion of the metal substrate. In electroless plating, the electrons required for the reduction of the metal ions are supplied by the oxidation of the chemical compound, (reducing agent) present in the solution.

When compared from the thermodynamic point of view, the driving force for immersion plating or galvanic displacement is the potential difference between half-cell (single electrode) reactions, the addition of which is the total reaction. Let's consider copper deposition on steel when immersed in an acidic sulfate solution containing copper ions. In this case, there are two half-cell cathodic reactions and one half-cell anodic reaction; i.e., hydrogen evolution reaction and reduction of copper ions are the two half-cell cathodic reactions, both of which gets the electrons required for the reduction from the dissolution reaction of iron (anodic half-cell reaction). Thus, the reduction of copper ions in the solution is only possible when the iron surface is exposed; once the surface is covered with copper, the reaction stops. Deposition process starts all over the surface that is in contact with the copper ion-containing solution.

Electrolytic metal plating on the other hand does not choose substrate material as long as it is conductive; for example, it is possible to electrodeposit zinc on iron, although electrode potential of zinc is far more electronegative than iron, because the energy required for the reduction of the metal ions in the solution is supplied externally with the help of a power supply. In these systems, a positive electrode (anode) is required to complete the circuit. The need for an anode in the system turns it into a line of sight process, which means the deposition process shows a strong dependence on its position relative to the anode. On the edges, we have thicker coatings, and so we cannot deposit inside holes, although the solution containing the metal ions to be deposited is in contact with the substrate metal. In electroless plating, the deposition starts and proceeds at the same rate all over the substrate as long as metal ions and reductant are present, which is one of the major advantages of the technique. However, a certain combination of substrate metal and reductant is required for the realization of the process. The reasons for this selectivity will be explained in the following sections.

1.4 THERMODYNAMIC ASPECTS OF ELECTROLESS DEPOSITION

Electroless deposition of metals is an electrochemical process that proceeds according to the principles of thermodynamics and kinetics of electrochemical reactions. Today, it is well known that the driving force of this spontaneous electroless deposition process is the potential difference between minimum two half-cell (single electrode) reactions, namely anodic half-cell and cathodic half-cell reactions and proceeds under mixed potential control.

The reduction of metal ions constitutes the main cathodic half-cell of the overall reaction. For the realization of the reduction reaction spontaneously, an anodic reaction (anodic half-cell) that will supply the electrons needed for the reduction of metal ions is required. The number of reducing agents suitable for electroless plating is rather limited. The major reducing agents used in electroless deposition and their standard potentials are given in Table 1.2.

The standard electrode potentials of the cathodic half-cell reactions can be extracted from electromotive force (EMF) series. Simply, for the standard electrode potential (E_0 or E_{rev}) for the cathodic half-cell reaction for nickel is presented in Equation 1.1:

$$\mathrm{Ni}^{+2} + 2\mathrm{e} \leftrightarrow \mathrm{Ni}^0 \quad E_0 = -0.26\ \mathrm{V}$$

Equation 1.1: Standard electrode potential of Ni_{+2}/Ni_0 reaction

This potential is valid for standard conditions, namely for nickel ion activity of 1, 298 K and 1 atm. As very well known, when the conditions are away from standard conditions, new values can be calculated by using Nernst equation. The potential E of the Me^{+n}/Me electrode is a function of the activity (a) of metal ions in the solution (Equation 1.2):

$$E = E_0 + \frac{RT}{nF} \ln a_{\mathrm{Me+r}}$$

Equation 1.2: Nernst equation

TABLE 1.2
Main Reducing Agents Used in Electroless Deposition and Their Standard Electrode Potentials

Name	Chemical Formula	Available Electrons	Standard Electrode Potential (V)	Used for deposition of
Sodium hypophosphite	$NaH_2PO_2{\cdot}H_2O$	2	−1.57 in alkaline/−0.5 acidic	Ni, Co
Sodium borohydride	$NaBH_4$	8	−1.24 alkaline	Ni, Co
Dimethylamine borane	$(CH_3)_2NH{\cdot}BH_3$	6	−1.18 alkaline	Ni, Co, Ag
Formaldehyde	HCHO	2	−1.11 alkaline	Cu, Ag
Hydrazine	N_2H_4	4	−1.16 alkaline	Cu

Source: Riedel, W., *Electroless Nickel Plating*, Finishing Publication Ltd, London, UK, 1989.

When we write half-cell electrode reactions (Equations 1.3 and 1.4) that constitute the total reaction, we can calculate the free energy or electrode potential of the total reaction (Equation 1.5) that will tell us whether these reactions are prone to proceed.

$$Me^{+n}_{soln} + ne \leftrightarrow Me \quad \text{(cathodic reaction)}$$

Equation 1.3: Cathodic half-cell reaction

$$Red_{soln} \leftrightarrow Ox_{soln} + ne \quad \text{(anodic half cell)}$$

Equation 1.4: Anodic half-cell reaction

$$Me^{+n}_{soln} + Red_{soln} \leftrightarrow Me + Ox_{soln} \ \text{(total reaction)}$$

Equation 1.5: Total of anodic and cathodic half cell reactions

Simply, if ΔG of the total reaction is negative, the reaction proceeds spontaneously in the written direction. Because the relation between standard free energy and the standard electrode potential is $\Delta G = -nFE_0$, the reaction proceeds in the written direction if ΔE of the total reaction is positive.

The thermodynamic calculations either by using chemical or electrochemical rules indicate that when the metal ions to be reduced are introduced into a solution that contains appropriate reductants, reactions are thermodynamically possible. However, it is fortunate that the rate of these reactions is very low and they require overcoming of large activation energy barriers to proceed at an appreciable rate. Thus, from the thermodynamic point of view, the electroless metal plating solution can be described as metastable that can be made to proceed at an appreciable rate by decreasing the activation energy barrier for the reaction. The decrease of the activation energy barrier for the reaction is achieved by the introduction of a catalytic substrate surface into the bath (Ni, Pt, Pd, etc.). The role of catalytic processes on electroless plating will be further discussed in kinetics section.

Thermodynamics is also an important tool in designing or developing electroless baths. For the reasons explained further in this book, we may require electroless nickel plating baths with different pH values extending from acidic to basic. In these cases, for the deposition process to occur, nickel must be in ionic form. Potential-pH diagrams based on thermodynamic calculations can be used for the selection of proper complexing and chelating agents. For example,

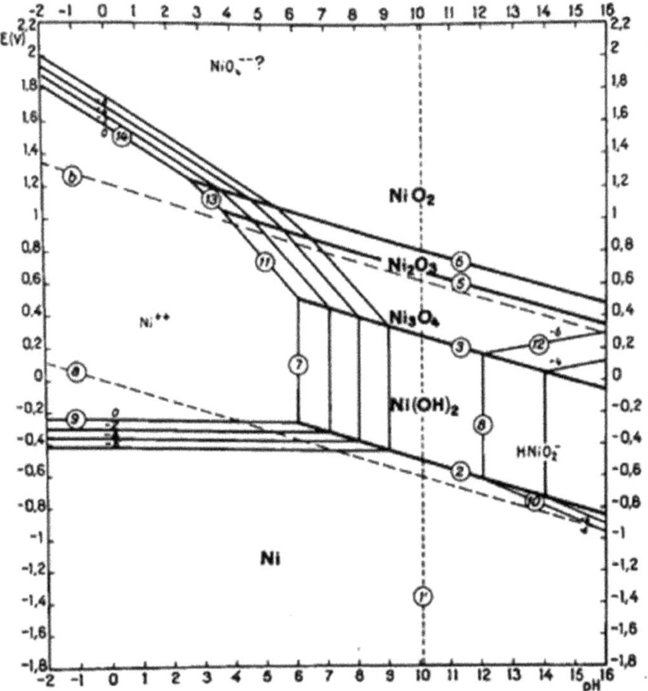

FIGURE 1.2 Potential-pH diagrams of Ni-H_2O.

the reductant (sodium borohydride) for electroless Ni–B plating require alkaline environments to function; thus, nickel should also be in ionic form under these conditions, which is not possible in water-based solutions without complexing and chelating agents (Figure 1.2). However, we can achieve the required ionic form that can be reduced with this reductant that contains nickel by forming nickel–amine complexes.

Similarly, the pH and environment oxidation potential dependence of the reductant agents can also be determined by using the potential-pH diagrams. For example, the stability region of reductants in Ni–P and Ni–B baths can be determined by using P-H_2O and B-H_2O potential-pH diagrams. An example of the usage of potential-pH diagrams for Ni–P system is presented in Figure 1.3. In this diagram, Ni-H_2O potential-pH diagram calculated for 0.6 g/L Ni^{+2} is superimposed on P-H_2O diagram (Pourbaix 1966). The hatched region of this diagram clearly indicates that it is thermodynamically possible to reduce nickel ions by the oxidation of hypophosphites to phosphites or phosphates.

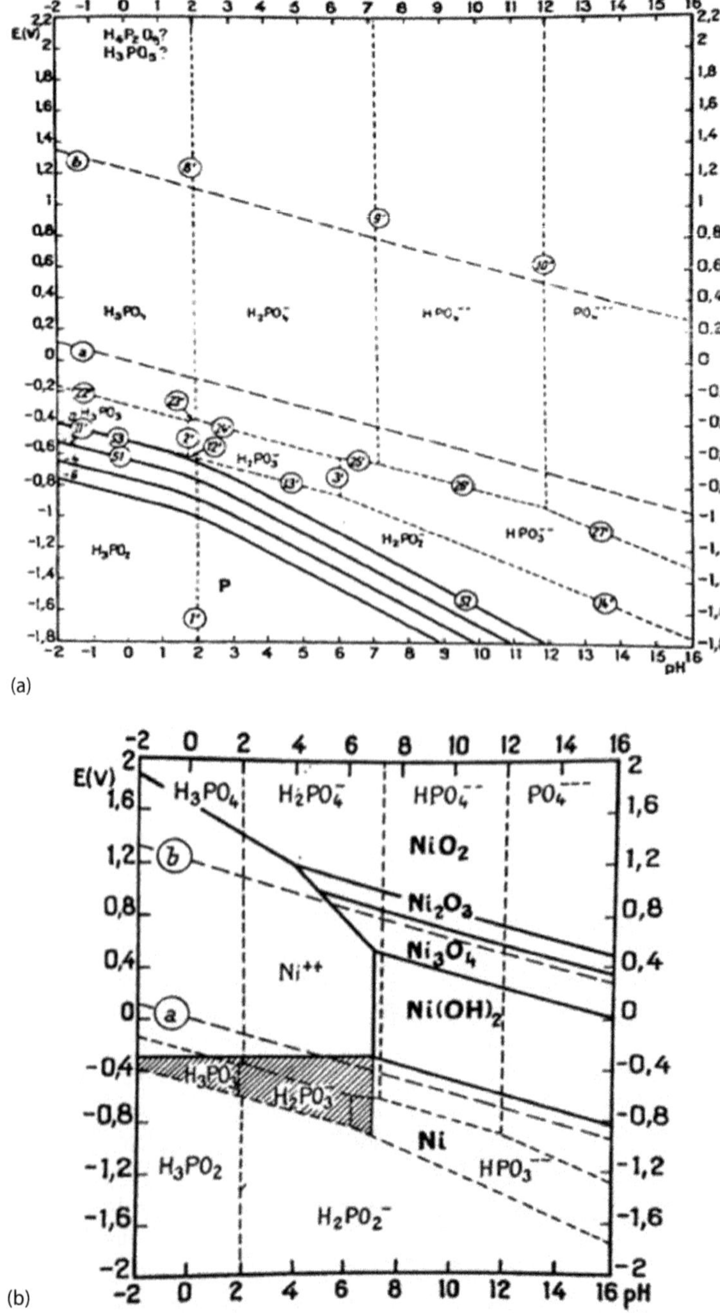

FIGURE 1.3 (a) P-H_2O potential-pH diagram. (b) Superimposed Ni-H_2O and P-H_2O diagram (for Ni^{+2} concentration of 0.6 g/L). (From Pourbaix, M., *Atlas of Electrochemical Equilibria in Aqueous Solutions*, Vol. 1, Pergamon Press, London, UK, 1966.)

1.5 CONTRIBUTION OF HYDROGEN EVOLUTION REACTION TO ELECTROLESS PLATING

Some interesting aspects of electroless deposition are:

- Only some substrate metals act as a catalyst,
- The restricted number of reductants, and
- Hydrogen evolution in almost all of them.

Among these, hydrogen evolution attracted a significant interest not only because of its generation mechanism but also its possible contribution to the catalytic activity of specific metals and also on the specificity of the reductants.

Five different mechanisms are proposed for the explanation of hydrogen evolution and its relation to metal and metalloid deposition in electroless metal plating. These are pure electrochemical (Elze 1960; Minjer 1975; Müller et al. 1960), metal hydroxide (Cavallotti and Salvago 1968; Salvago and Cavallotti 1972), hydride ion, atomic hydrogen (Hersch 1955; Lukes 1964), and modified atomic hydrogen or dehydrogenation mechanisms (Nikiforova and Sadakov 1967; Donahue 1972; Gorbunova et al. 1973). Among these mechanisms, two of them, namely dehydrogenation ("universal mechanism" as named by Meeraker (Van der Meeraker 1981)) and metal hydroxide mechanism (Cavallotti and Salvago 1968) are the most widely used and accepted mechanisms that have the ability to explain most of the observations in the electroless deposition.

Universal mechanism combines electrochemical reactions with the atomic hydrogen mechanisms and introduces additional dehydrogenation and recombination reactions to the anodic reaction sequences. The general scheme of anodic and cathodic reactions for electroless plating according to this mechanism is given below (Van Den Meerakker 1981), where RH is the reductant such as formaldehyde, hypophosphite or borohydride. *CatMe* in Equation 1.6 stands for catalytically active metal surface.

Anodic reactions:

$$\text{Dehydrogenation: } \mathrm{RH} \xrightarrow{\text{CatMe}} \dot{\mathrm{R}} + \dot{\mathrm{H}}$$

Equation 1.6: Dehyrogenation of the reductant during electroless metal deposition

$$\text{Oxidation: } \dot{\mathrm{R}} + \mathrm{OH}^- \rightarrow \mathrm{ROH} + \mathrm{e}$$

Equation 1.7: Oxidation of the reductant radical during electroless metal deposition

$$\text{Recombination: } \dot{\mathrm{H}} + \dot{\mathrm{H}} \rightarrow \mathrm{H_2}$$

Equation 1.8: Recombination of to give hydrogen gas molecules during electroless metal deposition

$$\text{Oxidation (alkaline solutions): } \dot{\mathrm{H}} + \mathrm{OH}^- \rightarrow \mathrm{H_2O} + \mathrm{e}$$

Equation 1.9a: Oxidation of in alkaline electroless metal deposition solutions

$$\text{Oxidation (acidic solutions): } \dot{H} \rightarrow H^{+} + e$$

Equation 1.9b: Oxidation of in acidic electroless metal deposition solutions

Cathodic reactions:

$$\text{Reduction to metal: } Me^{+n} + ne \rightarrow Me^{0}$$

Equation 1.10: Reduction of metal ions during electroless metal deposition

$$\text{Hydrogen evolution (alkaline solutions): } 2H_2O + 2e \rightarrow H_2 + 2OH^{-}$$

Equation 1.11a: Hydrogen evolution reaction in alkaline electroless metal deposition solutions

$$\text{Hydrogen evolution (acidic solutions): } 2H^{+} + 2e \rightarrow H_2$$

Equation 1.11b: Hydrogen evolution reaction in acidic electroless metal deposition solutions

In solutions where hypophosphite, aminoborane, and sodium borohydride are used as reductants, the co-deposition of phosphorus (Equation 1.12) or boron (Equation 1.13) is the other cathodic process.

$$H_2PO_2^{-} + e \rightarrow P + 2OH^{-}$$

Equation 1.12: Reaction that results in co-deposition of P in electroless nickel – phosphorous deposition solutions

$$B(OH)_4^{-} + 3e \rightarrow B + 4OH^{-}$$

Equation 1.13: Reaction that results in co-deposition of B in electroless nickel – boron deposition solutions

According to the reaction sequences given above, dehydrogenation step (Equation 1.6) that leads to the formation of reductant (R) and hydrogen radicals determines the catalytic nature of the process. Initiation of electroless deposition on a metal substrate that does not have the ability to catalyze this reaction is not possible. The metals (such as Pt, Pd, Ni, and Co) that act as catalytic surfaces for the electroless deposition process are also good hydrogenation and dehydrogenation catalysts and the poisons for this reaction, such as thiourea and mercaptobenzothiazole, act as stabilizers for almost all electroless deposition processes. When we look at the same reaction scheme from the reductant perspective, a proper reductant for the electroless process should involve dehydrogenation step on the catalytic metal surface (reaction 1.6) during its oxidation so that the reduction of metal ions occurs directly on the substrate surface. All of the reductants used in the electroless deposition processes possess this property.

Another evidence that supports the deterministic role of reaction 1.6 is the higher efficiencies of deposition for metals with higher catalytic activity (Pd, W, Au, Mo, and Re) for dehydrogenation reaction. For example, in the case of Pd, a higher efficiency (100%) can be obtained when compared to Ni–B (45%) efficiency because of the higher catalytic activity of Pd toward dehydrogenation reaction (Van Den Meerakker 1981).

The other mechanism that also found acceptance especially for the case of Ni–P deposition from acidic electrolytes is the metal hydroxide mechanism put forth by Salvago and Cavalotti (Cavallotti and Salvago 1968; Salvago and Cavallotti 1972) According to this mechanism on the catalytic surface of nickel, ionization of water occurs (Equation 1.14)

$$H_2O \xrightarrow{\text{catMe}} H^+ + OH^-$$

Equation 1.14: Ionization of water on catalytic nickel surface

Then, nickel ions at the surface hydrolyze to give nickel hydroxo complexes

$$Ni^{+2} + OH^- \rightarrow NiOH^+_{(ads)}$$

Equation 1.15: Hydrolysis reaction between nickel and hydroxyl ions to give hydroxo complexes

$$Ni^{+2} + 2OH^- \rightarrow Ni(OH)_{2(ads)}$$

Equation 1.16: Hydrolysis reaction between nickel and hydroxyl ions to give nickel hydroxide ad atoms

$$NiOH^+_{ads} + H_2O \rightarrow Ni(OH)_{2(aq)} + H^+$$

Equation 1.17: Hydrolysis of nickelhydroxo complexes to give nickel hydroxides and hydrogen ions.

As a result of the reaction of the hypophosphite ions with these hydrolyzed species, metallic nickel, atomic hydrogen, and orthophosphide ions are produced

$$Ni(OH)_{2(ads)} + H_2PO_2^- \rightarrow NiOH + H_2PO_3^- + H_{ad}$$

Equation 1.18: Reaction of hypophosphite ions with nickel hydroxide ad atoms to give orthophosphide ions and atomic hydrogen

$$NiOH^+_{(ads)} + H_2PO_2^- \rightarrow Ni + H_2PO_3^- + H_{ad}$$

Equation 1.19: Reaction of hypophosphite ions with nickel hydroxocomplexes to give metallic nickel, orthophosphide ions and atomic hydrogen

According to this mechanism, one of the sources of hydrogen evolution is the combination of atomic hydrogen that is formed in Equations (1.18 and 1.19). The other sources of hydrogen as put forth by this mechanism are the reaction of monovalent NiOH that is formed in accordance with reaction 1.18 with water (Equation 1.20) and/or direct reaction of hypophosphite ions with water (Equation 1.21).

$$NiOH + H_2O \rightarrow H_2PO_3^- + H_{ad}$$

Equation 1.20: Reaction of hypophosphite ions with mono valent nickel hydroxides complexes to give atomic hydrogen

$$H_2PO_2^- + H_2O \rightarrow H_2PO_3^- + H_2$$

Equation 1.21: Reaction of hypophosphite ions with water to give molecular hydrogen.

Accordingly, the requirement of catalytic metallic surface come into play not only for reactions 1.14 and 1.20 but also for the reduction of hypophosphite ions to elemental phosphorous (Equation 1.22).

$$Ni_{cat} + H_2PO_2^- \rightarrow P + NiOH + OH^-$$

Equation 1.22: Reduction reaction of hypophosphite ions on catalytic nickel surface to give elemental phosphorous.

The overall reaction with proper stoichiometry is given as (Djokic 2014):

$$Ni^{+2} + 4H_2PO_2^- + H_2O \rightarrow Ni + 3H_2PO_3^- + P + \frac{3}{2}H_2 + H^+$$

Equation 1.23: Overall reaction between nickel and hypophosphite ions to give Ni-P deposits and hydrogen gas.

1.6 KINETICS OF ELECTROLESS DEPOSITION PROCESSES

Kinetic aspects of electroless deposition are twofold: Kinetics of electrochemical reactions and nucleation and growth kinetics of the electroless deposited metals. In this section, the principles of electrochemical reaction kinetics relevant to electroless metal plating will be outlined. The second aspect is the kinetics of nucleation and growth will be briefly introduced in the section about the role of additives on electroless metal plating baths.

As explained in the previous sections, electroless deposition of metals is an electrochemical process under mixed potential control, which means that the process (at least the steps that involve electron transfer) proceeds, obeying electrochemical kinetic principles. Because of the electrochemical nature of the process, the rate of electroless deposition can be expressed as current and interrelated to the mass of electrodeposit by using Faraday's law (Equation 1.24). In this equation, w is the amount of deposited metal (in grams), A is the atomic mass (in grams), I is the deposition current (in amperes), t is the deposition time (in seconds), n is the number of electrons transferred, and 96487 is the charge in coulombs required for the deposition of equivalent of a metal.

$$w = \frac{A \cdot I \cdot t}{n \cdot 96487}$$

Equation 1.24: Faraday's law

Electroless deposition process consists of half-cell reactions (Equations 1.25 and 1.26) and an overall reaction (Equation 1.27), the rate of which depends on the properties of each half-cell reaction.

$$Me^{+n} + ne \rightarrow Me^0$$

Equation 1.25: Cathodic half-cell reaction

$$RH_{soln} \rightarrow Ox_{soln} + ne$$

Equation 1.26: Anodic half-cell reaction

$$Me^{+n} + RH_{soln} \rightarrow Me^0 + Ox_{soln}$$

Equation 1.27: Overall reaction between metal ions and the reductant.

The potential difference between the anodic and cathodic reactions is the driving force of the spontaneous overall reaction. Because the system consists of two electrochemical half-cell reactions with a potential difference that will polarize them accordingly, mixed potential theory can be applied. For a better description of the mixed potential theory, the kinetics of half-cell reactions will be summarized first.

1.6.1 Kinetics of Half-Cell Reactions

The coordinates of equilibrium conditions at a reversible potential can be shown graphically by using potential and current axis in linear or semi-logarithmic scales (Figure 1.4a and b). On the potential axis, the reversible potential extracted from thermodynamics (EMF series and Nernst equation) is marked, and on the current axis, the rate of the reaction at equilibrium (i_0) is marked. Because at E_{rev}, the rate of forward and reverse reaction are equal to each other and i_0 ($i_a = i_c = i_0$). The calculation of i_0, exchange current density, is not straight forward because, at these equilibrium conditions, there is no net charge transfer (hence, mass loss or gain); thus, these values can be calculated theoretically or by changing the equilibrium (polarization). By polarizing this half-cell reaction to either cathodic or anodic direction, a net current starts to pass showing dependence on the magnitude of the energy (overvoltage, η) applied for changing the equilibrium.

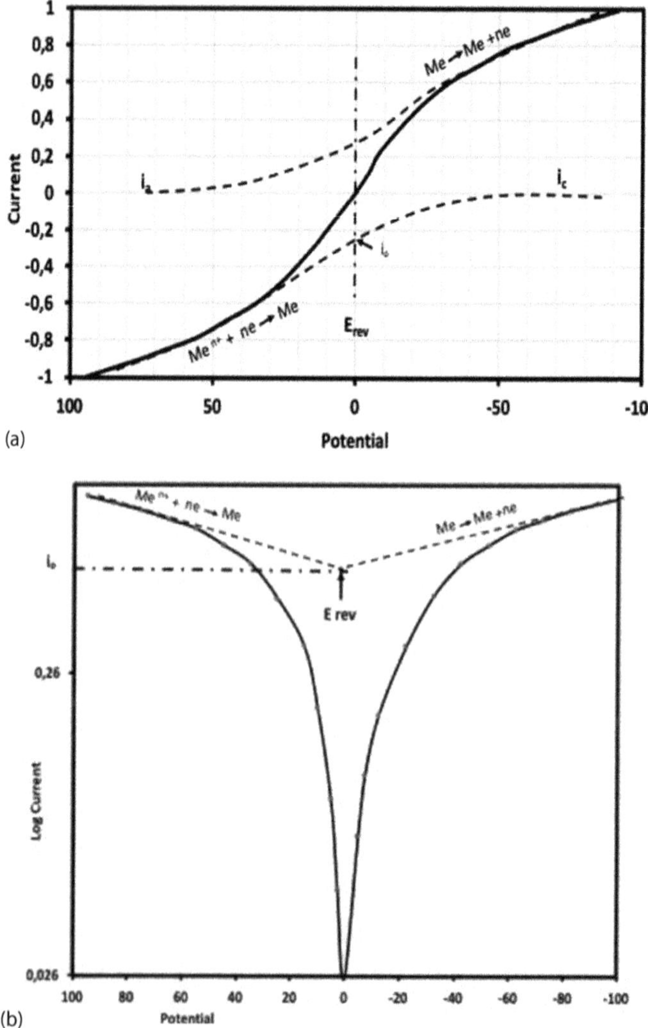

FIGURE 1.4 Schematic linear (a) and semi-logarithmic (b) representation of polarization curves.

$$\text{At equilibrium: } \eta = 0, \quad i_c = i_a = i_o$$

Equation 1.28: Definition of equilibrium conditions of a half-cell reaction.

$$\text{On cathodic polarization: } \eta_c = E_{rev} - E_p \;\; i_c > i_a \neq i_o \, i_{net} = i_c - i_a$$

Equation 1.29: Definition of the conditions for cathodic polarization of a half-cell reaction.

On anodic polarization: $\eta_a = E_{rev} - E_p \quad i_a > i_c \neq i_o \, i_{net} = i_a - i_c$

Equation 1.30: Definition of the conditions for anodic polarization of a half-cell reaction.

If the reaction is under activation control (in case where the slowest step is the electron transfer), the relation between the rate of the reaction (i_{net}) and overvoltage obeys the Butler-Volmer equation. Equation 1.31 represents the Butler-Volmer equation for anodic polarization. In this equation, η is the overvoltage (the difference between the polarized potential and reversible potential), b_a and b_c are the anodic and cathodic Tafel slopes, respectively.

$$i_{net} = i_a - i_c = i_o \exp\left(\frac{\eta}{b_a}\right) - i_o \exp\left(\frac{\eta}{|b_c|}\right)$$

Equation 1.31: Butler-Volmer equation for an anodically polarized half-cell.

As can be followed from the schematic (Figure 1.4b) for overvoltages in the range of 30–50 mV, a linear relation between E and log$i$ is started to be observed because the contribution of cathodic component (i_c) starts to become insignificant for anodic polarization and vice versa. For the anodic polarization case (Equation 1.31), the equation can be simplified by canceling the second exponential term (Equation 1.32). The same methodology can be applied for cathodic polarization case (Equation 1.33).

$$i_{net} = i_a - i_c = i_o \exp\left(\frac{\eta}{b_a}\right)$$

Equation 1.32: Simplified Butler-Volmer equation for an anodically polarized half-cell.

$$i_{net} = i_c - i_a = -i_o \exp\left(\frac{\eta}{|b_c|}\right)$$

Equation 1.33: Simplified Butler-Volmer equation for a cathodically polarized half-cell.

Taking logarithms of Equations (1.20 and 1.21) and solving them for η, the well-known Tafel equation is obtained:

$$\eta = a \pm b \log|i|$$

Equation 1.34: Tafel equation

where a and b are constants and $|i|$ is the absolute value of i_{net}, $\pm$ sign stands for anodic and cathodic processes, respectively. The theoretical values of these constants for activation-controlled cathodic reactions are:

$$a_c = \frac{2.303\,RT}{\alpha nF}\log i_o$$

Equation 1.35: Tafel equation constant *a*

$$b_c = \frac{2.303\,RT}{\alpha nF}$$

Equation 1.36: Tafel equation constant *b*

In these relations (Equations 1.35 and 1.36), α is the transfer coefficient R for gas constant, n is the number of electrons transferred, and F is the Faraday constant.

On the other hand, the derivative of Butler-Volmer equation (Equation 1.31) leads to the slope ($d\eta/di$) for $\eta \to 0$ V, and named as charge transfer resistance R_{ct}, that allows calculation of the rate of the half-cell reaction (exchange current density, i_0) without extension of cathodic or anodic polarization.

Besides activation (charge transfer) control, half-cell reactions can also be controlled by diffusion (concentration) polarization. In these cases, the slowest step during the realization of the reaction is the mass transport of the reacting species to the electrode solution interface. Butler-Volmer equation for combined polarization of a cathodic reaction is presented in Equation (1.37). In this equation, i_{lim} stands for limiting current.

$$i_c = \frac{i_0 \exp\left[\frac{-(1-\alpha)nF}{RT}\right]\eta_c}{1-\frac{i_0}{i_{lim}}\exp\left[\frac{-(1-\alpha)nF}{RT}\right]\eta_c}$$

Equation 1.37: Butler-Volmer equation for combined (activation + diffusion) polarization controlled cathodic half-cell reaction. In this equation i_{lim} stands for limiting current.

$$if\ i_{lim} < i_0 \exp\left[\frac{-(1-\alpha)nF}{RT}\right]\eta_c$$

denominator of equation 1.37 goes to 1 (activation control)

Equation 1.38: Definition of conditions for activation control in the combined Butler-Volmer equation

denominator goes to 1 - activation controlled

$$if\ i_{lim} > i_0 \exp\left[\frac{-(1-\alpha)nF}{RT}\right]\eta_c$$

$i_{lim} = i$ *reaction 1.37 becomes potential independent (diffusion control)*

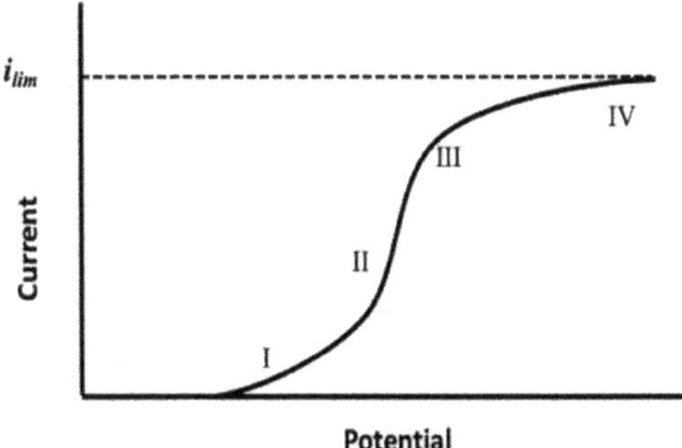

FIGURE 1.5 Four regions that can be observed during polarization of a half-cell reaction to one direction (anodic or cathodic). I: linear region (slope gives R_{ct}); II: exponential region (slope gives Tafel slope); III: mixed control region; IV: diffusion controlled region.

$i_{lim} = i$ and the reaction becomes independent of potential, diffusion control

Equation 1.39: Definition of conditions for diffusion control in the combined Butler-Volmer equation

$$i_{lim} = \frac{nFD}{\delta} C_b$$

Equation 1.40: Definition of limiting current.

Depending on the magnitude of i_{lim}, Equation 1.37 is converted into activation-controlled (condition 1.38) or diffusion-controlled state (condition 1.39). Limiting current by definition is a function of diffusion coefficient (D), number of electrons transferred (n), diffusion layer thickness (δ), and concentration of diffusing species (C_b).

A half-cell reaction that starts with activation control and proceeds with mass transport control on a linear scale upon polarization is schematically presented in Figure 1.5.

From the evaluations of the half-cell reactions that are under activation control, the most effective parameters on the kinetics of the reaction are exchange current density, i.e., the rate of the reaction at equilibrium, the number of electrons transferred, and α, transfer or symmetry coefficient that indicates the symmetry of the activation energy barrier for activation-controlled reactions. All these parameters reflect the values of the constants in Tafel equation. Simply, the lower the value of Tafel constants, the higher is the rate of the reaction upon polarization. On the other hand, the most effective parameters for transport-controlled reactions are the diffusion coefficients (D), the thickness of diffusion layer (δ), and the bulk concentration of diffusing species in the solution (C_b).

1.7 MIXED POTENTIAL THEORY

Mixed potential theory is first established by Wagner and Traud for the explanation of kinetics of corrosion in their classical paper "On the Interpretation of Corrosion Processes through the Superposition of Electrochemical Partial Processes and on the Potential of Mixed Electrodes," in 1938.

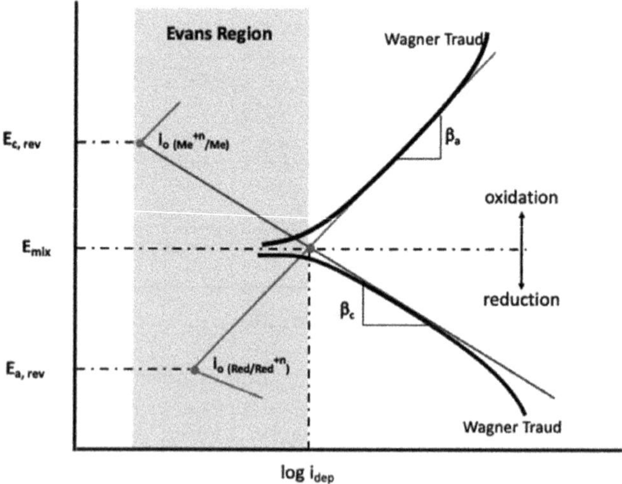

FIGURE 1.6 Schematic representation of spontaneous polarization (gray-shaded, Evans region) and forced polarization regions.

According to this theory, two half-cell reactions that occur in the same solution on the same electrode will be anodically and cathodically polarized because of the potential difference between them, obeying Butler-Volmer equation till a new "equilibrium" state is reached at E_{mix} (gray-shaded region in Figure 1.6). During this stage, some sites on the substrate metal will act as anodic sites and the others will act as cathodic sites, and the slower half-cell reaction will control the overall reaction rate (see Figure 1.8a–c).

This is not a real equilibrium because net changes are occurring at E_{mix}. The polarization that occurs spontaneously cannot be followed experimentally. However, it can be determined by further polarizing them externally from E_{mix} potential, assuming that they will still obey Butler-Volmer equation upon further polarization (region described by Wagner and Traud) or calculated by using relevant kinetic parameters.

Although this theory has been put forth for electrochemical corrosion reactions, it has been verified that most of the electroless deposition processes also obey the principles of mixed potential theory (Paunovic 1983; Flis and Duquette 1997). This is not an unexpected outcome because of the similarity between electrochemical corrosion and electroless metal deposition reactions. Both of them consists of minimum two half-cell reactions (cathodic and anodic), these reactions occur on the same substrate, both of them are spontaneous and they are driven by a potential difference between anodic and cathodic half-cells.

The self-polarized spontaneous polarization region can be calculated or schematically drawn for presenting roles of different parameters on the polarization behavior (Evans diagrams).The self-polarized spontaneous polarization region can be calculated by using Equations (1.19 and 1.20), if i_0 values, E_{rev} values, and Tafel slopes for both anodic and cathodic reactions are known.

Another way of determining the spontaneous region of polarization and the rate of reactions at E_{mix} is to use polarization curves obtained by polarization from E_{mix}. The rate of reactions at E_{mix} can be determined by extrapolating the linear regions of anodic and cathodic polarization curves (Tafel extrapolation method), assuming that they still obey Butler-Volmer equation upon polarization from E_{mix}. In cases where E_{rev} values of anodic and cathodic reactions are known by further extrapolating anodic and cathodic polarization curves to E_{rev}, one can determine one of the most important kinetic parameters, the exchange current density (i_0).

Another way of determining the reaction rate at E_{mix} is to use low potential approximation. The same approach for the determination of exchange current density (Equation 1.36) can be used simply by replacing i_0 with deposition rate (i_{dep}).

During the application of mixed potential theory concepts to electroless deposition, polarization curves for each half-cell reaction can be obtained separately that allows the determination of kinetics of each half-cell reaction that constitutes total reaction. This is realized by investigating the polarization behaviors of cathodic and anodic half-cell reactions in the absence of reductants and in the absence of metal ions, respectively (Figure 1.7) (Paunovic 1968). The possibility of polarizing each half-cell electrode reaction is the advantage of electroless deposition systems when compared with corrosion. In this manner, the polarization of each half-cell reaction that occurs spontaneously, when the reductants and oxidants are present in the same solution, can be followed and determined (Figure 1.7). By using this method of obtaining polarization behavior of the half-cells that constitutes the electroless deposition process, factors affecting the kinetics of the process can be determined. For example, the role of additives, exchange current densities of different catalytic substrates, concentration of reductants, and so on, in the kinetics of the half-cell reactions can be successfully evaluated by this methodology.

Evans diagrams (the schematic diagrams representing the spontaneous region) are a very effective way of presenting the role of several parameters on the kinetics of electroless deposition processes, such as the rate of controlling half-cell reaction,

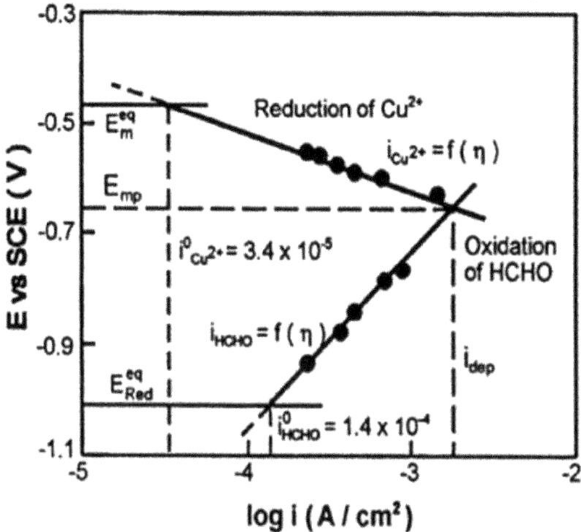

FIGURE 1.7 Overlayed Cu^{+2} reduction and formaldehyde oxidation polarization curves (Evans diagram). Cu^{+2} reduction curve obtained for 0.1 M $CuSO_4$ and 0.175 EDTA solution with pH 12.5 at 25° C. Formaldehyde oxidation curve obtained for 0.05 M HCHO and 0.175 M EDTA solution at 25° C. (From Paunovic, M., *Plating*, 51, 1161–1167, 1968.)

the importance of i_0 in kinetics, and type of control (activation or diffusion control) and concentration changes in the electrolyte. Several examples of Evans diagrams are presented in Figure 1.8 that can be used for understanding the role of kinetic and thermodynamic parameters on electroless deposition.

In studies that investigate the validity of the mixed potential theory for electroless deposition, the deposition rates calculated from i_{dep} are in accordance with the rates determined by weight gain measurements especially for electroless copper deposition in formaldehyde (Paunovic 1968) and Ni–dimethylamine borane (DMAB; Paunovic 1983; Flis and Duquette 1997) solutions and copper and nickel in hypophosphite solutions (Martins and Nunes 2016). However, for electroless nickel coatings conducted by using hypophosphite's and reductants, the rates attained by mixed potential theory are much lower than the gravimetric ones. This special behavior observed in hypophosphite baths is generally attributed to chemical reduction of nickel ions by the active adsorbed hydrogen that forms on

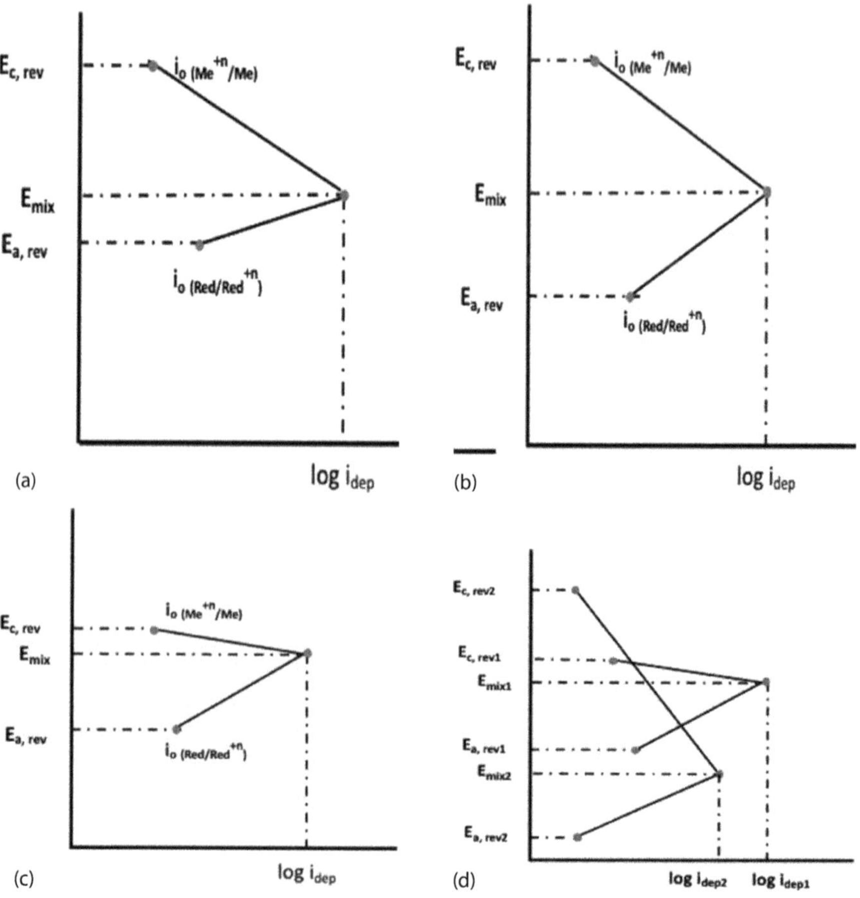

FIGURE 1.8 Evans diagram describing: (a) cathodic control; (b) mixed control; (c) anodic control; (d) a reaction with higher reaction tendency (differences between $E_{a,rev}$ and $E_{c,rev}$) that can result in lower deposition rate; *(Continued)*

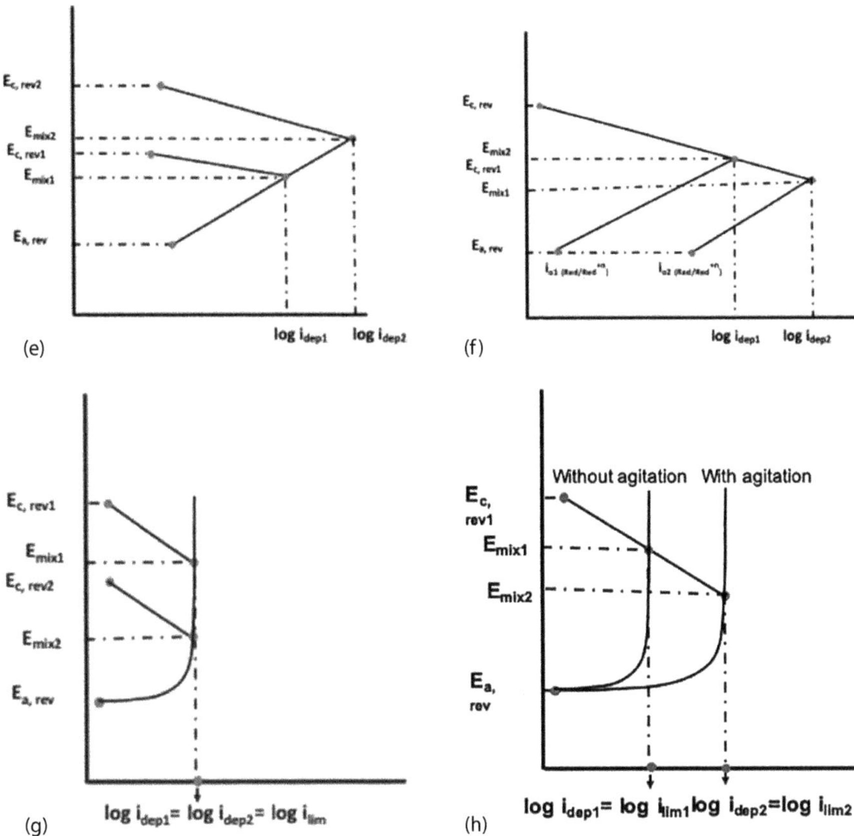

FIGURE 1.8 (Continued) Evans diagram describing: (e) effect of increased cathodic reactant concentration on deposition rate by shifting E_{rev} to more positive potentials; (f) effect of exchange current density on deposition rate (higher i_0 means easier polarization, higher deposition rate); (g) deposition rate does not change with the change of $E_{c,rev}$, because of diffusion-controlled anodic reaction; (h) the role of agitation on the deposition rate of diffusion controlled reactions.

the catalytic surface through adsorption and dehydrogenation of hypophosphite ions and/or to the multistep complexity of reduction and oxidation reactions (Magagnin et al. 2011).

1.8 THE ROLE OF DEPOSITION PARAMETERS AND CHEMICAL COMPONENTS ON ELECTROLESS NICKEL COATING PROCESS

1.8.1 The Role of Deposition Parameters

Before going into details of bath components, the role of deposition parameters on the properties of the electroless nickel coatings will be summarized. The deposition parameters other than the composition of the solution are pH, temperature of the solution and agitation. Thus, when compared to electrolytic metal plating, the variables are limited. In electrolytic metal plating, the rate of deposition can additionally be easily controlled by changing the magnitude and shape (e.g., pulse plating) of the applied voltage.

Generally, for Ni–P coatings, a decrease in the pH leads to a decrease in the deposition rate and an increase in the P content of the coating. This effect is more pronounced in acidic Ni–P baths. In acidic baths, the pH of the solutions is not generally allowed to go below 3.5–4 because the deposition rate decreases very dramatically (Figure 1.9). The reason for the decrease in deposition rate can be explained with Evans diagrams (see Figure 1.8). With the decrease of pH, the contribution of hydrogen evolution cathodic reaction increases and oxidation rate of the reductant decreases (Equation 1.7), thus leading to a decrease in the deposition rate. Generation of hydrogen ions during deposition leads to a decrease in pH during deposition. The decrease in pH not only results in lowering of the deposition rate and the increasing of P content (Figure 1.9) but also increases the stability of the bath. Additionally, by lowering the pH value, the adhesion of coating to the substrate increases, internal stress shifts to compressive direction, porosity decreases in the coatings, and the solubility of the phosphite increases (reaction product of the reductant), thus extending the life of the bath. However, as explained above, because incremental changes in pH dramatically change P content (e.g., a change in pH from 4.5 to 4 increases the P content from 9–14), very good complexing agents (buffers) are needed to keep the pH of the acidic Ni–P baths constant. This class of Ni–P electroless deposition (acidic) baths are suitable

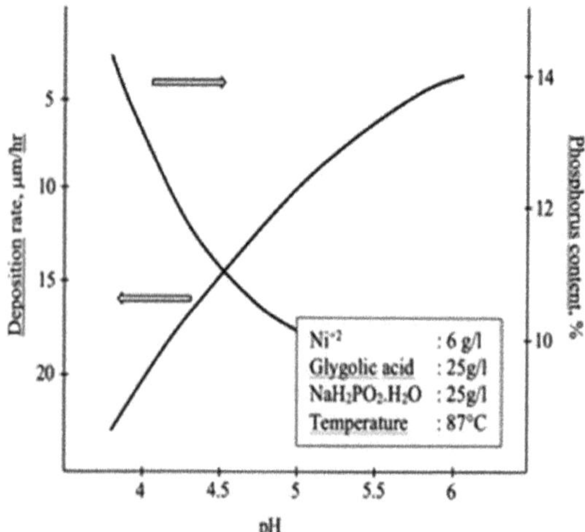

FIGURE 1.9 Effect of solution pH on deposition rate and P content of the coatings. (Adapted from Mallory, G.O., The Fundamental Aspects of Electroless Nickel Plating, In *Electroless Plating: Fundamentals and Applications*, American Electroplaters and Surface Finishing Society, New York, pp. 1–56, 1990.)

for deposition of low (3%–5%), medium (6%–9%), and high (10%–14%) phosphor-containing deposits by adjusting the deposition parameters and composition of the solution. Because the main parameter defining the P content is the pH of the solution, the use of different complexing agents (buffers) may be required for different P contents. As will be given in the other chapters of the book, P content of the coatings has a strong influence on the physical (conductivity, hardness, magnetic, thermal expansion) and chemical (corrosion resistance) properties. Ni–B baths based on DMAB show similar features as in Ni–P baths with respect to changes in pH.

The other types of Ni–P baths are alkaline; these baths are generally preferred for the deposition on temperature- and acid-sensitive materials (polymers) because it is possible to use these baths at lower temperatures (40°C–75°C) than acidic baths (85°C–90°C). The stability of these baths is lower than the acidic baths because the potential of the reductant reaction is higher compared to the ones in acidic solutions (Table 1.2). The lowest operating pH of these baths is approximately 8 baths and generally does not allow to deposit coating with high P and B content (DMAB). Contrary to acidic baths, the decrease in the pH of the solution does not dramatically increase the P content in these baths but lowers the deposition rate. On the other hand, Ni–B baths based on $NaBH_4$ is a special case of baths that should be treated separately because they can only be kept stable at high pH values.

Temperature is another key parameter in electroless nickel plating. As high activation energy is required for the initiation of chemical and/or electrochemical reactions on catalytic metal surface, high temperatures are required to carry out the deposition process. The rate of deposition shows an exponential relation with temperature (Mallory 1990), indicating the necessity of overcoming energy barrier for the initiation of the reaction. Generally, in acidic baths, deposition temperatures are higher than the alkaline baths. The control of deposition temperature for alkaline baths is more critical than acidic ones because increased temperatures may lead to bath decomposition.

The solution mixing or agitation is one of the parameters that have not attracted necessary attention in electroless deposition research. There are very limited studies in the literature that systematically investigated the role of transport in the kinetics of electroless nickel deposition processes. In this book, because a section is specifically devoted to ultrasonic-aided electroless deposition, the possible role of agitation on electroless deposition will be briefly discussed.

Agitation of the solution has direct effects on the realization of oxidation and reduction reactions on the metal surface through:

1. Transport of reactants and reaction products from and away from the reaction interface (avoiding the local pH changes, removal of hydrogen bubbles from the surfaces, and easier chemistry equilibration in recesses and cavities).
2. Increasing the reaction rate if reactions involve diffusion of ions (transport-controlled) by increasing the limiting current.

1.8.2 Chemical Components of Electroless Nickel Deposition Baths and Their Functions

An electroless nickel bath is basically composed of nickel ion sources and reductants (Table 1.4). Other than these components stabilizers, complexing agents and wetting agents are required for a properly functioning electroless bath. The classification of electroless nickel bath and their main constituents are presented in Table 1.3.

1.8.2.1 Main Components, Nickel Ion Sources, and Reductants

The main components of electroless nickel baths are nickel ion sources and reductants. Basically, any nickel salt that can dissolve in water-based solution to give nickel in ionic form can be used as nickel ion source. However, for reasons of availability, solubility, and the knowledge accumulated about their behavior because of their extensive use in electrodeposition, nickel sulfate and nickel chloride are the most widely used nickel ion sources. When depositing on aluminum and steel, chloride-containing baths are not preferred because of the risk of inducing localized corrosion on these substrates. Nickel hypophosphite is also used as nickel ion source; however, its cost is the major drawback. This nickel salt can be directly used in acidic electrolytes by adjusting the solution pH to not allow them to precipitate as hydroxides. For using these baths at high pH values (alkaline baths), complexing agents such as amines are needed.

TABLE 1.3
Classification of Electroless Nickel Bath and Their Main Constituents

Bath Type	Pure Ni	Acid Ni–P and Ni–B	Alkaline Ni–P and Ni–B
pH	10–11	4–5.5 (low to medium P and B) 5.5.–6.5 (low P and B)	8.5–14
Temperature	85–90	75–95	25–100
Deposition rate (micron/hr)	6–12	10–25	10–15
Metal source	Nickel acetate	Nickel sulfate, nickel chloride	Nickel sulfate, nickel chloride
Reductant	Hydrazine	Na hypophosphite for Ni–P Dimethylamine borane for Ni–B	Na hypophosphite for Ni–P Dimethylamine borane for Ni–B Sodium borohydride for Ni–B
Complexing agents	EDTA glycolic acid	Citric, lactic, glycolic, propionic acids, sodium citrate, succinic acid	Citric, lactic, glycolic and propionic acids, sodium citrate, sodium acetate, sodium pyrophosphate, ammonium and amine salts
Stabilizers	—	Thiourea, lead acetate, heavy metal salts, mercaptotriazoles	Thiourea, lead acetate, heavy metal salts, thio-organic salts, thallium nitrate, selenium salts
Wetting agents	Ionic, non-ionic surfactants	Ionic, non-ionic surfactants	Ionic, non-ionic surfactants

The number of reductants that can be used during electroless nickel deposition is limited. The reductants are hydrazine, hypophosphites, DMAB, and sodium borohydride. Although there are a number of redox couples that have the ability to reduce nickel ions, the above listed are the only reductants that can be used for electroless nickel plating. Different from other redox couples, these reductants require a catalytic surface for the initiation of their function, which makes them special and very suitable for coating applications. This property of the reductants allows deposition only on the substrates without homogeneous nucleation reactions in the solution. The reason for this selectivity is explained by the ability of the surfaces to catalyze dehydrogenation reactions, which is an important step for the proper functioning of the reductants.

Pure electroless nickel coatings can only be produced by using hydrazine as the reductant. The share of this type of electroless plating is very low compared to the Ni–P and Ni–B-based coatings. The Ni–P and Ni–B coatings can be described as metastable, saturated alloys in their as-deposited amorphous or nanocrystalline state. The most widely used electroless nickel coatings are produced with the use of hypophosphites as reductants. As discussed in the introduction of this section, it is possible to produce a wide range of amorphous-nanocrystalline Ni–P alloy coating with varying P contents. These alloys can be used in as-deposited amorphous state (usually the ones with high P content) for better corrosion protection. Ni–P coatings with lower P content are generally heat-treated for increased hardness by the precipitation of nickel phosphides.

The third group of electroless nickel coatings is Ni–B alloy coatings. For the production of these coatings, two different types of reductants are used. These are DMAB and sodium borohydride. These coatings are harder than the phosphorus-containing coatings and specifically used for tribological applications.

1.8.2.2 Complexing Agents

The functions of complexing agents are to keep the pH level stable (buffering), to prevent the precipitation of nickel as phosphides/or borides and to reduce the concentration of free nickel ions. The majority of the complexing agents in electroless nickel plating are organic acids or their salts. Pyrophosphates and ammonia or ammonium salts are two of the few inorganic complexing agents. Pyrophosphates are mainly used in alkaline nickel baths; on the other hand, ammonia and ammonium salts are mainly used for pH control in alkaline baths. Other acids used for buffering action are acetic, propionic, glutaric, succinic, and adipic acids.

As well known, when the metal ions are dissolved in water-based solution, they are surrounded with water molecules, the number of molecules that are attached to the ion is named as coordination number. For nickel, this number is 4 or 6. Metal ion complexes form in cases when these water molecules are replaced with another ion or molecule. The molecule or ion that replaces water molecule is named as complexing agents. Complexing of the ion changes its chemical and electrochemical properties. With increasing stability of the complex, it becomes more difficult to remove the solvation sheath to free and make ready the nickel ion for reduction. As a result, in baths where a complexing agent with higher stability constant is used, the rate of nickel ion reduction will be lower. Accordingly, in a bath with citrates, the rate of deposition will be lower but the stability of the bath will be higher (Table 1.4). This is the main reason why citrates are commonly used as complexing agents in relatively less stable alkaline baths.

TABLE 1.4
Commonly Used Complexing Agents in Electroless Nickel Baths and Their Stability Constants

Anion	Acid	Stability pK = −logK
Acetate	CH_3COOH	1.5
Succinate	$HOOCCH_2CH_2COOH$	2.2
Aminoacetate	NH_2CH_2COOH	6.2
Malonate	$HOOCHCH_2COOH$	4.2
Pyrophosphate	$H_2O_3POPO_3H_2$	5.3
Malate	$HOOCCH_2(OH)COOH$	3.4
Citrate	$HOOCCH_2(OH)C(COOH)_2$	6.9

Source: Scliesinger, M., Electroless Deposition of Nickel, In *Modern Electroplating*, Paunovic, M. and Schlesinger, M. (Eds.), 5th ed., Vol. 111, The Electrochemical Society Series, John Wiley & Sons, Hoboken, NJ, 2010.

Most of the complexing agents for electroless nickel are effective in a pH range of 4–6. In cases of alkaline baths, the choice of complexing agents should be made accordingly. In alkaline Ni–P baths, citrates and pyrophosphates are mainly used as complexing agents along with ammonium chloride as pH level stabilizer (Table 1.5). In the case of Ni–B coatings from borohydride baths which are highly alkaline, EDTA or tartrate salts are used as complexing agents.

1.8.2.3 Stabilizers

As mentioned in the previous sections, electroless deposition baths are metastable, which are prone to decompose by themselves as a result of homogeneous reactions within the solution. Bath decomposition starts with extensive gas evolution in the bath followed by the precipitation of phosphides or borides of nickel depending on the type of reductant used. Although the use of stabilizers is not compulsory for the functioning of the electroless deposition baths, in all industrial-scale baths, these components are added to the bath for minimizing the risk of self-decomposition. Due to the local overheating or the homogeneous localized reduction of nickel ions with hydroxyl ions, very fine nickel metal particles form within the solution acting as a very efficient heterogeneous reaction center. The other sources of solid particles that may lead to decomposition are the ones that are dragged into solution in the form of dust, fumes or other contaminants. There are four classes of stabilizers (Mallory and Hajdu 1991):

- Class I: Compounds of group VI elements, Se, S, and Te;
- Class II: Oxyanions: AsO^{-2}, IO^{-3}, and MoO_4^{-2};
- Class III: Heavy metal ions: Sn_{+2}, Pb_{+2}, Hg_{+2}, Sb_{+2}, and Tl_{+2};
- Class IV: Unsaturated organic acids: Maleic and itaconic acids.

TABLE 1.5
Some Exemplary Electroless Nickel Bath Compositions That Use Hypophosphite as Reducing Agent

	Acid Baths				Alkaline Baths			
Bath Constituents (g L^{-1})	**1**	**2**	**3**	**4**	**5**	**6**	**7**	**8**
Nickel chloride hexahydrate	30	30	—	21	26	30	20	—
Nickel sulfate hexahydrate	—	—	25	—	—	—	—	25
Sodium hypophosphite monohydrate	10	10	23	24	24	10	20	25
Hydroacetic acid	35	—	—	—	—	—	—	—
Sodium citrate	—	12.6	—	—	—	84	10	—
Succinic acid	—	—	—	7	—	—	—	—
Sodium flouride	—	—	—	5	—	—	—	—
Lactic acid	—	—	—	—	27	—	—	—
Propionic acid	—	—	—	—	2.2	—	—	—
Ammonium chloride	—	—	—	—	—	50	35	—
Sodium pyrophosphate	—	—	—	—	—	—	—	50
Lead ion (ppm)	—	—	1	—	2	—	—	—
pH	4–6	4–6	4–8	6	4–6	8–10	9–10	10–11
Temperature (°C)	100	100	85	100	100	95	85	70

Source: Minjer, C.H., *Electrodepos. Surf. Treat.*, 3, 261, 1975.

The main function of these materials is to prevent the homogeneous reactions that lead to the decomposition of the bath in an uncontrolled manner. The general character of stabilizers as mentioned earlier is that they are poisons for hydrogenation/dehydrogenation catalysts.

The functioning mechanism of different stabilizers is expected to be different. The studies conducted on Class I and II stabilizers indicate that they have a strong adsorption tendency on the catalytic surface, which is verified by the increase in the deposition potential (E_{mix}) with stabilizer concentration (Mallory and Hajdu 1991). Sulfur-based Class I stabilizers such as thiourea adsorb very strongly to the surface, leading to the formation of microporosity and integration of S in the coating. Another indication of the adsorptive character of Class I and II stabilizers is the decrease in the deposition rate with the increase in their concentration. By the agitation of the electrolyte, stabilizers start to compete with the reductants for adsorbing on the catalytic surface that results in the decrease in the deposition rate. The concentration of stabilizer in the case of agitated solution may be decreased from 10 to 2 ppm.

Contrary to Class I and II type stabilizers, Class III type stabilizers do not exert an appreciable effect on the deposition potential (E_{mix}). For example, although an increase in the lead concentration from 1 to 10 ppm results in a decrease of deposition rate from 15 to 2 micron/hr, the shift in the deposition potential was observed to be only 16 mV (Mallory and Hajdu 1991). These experimental results clearly show that a mechanism other than adsorption is operational on the catalytic surface in the case of these stabilizers. The working mechanism put forth by the researchers for these stabilizers is based on the model proposed by Cavalotti and Salvago (Cavallotti and Salvago, 1968; Salvago and Cavallotti 1972). According to this model, hydrolysis of nickel ion is an important step during deposition; some of these hydrolyzed species can desorb from the catalytic surface into the double layer and form colloidal particles. The role of heavy metal ions comes into play in this stage; these ions by adsorbing on the colloidal nickel particles act as a repellent by changing the surface charge of them. Thus, these colloidal particles cannot grow to critical sizes that will lead to the initiation of spontaneous decomposition on them.

The action of Class IV of stabilizers is suggested to be similar to the Class I or II stabilizers, i.e., by adsorbing to the catalytic surface. However, there is also experimental evidence that the double bond combining these stabilizers to the colloidal particles can also be an acting mechanism of them.

The stabilizers and their action mechanisms are similar for Ni hypophosphite and Ni-DMAB baths; however, borohydride-based EN baths should again be considered separately because of their very high pH and operating temperature. Thallium salts within the concentration range of 40–50 ppm are widely used as stabilizers in this bath.

The use of toxic chemicals as stabilizers (Class I and II) has become a serious environmental concern in electroless plating. The ongoing efforts for replacing toxic stabilizers from the electroless bath will be considered in the coming chapters of this book.

REFERENCES

Brenner, A. 1984. "Reminiscences of Early Electroless Plating." *Plat. Surf. Finish.* 71 (7): 24–27.

Brenner, A., and G. E. Riddell. 1946. "Nickel Plating on Steel by Chemical Reduction." *J. Res. Natl. Bur. Stand.* 37 (1): 31. doi:10.6028/jres.037.019.

Breteau, P. 1911. "Nickel Spontaneous Decomposition by Hypophosphite." *Bull. Soc. Chim.* 9 (4): 515.

Cavallotti, P., and G. Salvago. 1968. "Studies on Chemical Reduction of Nickel and Cobalt by Hypophosphite." *Electrochim. Metallorum.* 3: 239.

Djokic, S. 2014. *Modern Aspects of Electrochemistry*, Vol. 57. New York: Springer. doi:10.1007/978-1-4939-0289-7.

Donahue, F. M. 1972. "Interactions in Mixed Potential Systems." *J. Electrochem. Soc.* 119: 72–74.

Elze, J. 1960. "Studien über die Inhibitorwirkung bei elektrolytischerundgesamtstromloser Nickelabscheidung." *Metall* 14: 104.

Flis, J., and D. J. Duquette. 1997. "Nucleation and Growth of Electroless Nickel Deposits on Molybdenum Activated with Palladium." *J. Electrochem. Soc.* 37 (1): 85–90. doi:10.1149/1.2115541.

Gorbunova, K. M., M. V. Ivanov, and V. P. Moiseev. 1973. "Electroless Deposition of Nickel-Boron Alloys." *J. Electrochem. Soc.* 120 (5): 613–618.

Hersch, P. 1955. "No Title." *Trans. Inst. Met. Finish.* 33: 417.

Kanani, N. 2004. *Electroplating Basic Principles.* Burlington, VT: Elsevier.

Lukes, R. M. 1964. “The Mechanism for the Autocatalytic Reduction of Nickel by Hypophosphite Ion.” *Plating* 51: 969.

Magagnin, L., P. Cojocaru, and P. L. Cavallotti. 2011. “Electrochemical Behaviour of Nickel ACD-Electroless Deposition.” *ECS Trans.* 33 (18): 1–6.

Mallory, G. O., and J. B. Hajdu. 1991. *Electroless Plating: Fundamentals and Applications.* New York: American Electroplaters and Surface Finishing Society.

Mallory, G. O. 1990. “The Fundamental Aspects of Electroless Nickel Plating.” In *Electroless Plating: Fundamentals and Applications*, pp. 1–56. New York: American Electroplaters and Surface Finishing Society.

Martins, J. I., and M. C. Nunes. 2016. “On the Kinetics of Copper Electroless Plating with Hypophosphite Reductant.” *Surf. Eng.* 32 (5): 363–371. doi:10.1179/17432944 15Y.0000000066.

Meeraker, J. Van der. 1981. “On the Mechanism of Electroless Plating: II. One Mechanism for Different Reductants.” *J. Appl. Electrochem.* 11: 395–400.

Minjer, C. H. de. 1975. “Some Electrochemical Aspects of the Electroless Nickel Process with Hypophosphite.” *Electrodepos. Surf. Treat.* 3: 261. doi:10.1016/0300-9416(75)90004-8.

Müller K., 1960, “Elektrochemishe Thermodynamik der Stromlosen Metallabscheidung” *Metalloberflache*, 14, 65.

Nikiforova, A. A., and G. A. Sadakov. 1967. “Physicochemical Principles of Nickel Plating.” *Sov. Electro. Chem.* 3: 1076.

Paunovic, M. 1968. “Electrochemical Aspects of Electroless Deposition of Metals.” *Plating* 51: 1161–1167.

Paunovic, M. 1983. “Electrochemical Aspects of Electroless Nickel Deposition.” *Plat. Surf. Finish.* 70: 62–66.

Popov, K. I., S. Djokic and B. Grgur. 2002. “Metal Deposition without an External Current.” In *Fundamental Aspects of Electrometallurgy*, pp. 249–270. New York: Kluwer Academic Publishers.

Pourbaix, M. 1966. *Atlas of Electrochemical Equilibria in Aqueous Solutions*, Vol. 1. London, UK: Pergamon Press. doi:10.1039/c5cc08796a.

Riedel, W. 1989. *Electroless Nickel Plating.* London, OH: Finishing Publication Ltd.

Riedel, W. 1991. *Electroless Nickel Plating Riedel.* Metals Park, OH: ASM International.

Salvago, G., and P. L. Cavallotti. 1972. “Characteristics of the chemical reduction of Ni alloys with hypophosphite.” *Plating* 59: 665.

Scliesinger, M. 2010. “Electroless Deposition of Nickel.” In *Modern Electroplating*, Paunovic, M., and Schlesinger, M. (Eds.), 5th ed., Vol. 111. The Electrochemical Society Series. Hoboken, NJ: John Wiley & Sons. doi:10.1149/1.2425993.

Shipley, C.R. 1984. “Historical Highlights of Electroless Plating” *Plating and Surface Finishing* 71 (6), 24–27.

Wurtz, A. 1844. “Compt Rendus Academie of Science Paris.” *Ann. Chim. et Phys.* 3 (11): 18.

Zhang, B. 2016. *History–From the Discovery of Electroless Plating to the Present.* doi:10.1016/B978-0-12-802685-4/00001-7.

Zhang, B. W., S. Z. Liao, H. W. Xie, and H. Zhang. 2014. “Progress of Electroless Amorphous and Nano Alloy Deposition: A Review–Part 2.” *T I Met. Finish.* 92 (2): 74–80. doi:10.1 179/0020296714Z.000000000168.

2 Activation, Initiation, and Growth of Electroless Nickel Coatings

Esteban Correa, Alejandro Alberto Zuleta Gil, Juan G. Castaño, and Félix Echeverría

CONTENTS

2.1 SUBSTRATE CLEANING

By definition, electroless coatings can only be formed on catalytic surfaces, that is, a surface capable of providing the necessary conditions for the oxidation–reduction reactions to occur. Therefore, substrate surface preparation is an essential primary treatment before coating any substrate. The coating performance is significantly influenced by its ability to adhere properly to the substrate material. Usually, the correct preparation of the substrate surface is considered the most important factor that affects the ultimate success of the surface treatment (Rao et al. 2005). The presence of surface pollutants such as waxes, greases, abrasives, impurities from polishing or shot blasting, machining oils, fingerprints, oxide layers, among others, (ASTM B319-91 2014) may affect the nucleation and growth of nickel coating. Such a presence will favoring the formation of pores and reducing the adhesion to the substrate, which will have a detrimental effect on its resistance to corrosion and wear (National Physical Laboratory 1982; Kerr et al. 1996). In brief, substrate cleaning is mainly associated with the removal or elimination of particles (organic or inorganic) from its surface; but also refers to leaving the substrate surface in a totally known and controlled condition to guarantee the success of the surface treatment.

2.1.1 Ferrous Substrates

The physical and chemical characteristics of the substrate surface, as well as the composition of the base metal, influence the selection of the cleaning process prior to the electroless nickel coating deposition process (Mandich 2003). For this reason, cleaning processes have been established for the various ferrous substrates, such as low-, medium-, and high-carbon steels and stainless steels.

Except for stainless steels, most ferrous substrate surfaces exhibit little resistance to corrosion. This means that the cleaning process must go beyond the removal of dirt, oils, and greases to include the removal of sediments and oxide layers formed on the substrate. For these purposes, procedures that involve cleaning surfaces through mechanical polishing with abrasive paper, shot peening, girt-blasting, or sandblasting are commonly used.

Following this procedure, dirt, oils, and greases are usually removed from the surface of ferrous substrates (degreasing) using organic solvents such as ethanol (Belakhmima et al. 2017; Lee et al. 2018; Oliveira et al. 2017) or acetone (Vitry et al. 2017; Zuleta et al. 2009; Das and Sahoo 2012). Then, the removal of impurities and additional oxide layers present on the surface is carried out. For this purpose, two cleaning processes are used: (i) alkaline cleaning and (ii) acid pickling. These procedures can be performed independently or consecutively, depending on the amount of pollutants found on the surface and the nature of the substrate.

Alkaline cleaning is a procedure that involves the use of alkaline solutions (usually pH > 11) containing sodium (Staia et al. 1997; Czagány et al. 2017; Belakhmima et al. 2017; Oliveira et al. 2017) or potassium hydroxide (Goettems et al. 2017). Such alkaline solutions are used in compositions that vary from 30 to 120 g/L NaOH or KOH (ASTM B322-99 2014) at temperatures ranging between 60°C and 90°C (Olarewaju Ajibolam et al. 2015; Oliveira et al. 2017; Staia et al. 1997;

Czagány et al. 2017). Solutions containing mixtures of sodium hydroxide (50 g/L), sodium carbonate (30 g/L), sodium phosphate (30 g/L) (Wang et al. 2018; Fayyad et al. 2018), and sodium silicate (Xie et al. 2016) are also commonly used. Due to the pH of the solutions, alkaline cleaning has the advantage of preventing the dissolution of the substrate during the process, which is why it is one of the most commonly used methods for cleaning low-, medium-, and high-carbon steels. Some of the conventional cleaning methods for this type of substrates are specified in ASTM B183 (ASTM B183-18 2018).

On the other hand, high-carbon steels (C weight > 0.35%) and hardened low-carbon steels require a variation in the cleaning process (ASTM B319-91 2014), which involves additional processes for the alkaline cleaning process. This is due to the presence of a dark layer (smut) after the initial cleaning, which must be removed before the deposition process because it can decrease the adherence of the electroless coating to the substrate. This process is performed by immersing the substrate in HCl (Karrab et al. 2013; Contreras et al. 1999; Lin and He 2005) or H_2SO_4 (Voorwald et al. 2007) solution, in compositions that vary between 5% and 20% (V/V). The immersion time in this type of acidic solutions must be short due to the high susceptibility that such substrates exhibit to hydrogen fragilization (ASTM B319-91 2014). In cases where smut is strongly attached to the surface, the substrate must be dipped in a sodium cyanide solution (26 g/L) (ASTM B319-91 2014).

Stainless steels require a cleaning process different from those described above, due to the presence of a thin Cr_2O_3 film (from 3 to 5 nm) on its surface (Kerber and Tverberg 2000), which prevents the initial nickel reduction reaction from occurring. For this reason, after cleaning and before immersion in the electroless nickel bath, the chromium oxide film must be removed (Nikitasari and Mabruri 2016; ASTM B254-92 2014). For this purpose, the following procedure is used (Zuleta 2008; Zuleta et al. 2009): (i) mechanical cleaning of the surface with either abrasive paper, shot peening, gritblasting, or sandblasting, (ii) washing with distilled water, (iii) alkaline cleaning at 70°C for 15 min in a solution containing 25 g/L NaOH, 25 g/L Na_2CO_3, 30 g/L Na_3PO_4, and 8 g/L Na_2SiO_3, (iv) immersing test pieces in H_2SO_4 (24%) solution for 60 s, and (v) rinsing with distilled water.

The steel substrates used to generate electroless nickel coatings include SAE 1018 (Belakhmima et al. 2017), SAE 1020 (Karabulut et al. 2017; Staia et al. 1996), SAE 1040 (Kaya et al. 2008; Mukhopadhyay, Kumar, and Sahoo 2017; Das and Sahoo 2012), SAE 1045 (Dong et al. 2009; Contreras et al. 1999; Lin and He 2005), SAE 4140 (Soares et al. 2017; Vite-Torres et al. 2015), SAE 4340 (Garcés et al. 1999; Nascimento et al. 2001), SAE 304 (Tang and Zuo 2008; Zuleta et al. 2009), SAE 316 (Abdeli et al. 2009; Wang et al. 2015), and gray cast iron (Park and Kim 2018; Yamamoto et al. 1989; Pan et al. 2013a).

2.1.2 Aluminum Alloys

Aluminum is a fairly reactive light metal, and thus, a thin oxide layer (between 40 and 100 Å) forms on its surface when the material is exposed to the atmosphere (Fuente et al. 2007). In addition, traces of dirt may also be commonly found on aluminum or aluminum alloy surfaces after their manufacturing processes. Both the thin oxide layer and the dirt found on the surface isolate the metal, preventing the metal anchor between the coating and the aluminum substrate and causing adhesion issues, blistering, or coating failure (Court et al. 2017; Court et al. 2000).

In this sense, the cleaners most commonly used to remove dirt and the natural oxide layers present on the surfaces of aluminum-based materials are organic solvents such as: (i) acetone or alcohols (Vitry et al. 2012b; Dadvand et al. 2003; Farzaneh et al. 2016); alkaline mixtures of silicates, borates, and phosphates (Court et al. 2017; Court et al. 2000); (ii) alkaline solutions of sodium or potassium hydroxide (Shirmohammadi Yazdi et al. 2013; Ebrahimi et al. 2015); (iii) acid solutions (Akyol et al. 2018) with the presence of fluoridated species (Delaunois and Lienard 2002; Delaunois et al. 2000; Vijayanand and Elansezhian 2014a, 2014b); and/or (iv) mechanical surface polishing with abrasive paper (Sudagar et al. 2013; Sudagar et al. 2010) following the typical metallographic preparation procedures with silicon carbide paper. It is also common to use ultrasonic baths, which help to remove organic debris or particles embedded in the substrate surface (Tsai et al. 2012; Hsu et al. 2004; Ebrahimi et al. 2015). It should be noted that, owing to the increasingly stringent environmental legislation, cleaning solutions containing silicates, borates, phosphates, and fluorinated species have been restricted and replaced by hydroxide alkaline cleaners. In summary, substrate cleaning includes two main aspects: removal of the natural oxide layer by mechanical (mechanical grinding or polishing) or chemical (detergents, organic solvents, or inorganic mixtures) methods and dirt removal with detergents or alkaline solutions. The aluminum substrates used most frequently to generate electroless nickel coatings are the AA6061 alloys (Hamid and Elkhair 2002; Kinast et al. 2014; Sheng 2016; Unigovski et al. 2013; Dadvand et al. 2003; Farzaneh et al. 2016), AA1050 (Khan et al. 2007; Rudnik et al. 2011; Delaunois et al. 2000), LM24 (Franco et al. 2015, 2013), A356 (Hashemi and Ashrafi 2018; Pancrecious et al. 2015), AA2017 (Hino et al. 2005), AA7075 (Kumar et al. 2014; Liew et al. 2013; Oskouei et al. 2016; Rahmat et al. 2013; Hsu et al. 2004; Sudagar et al. 2010), Al-Si (Rajendran et al. 2010), $AlMg_2$ (Szirmai et al. 2010; Takács et al. 2007), AA2024 (Vitry et al. 2008; Vitry et al. 2012b), AA3004 (Shirmohammadi Yazdi et al. 2013), and commercially pure aluminum (Al100) (Arakawa et al. 2014; Akyol et al. 2018; Court et al. 2017; Ebrahimi et al. 2015; Hino et al. 2005).

2.1.3 Magnesium Alloys

The cleaning process of such substrates uses acetones (Baskaran et al. 2006a, 2009, 2006b) or alcohols (Pan et al. 2013b; Stremsdoerfer et al. 2008; Brunelli et al. 2009), which remove the organic substances present on the surface (Das and Sahoo 2011b; Sahoo and Das 2011; Das and Sahoo 2011a, 2012). In addition, the hot mixture of sodium hydroxide and trisodium phosphate (Zuleta et al. 2012; Gu et al. 2005; Song et al. 2006; Zuleta et al. 2011) or other acid solutions (Zhang et al. 2009) helps to remove traces of dirt or small rust layers formed owing to the substrates being exposed to the environment. As with aluminum substrates, magnesium surfaces are usually subject to metallographic preparation processes using silicon carbide paper to remove oxide layers and standardize the surface (Correa et al. 2013a). Gritblasting or sandblasting process is also commonly performed on magnesium substrates

(Correa et al. 2012; Correa et al. 2013a) to remove oxide layers or to generate a known roughness, as electroless nickel coatings exhibit better properties on rough substrates (Vitry et al. 2010; Zuleta et al. 2012). The most commonly used magnesium substrates to generate electroless nickel coatings are the AZ91D (Liu et al. 2010; Correa et al. 2017; Wang et al. 2012a, 2012b; Gu et al. 2005), AZ31B (Zuleta et al. 2017; Chang et al. 2011; Fan et al. 2010), and AM50 alloys (Petro and Schlesinger 2011), as well as commercially pure magnesium (Correa et al. 2012).

2.1.4 Copper

For copper and its alloys, the most widely used procedure is to use acetone at room temperature to remove any traces of dirt, dust, or grease (Tang et al. 2013, 2014; Lin et al. 2016). However, other studies have reported heating acetone at 40°C (Xu 2016), using a diluted HCl solution (De et al. 2016), utilizing a cleaning solution with deionized soap (Contreras-Lopez et al. 2016), or submerging the substrates for 5 min in an unspecified alkaline medium at a temperature between 70°C and 90°C (Ahmad et al. 2015; Rabiatul Adawiyah and Saliza Azlina 2018). For the removal of the thin oxide layer formed on copper and copper alloy surfaces, some authors have reported the immersion of the substrate in H_2SO_4 solution (5 wt%) for 2 min (Tian et al. 2013; Lin et al. 2016), in a solution containing 10 g of $Na_2S_2O_8$, 10 g of NaOH, and 200 mL of distilled water for 5 min at a temperature between 50°C and 60°C (Ahmad et al. 2015), or in a concentrated NaOH solution with no given time or temperature specifications (Rabiatul Adawiyah and Saliza Azlina 2018; Contreras-Lopez et al. 2016). Another study has reported the use of gritblasting and sandblasting to remove the abovementioned oxide layer, but it establishes neither the size of the particles used nor the air pressure or operation time (Ghaderi et al. 2015). However, other studies have simply reported that substrate cleaning (removal of dirt and oxide layer) was performed with acidic cleaning solutions (Strandjord et al. 2002; Fukuda et al. 2018). The most commonly used copper substrates for generating electroless nickel coatings are bronze (Tang et al. 2014; Contreras-Lopez et al. 2016), Cu-DHP (Deoxidation High Phosphorous) alloy (Ghaderi et al. 2015), copper-coated laminated coupons (Oda et al. 2018), and commercially pure copper (Nakao et al. 2003; Yee et al. 2014; Tian et al. 2013; Ahmad et al. 2015; Xu 2016; De et al. 2016; Rabiatul Adawiyah and Saliza Azlina 2018).

2.1.5 Titanium

Information on cleaning procedures for titanium substrates and their alloys is limited. All studies mentioned herein first report using metallographic preparation with silicon carbide paper to standardize the substrate surfaces and then a cleaning process for the removal of dirt or grease. Dabalá et al. (Dabalà et al. 2001, 2004; Brunelli et al. 2009) reported the use of alcohol to degrease the surface of the titanium alloy Ti6Al4V, while another study (Mahmoud 2009) reported the use of acetone for the same purpose, but supplementing the cleaning procedure by immersing the substrates in an alkaline cleaning solution containing 50 g/L of NaOH, 20 g/L of Na_2CO_3, 20 g/L Na_3PO_4, and 2 g/L sulfonic acid at 80°C for 10 min (Gao et al. 2017) or at 50°C for 5 min (Mahmoud 2009). On the other hand, Uma Rani et al. (2010) reported substrate cleaning using a solvent with ultrasonic agitation and degreasing the surface with butanone vapor ($CH_3COC_2H_5$) for 10 min. Now, to remove the oxide layer and scales from the substrate surface, substrates are commonly submerged in a solution containing 400 g/L of HNO_3 and 50 g/L of HF at room temperature (approximately 25°C) for 3 min (Gao et al. 2017; Mahmoud 2009), or first, in a solution containing 500 g/L of NaOH and 100 g/L of $CuSO_4{\cdot}5H_2O$ at 90°C between 15 and 20 min, and, then, in a solution containing 275 mL/L of HNO_3 and 225 mL/L of HF between 20 and 30 s (Uma Rani et al. 2010). The most commonly used titanium substrates for generating electroless nickel coatings are the Ti6Al4V alloy (Brunelli et al. 2009; Dabalà et al. 2001, 2004; Uma Rani et al. 2010) and commercially pure titanium (Gao et al. 2017).

2.1.6 Polymeric Substrates

Due to their nature and a great variety of functional groups they possess, different processes are used for cleaning polymeric substrates. Some of them are quite simple, while others are complex and require the use of highly specialized equipment. In this section, only the simple procedures will be discussed, because the others involve, simultaneously, the following processes: etching, sensitizing, and activation, which will be later discussed in a different section. For the removal of dust particles, dirt, or grease found on polypropylene (PP) or polycarbonate (PC) substrates, ethanol-based ultrasonic cleaning has been reported to be used (Charbonnier and Romand 2003), while for acrylic resin, immersion in a 50% NaOH solution has been reported to be used (Huang et al. 2014). Li et al. (2014) first used NH_4OH solution at 50°C for 10 min. Then, an HCl solution was used at 50°C for 5 min to clean polyethylene terephthalate substrates. On the other hand, carbon fiber-reinforced polyether ether ketone was sanded with SiC paper no. 2000, followed by sonication in acetone (Su et al. 2015) solution. Polyimide was submerged sequentially in a 10 wt% solution of NaOH and 10 wt% solution of HCl (Woo et al. 2009). Substrates manufactured with the KMPR® photoresist polymer were rinsed in deionized water and isopropyl alcohol and then dried with nitrogen stream (Zeb et al. 2017). These processes were performed for removing dirt or increasing the roughness of the substrates and thus, improving the coating adhesion to the substrate.

2.1.7 Ceramic Substrates, Ceramic, and Metallic Powders

As discussed above, alkaline cleaners are used to remove dirt or grease from substrates, while acidic cleaners are used to remove thin oxide layers. Arakawa et al. (2014) reported the cleaning of silicon-nitride substrates using an alkaline solution (chemical reagents were not specified) at 80°C for 5 min and their subsequent immersion in an acidic solution containing sulfuric acid at room temperature for 1 min. On the other hand, acetone-based ultrasonic cleaning for 10 min has been reported to remove traces of grease or dirt present on the surface of carbon fiber (Balaraju et al. 2016) or silicon flake (Bernasconi et al. 2016) substrates. However, some studies have also reported the cleaning of wafer-type silicon substrates using piranha solution (H_2O_2:H_2SO_4, ratio 1:2) for 20 min at 80°C (Delbos et al. 2015) or consecutive immersion in concentrated hydrofluoric (HF) acid and dilute nitric acid (Karmalkar 1997) to remove dirt, grease, or oxides present on the surface. To cover particles through the electroless nickel technique, the use of alkaline cleaners has been reported, without specifying the temperature, time, or nature of the chemical reagents, to remove dirt or boron carbide grease (Deepa et al. 2014). On the other hand, silicon carbide particles are cleaned with acetone for 15 min (Dikici et al. 2011; Mohandas and Radhika 2017) or 25 min (Pázmán et al. 2012a, 2012b) or with isopropyl alcohol (2-propanol) (Umasankar et al. 2014). Alumina nanoparticles have a somewhat more complex process, because, to clean their surface from the adsorbed chemical species, a vacuum oven was used to dry the nanoparticles at 110°C for 24 h (Zheng et al. 2012). Acetone was also used to clean ceramic vanadium carbide and tungsten carbide powders (Giampaolo et al. 1997), as well as mixtures of boron carbide and nano-titanium boride powders (Saeedi Heydari et al. 2018). Regarding metallic powders, Ramaseshan et al. (2001) reported the cleaning of a mixture of titanium and aluminum powders (ratio 80:20) by using a solution containing chemical reagents such as ammonium chloride, sodium citrate, and sodium fluoride. The mixture of metallic powders was immersed in a glass beaker with the cleaning solution for 1 h at room temperature to remove the oxide layer present on the surface of the powders. The same procedure was used to remove the oxide layer from the surface of titanium powders elsewhere (Zangeneh-Madar and Jafari 2012). However, titanium powders have also been cleaned by immersion in an alkaline bath (Barbat and Zangeneh-Madar 2014).

2.2 SUBSTRATE ACTIVATION

In general, for the formation of electroless coatings, the substrate surface must be catalytic because this is a mixed potential process in which the cathodic and anodic electron transfer reactions occur at equal speeds on an electronically conductive surface (Gafin and Orchard 1993; Paunovic and Schlesinger 2006). On this surface, which is simply an electroless substrate–solution interface, coating formation occurs and involves the adsorption of the reducing agent. This agent reacts with the water from the electroless solution or bath, generating atomic hydrogen on the surface and thus, favoring the reduction in nickel ions and the subsequent coating growth (Mallory and Hajdu 1990; Babhale et al. 2003). These surfaces, called catalytic surfaces (Abrantes and Correia 1994), are, for example, the surfaces of some metallic substrates or metal particles dispersed on non-catalytic substrates (Paunovic and Schlesinger 2006), such as the surfaces of ceramic or polymeric materials. Although some metallic substrates have catalytically active surfaces, in most cases, they have oxide layers that can interfere with the electroless coating formation process. Therefore, besides the pretreated cleaning and pickling processes, a previous "activation" process must be performed to improve the catalytic characteristics of the surface and guarantee the formation of an adherent coating (Pancrecious et al. 2018; Lee et al. 2018). In the present work, the "activation" process may include etching, sensitization, pickling, or conversion coating processes, depending on the nature of the substrate.

2.2.1 Ferrous Substrates

Although some metallic substrates have catalytically active surfaces, in most cases, they have oxide layers that can interfere with the electroless coating formation process. Thus, besides the cleaning and pickling processes described in the preceding section, a previous "superficial activation" process must be performed to improve the catalytic characteristics of the surface and guarantee the formation of an adherent coating on the substrate (Lee et al. 2018). This must also be done on the surfaces of passivable ferrous materials, which do not favor the deposition process due to the formation of a non-catalytic ceramic film. Therefore, their surface must be activated through the generation of catalytically active nuclei. To date, several superficial activation procedures have been reported, for example, (i) photochemical activation, (ii) mechanical activation, and (iii) electrochemical activation (Paunovic and Schlesinger 2006).

2.2.1.1 Photochemical Activation

It is based on chemical photoelectron kinetics and the formation of nanoparticles with high catalytic activity (Eliaz and Gileadi 2018). Usually, such activation involves a photochemical reaction that precedes the electrochemical reduction process. However, such activation is implemented on a small scale at the industrial level, and therefore, will not be detailed in this paper. Some of the photochemical activation processes have been described by Paunovic and Schlesinger (2006).

2.2.1.2 Mechanical Activation

The mechanical activation process is one of the most commonly used procedures at the industrial level, because, besides removing pollutants and oxide layers present on the substrate surfaces, it fosters a surface with greater roughness and improves the mechanical adhesion of the coating to the substrate (Correa et al. 2012). The usual method is sandblasting or gritblasting, in which a pressurized jet of sand or alumina particles collides with the substrate surface. During this surface process, there is a severe plastic deformation of the substrate surface and an increase in the amount of microstructural defects, which in turn increase the amount of energy stored on the surface (Guo et al. 2012; Harris and Beevers 1999; Yunes and Azadbeh 2013). This favors the presence of active sites suitable for the adsorption of nickel ions and the reduction in agents responsible for the autocatalytic reaction. Figure 2.1 shows the scanning electron micrographs of a steel St-37 surface with different roughness values, which was immersed in an electroless NiB solution for different times. As can be seen, the amount of nickel nodules present increases with the steel roughness, while the nodule size decreases. After 30 s of immersion, the roughest sample is almost completely covered by nickel nodules. After 60 s, the surfaces are completely covered. However, the rougher sample seems to be homogeneously more covered. This implies that the formation of electroless nickel coatings is favored when substrates exhibit higher roughness (Vitry et al. 2010).

2.2.1.3 Electrochemical Activation

The electrochemical activation process involves the reduction of metallic particles on the substrate surface to favor the creation of catalytically active sites that support

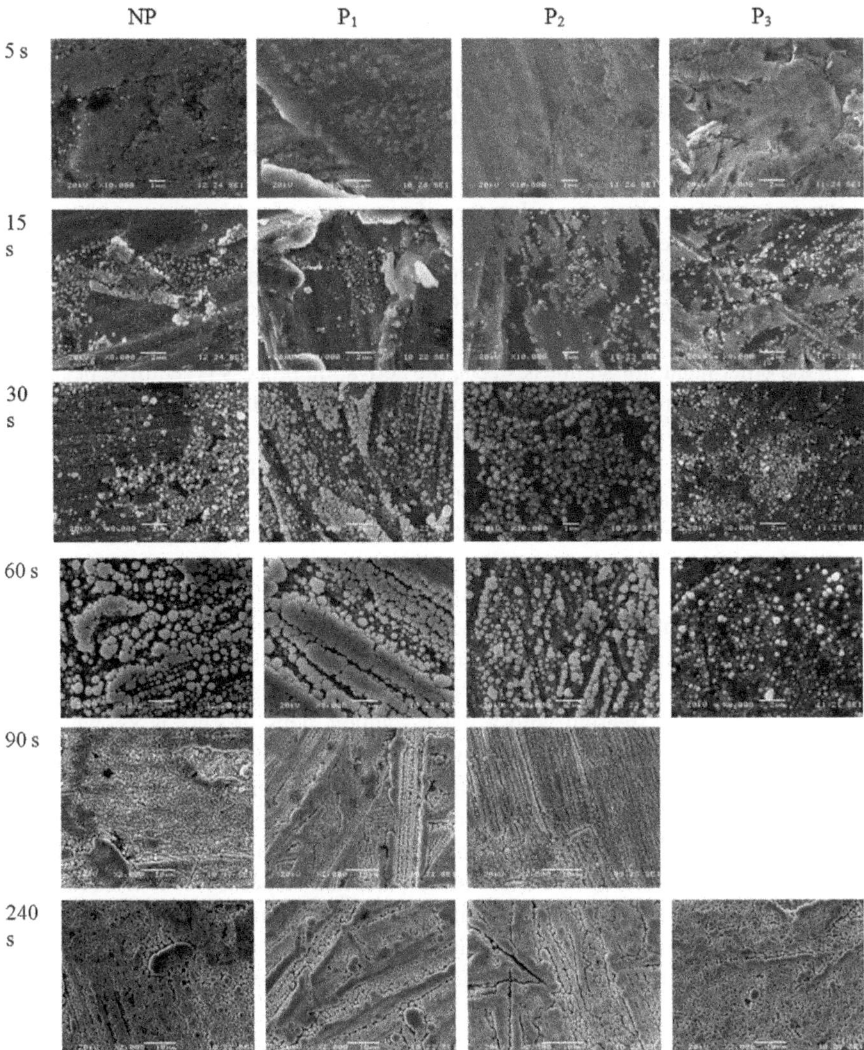

FIGURE 2.1 Scanning electron micrographs (secondary electrons) for a St-37 steel surface with different roughness values, immersed for different times in an electroless Ni–B bath. NP: $R_a = 0.674 \pm 0.111$, P_1: $R_a = 0.559 \pm 0.144$, P_2: $R_a = 0.294 \pm 0.084$, and P_3: $R_a = 0.476 \pm 0.120$. (Reprinted from *Mater. Sci. Eng. B Solid State Mater. Adv. Technol.*, 175, Vitry, V. et al., Initiation and Formation of Electroless Nickel-Boron Coatings on Mild Steel: Effect of Substrate Roughness, 266–273, Copyright 2010, with permission from Elsevier.)

the initiation of oxidation and reduction reactions. For the latter, one of the methods used is the immersion of the substrate in an acidified solution with Pd^{2+} ions, which provides this activity to the substrate surfaces of various materials. However, the use of palladium increases the production costs because this material is expensive (19,000–26,000 USD/kg) (Blickensderfer 2018). For this reason, industries seek to reduce or eliminate the use of this chemical element from the electroless process. In addition, the use of palladium during the activation process may compromise the stability of the electroless bath because palladium may not be completely removed after the activation process (Krishnan et al. 2006).

Overall, for non-passivable ferrous substrates, an activation process prior to the electroless coating deposition process is not required because the surface of these substrates is catalytically active. However, if some type of oxide or corrosion product is present on the substrate surface, the deposition process may be inhibited. Consequently, chemical attack activation-based treatments are frequently adopted, using hydrochloric or sulfuric acid solution to remove the corrosion product layers (Liu et al. 2015; Lee et al. 2018).

On the other hand, for stainless steels, the activation process is required because the oxide layers formed on the surface of these substrates are enriched with chromium, which, due to its high affinity with oxygen, promotes the formation of a thin Cr_2O_3 layer on their surface. This layer is usually dense and adheres strongly to the substrate (Li et al. 2005). This oxide layer, being nonconductive or catalytic, requires an activation treatment similar to that used for ceramic or polymeric substrates. Three procedures are generally used to activate stainless-steel surfaces: (i) palladium activation, (ii) nickel strike, and (iii) chemical attack. The first two procedures are also commonly used for the activation of ceramic or polymeric substrates. After completing the cleaning and degreasing processes, the first activation process involves sensitizing the substrate surface by immersing the substrate in a $SnCl_2$ solution before treating with a $PdCl_2$ solution. This process is known as a two-step activation process. A typical surface sensitization solution contains 10 g/L of $SnCl_2$ and 40 mL/L of HCl (37%) (Paunovic and Schlesinger 2006). During this process, the substrate treated in tin chloride is sensitized to adsorb Pd^{2+} ions on its surface. This occurs due to the formation of tin chloride films in island form, which are hydrolyzed and adhered to the substrate (Wei and Roper 2014; Sviridov et al. 2003). The second step consists of immersing the piece in a nucleation solution, which generally contains 0.1–1.0 g/L $PdCl_2$ and 5–10 mL/L HCl (37%), where the dispersed palladium nuclei acting as catalytic sites are formed according to the following simplified reaction (2.1):

$$Pd^{2+} + Sn^{2+} \rightarrow Pd^0 + Sn^{4+} \tag{2.1}$$

The palladium nuclei formed are distributed as islands of less than 10 Å and as high as 40 Å in diameter (Paunovic and Schlesinger 2006). After these nuclei are present on the substrate surface, the nickel ions are reduced at the activated sites, around the palladium atoms, which are subsequently dissolved during the initial induction period. In addition, this process is also usually done in a single-step activation process, performed through the immersion of the substrate in a solution where $SnCl_2$ and $PdCl_2$ are mixed, forming Sn-Pd complexes, which subsequently transform into colloidal palladium or Sn/Pd alloy particles (Paunovic and Schlesinger 2006; Wei and Roper 2014). It has been found that the beginning of the deposition process on single-step activated substrates requires lesser time as compared to the two-step process

(Wei and Roper 2014). In this case, and similarly, for the two-step activation process, the palladium complexes reduced on the substrate surface provide catalytic nucleation sites at which deposition will occur. In the initial coating formation stages, the deposition on an active palladium surface is characterized by three-dimensional nucleation, growth, and coalescence of the crystals. This nucleation mechanism results in the formation of a continuous coating on the substrate surface (Petro 2014).

Surface activation by immersing the substrate in nickel solutions is commonly referred to as nickel strike. This procedure involves immersing substrates in an electrolytic solution containing nickel ions, which may originate from chemical species such as nickel chloride ($NiCl_2$) or nickel sulfamate ($H_4N_2NiO_6S_2$). The immersion of the substrates in these solutions supports, through chemical deoxidation, the deposition of a Ni^{2+} ion layer on the substrate surface, which is catalytic and favors the subsequent deposition of the electroless coating on its surface (Dai et al. 2014; Mallory and Hajdu 1990). The most widely reported procedure in the literature uses the chemical solution, known as "Woods Nickel Strike," to generate the Ni^{2+} ion layer absorbed on the substrate surfaces. Table 2.1 shows the chemical composition and operating conditions of "Woods Nickel Strike."

The third procedure involves the removal of the chromium oxide layer from the substrate surface. Substrates are immersed in H_2SO_4 (5%) (Zuleta et al. 2009; Lee et al. 2018), HCl (5%) solution (Vitry and Delaunois 2015), or a mixture of both (Khosravipour 2004), for approximately 60 s (Homjabok et al. 2010). Subsequently, the samples are rinsed before the electroless bath. The time interval between the complete removal of the oxide layer and the deposition process must be short to avoid the regeneration of the chromium oxide layer. It has been found that the time required to perform the complete removal of oxide layers on AISI 304 steel using a solution containing 5.5 M of HCl at 55°C is 42 s (Li et al. 2005). However, some authors argue that this process should use H_2SO_4 instead of HCl, because the latter promotes intergranular substrate corrosion (Homjabok 2010). Some of the compositions used for the activation of stainless steels are shown in Table 2.2.

TABLE 2.1
Chemical Composition and Operating Conditions of the Chemical Solution Used to Perform the Nickel Strike Procedure

Woods Nickel Strike	
Nickel chloride	240 g/L
Hydrochloric acid (32% vol.)	320 mL/L
Anodic current density	1 A/dm^2, 30–60 s
Cathodic current density	2 A/dm^2, 2–60 s

Source: Mallory, G.O. and Hajdu, J.B., *Electroless Plating: Fundamentals and Applications*, American Electroplaters and Surface Finishers Society, Orlando, FL, 1990.

TABLE 2.2
Activation Processes for Some Stainless-steel Substrates

Substrate	Solution	Operation Conditions	References
304	HCl (30% vol.)	Temperature: Room Time: 1–5 min	Vitry and Delaunois (2015)
	HCl (20% vol.)	Temperature: 50°C	Khosravipour (2004)
	H_2SO_4 (25% vol.)	Temperature: Room Time: 60 s	Zuleta et al. (2009)
316L	150 mL/L H_2SO_4 (98%) and 20 g/L NaCl	Temperature: Room Time: 1–3 min	Fang et al. (2015)
SAE HNV3	H_2SO_4 (5%)	Temperature: Room Time: 2 min	Goettems et al. (2017)
Stainless steel foam	10 g/L $SnCl_2$ and 0.2 g/L $PdCl_2$	—	Abdel Aal and Shehata Aly (2009)

2.2.2 Aluminum Alloys

The most commonly used aluminum alloys, technologically speaking, have considerable silicon percentages in their chemical composition. The presence of silicon causes some superficial areas of aluminum alloys to be non-catalytic. Therefore, after the cleaning stage, it is almost mandatory to perform procedures to prevent spontaneous aluminum oxidation and remove silicon from the surface. To remove silicon from aluminum alloy surfaces, solutions rich in fluorinated species mixed with concentrated nitric acid are commonly used. Fluoride removes silicon particles, while nitric acid blocks the oxidative action of fluorine on aluminum (Bibber 2009; Delaunois et al. 2000; Delaunois and Lienard 2002). As mentioned above, the presence of non-catalytic sites or traces of oxides inhibits the subsequent formation of electroless coatings because metallic bonds are not generated between the coating and the substrate, thus generating poor coating adhesion, blistering, and failures (Court et al. 2017). Once the oxide layer and the silicon particles have been removed from aluminum alloy surfaces, efforts must be made to keep the surface free for oxide layers and silicon particles. For these purposes, aluminum alloys are subjected to a procedure known as zincating. Zincating is a process that seeks to keep the aluminum surface completely free of aluminum oxides and is carried out by immersing the substrates in alkaline solutions containing zinc (typically zinc oxide and sodium hydroxide) (Takács et al. 2007). This process is governed by the following chemical reactions (Court et al. 2000):

$$Al + 3OH^- \rightarrow Al(OH)_3 + 3e \quad (2.2)$$

$$Al(OH)_3 \rightarrow H_2AlO_3^- + H^+ \quad (2.3)$$

$$Zn(OH)_4^{2-} + 2e \rightarrow Zn + 4OH^- \quad (2.4)$$

$$H^+ + e \rightarrow \frac{1}{2}H_2 \tag{2.5}$$

Chemical reactions (2.2) and (2.3) show the dissolution of aluminum in the alkaline medium, while chemical reactions (2.4) and (2.5) show the deposition of zinc on the substrate surface and the respective hydrogen evolution that accompanies the process.

Although some studies have reported zincate treatments in one step (Franco et al. 2013; Hashemi and Ashrafi 2018), this process is usually performed in two steps (Court et al. 2017; Hino et al. 2005; Vitry et al. 2012c), which is known as a double zincate treatment. In this process, the alloys are first submerged in a zinc bath for a certain period, and then extracted and washed with deionized water. Then, they are submerged in a solution containing nitric acid at 50% v/v for 15 s. Finally, aluminum substrates are submerged again in the zinc solution for a given period (Dadvand et al. 2003). The process of immersion in nitric acid is known as acid stripping and is twofold: (i) to dissolve the alloying elements exposed during the first zincate treatment and (ii) to develop a surface rich in aluminum again (Murakami et al. 2013a). Then, during the second zincate treatment, a softer surface is formed. Figure 2.2 shows the electron micrographs of the aluminum surfaces after having been subjected to a one-step and double zincate processes. It is easy to observe that during the one-step zincate treatment (Figure 2.2a), the surface is fully covered. However, the polishing lines of the substrate can also be observed. An inspection at higher magnification reveals the presence of small zinc grains. Figure 2.2b shows the aluminum surface after the double zincate treatment process, which reveals a smoother surface because the contour surfaces of the substrate are not visible. The polishing lines of the substrate were removed during the acid stripping process. Some pits may also be observed on the surface, which is attributable to the removal of aluminum or intermetallic particles during acid stripping.

Figure 2.3 shows cross sections of an aluminum alloy subjected to the zincating process. This diagram suitably and illustratively summarizes what happens during the double zincate treatment process.

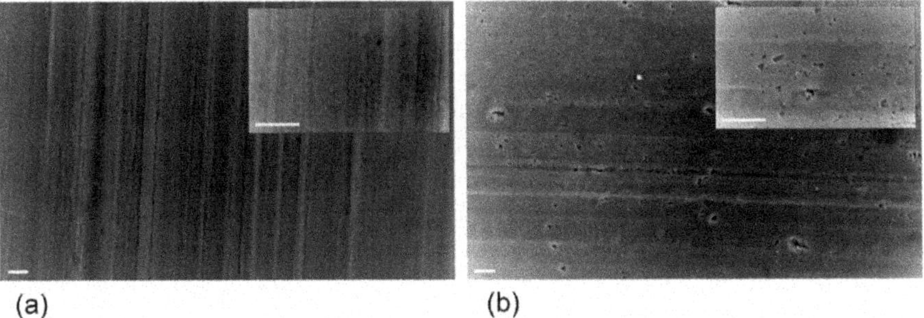

FIGURE 2.2 Electron micrographs of aluminum surfaces after (a) single zincate treatment for 60 s and (b) double zincate treatment for 60 s. The magnification is 500×. The magnification of images in the boxes is 2500×. All scales represent 10 μm. (With kind permission from Taylor & Francis: *Trans. IMF*, Monitoring of Zincate Pre-Treatment of Aluminium Prior to Electroless Nickel Plating, 95, 2017, 97–105, Court, S. et al.)

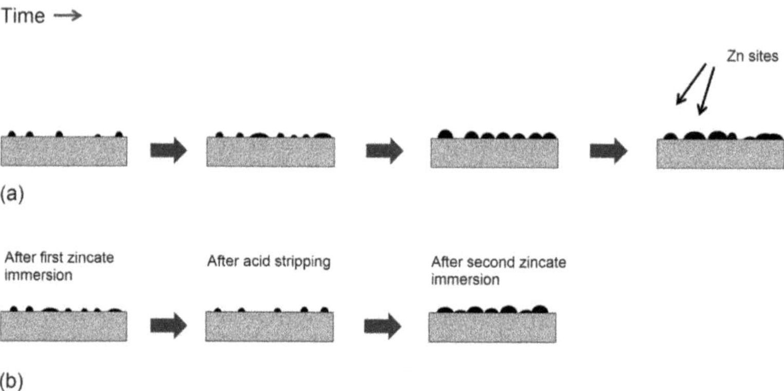

FIGURE 2.3 Cross section of the surface layers formed on an aluminum substrate after several previous zincate treatments: (a) single zincate treatment in a synthetic laboratory solution and (b) double zincate treatment in a synthetic laboratory solution. (With kind permission from Taylor & Francis Group: *Trans. IMF*, Monitoring of Zincate Pre-Treatment of Aluminium Prior to Electroless Nickel Plating, 95, 2017, 97–105, Court, S. et al.)

The role played by zinc in the process of forming electroless nickel coatings on aluminum alloys is quite noble. When aluminum alloys are immersed directly in the electroless nickel bath, the displacement reaction between aluminum and nickel ions has a strong tendency to occur, because there is a large potential difference between aluminum $\left(\varphi_{Al^{3+}/Al} = -1.662V\right)$ and nickel $\left(\varphi_{Ni^{2+}/Ni} = -0.250V\right)$. This displacement reaction occurs quickly and strongly, resulting in poor adhesion and subsequent coating failure (Hashemi and Ashrafi 2018). Now, the standard potential of zinc $\left(\varphi_{Zn^{2+}/Zn} = -0.763V\right)$ is closer to aluminum than to nickel. Therefore, when the zincated aluminum substrate is immersed in the electroless nickel bath, zinc dissolution occurs first. Then, a nickel coating is formed on the treated aluminum surface and not on the zinc surface (Dadvand et al. 2003; Hashemi and Ashrafi 2018).

However, some previous studies (Takács et al. 2007; Murakami et al. 2013a) have shown that, if a single zincate treatment is performed, zinc is vigorously dissolved in the electroless nickel bath because the zinc particles formed are thick. The bath generates electrons that are consumed by the reduction of the reducing agent (hydrazine, sodium hypophosphite, sodium borohydride, or dimethylamine borane, depending on the case) and water, generating appreciable quantities of hydrogen gas evolving from the substrate surface. The presence of appreciable quantities of gaseous hydrogen generates coatings with porosity, homogeneity, and adhesion issues (Correa et al. 2012; Correa et al. 2013a). Consequently, zincating in two stages is preferable, because the zinc particles formed on the surface are finer. Then, it would be possible to control the amount of hydrogen that evolves during zinc dissolution in the electroless nickel bath, thus generating a coating with fewer defects (Murakami et al. 2006, 2013a, 2013b; Hino et al. 2005).

However, the double zincate treatment process may not be performed on any aluminum alloys. For example, it has been reported that the presence of magnesium, manganese, iron, or silicon as aluminum alloying elements generates an excess of electrons during the zincate treatment process because these elements tend to dissolve preferentially. The presence of these electrons increases zinc deposition on aluminum surfaces, and as discussed above, the presence of coarse zinc particles causes defects in the final electroless nickel coating (Hino et al. 2005; Murakami et al. 2006, 2013a, 2013b). For this reason, the scientific community has begun seeking alternative aluminum surface activation procedures. Sudagar et al. (2010) reported the activation of aluminum substrates by immersion in solutions containing hydrogen fluoride, nickel salts, and boric acid for 1 min at room temperature. It was evidenced that a thin layer of nickel is formed on the aluminum surface through the contact reduction reaction between the substrate and the nickel salt solution. This small layer acts as a catalytic surface for the electroless nickel coating. This procedure is called nickel strike. Another alternative procedure reported is the hypophosphite absorbed layer (Takács et al. 2007), which involves activating the aluminum substrates by immersion in solutions containing sodium hypophosphite and lactic acid, with a pH of 4.6. This immersion produces a molecular hypophosphite absorbed layer on the aluminum surface, which ultimately acts as a catalytic surface for the electroless nickel coating. Azumi et al. (2003) reported a more sophisticated procedure, which involves forming Al-Ni alloy layers through the magnetron sputter deposition technique. The presence of nickel in the alloy acts as catalytic sites for the formation of electroless nickel coating. These procedures work quite well with acid or neutral electroless nickel baths. However, when an alkaline electroless nickel bath is required, i.e., the formation of an electroless nickel-boron coating reduced with sodium borohydride on an aluminum surface, special care must be taken because that alkalinity corrodes vigorously the substrate. Therefore, and despite the aforementioned disadvantages of zincating, a double zincating followed by the deposition of an intermediate layer of an electroless 3–4 μm thickness NiP layer is required as recommended by Delaunois et al. (Delaunois et al. 2000; Delaunois and Lienard 2002).

2.2.3 Magnesium Alloys

To coat commercially pure magnesium and its alloys through the electroless technique, previous surface pretreatments are required. This is because magnesium and its alloys have high chemical activity, making it a difficult material to coat. In addition, it is difficult to coat magnesium alloys by the electroless technique because (i) the presence of magnesium oxide or hydroxide on the surface decreases the adhesion between the substrate and the coating, (ii) magnesium alloys may have different microstructural phases, which causes different growth electroless coating rates, (iii) the electroless bath, regardless of its nature, is an aggressive medium for magnesium alloys, and (iv) the existence of pores in the coating favors the formation of a strong galvanic pairing between magnesium and nickel, which causes an increase in the corrosion rate of the substrate. The most commonly used pretreatment starts with acid etching in chromium ion solution, followed by activation in HF acid solution (Ambat and Zhou 2004; Liu and Gao 2006a, 2006b, 2006c; Song et al. 2006). The treatment of magnesium surfaces with HF acid promotes the formation of a thin layer composed of $Mg(OH)_2/MgF_2$, which facilitates the deposition of electroless coatings by preventing substrate dissolution within the electroless solution (Xiang et al. 2001; Liu

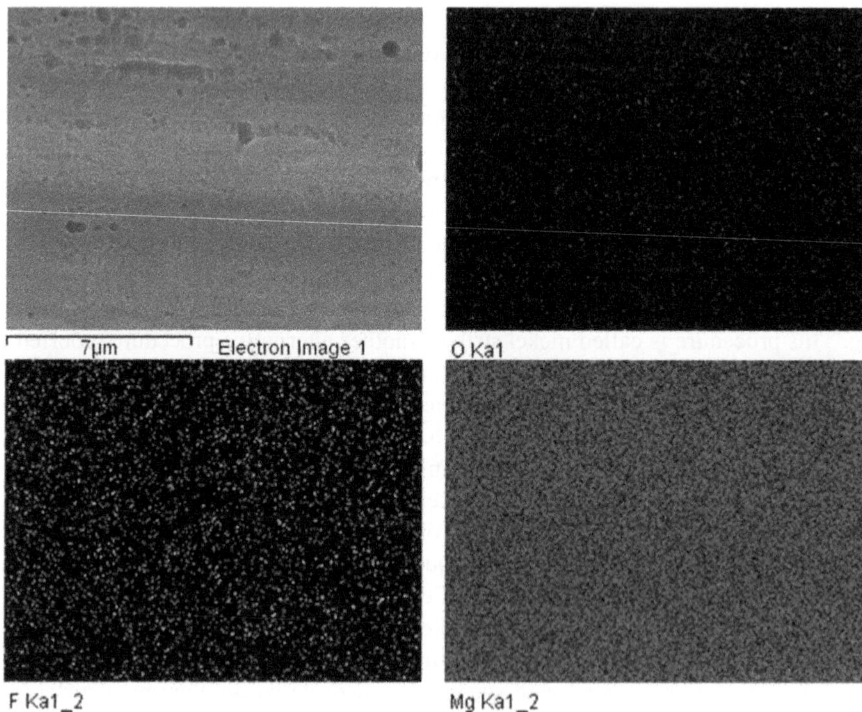

FIGURE 2.4 Scanning electron micrograph (backscattered electrons) and energy-dispersive X-ray maps showing the surface of commercially pure magnesium treated in an aqueous hydrofluoric acid solution.

et al. 2010; Liu and Gao 2006a). Figure 2.4 shows the morphology of this Mg $(OH)_2$/MgF_2 layer. At both high and low resolutions, the layer morphology is porous, with cavities of a few nanometers and thickness between 1.6 and 3.2 μm. Furthermore, the abovementioned species is stable and dissolves little in the electroless bath, thus protecting the magnesium substrates (Liu et al. 2010; Gu et al. 2005).

Now, if the pretreatment with fluorinated chemical species includes an excess of sodium ions (Na^+), the partial transformation of the $Mg(OH)_2$/MgF_2 layer to $Mg(OH)_2$/$NaMgF_3$ is favored, as shown in Figure 2.5. The $NaMgF_3$ species is known as neighborite (Sevonkaev et al. 2008; Zuleta et al. 2017). Regarding magnesium substrates, once they are in a humid environment (such as the electroless solution), they tend to form a thin magnesium hydroxide layer. If fluorinated species are added to the same humid environment (small amounts of HF or NH_4HF_2), $Mg(OH)_2$ is partially transformed into MgF_2, either by hydroxide dissolution or by the substitution or exchange of hydroxyl ions by fluoride ions (Bradford et al. 1976). In addition, if an excess of sodium ions (reducing agents such as sodium hypophosphite or sodium borohydride) is added to the humid environment in the presence of fluoride ions, MgF_2 is partially transformed into $NaMgF_3$ (Zuleta et al. 2017).

The modified pretreatments have been proposed to prevent the use of chromium and HF species, which are harmful to human health. For example, an etching

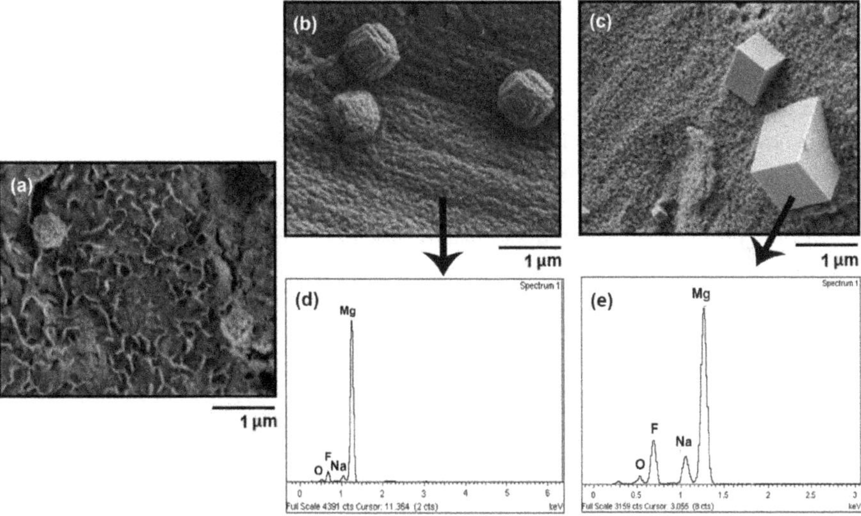

FIGURE 2.5 Scanning electron micrographs (secondary electrons) and energy-dispersive X-ray point analysis of the MgF_2 transformation process into $NaMgF_3$ after (a) 15 s, (b) 30 s and (c) 60 s of immersion in the alkaline electroless Ni-P plating bath. (d,e) EDX spectra of the magnesium surface after 30 s of immersion at locations indicated in (b) and (c), respectively. (Reprinted from *Surf. Coat. Technol.*, 321, Zuleta, A.A. et al., Study of the Formation of Alkaline Electroless NiP Coating on Magnesium and AZ31B Magnesium Alloy, 309–320, Copyright 2017, with permission from Elsevier.)

pretreatment in an oxalic acid solution, followed by activation in a sodium pyrophosphate and potassium fluoride solution, has been reported. Then, the samples were treated in a zinc bath, and finally, submerged in the electroless bath (Chen et al. 2006). Another fairly elaborate pretreatment consists of anodizing the samples through the DOW 17 process, followed by immersion in a TiB_2 catalytic powder mixture, epoxy resin, curing agent, and an organic solvent to form an approximately 10–15 μm film on the anodized surface (Sun et al. 2008). Another common practice is to perform a surface chemical attack by immersing the magnesium alloy in a complex mixture of different chemical reagents, such as magnesium dihydrogen phosphate, phosphoric acid, acetic acid, ethanol, and nitric acid. subsequently, an electroless NiP pre-coating is performed to finally submerge the piece in the electroless NiB bath (Zhang et al. 2008).

Other studies have reported processes where a zinc layer is formed on the surface of the magnesium alloys by immersion in a bath of liquid zinc (Wu et al. 2010) or in a zinc sulfate solution (Wang et al. 2012a, 2012b). The abovementioned pretreatment is generally applied to remove oxide layers and to prevent re-oxidation of the metal surface. In other studies (Cai et al. 2011; Chang et al. 2011), the formation of a NiB coating is reported using the electroless technique on the AZ31B magnesium alloy. The process employed includes surface activation (unspecified procedure) followed by zinc galvanizing, and finally, the formation of a thin NiP layer. The electroless NiB coating was formed on the treated surface of the AZ31B alloy. The results show the successful formation of a NiP/NiB duplex layer. Similarly, Fan et al. (2010) formed NiP/NiB duplex coatings on the AZ31B magnesium alloy. Before the deposition process, the substrates were etched in a chromium solution and then passivated in another HF solution. After all this and due to the aggressiveness of the electroless NiB bath, the magnesium substrates were coated with an electroless NiP layer with the aim of isolating the substrate from the electroless NiB bath. An inspection of the cross section shows that the thickness of the NiP layer is approximately 5 μm and no evidence of the presence of the pretreatments is observed. That is, the substrate/coating interface is smooth and does not reflect the roughness generated by the etching and passivation treatments. The electroless NiB layer is homogeneous and continuous and exhibits a thickness of 10 μm. The scientific and technical community that works with electroless coatings is well aware that the state of the substrate surface has a significant influence on the deposition process and on the coating properties (Vitry et al. 2010). However, the influence of the substrate on the coating properties and structure is not determined in any of the four previous studies (Cai et al. 2011; Chang et al. 2011; Fan et al. 2010; Vitry et al. 2010), because the NiB coating was not formed directly on the magnesium alloy surfaces.

Therefore, the scientific community seeks to obtain electroless nickel coatings directly on magnesium and its alloys, trough procedures that do not include chromium, fluorine, or any other chemical agent harmful to the human health or the environment. According to Petro and Schlesinger (2011), electroless NiB coatings were formed directly on the AZ91D and AM50 magnesium alloys. For these purposes, the substrates were degreased and polished with silicon carbide paper no. 240. The result is a coating that has some discontinuities, which were attributed to localized alloy oxidation, which inhibits coating nucleation. It was suggested that the oxidation localized on the aluminum intermetallic compounds occurred either during the electroless process or during the dry mechanical polishing before the pretreatment.

Wang et al. (2012a, 2012b) deposited an electroless NiB coating directly on the AZ91D magnesium alloy. The alloy surface was etched in acetic acid. The pretreatment generated a rough surface, suitable for the nucleation of the NiB coating, even when corrosion products, not identified, were observed on the treated alloy surface. After the acetic acid pretreatment, it was observed that the surface exhibited a uniform honeycomb structure that favored the formation of the coating. The results showed that the NiB coating is compact and dense. However, cracks were observed along the cell boundaries or through the nodules. Despite this, resistance to corrosion was improved for the substrates.

Recently, the present authors have developed an environmentally friendly procedure that allows the formation of electroless nickel coatings (alloyed with phosphorus or boron) directly on the surfaces of pure magnesium and magnesium alloys (AZ31B and AZ91D) (Correa et al. 2013c, 2013b; Calderón et al. 2016; Correa et al. 2012, 2013a, 2017; Zuleta et al. 2012, 2017). The procedure includes degreasing of the substrates through ultrasonic ethanol baths. Then, substrate surfaces are grit-blasted for 5 min using alumina particles of 150 μm diameter and air pressure of 60 psi. Finally, the substrates are subjected to an alkaline cleaning procedure in a solution containing 37 g/L NaOH and 10 g/L Na_3PO_4. The substrates are washed with distilled water and

FIGURE 2.6 Magnesium surface after gritblasting process for 5 min using alumina particles with a size of 150 μm under air pressure of 60 psi.

dried with a stream of hot air between each step. After the cleaning and drying processes, the substrates are immediately transferred to the electroless nickel solution.

Gritblasting of the magnesium surfaces is essential to prevent the use of chemical agents, such as HF acid, chromium, silver, and palladium, when activating surfaces and nucleating coatings. During gritblasting, the magnesium surfaces suffer erosion and high plastic deformation due to the impact of the alumina particles, which generates highly irregular and rough surfaces, as observed in Figure 2.6. This increase in substrate roughness is beneficial for the mechanical anchoring of the electroless coating (Vitry 2010a; Das and Sahoo 2011b). In addition, the severe plastic deformation generates an increase in localized energy in the grain boundaries due to the high density of the caused point or line defects (Lee and Kim 2011; Wang and Li 2003). This can be evidenced more clearly in Figure 2.7, which shows optical images of the magnesium substrates with different R_a values before and after immersion in the NiP electroless coating bath. The surface that was grit-blasted with alumina was uniformly covered by the electroless NiP coating. In contrast, substrates polished with silicon carbide abrasive paper exhibited dark and light regions, labeled "1" and "2," respectively, due to a nonuniform coating coverage on surfaces of reduced roughness. A closer inspection of the specimens by scanning electron microscopy (Figure 2.8) revealed that the coating formed on the grit-blasted surface was uniform on the magnesium surface, formed by fine nodules and with a cauliflower-like surface morphology, which is typical of electroless coatings. On the contrary, the substrates polished with silicon carbide abrasive paper showed different regions where nodular or cubic morphologies prevailed. The presence of cubic morphologies will be explained in detail later. Gritblasting with alumina particles produces rougher surfaces and the highest possible rate of electroless coating formation. In contrast, polishing the surfaces

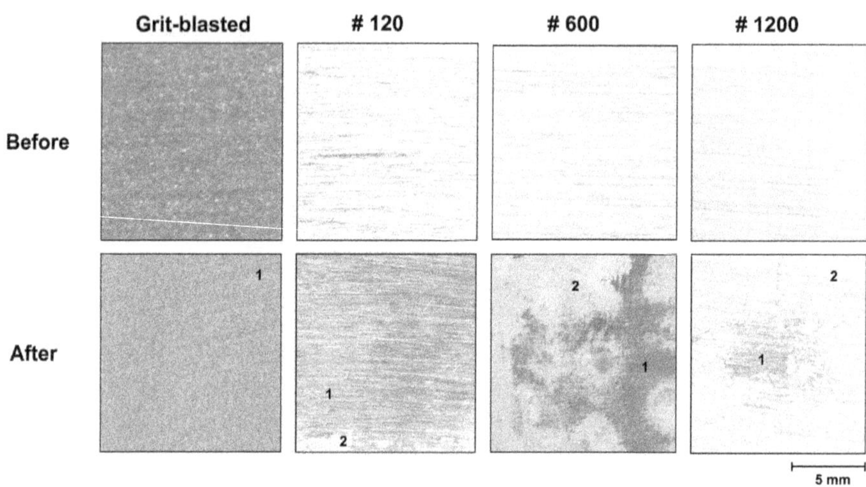

FIGURE 2.7 Optical images of commercially pure magnesium surface with different surface finishes before and after immersion for 30 min in an electroless NiP bath. The roughness values (R_a) are 2.1 ± 0.22, 1.33 ± 0.04, 0.62 ± 0.10, and 0.30 ± 0.01 for the grit-blasted surface and surfaces polished with abrasive paper 120, 600, and 1200, respectively. (Reprinted from *Surf. Coat. Technol.*, 321, Zuleta, A.A. et al., Study of the Formation of Alkaline Electroless NiP Coating on Magnesium and AZ31B Magnesium Alloy, 309–320, Copyright 2017, with permission from Elsevier.)

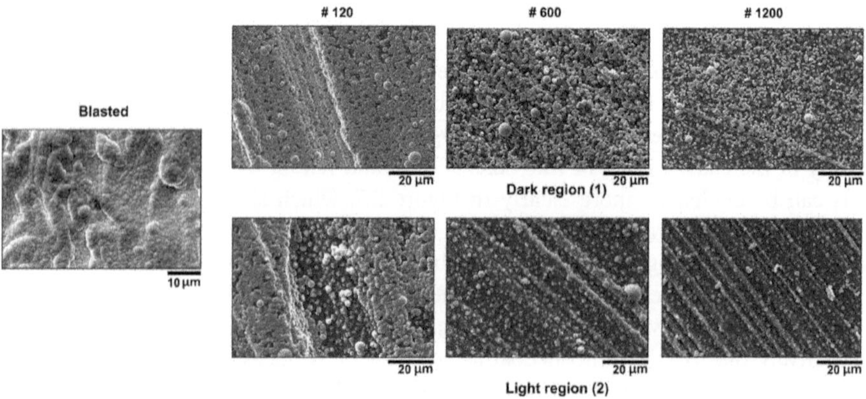

FIGURE 2.8 Scanning electron micrographs (secondary electrons) of the magnesium surfaces with variable roughness after immersion in an alkaline electroless Ni-P bath for 30 min. The upper and lower rows present details of the dark and light regions, respectively, which are evident in the optical images shown in Figure 2.7. (Reprinted from *Surf. Coat. Technol.*, 321, Zuleta, A.A. et al., Study of the Formation of Alkaline Electroless NiP Coating on Magnesium and AZ31B Magnesium Alloy, 309–320, Copyright 2017, with permission from Elsevier.)

with silicon carbide abrasive paper results in the progressive reduction of coating growth rates as the size of the carbide particle is reduced. The results are explained, as already mentioned above, by the effect of mechanical treatments on the magnesium surface topography and by the amount of plastic substrate deformation, with the rougher surface increasing the number of favorable sites (catalytically active) for nickel coating nucleation (Correa et al. 2013b; Zuleta et al. 2017).

However, the alumina gritblasting process alone is not enough to generate an electroless nickel coating with suitable properties on magnesium substrates. As mentioned earlier, magnesium substrates exhibit high reactivity when submerged in electroless nickel solutions. This is prevented, usually, with the help of HF, which is harmful to human health and to the environment. As an alternative, small amounts of ammonium bifluoride acid (NH_4HF_2) can be used in the electroless bath. The NH_4HF_2 and HF toxicity levels are similar, but NH_4HF_2 is easier to handle (it is solid), less volatile, and cheaper than HF (Zuleta et al. 2012; Correa et al. 2012). The presence of small amounts of ammonium bifluoride in electroless nickel baths (using reducing agents such as sodium hypophosphite or sodium borohydride) favors the formation of coatings with adequate properties on magnesium substrates. This is due to the presence of fluorine decreases the corrosion rate by acting as a corrosion inhibitor for magnesium substrates within the electroless nickel bath by forming a small porous MgF_2 layer. This layer is partially transformed into neighborite due to the sodium (coming from the reducing agents) present in the electroless nickel bath. Chronologically, the MgF_2 layer is formed after a few seconds of immersion in the electroless nickel bath passivating the surface. In specific places, a relatively low population density of cubic $NaMgF_3$ particles is developed from MgF_2 particles. The size of these $NaMgF_3$ particles is approximately 1 μm. While MgF_2 and $NaMgF_3$ particles are being formed, spherical nickel particles are deposited. These particles subsequently increase in number and size and extend through the substrate surface wrapping the $NaMgF_3$ crystals. The crystals are then covered by additional growth of the electroless nickel coating remaining at the base of the coating. The presence of the fluorine species in the electroless nickel bath allows the coating to form without significant magnesium substrate corrosion and prevents the decomposition of the electroless solution due to the presence of Mg^{2+} ions released by substrate corrosion (Zuleta et al. 2012). The process previously described will be further evidenced and explained in a later section.

2.2.4 Copper

Copper is widely used in microelectronic applications due to its excellent electrical conductivity and simple processing. However, this material and its alloys easily corrode in the atmosphere due to the presence of water and some chemical species (Rice 1981). Therefore, copper surfaces must be protected through coatings that preserve their excellent properties and limit their deterioration by oxidation. Among the various protection options (Salahinejad et al. 2017), electroless nickel technology emerges as the technique preferred by industry to protect copper from atmospheric corrosion while maintaining its electrical conductivity and weldability characteristics. As mentioned (Loto 2016; Krishnan et al. 2006), electroless nickel coatings can be deposited using sodium hypophosphite, sodium borohydride, dimethylamine borane, or hydrazine. The first one generates NiP deposits, and the second and third generate NiB deposits, while the last one generates nickel alloy coatings. Electroless NiB coatings have better mechanical characteristics than electroless NiP coatings. However, the latter exhibits better performance against corrosion than the former (Xu 2016). For this reason, copper substrates are usually protected by electroless NiP coatings. Still, sodium hypophosphite oxidation may not be performed on copper surfaces due to the potential difference between them (Kunimoto et al. 2011a, 2011b). In other words, copper surfaces are not catalytically active to oxidize the hypophosphite, and consequently, reduce nickel (Homma et al. 2001). Therefore, copper surfaces that require protection by electroless NiP coatings must be previously activated to generate active sites that allow for hypophosphite oxidation (Huh et al. 2016; Tian et al. 2013; Lin et al. 2016).

The most widely used process for the activation of a copper surface is the immersion of substrates in solutions containing palladium ions (Pd^{2+}) for 1 or 2 min (Ahmad et al. 2015; De et al. 2016; Lin et al. 2016; Rabiatul Adawiyah and Saliza Azlina 2018; Strandjord et al. 2002; Tang et al. 2014; Tian et al. 2013; Homma et al. 2001; Yee et al. 2014). This dip generates small Pd^{2+} nuclei adhered on a copper surface. Huh et al. (2016) precisely explained the activation mechanisms using palladium. There, a solution containing 30 g/L sulfuric acid, 25 mg/L palladium, and 15 g/L oxycarboxylic acid was used. Copper substrates were submerged for different periods, starting with 15 s and ending with 120 s. The number of Pd^{2+} nuclei on the copper surface increased with the activating solution immersion time. The number of Pd^{2+} nuclei increased until reaching 60 s. After this time, and up to 120 s, it remained constant. However, nuclei sizes remained constant until the first 60 s of immersion. Thereafter, and up to 120 s, nuclei sizes increased. The Pd^{2+} nuclei size increases were caused by the coalescence of two or more Pd^{2+} nuclei. Once the Pd^{2+} nuclei were on the copper surface, the surface was transferred to an electroless nickel solution. Within the electroless solution, the Pd^{2+} nuclei provided the electrons necessary for performing the nickel reduction. In this way, the first nickel clusters were deposited on the copper surface and not on the surface of the palladium nuclei. This is possible because palladium electrons can travel through the copper substrate given its high conductivity. Then, the subsequent coating formation occurred on the first nickel clusters through the oxidation–reduction reactions typical of this process. The palladium nuclei were embedded in the substrate-coating interface. However, the coating was not deposited on the palladium surface, but rather the palladium nuclei were covered by the lateral growth of the electroless nickel coating. Figure 2.9 presents a schematic of the deposition of electroless nickel coatings on the copper surfaces activated with palladium.

Furthermore, palladium is part of the platinum-group metals. The commercial value of these metals varies according to their durability, resistance to corrosion, and catalytic properties. For this reason, the price of palladium has been increasing with its application in various industrial sectors (mainly as a catalyst in vehicle gas exhaust pipes), while its reserves constantly decrease. (Vaškelis et al. 2005; Chen et al. 2012). This constant increase in palladium prices prompts the academic, scientific, and industrial communities to search for copper surface activation alternatives. In this sense, it is known as an alternative procedure that effectively activates copper surfaces, allowing

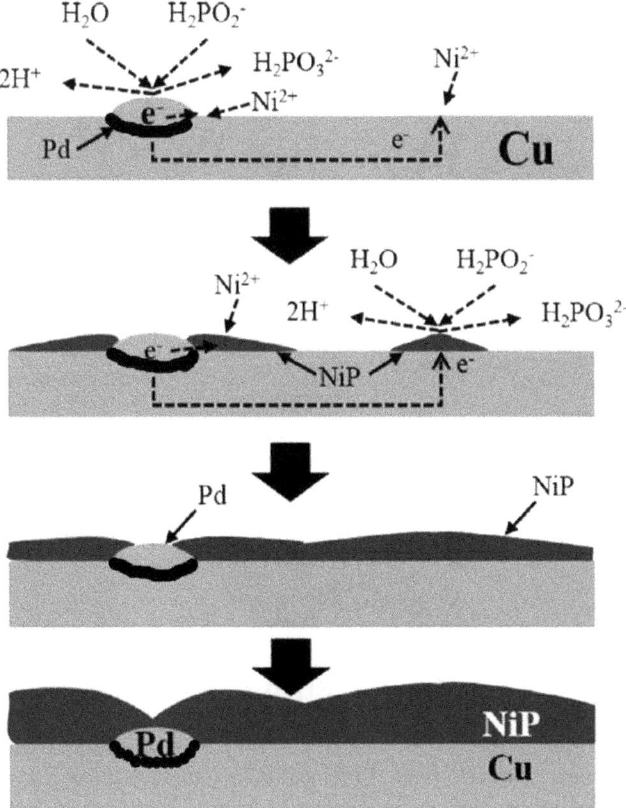

FIGURE 2.9 Schematic diagram of the deposition of electroless nickel coatings on copper surfaces activated with palladium. Note that nickel does not grow on Pd^{2+} nuclei. (Huh, S.-H. et al., *J. Electrochem. Soc.*, 163, D49–D53, 2016. Reproduced by permission from the Electrochemical Society.)

for the subsequent formation of electroless nickel coatings. This procedure consists of activating copper surfaces with nickel (Ni-activation). This facilitates the formation of a very thin layer (scarce nanometer thick) of metallic nickel through the reduction of the Ni^{2+} species. To achieve the thin nickel layer, an excess of thiourea (Tian et al. 2013), Ti^{3+} ions (Yagi et al. 2005), or formaldehyde (Lin et al. 2016) may be used as a reducing agent. Once the thin layer of metallic nickel is formed, the treated substrate is coated with the desired electroless nickel coating.

The activation with excess thiourea is performed by immersing the cleaned copper substrates in a solution containing 40 g/L $NiSO_4 \cdot 6H_2O$, 170 g/L $(NH_2)_2CS$, and 30 g/L H_3BO_3 at a temperature of 60°C and pH of 1.0. The excess of thiourea in the solution is adsorbed on the copper surface, causing a negative displacement in its electrochemical potential, which leads to the generation of a replacement reaction between the Ni^{2+} ions and copper. The copper on the surface dissolves and corrodes while the nickel ions are reduced. At the same time, a sulfur reduction occurs.

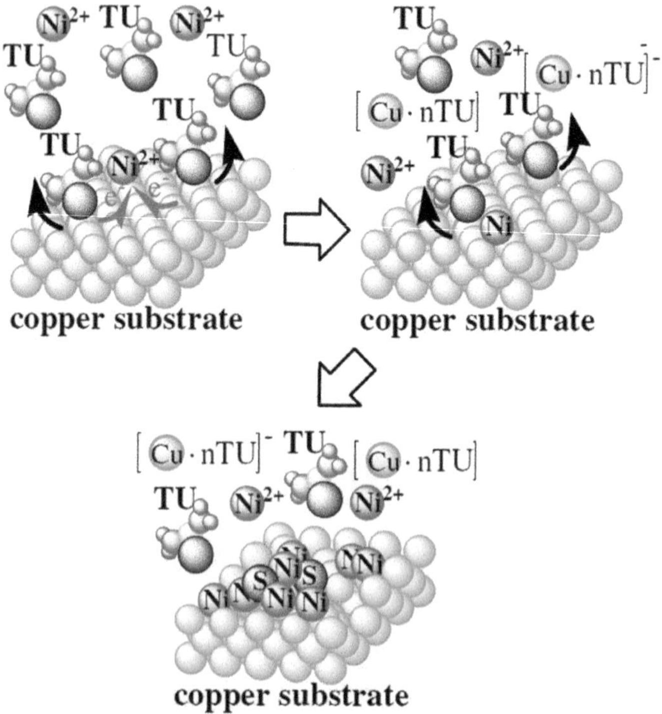

FIGURE 2.10 Outline of nickel activation process for copper substrates in a solution containing a high concentration of thiourea. (Reprinted from Tian, D. et al., *Surf. Coat. Technol.*, 228, A Pd-Free Activation Method for Electroless Nickel Deposition on Copper, 27–33, Copyright 2013, with permission from Elsevier.)

In other words, the thin layer of metallic nickel contains sulfur inclusions in its structure, as well as the presence of absorbed thiourea. Figure 2.10 shows an outline of the nickel activation process for copper substrates in a solution containing a high concentration of thiourea (Tian et al. 2013).

2.2.5 Titanium

Titanium and its alloys are quite difficult to coat because, due to their thermodynamic behavior, they tend to form a passive and tenacious oxide layer (Uma Rani et al. 2010). The section in which the substrate cleaning procedures are explained includes various procedures to eliminate the oxide layer. However, it regenerates so quickly that forming the coating before the oxide layer blocks the access to nickel ions is quite complex. If electroless nickel coatings are formed on dirty or oxidized surfaces, their properties will not be as desired. Some authors reported that titanium substrates, after being cleaned, are subjected to activation processes using solutions containing HF acid (Dabalà et al. 2001, 2004; Brunelli et al. 2009) or tetrafluoro-sodium borate (Gao et al. 2017). The role played by fluorine ions in the activation process of titanium substrates is not clear, and therefore, it is not yet known whether titanium substrates can be catalytically activated. The information available (Lamolle et al. 2009; Straumanis and Chen 1951; Liu et al. 2008; Zahran et al. 2016; Korotin et al. 2012) until now establishes that, once the titanium surfaces (which are completely or partially covered by the passive oxide layer) come into contact with the solution containing fluorine ions, there are two possibilities: (i) if the immersion time is short (and also depending on the fluorine reagent concentration), the oxide layer is dissolved in some places, on which titanium trifluoride is formed (TiF_3), or (ii) if the immersion time is extensive (and also depending on the fluorine reagent concentration), the oxide layer is completely dissolved and there is no presence of fluorinated species. Whatever the case may be, the titanium surface treated with fluorine ion solutions will end up being rougher than the original substrate surface. The information available is not clear as to whether the TiO_2 layer was completely removed or not before the electroless nickel process. However, the studies in which fluorine ions were used in the activation process reported a successful formation of electroless nickel coatings. The role played by TiF_3 or the roughness in the formation of electroless nickel coatings on titanium substrates treated with fluorine solution is not yet clear and should be further studied in more detail.

Other authors have preferred to use well-known technological processes to successfully form electroless nickel coatings, for example, using double zincating pretreatments to form electroless nickel coatings on Ti_6Al_4V alloy surfaces (Uma Rani et al. 2010). In this study, the alloy was cleaned and zincated twice by immersion for 3–4 min in a solution containing 100 g/L $Na_2Cr_2O_7$, 65 mL/L HF (40%), and 12 g/L $ZnSO_4{\cdot}5H_2O$ at a pH of 2.0 and a temperature of approximately 116°C. After double zincating, the substrates were transferred to the electroless nickel solution. The coating was successfully formed. Another alternative is to anodize the titanium and then coat it using the electroless nickel technique (Mahmoud 2009). Anodizing forms a porous and conductive oxide layer that allows the subsequent nucleation of the coating. For these purposes, a solution containing 100 g/L of oxalic acid at room temperature was used. The current density used for anodizing was between 60 and 95 mA/cm^2 using a stainless-steel counter-electrode. After anodizing, the substrates were transferred to the electroless nickel solution. Thus, the coating was successfully formed.

2.2.6 Polymeric Substrates

Figure 2.11 shows a schematic of the conventional polymer plating process using the electroless technique. This process usually involves various stages: (i) chemical etching, using chromic acid to increase roughness and surface area of substrates (Charbonnier et al. 2001); (ii) surface sensitization with tin chloride to promote chemisorption of the catalytic agent (Schilling et al. 1996); (iii) chemical activation of the surface using palladium chloride as a catalytic agent, which fosters the initial reduction of nickel ions (Charbonnier and Romand 2003); and finally, the acceleration stage, using either an acid or a strong base. The goal of the acceleration stage is to convert the hydrolyzed adsorbed catalyst into active metallic palladium (Rantell and Holtzman 1974; Teixeira and Santini 2005).

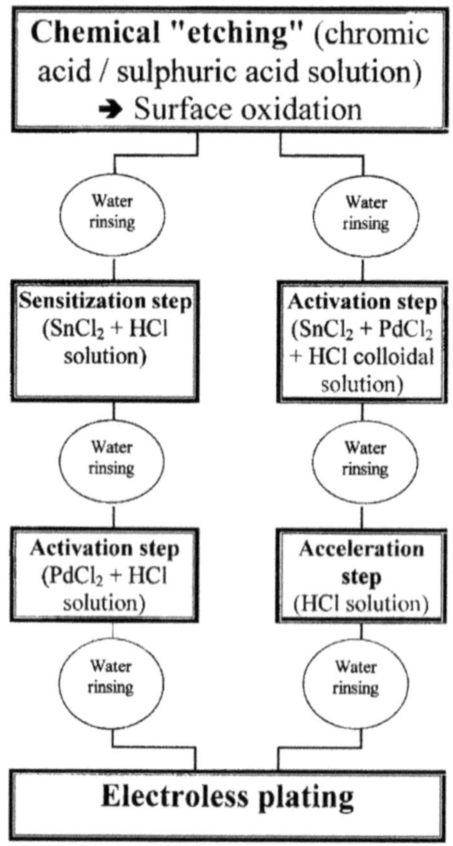

FIGURE 2.11 Schematic of the conventional polymer plating process using the electroless technique. (Reprinted from *Int. J. Adhes. Adhes.*, 23, Charbonnier, M., and Romand, M., Polymer Pretreatments for Enhanced Adhesion of Metals Deposited by the Electroless Process, 277–285, Copyright 2003, with permission from Elsevier.)

However, given the environmental regulations that limit the use of chromium species and the cost of some chemical reagents (such as palladium), several alternative methods for metallizing substrates have been used. First, to prevent the chemical etching stage from using chromic acid, two techniques have been used: modification by plasma and surface grafting. According to Kang and Neoh (2009), plasma modification is probably the most versatile polymer surface modification technique. The plasmas clean the surface, perform surface ablation or etching, reticulate the surface, and modify the chemical structure of the surface. All the above without using substances that are toxic to humans or harmful to the environment. On the other hand, surface grafting supports molecular designs and the redesign of polymeric surfaces. This is achieved through the implementation of functional groups that allow for the subsequent activation of polymeric surfaces. Additional details on surface modification techniques of polymeric materials can be found elsewhere (Kang and Neoh 2009).

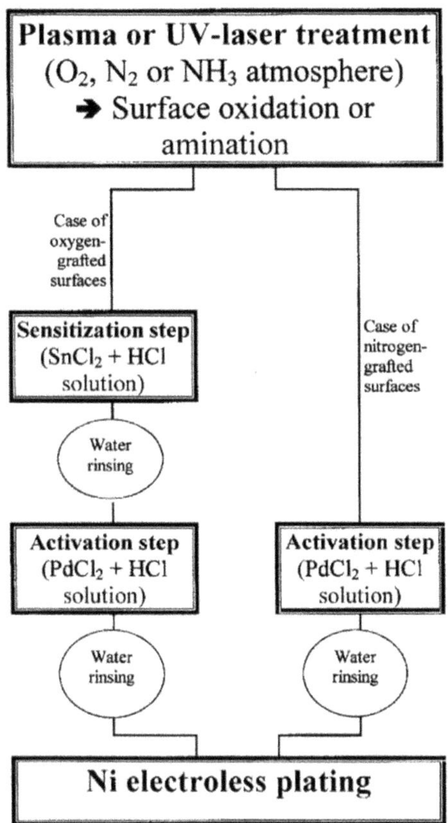

FIGURE 2.12 Polymer plating process using the electroless technique without chromic acid in the chemical etching stage. (Reprinted from *Int. J. Adhes. Adhes.*, 23, Charbonnier, M., and Romand, M., Polymer Pretreatments for Enhanced Adhesion of Metals Deposited by the Electroless Process, 277–285, Copyright 2003, with permission from Elsevier.)

Figure 2.12 shows the polymer plating process using the electroless technique without chromic acid in the chemical etching stage. In this process, as reported by Charbonnier and Romand (2003), the surface sensitization stage may also be eliminated. This deletion will depend on the type of atmosphere in which the plasma surface modification procedure is performed. According to Żenkiewicz et al. (2015), the most commonly used catalytic agent is palladium. This is due to its effectiveness in the catalyzation of the oxidation reaction for most of the reducing agents used in electroless metallization, although it is also possible to use copper, gold, silver, and aluminum as catalysts (Kim et al. 2016; Kimura et al. 2010; Petukhov et al. 2014; Li et al. 2014). However, the price of palladium is considerable, which means that alternatives must be considered for the activation stage. Wang et al. (2013) reported a simple method for coating polyvinyl chloride (PVC) sheets through the electroless technique without using chemicals containing tin or palladium. The substrates were functionalized in a solution containing chitosan and polyethylene glycol at 80°C for 30 min.

Then, the functionalized PVC was immersed in a copper sulfate solution for 1 h. Finally, the substrates were immersed in a sodium borohydride solution to reduce the copper found on the surface. The metallic copper acts as a catalytic site for the formation of the electroless nickel coating. Something similar was reported by Garcia et al. (2010) for the metallization of acrylonitrile-butadiene-styrene substrates, the difference lying in the fact that the functionalization of the polymer surface was performed by polyacrylic acid covalent grafting. This means that the copper was adsorbed on the surface of the polyacrylic acid and then reduced to serve as a catalytic site for the formation of electroless nickel coating. Another study (Charbonnier et al. 2006) used plasma modifications in an ammonium atmosphere to condition and functionalize the surface of different polymeric substrates (e.g., PC, PP, polytetrafluoroethylene, polybutylene terephthalate, or liquid crystal polymers). Next, the functionalized substrates were immersed in an ethanol and nickel acetate solution. Then, the nickel on the functionalized surface was reduced using sodium borohydride. The reduced nickel acts as a catalytic site for the formation of the electroless nickel coating.

2.2.7 Ceramic Substrates, Ceramic, and Metallic Powders

The conventional procedure to activate the surfaces of the ceramic substrates involves immersion in a colloidal solution, which facilitates the chemisorption of noble metal atoms, such as palladium. According to several previous studies (Karmalkar and Kumar 2004; Karmalkar 1999; Carraro et al. 2007; Singh et al. 2014; Karmalkar 1997; Balaraju et al. 2016; Bernasconi et al. 2016; Delbos et al. 2015), the colloidal solutions typically used include $PdC1_2$, $SnCl_2$, or $PdCl_2$-strong acid. However, when substrates require metallization for use in the microelectronics industry (e.g., silicon substrates), such solutions are too aggressive, causing undesired damages, and thus, restricting their use. For this reason, modifications have been proposed for these colloidal systems to mitigate the aggressiveness. One of these modifications involves adding NH_4OH or NH_4F_2 to the colloidal system. Then, the colloidal system for activation would compound by $PdC1_2$-strong acid-NH_4F or $PdC1_2$-strong acid-NH_4OH (Karmalkar 1999). The presence of ammonia generates a complex with palladium, which has greater adhesion to the substrate than when it is not in its complex form. This improves the substrate activation, resulting in electroless coating with more suitable properties.

As for the ceramic powders, two types of surface activations can be used. The first involves immersing particles in a solution containing 30 g/L of sodium hypophosphite and 20 mL/L of lactic acid for 30 min at 85°C. This immersion generates a thin layer of absorbed hypophosphite on the surface of the particles. This layer helps the subsequent nucleation of electroless coating, and the procedure is known as hypophosphite absorbed layer. The second activation is the "conventional activation" through surface sensitization using tin chloride and later activation with palladium chloride. As mentioned above, tin facilitates the adhesion of palladium, with the latter serving as a catalyst for nickel reduction reactions (Huang and Chen 2017; Pázmán et al. 2012a, 2012b; Umasankar et al. 2014; Mohandas and Radhika 2017; Saeedi Heydari et al. 2018). Some ceramic powders, such as tungsten and vanadium carbides, may be directly coated without using surface activation methods (Giampaolo et al. 1997).

For metallic powders, the most commonly used activation procedure is known as nickel strike, which involves immersing the metallic particles in a nickel salt solution (typically, nickel chloride). As a result of this immersion, the nickel particles are adsorbed on the metal surface, acting as a catalytic surface for electroless nickel coating (Beygi et al. 2012; Barbat and Zangeneh-Madar 2014; Ramaseshan et al. 2001; Zangeneh-Madar and Jafari 2012).

2.3 MORPHOLOGY EVOLUTION

The electroless nickel coating properties are influenced, among other characteristics, by the deposit formation process in the initial stages (Homma et al. 1997); thus, the understanding of the initiation and growth stages is of high interest for their control. Electroless nickel coatings that have not been subjected to any type of heat treatment or that are found in the as-plated condition are a supersaturated metastable alloy (Mallory and Hajdu 1990). This occurs because the energy available during the deposition process is not sufficient to retain the intermetallic structures predicted by the corresponding phase diagram (Vitry et al. 2012a). Therefore, in their as-plated condition, the alloying element (phosphorus or boron) is trapped between the nickel atoms, generating supersaturation (Mallory and Hajdu 1990), and the predicted structure is only reached after applying a suitable heat treatment (Vitry et al. 2012a).

Now, regardless of the type of surface where the coating is formed, or of the chemical composition or conditions of the electroless bath, the morphology of electroless nickel coatings is quite similar. The coatings are uniform in thickness, follow the shape of the piece to be coated (Contreras et al. 2006), and comprise nickel agglomerates or nodules that are randomly distributed, revealing a cauliflower-like appearance (Anik et al. 2008; Baskaran et al. 2006a; Vitry et al. 2010). The preferential nodule growth direction is vertical with respect to the surface of the substrate, which explains the columnar structure reported for such coatings (Riddle and Bailer 2005; Rao et al. 2005). In addition, small pores (Hamid et al. 2010) or cracks (Contreras et al. 2006) may be found on nodule boundaries or surfaces due to the generation of gaseous hydrogen or the release of residual stress, respectively. The size of the nickel nodules is affected by several factors. For example, increasing the deposition time (Dominguez-Rios et al. 2007), bath temperature (Bulbul 2011), substrate type, surface characteristics of the substrate, or the concentration of the reducing agent (Baskaran et al. 2006a) decreases the nodule size, while the concentration of the nickel salt does not exhibit any significant effect thereon (Bulbul 2011). However, this matter will be further in Chapters 3 and 4 of this book.

2.3.1 Ferrous Substrates

In most cases, the stages of electroless nickel coating formation on ferrous substrates proceed as follows: (i) preferential grain growth with a favorable orientation, (ii) restriction of vertical grain growth, (iii) lateral grain binding, (iv) initial grain growth cessation, and (v) nucleation of new grain layers (Petro 2014; Paunovic and Schlesinger 2006). The growth mechanism described below was performed on a carbon steel (St-37) surface, as reported by Vitry et al. (Vitry et al. 2012c; Bonin

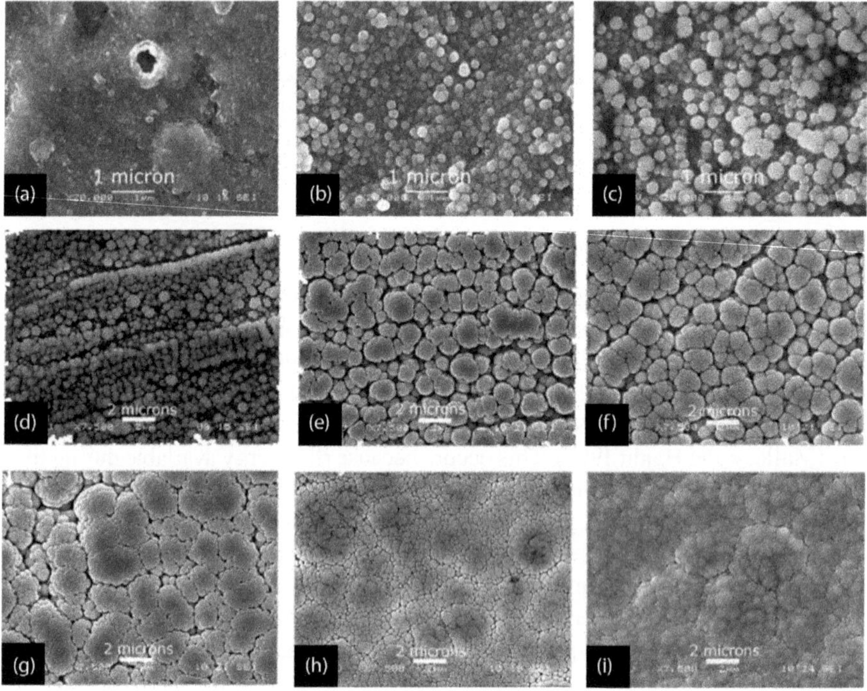

FIGURE 2.13 Scanning electron micrographs (secondary electrons) of the St-37 steel surface after (a) 15, (b) 30, (c) 60, (d) 90, (e) 240, (f) 420, (g) 600, (h) 1800, and (i) 3600 s of immersion in an electroless NiB bath. (Image reproduced and adapted with permission from Elsevier Vitry et al. 2012.)

and Vitry 2016; Vitry and Delaunois 2015). The ferrous surface was mechanically polished with SiC abrasive paper no. 4000, degreased with acetone, activated in an HCl solution (30% vol.), and extracted and washed with deionized water. Following this, the steel sample was immersed in the NiB electroless bath. Figure 2.13 shows the scanning electron micrographs that describe the morphological evolution during 3600 s (60 min) of immersion in an electroless NiB bath using sodium borohydride as a reducing agent. For the steel surface in its initial state (0–15 s), the presence of nickel nodules is not observed on the ferrous surface. After 15 s of immersion in the electroless bath (Figure 2.13a), a microelemental analysis by EDX (energy-dispersive X-ray analysis) showed approximately 6% nickel on the surface. The nodules were formed preferably on the polishing lines and the surface defects of the substrate. Subsequently, after a 30-s deposition (Figure 2.13b), the presence of nickel became even more evident as nodules of 0.1–0.2 μm in diameter were distributed over the substrate surface, which is currently not completely covered. The nodule size increased with the process duration. For example, after 90–240 s of electroless bath immersion (Figure 2.13d and e), a greater coverage was observed. In addition, the surface was leveled and nodule diameters showed growth. After immersion for 420–600 s (7–10 min) (Figure. 2.13f and g), the nodule sizes started refining, but the morphology remained quite similar to that already observed previously. After 1800 s

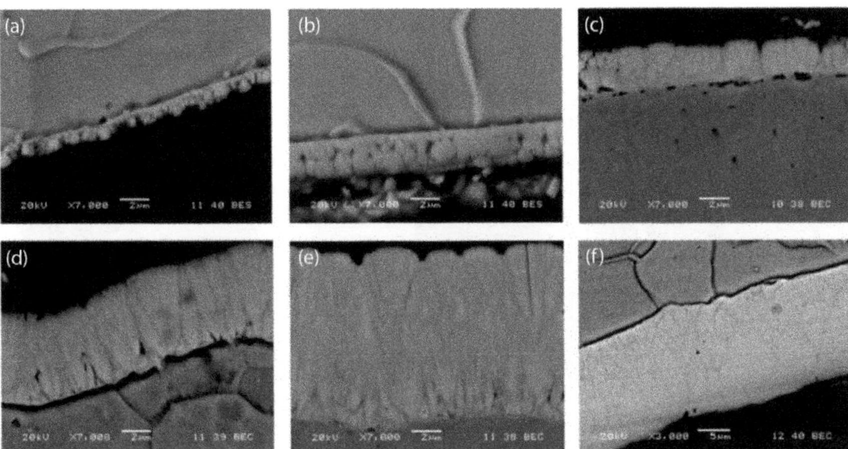

FIGURE 2.14 Scanning electron micrographs of cross sections of the coating formed on the St-37 steel surface after (a) 90, (b) 240, (c) 420, (d) 600, (e) 1800, and (f) 3600 s of immersion in an electroless NiB bath. (Image reproduced with permission from Elsevier Vitry et al. 2012.)

(30 min) of immersion (Figure 2.13h), the columns thickened, while, after 3600 s (60 min) (Figure 2.13i), the intercolumnar space appeared to have disappeared, or rather, to have been filled. The nodule size was smaller, but the cauliflower-like morphology of these coatings was still observed.

Figure 2.14 shows the scanning electron micrographs of the cross sections of an electroless nickel coating deposited on a St-37 steel surface. In the cross sections, for the initial 90 s (Figure 2.14a), two nickel layers were visible: a continuous layer adjacent to the substrate surface on which the other grew, which comprised nickel nodules of different sizes. After 240 s (Figure 2.14b), the nodules grouped, forming an approximately 1 μm thick layer. After 420 s (Figure 2.14c), the nodules were no longer able to grow laterally and the columnar growth of the coating began to become evident. The upper columns were a mixture of larger and smaller columns, which indicates the start of the grain refinement process. After 600 s (Figure 2.14d), a phenomenon similar to secondary germination occurred, inducing the separation of the column into several smaller branches and leading to the refinement of the upper part of the coating. This process continues until 1800 s (Figure 2.14e) alongside coating densification, which becomes much more evident after 3600 s (Figure 2.14f).

Through observations made from gravimetric mass gain curves during the coating formation process, it was observed that, during the first 240 s, an initial rapid growth phase in which spherical nodules were formed. Then, there was a deceleration in the growth rate because the thickness and quantity of nickel deposited increased very slowly during the densification phase (240–420 s). After this time and as soon as the columnar morphology was formed, both the mass gain and thickness increased until the end of the process.

As mentioned earlier, the morphological evolution of electroless nickel coatings is quite similar, regardless of the substrate where they are deposited. Figure 2.15 shows the surface and cross section of an electroless NiB coating formed on the surface of

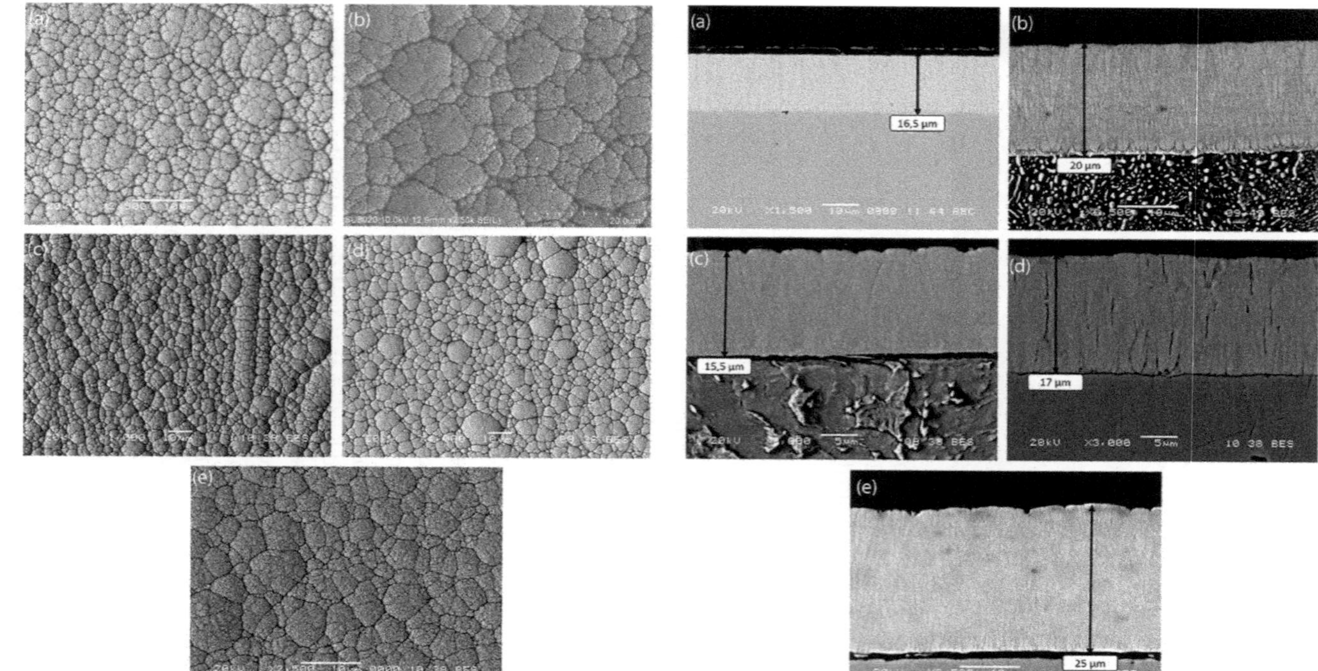

FIGURE 2.15 Scanning electron micrographs (secondary electrons) of surfaces (left) and cross sections (right) of an electroless NiB coating after 60 min of immersion. (a) St 37 steel (mild steel), (b) AISI 1070 steel (low-alloy carbon steel), (c) STM A353 (cryogenic steel), (d) AISI 304 (austenitic stainless steel), and (e) ASTM A182 (Grade F55) (duplex Austeno-ferritic stainless steel). (Reproduced from *Appl. Surf. Sci.*, 359, Vitry, V., and Delaunois, F., Applied Surface Science Formation of Borohydride-Reduced Nickel–Boron Coatings on Various Steel Substrates, 692–703, Copyright 2015, with permission from Elsevier.)

different ferrous substrates during 60 min of immersion. It can be easily seen that all coatings have typical cauliflower-like morphology, with the difference being in nodule sizes. The size of the nodules formed on St-37, AISI 304, and ASTM A182 steel surfaces was close to 0.3 μm, while that of the nodules formed on AISI 1070 and ASTM A353 steel surfaces was close to 0.5 μm. The difference in the sizes may be because the coatings were in different nodule refinement stages. The cross-section images evidence the separation of the columns into several smaller branches. This suggests that the growth of columns is the result of the competition between column selection (lateral growth of some columns) and refining (separation of columns into smaller ones).

2.3.2 Aluminum Alloys

The morphological evolution of electroless nickel coatings formed on aluminum and its alloys is quite similar, regardless of the activation method used (double zincating, nickel strike, hypophosphite absorbed layer, etc.). Figure 2.16 shows the scanning electron micrographs for aluminum alloy surfaces after a double zincating process. As discussed in a previous section, the zinc film was developed by the nucleation of crystals that overlapped and multiplied until a soft zinc layer was formed on the aluminum (Khan et al. 2007).

Figure 2.17 shows the scanning electron micrographs for an AA 1050 aluminum alloy surface after immersion in an electroless NiP bath for different periods. Here, some small nickel particles began to nucleate in the sites where the zinc from the pretreatment had dissolved (Figure 2.17a). Nickel nuclei became catalytic sites for coating. Over time, the nickel nuclei grew and coalesced until the substrate surface was covered almost completely by the coating (Backović et al. 1979; Li et al. 2006; Khan et al. 2007). However, some empty spaces could be observed (Figure 2.17d), which will be filled throughout the coating formation process given the autocatalytic characteristic of the nickel particles.

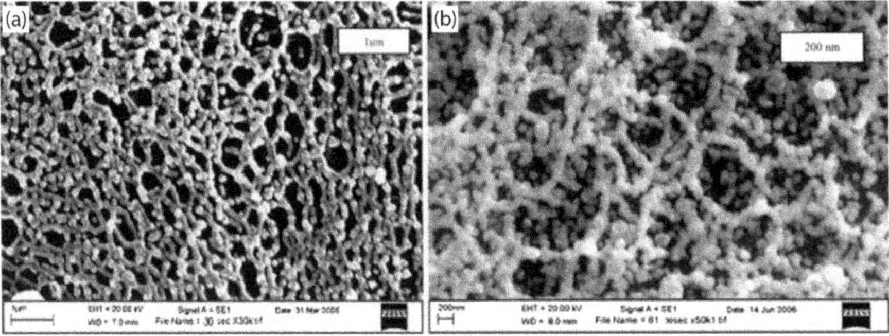

FIGURE 2.16 Scanning electron micrographs (secondary electrons) for aluminum alloy surfaces: (a) AA1050 and (b) AA6061 after a 30 s double zincating procedure before immersion in an electroless NiP bath. (With kind permission from Springer Science+Business Media: *J. Appl. Electrochem.*, Surface Characterization of Zincated Aluminium and Selected Alloys at the Early Stage of the Autocatalytic Electroless Nickel Immersion Process, 37, 2007, 1375–1381, Khan, E. et al.)

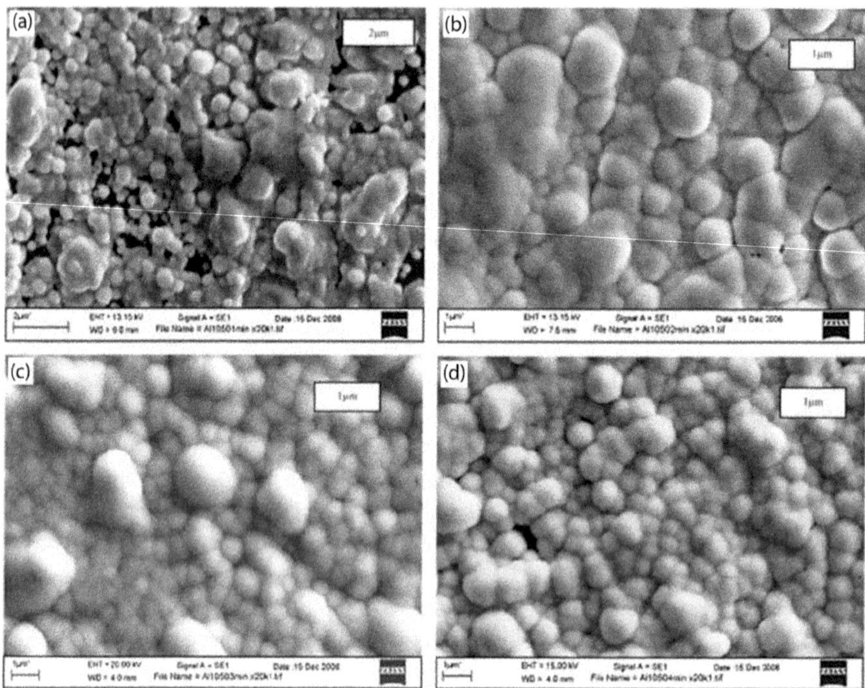

FIGURE 2.17 Scanning electron micrographs (secondary electrons) of the 1050 aluminum alloy surfaces following immersion in an alkaline electroless NiP bath for (a) 1, (b) 2, (c) 3, and (d) 4 min. (With kind permission from Springer Science+Business Media: *J. Appl. Electrochem.*, Surface Characterization of Zincated Aluminium and Selected Alloys at the Early Stage of the Autocatalytic Electroless Nickel Immersion Process, 37, 2007, 1375–1381, Khan, E. et al.)

Figure 2.18 shows the scanning electron micrographs for aluminum alloy (AA1050 and AA6061) surfaces following immersion in the NiP electroless bath. It is clearly observed that the size of the developed nickel nodules was larger for the AA1050 alloy than that for the AA6061 alloy, although the typical electroless nickel coating morphology was retained. After 4 min of immersion in the electroless bath, both surfaces were covered by the nickel coating. Therefore, the deposition mechanism of the coating was quite similar for both substrates. Still, the phosphorus percentage into the coating was 9% for the AA6061 alloy but only 3% for the AA1050 alloy (Khan et al. 2007).

Figure 2.19 shows the scanning electron micrographs of the cross sections of an electroless nickel coating deposited on a St-37 steel surface. Coatings with consistent thickness were clearly observed. For the electroless NiB coating (Figure 2.19, left), the coating formation rate was approximately 16 μm/h and its columnar structure was easily noticeable. The coating formation rate of the electroless NiP bath

FIGURE 2.18 Scanning electron micrographs (secondary electrons) of the aluminum alloy surface: (a) AA1050 and (b) AA6061 following immersion in the electroless NiP bath for 4 min. (With kind permission from Springer Science+Business Media: *J. Appl. Electrochem.*, Surface Characterization of Zincated Aluminium and Selected Alloys at the Early Stage of the Autocatalytic Electroless Nickel Immersion Process, 37, 2007, 1375–1381, Khan, E. et al.)

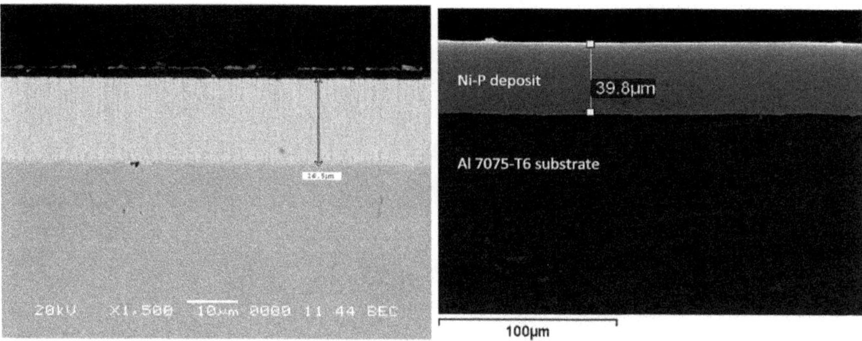

FIGURE 2.19 Scanning electron micrographs of cross sections of an Al-Cu-Mg aluminum alloy (left) after immersion in an electroless NiB bath for 60 min and an AA7075-T6 aluminum alloy (right) after immersion in the electroless NiP bath for 60 min. (Reproduced from *Int. J. Fatigue*, 52, Rahmat, M.A. et al., The Effect of Electroless NiP Coatings on the Fatigue Life of Al 7075-T6 Fastener Holes with Symmetrical Slits, 30–38, Copyright 2013, with permission from Elsevier; Reproduced from *Surf. Coat. Technol.*, 202, Vitry, V. et al., Mechanical Properties and Scratch Test Resistance of Nickel-Boron Coated Aluminium Alloy after Heat Treatments, 3316–3324, Copyright 2008, with permission from Elsevier.)

(Figure 2.19, right) was approximately 39 μm/h. In neither case, there were appreciable flaws in the substrate/coating interface, while in the free surface, a typical cauliflower-like morphology was observed (Rahmat et al. 2013; Vitry et al. 2008).

The differences in coating formation rates, as well as in alloying element (P or B) content, were linked to the different conditions (concentration of chemical reagents, temperature, pH, etc.) of the electroless baths used.

2.3.3 Magnesium Alloys

The substrates shown below are commercially pure magnesium and its AZ31B alloy. The samples, with dimensions 10 × 10 × 2 mm^3, were mechanically dry-polished with abrasive paper no. 100. Then, they were grit-blasted with alumina (150 μm), washed with deionized water, sequentially submerged in ultrasonic ethanol baths, and hot air-dried. Then, a final cleaning was performed in a solution containing 37 g/L of NaOH and 10 g/L of Na_3PO_4 for 10 min at 65°C. The samples were then transferred immediately to the electroless NiP solution (Correa et al. 2013a; Zuleta et al. 2017). Figure 2.20 shows the scanning electron micrographs that describe the morphological evolution during the first 60 s of immersion in an alkaline electroless NiP bath. For the magnesium surface in its initial state (0 s),

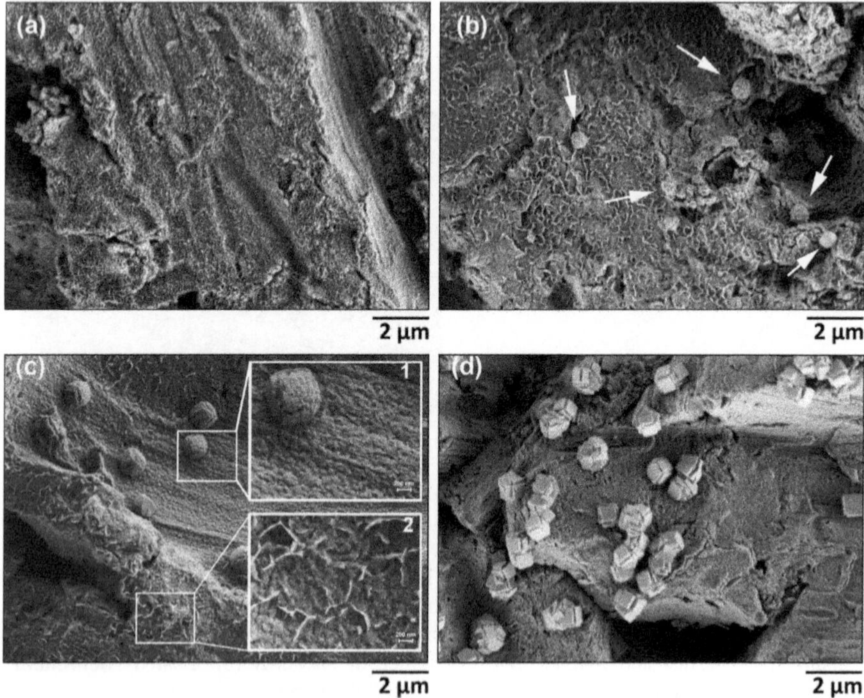

FIGURE 2.20 Scanning electron micrographs (secondary electrons) of pure magnesium surface after (a) 0, (b) 15, (c) 30, and (d) 60 s of immersion in an alkaline electroless NiP bath. (Reproduced from *Surf. Coat. Technol.*, 321, Zuleta, A.A. et al., Study of the Formation of Alkaline Electroless NiP Coating on Magnesium and AZ31B Magnesium Alloy, 309–320, Copyright 2017, with permission from Elsevier.)

a typical $Mg(OH)_2$ morphology was observed (Merino et al. 2010; Volovitch et al. 2009), which was mainly composed of plates of approximately 10–20 nm thickness located perpendicularly to the surface. The $Mg(OH)_2$ originated from the cleaning process (Song et al. 2006), because this was carried out in a highly alkaline environment (pH ~ 10.5). After 15 s, some particles became spherical (see arrows), with an approximate diameter of 0.5 μm. Afterward and after 30 s, the presence of two different regions is observed. One region where the previously observed morphology typical of the compound $Mg(OH)_2$ is maintained (labeled as 2) accompanied by another region exhibiting a totally different morphology composed of a structure of nanometric texture (labeled as 1), which suggests the formation of a film on the surface. According to Sevonkaev et al. (2008) and Lui et al. (2010), where similar morphologies were found, this film was generated by the MgF_2 compound. The presence of both morphologies (1 and 2) suggested that $Mg(OH)_2$ began its transformation into MgF_2 after 30 s of immersion in the electroless bath. This transformation has already been explained above in a previous section. After 60 s, $Mg(OH)_2$ has fully transformed into MgF_2 because the lamellar hydroxide structure was no longer observed (Zuleta et al. 2017).

Figure 2.21 shows high-resolution scanning electron micrographs of the pure magnesium surface after 60 s of immersion in an alkaline electroless NiP bath. The backscattered electron image shows light NiP particles. The pseudo-cubic crystals observed at 30 s were transformed into well-defined cubes with sizes ranging between 0.5 and 1 μm. In a previous section, these cubes were identified as $NaMgF_3$ (Zuleta et al. 2017).

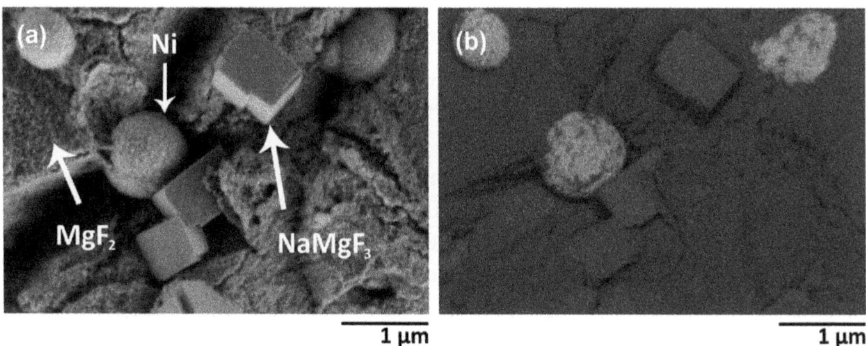

FIGURE 2.21 High-resolution scanning electron micrographs (a) secondary electrons and (b) backscattered electrons of a pure magnesium surface after 60 s of immersion in an alkaline electroless NiP bath. (Reproduced from *Surf. Coat. Technol.*, 321, Zuleta, A.A. et al., Study of the Formation of Alkaline Electroless NiP Coating on Magnesium and AZ31B Magnesium Alloy, 309–320, Copyright 2017, with permission from Elsevier.)

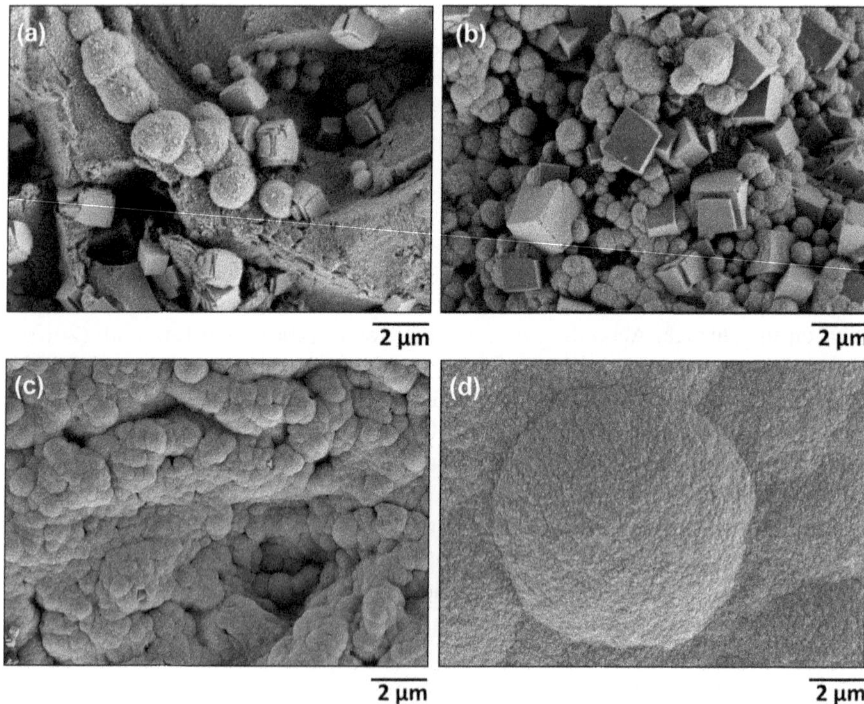

FIGURE 2.22 Scanning electron micrographs (secondary electrons) of pure magnesium surface after (a) 120, (b) 300, (c) 600, and (d) 1800 s of immersion in an alkaline electroless NiP bath. (Reproduced from *Surf. Coat. Technol.*, 321, Zuleta, A.A. et al., Study of the Formation of Alkaline Electroless NiP Coating on Magnesium and AZ31B Magnesium Alloy, 309–320, Copyright 2017, with permission from Elsevier.)

Figure 2.22 shows the scanning electron micrographs that describe the morphological evolution after 120 s of immersion in an alkaline electroless NiP bath. After 120 s, more NiP nodules were formed around the preexisting nodules and coating formation continued autocatalytically. After 600 s, the magnesium surface was almost completely covered by the coating, while the cubic $NaMgF_3$ particles were embedded by NiP nodules. After 1800 s, the magnesium surface was completely covered by the electroless NiP coating (Zuleta et al. 2017).

Regarding the AZ31B magnesium alloy, Figure 2.23 shows the morphological evolution of the magnesium alloy surface after immersion in an alkaline electroless

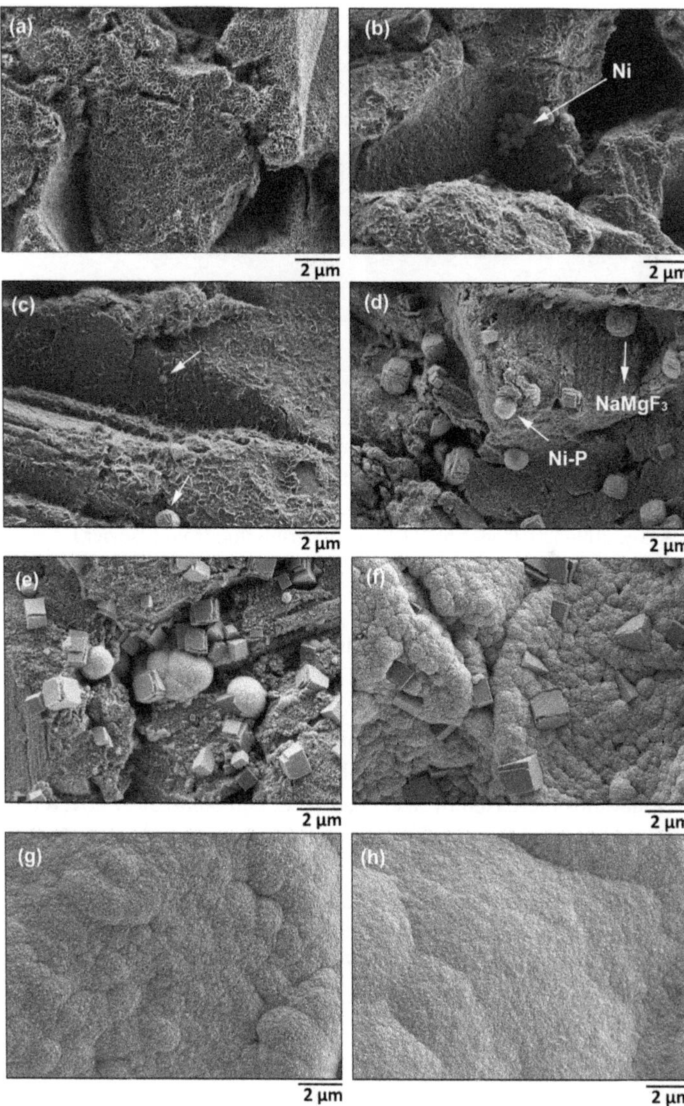

FIGURE 2.23 Scanning electron micrographs (secondary electrons) of an AZ31B magnesium alloy surface after (a) 0, (b) 15, (c) 60, (d) 120, (e) 300, (f) 600, (g) 1200, and (h) 1800 s of immersion in an alkaline electroless NiP bath. (Reproduced from *Surf. Coat. Technol.*, 321, Zuleta, A.A. et al., Study of the Formation of Alkaline Electroless NiP Coating on Magnesium and AZ31B Magnesium Alloy, 309–320, Copyright 2017, with permission from Elsevier.)

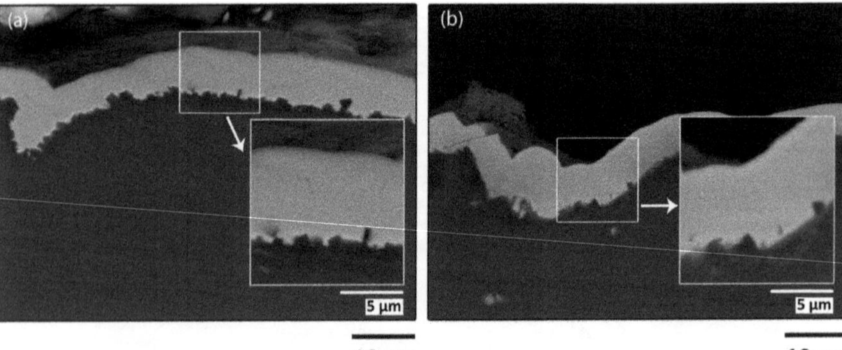

FIGURE 2.24 Scanning electron micrographs (secondary electrons) of the cross sections of pure magnesium (a) and AZ31B magnesium alloy (b) surfaces after immersion for 1800 s in an alkaline electroless NiP bath. (Reproduced from *Surf. Coat. Technol.*, 321, Zuleta, A.A. et al., Study of the Formation of Alkaline Electroless NiP Coating on Magnesium and AZ31B Magnesium Alloy, 309–320, Copyright 2017, with permission from Elsevier.)

NiP bath for 1800 s. After 15 s of immersion, a change in the morphology and the presence of NiP nodules were observed. The morphological changes were similar to those described for pure magnesium. After 60 s, some $NaMgF_3$ particles were observed, and it became evident that the oxide/hydroxide layer had been dissolved or transformed. The number of NiP nodules increased after 120 s of immersion. At 1800 s, the nodules increased in size, coalesced, and completely covered the alloy surface. It should be noted that the formation of the electroless coating NiP occurred more quickly on an alloy surface than on a pure magnesium surface. Figure 2.24 shows the cross sections of electroless NiP alkaline coatings formed on pure magnesium and AZ31B alloy surfaces. The images revealed that the coating thickness was approximately 6.3 µm. Some cubic $NaMgF_3$ crystals might have been observed in a layer that might be up to 2.5 µm thick. The crystals were embedded in the electroless coating (Zuleta et al. 2017).

Figure 2.25 schematically summarizes, from the discussion carried out in this section, the process of forming the alkaline electroless NiP coating on magnesium and its AZ31B alloy.

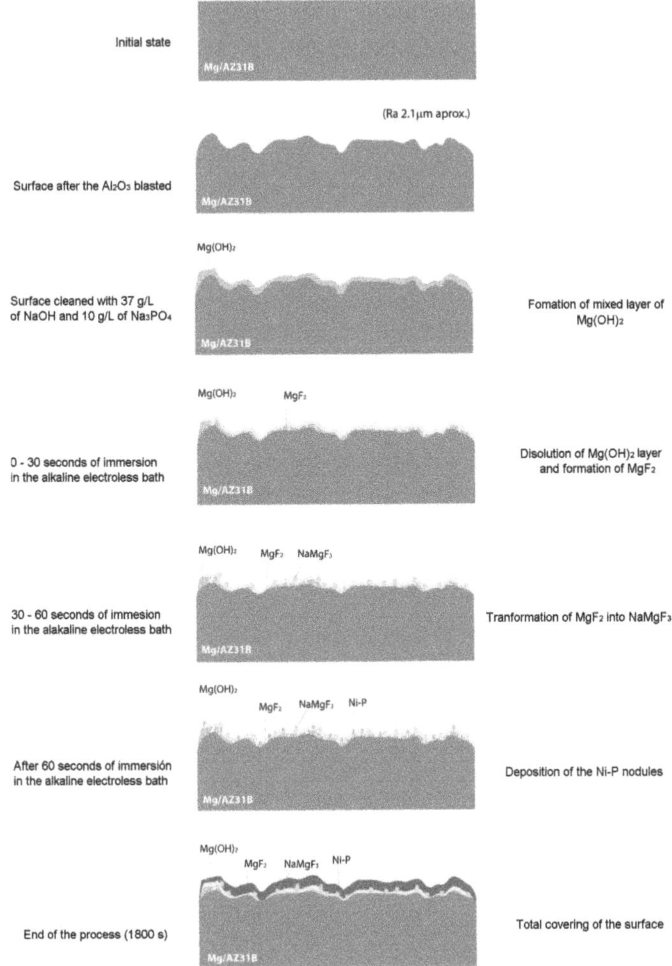

FIGURE 2.25 Schematic summary of the electroless NiP alkaline coating formation process on pure magnesium and AZ31B magnesium alloy surfaces. (Reproduced from *Surf. Coat. Technol.*, 321, Zuleta, A.A. et al., Study of the Formation of Alkaline Electroless NiP Coating on Magnesium and AZ31B Magnesium Alloy, 309–320, Copyright 2017, with permission from Elsevier.)

2.4 GENERAL CONCLUSIONS

The electroless nickel technology is quite versatile because it supports coating formations on several substrates of different nature. Therefore, the technology is widely used in different industries to improve the properties of the treated substrates. The number of variations found for sensitizing and activating the surfaces of different substrates evidences the interest of the scientific community in generating environmentally friendly and economically efficient procedures, which allow for a greater use of the electroless nickel technology.

The substrate cleaning process before the electroless process significantly improves the characteristics of the deposit formed, in contrast to the untreated surfaces. Furthermore, the process of removing oxide layers, grease, and dirt has proven to be more efficient when using gritblasting, which favors the rapid formation of more compact and adherent electroless coatings.

The activation of nonconductive surfaces, such as non-passivable metals, polymers, and ceramics, is performed through similar processes, involving the use of palladium and tin. However, there is a constant search for less expensive alternatives to obtain similar or better coating characteristics than those obtained with this metal.

The morphological evolution of electroless coatings on ferrous, aluminum, and magnesium surfaces has been widely studied, although some key aspects are yet to be elucidated, such as the influence of substrates on the size of nickel nodules. Conversely, the morphological evolution of the electroless coatings on copper and titanium surfaces, as well as the activation process for titanium surfaces, requires further systematic and detailed studies to better observe and understand the coating formation stages.

REFERENCES

Abdel Aal, A., and M. Shehata Aly. 2009. "Electroless Ni-Cu-P Plating onto Open Cell Stainless Steel Foam." *Applied Surface Science* 255 (13–14): 6652–6655. doi:10.1016/j.apsusc.2009.02.073.

Abdeli, M., P. N. Ahmadi, and R. A. Khosroshahi. 2009. "Study of Chemical Composition and Heat Treatment of Electroless Ni–B–Tl Alloy Coating on AISI 316 Stainless Steel." *Surface Engineering* 25 (2): 127–130. doi:10.1179/026708408X356830.

Abrantes, L. M., and J. P. Correia. 1994. "On the Mechanism of Electroless NiP Plating." *Journal of the Electrochemical Society* 141 (9): 2356. doi:10.1149/1.2055125.

Ahmad, N. B., N. A. B. Fadil, and H. B. Hashim. 2015. "The Effect of Organic Solvent (Ethanol) on Electroless NiP Deposition." In *Proceedings of the IEEE/CPMT International Electronics Manufacturing Technology (IEMT) Symposium*, June 1–4. IEEE. doi:10.1109/IEMT.2014.7123117.

Akyol, A., H. Algul, M. Uysal, H. Akbulut, and A. Alp. 2018. "A Novel Approach for Wear and Corrosion Resistance in the Electroless NiP-W Alloy with CNFs Co-Depositions." *Applied Surface Science* 453 (September): 482–492. doi:10.1016/j.apsusc.2018.05.152.

Ambat, R., and W. Zhou. 2004. "Electroless Nickel-Plating on {AZ91D} Magnesium Alloy: Effect of Substrate Microstructure and Plating Parameters." *Surface and Coatings Technology* 179 (2–3): 124–134. doi:10.1016/S0257-8972(03)00866-1.

Anik, M., E. Körpe, and E. Şen. 2008. "Effect of Coating Bath Composition on the Properties of Electroless Nickel–Boron Films." *Surface and Coatings Technology* 202 (9): 1718–1727. doi:10.1016/j.surfcoat.2007.07.031.

Arakawa, T., N. Watanabe, T. Nakada, J. Oshikiri, H. Umemoto, A. Hashimoto, and I. Koiwa. 2014. "Influence of Complexing Agents on Adhesion Strength of Electroless Nickel–Phosphorus Plating to Silicon Nitride–Aluminum–Polyimide Mixed Substrates." *Bulletin of the Chemical Society of Japan* 87 (5): 626–630. doi:10.1246/bcsj.20130282.

ASTM B183-18. 2018. "Standard Practice for Preparation of Low-Carbon Steel for Electroplating." West Conshohocken, PA: ASTM International. doi:10.1520/B0183-79R09.2.

ASTM B254-92. 2014. "Standard Practice for Preparation of and Electroplating on Stainless Steel." West Conshohocken, PA: ASTM International. doi:10.1520/B0254-92R14.

ASTM B319-91. 2014. "Standard Guide for Preparation of Lead and Lead Alloys for Electroplating." West Conshohocken, PA: ASTM International. doi:10.1520/B0319-91R14.

ASTM B322-99. 2014. "Standard Guide for Cleaning Metals Prior to Electroplating." West Conshohocken, PA: ASTM International. doi:10.1520/B0322-99R14.

Azumi, K., T. Yugiri, T. Kurihara, M. Seo, H. Habazaki, and S. Fujimoto. 2003. "Direct Plating of Electroless NiP Layers on Sputter-Deposited Al-Ni Alloy Films." *Journal of the Electrochemical Society* 150 (7): C461. doi:10.1149/1.1576770.

Babhale, R. A., R. S. Sonawane, S. S. Bodhale, S. K. Apte, and B. B. Kale. 2003. "Electroless Nickel Deposition on Mild Steel by Using a New Bath Formulation and Its Characterization." *Indian Journal of Chemical Technology* 10: 154–158.

Backović, N., M. Jančić, and LJ Radonjić. 1979. "Study of Electroless NiP Deposition on Aluminium." *Thin Solid Films* 59 (1): 1–12. doi:10.1016/0040-6090(79)90358-4.

Balaraju, J. N., P. Radhakrishnan, V. Ezhilselvi, A. Anil Kumar, Z. Chen, and K. P. Surendran. 2016. "Studies on Electroless Nickel Polyalloy Coatings over Carbon Fibers/CFRP Composites." *Surface and Coatings Technology* 302 (September): 389–397. doi:10.1016/J.SURFCOAT.2016.06.040.

Barbat, N., and K. Zangeneh-Madar. 2014. "Preparation of Ti–Ni Binary Powder via Electroless Nickel Plating of Titanium Powder." *Powder Metallurgy* 57 (2): 97–102. doi:10.1179/1743290113Y.0000000070.

Baskaran, I., R. Sakthi Kumar, T. S. N. S. Narayanan, and A. Stephen. 2006a. "Formation of Electroless Ni–B Coatings Using Low Temperature Bath and Evaluation of Their Characteristic Properties." *Surface and Coatings Technology* 200 (24): 6888–6894. doi:10.1016/j.surfcoat.2005.10.013.

Baskaran, I., T. S. N. S. Narayanan, and A. Stephen. 2006b. "Effect of Accelerators and Stabilizers on the Formation and Characteristics of Electroless Ni–P Deposits." *Materials Chemistry and Physics* 99 (1): 117–126. doi:10.1016/j.matchemphys.2005.10.001.

Baskaran, I., T. S. N. S. Narayanan, and A. Stephen. 2009. "Corrosion Resistance of Electroless Ni–Low B Coatings." *Transactions of the IMF* 87 (4): 221–224. doi:10.1179/174591909X438848.

Belakhmima, R. A., N. Errahmany, M. Ebn Touhami, H. Larhzil, and R. Touir. 2017. "Preparation and Characterization of Electroless Cu–P Deposition Protection for Mild Steel Corrosion in Molar Hydrochloric Solution." *Journal of the Association of Arab Universities for Basic and Applied Sciences* 24 (1): 46–53. doi:10.1016/j.jaubas.2017.01.001.

Bernasconi, R., A. Molazemhosseini, M. Cervati, S. Armini, and L. Magagnin. 2016. "Application of Self-Assembled Monolayers to the Electroless Metallization of High Aspect Ratio Vias for Microelectronics." *Journal of Electronic Materials* 45 (10): 5449–5455. doi:10.1007/s11664-016-4753-5.

Beygi, H., H. Vafaeenezhad, and S. A. Sajjadi. 2012. "Modeling the Electroless Nickel Deposition on Aluminum Nanoparticles." *Applied Surface Science* 258 (19): 7744–7750. doi:10.1016/j.apsusc.2012.04.132.

Bibber, J. W. 2009. "Zincate- or Stannate-Free Plating of Magnesium, Aluminum, and Titanium." *Metal Finishing* 107 (7–8): 28–30. doi:10.1016/S0026-0576(09)80202-X.

Blickensderfer, J. K. 2018. *Electroless Deposition of Amorphous Iron-Alloy Coatings.* Case Western Reserve University, Cleveland, OH.

Bonin, L., and V. Vitry. 2016. "Mechanical and Wear Characterization of Electroless Nickel Mono and Bilayers and High Boron-Mid Phosphorus Electroless Nickel Duplex Coatings." *Surface and Coatings Technology* 307 (December): 957–962. doi:10.1016/j.surfcoat.2016.10.021.

Bradford, P. M., B. Case, G. Dearnaley, J. F. Turner, and I. S. Woolsey. 1976. "Ion Beam Analysis of Corrosion Films on a High Magnesium Alloy (Magnox Al 80)." *Corrosion Science* 16 (10): 747–766. doi:10.1016/0010-938X(76)90007-X.

Brunelli, K., M. Dabalà, F. Dughiero, and M. Magrini. 2009. "Diffusion Treatment of Ni–B Coatings by Induction Heating to Harden the Surface of Ti–6Al–4V Alloy." *Materials Chemistry and Physics* 115 (1): 467–472. doi:10.1016/j.matchemphys.2009.01.016.

Bulbul, F. 2011. "The Effects of Deposition Parameters on Surface Morphology and Crystallographic Orientation of Electroless NiB Coatings." *Metals and Materials International* 17 (1): 67–75. doi:10.1007/s12540-011-0210-4.

Cai, Y., Z. Z. Chang, C. M. Chen, Y. T. Bai, Y. W. Lin, and Z. Q. Li. 2011. "Complexing Agent on the Surface of Magnesium Alloy Plating of NiB Alloy." In *Advanced Materials Research* 311–313: 327–330.

Calderón, J. A., J. P. Jiménez, and A. A. Zuleta. 2016. "Improvement of the Erosion-Corrosion Resistance of Magnesium by Electroless NiP/Ni$(OH)_2$-Ceramic Nanoparticle Composite Coatings." *Surface and Coatings Technology* 304: 167–178. doi:10.1016/j.surfcoat.2016.04.063.

Carraro, C., R. Maboudian, and L. Magagnin. 2007. "Metallization and Nanostructuring of Semiconductor Surfaces by Galvanic Displacement Processes." *Surface Science Reports* 62 (12): 499–525. doi:10.1016/j.surfrep.2007.08.002.

Chang, Q. H., Z. Z. Chang, Y. T. Bai, X. Y. Chen, and D. Peng. 2011. "The Effect of NaOH on the Surface of Magnesium Alloy NiB Alloy Plating Properties." *Advanced Materials Research* 189–193: 347–50.

Charbonnier, M., and M. Romand. 2003. "Polymer Pretreatments for Enhanced Adhesion of Metals Deposited by the Electroless Process." *International Journal of Adhesion and Adhesives* 23 (4): 277–285. doi:10.1016/S0143-7496(03)00045-9.

Charbonnier, M., M. Romand, and Y. Goepfert. 2006. "Ni Direct Electroless Metallization of Polymers by a New Palladium-Free Process." *Surface and Coatings Technology* 200: 5028–5036.

Charbonnier, M., M. Romand, E. Harry, and M. Alami. 2001. "Surface Plasma Functionalization of Polycarbonate: Application to Electroless Nickel and Copper Plating." *Journal of Applied Electrochemistry* 31 (1): 57–63. doi:10.1023/A:1004161707536.

Chen, C.-H., H.-L. Yang, H.-R. Chen, and C.-L. Lee. 2012. "Activity on Electrochemical Surface Area: Silver Nanoplates as New Catalysts for Electroless Copper Deposition." *Journal of the Electrochemical Society* 159 (9): D507–D511. doi:10.1149/2.025209jes.

Chen, J., G. Yu, B. Hu, Z. Liu, L. Ye, and Z. Wang. 2006. "A Zinc Transition Layer in Electroless Nickel Plating." *Surface and Coatings Technology* 201 (3–4): 686–690. doi:10.1016/j.surfcoat.2005.12.012.

Contreras, A., C. León, O. Jimenez, E. Sosa, and R. Pérez. 2006. "Electrochemical Behavior and Microstructural Characterization of 1026 Ni–B Coated Steel." *Applied Surface Science* 253 (2): 592–599. doi:10.1016/j.apsusc.2005.12.161.

Contreras, G., C. Fajardo, J. A. Berríos, A. Pertuz, J. Chitty, H. Hintermann, and E. S. Puchi. 1999. "Fatigue Properties of an AISI 1045 Steel Coated with an Electroless NiP Deposit." *Thin Solid Films* 355–356: 480–486. doi:10.1016/S0040-6090(99)00672-0.

Contreras-Lopez, L. F., E. M. Hernandez-Hernandez, C. A. Cortes-Escobedo, J. Santa Ana-Tellez, and W. Martinez-Velazco. 2016. "Electroless Nickel Plating Process in Electrodes for Use in Oxi-Hydrogen Reactors." In *2016 XVI International Congress of the Mexican Hydrogen Society (CSMH)*, pp. 1–5. IEEE. doi:10.1109/CSMH.2016.7947662.

Correa, E., A. A. Zuleta, L. Guerra, J. G. Castano, F. Echeverria, A. Baron-Wiechec, P. Skeldon, and G. E. Thompson. 2013a. "Formation of Electroless NiB on Bifluoride-Activated Magnesium and AZ91D Alloy." *Journal of the Electrochemical Society* 160 (9): D327–D336. doi:10.1149/2.012309jes.

Correa, E., A. A. Zuleta, L. Guerra, M. A. Gómez, J. G. Castaño, F. Echeverría, H. Liu et al. 2013b. "Coating Development during Electroless Ni–B Plating on Magnesium and AZ91D Alloy." *Surface and Coatings Technology* 232: 784–794. doi:10.1016/j.surfcoat.2013.06.100.

Correa, E., A. A. Zuleta, L. Guerra, M. A. Gómez, J. G. Castaño, F. Echeverría, H. Liu, P. Skeldon, and G. E. Thompson. 2013c. "Tribological Behavior of Electroless Ni–B Coatings on Magnesium and AZ91D Alloy." *Wear* 305 (1–2): 115–123. doi:10.1016/j.wear.2013.06.004.

Correa, E., A. A. Zuleta, M. Sepúlveda, L. Guerra, J. G. Castaño, F. Echeverría, H. Liu, P. Skeldon, and G. E. Thompson. 2012. "Nickel–Boron Plating on Magnesium and AZ91D Alloy by a Chromium-Free Electroless Process." *Surface and Coatings Technology* 206 (13): 3088–3093. doi:10.1016/j.surfcoat.2011.12.023.

Correa, E., J. F. Mejía, J. G. Castaño, F. Echeverría, and M. A. Gómez. 2017. "Tribological Characterization of Electroless NiB Coatings Formed on Commercial Purity Magnesium." *Journal of Tribology* 139 (5). doi:10.1115/1.4036169.

Court, S. W., B. D. Barker, and F. C. Walsh. 2000. "Electrochemical Measurements of Electroless Nickel Coatings on Zincated Aluminium Substrates." *Transactions of the Institute of Metal Finishing* 78 (4): 157–162. doi:10.1080/00202967.2000.11871330.

Court, S., C. Kerr, C. Ponce de León, J. R. Smith, B. D. Barker, and F. C. Walsh. 2017. "Monitoring of Zincate Pre-Treatment of Aluminium Prior to Electroless Nickel Plating." *Transactions of the Institute of Metal Finishing* 95 (2): 97–105. doi:10.1080/00202967.2016.1236573.

Czagány, M., P. Baumli, and G. Kaptay. 2017. "The Influence of the Phosphorus Content and Heat Treatment on the Nano-Micro-Structure, Thickness and Micro-Hardness of Electroless NiP Coatings on Steel." *Applied Surface Science* 423: 160–169. doi:10.1016/j.apsusc.2017.06.168.

Dabalà, M., A. Variola, and M. Magrini. 2001. "Surface Hardening of Ti–6Al–4V Alloy Using Combined Electroless Ni–B Plating and Diffusion Treatments." *Surface Engineering* 17 (5): 393–396. doi:10.1179/026708401101518079.

Dabalà, M., K. Brunelli, R. Frattini, and M. Magrini. 2004. "Surface Hardening of Ti–6Al–4V Alloy by Diffusion Treatment of Electroless Ni–B Coatings." *Surface Engineering* 20 (2): 103–107. doi:10.1179/sur.2004.20.2.103.

Dadvand, N., G. J. Kipouros, and W. F. Caley. 2003. "Electroless Nickel Boron Plating on AA6061." *Canadian Metallurgical Quarterly* 42 (3): 349–364. doi:10.1179/cmq.2003.42.3.349.

Dai, K. J., Y. Xiong, and X. L. Zhang. 2014. "Preparation of Electroless Ni–P Coating on ZnS Substrate by Strike Nickel Activation Process." *Materials Technology* 29 (4): 241–244. doi:10.1179/1753555714Y.0000000143.

Das, S. K., and P. Sahoo. 2011a. "A Parametric Investigation of the Friction Performance of Electroless NiB Coatings." *Lubrication Science* 23 (2): 81–97. doi:10.1002/ls.145.

Das, S. K., and P. Sahoo. 2011b. "Tribological Characteristics of Electroless NiB Coating and Optimization of Coating Parameters Using Taguchi Based Grey Relational Analysis." *Materials & Design* 32 (4): 2228–2238. doi:10.1016/j.matdes.2010.11.028.

Das, S. K., and P. Sahoo. 2012. "Influence of Process Parameters on Microhardness of Electroless NiB Coatings." *Advances in Mechanical Engineering* 2012: 1–11. doi:10.1155/2012/703168.

De, J., T. Banerjee, R. S. Sen, B. Oraon, and G. Majumdar. 2016. "Multi-Objective Optimization of Electroless Ternary Nickel–Cobalt–Phosphorus Coating Using Non-Dominant Sorting Genetic Algorithm-II." *Engineering Science and Technology, an International Journal* 19 (3): 1526–1533. doi:10.1016/j.jestch.2016.04.011.

Deepa, J. P., T. P. D. Rajan, C. Pavithran, and B. C. Pai. 2014. "Studies on Electroless Nickel Boride Coating on Boron Carbide Particles." *Surface Engineering* 30 (10): 702–708. doi:10.1179/1743294414Y.0000000279.

Delaunois, F., and P. Lienard. 2002. "Heat Treatments for Electroless Nickel–Boron Plating on Aluminium Alloys." *Surface and Coatings Technology* 160 (2–3): 239–248. doi:10.1016/S0257-8972(02)00415-2.

Delaunois, F., J. P. Petitjean, M. Jacob-Dulière, and P. Liénard. 2000. "Autocatalytic Electroless Nickel-Boron Plating on Light Alloys." *Surface and Coatings Technology* 124: 201–209.

Delbos, E., D. A., H. E. Belghiti, J. Vigneron, M. Bouttemy, and A. Etcheberry. 2015. "Localised Metallisation Process for Silicon Solar Cells." *Physica Status Solidi (C) Current Topics in Solid State Physics* 12 (12): 1427–1432. doi:10.1002/pssc.201510106.

Dikici, B., C. Tekmen, M. Gavgali, and U. Cocen. 2011. "The Effect of Electroless Ni Coating of SiC Particles on the Corrosion Behavior of A356 Based Squeeze Cast Composite." *Strojniski Vestnik/Journal of Mechanical Engineering* 57 (1): 11–20. doi:10.5545/sv-jme.2010.111.

Dominguez-Rios, C., R. Torres-Sanchez, and A. Aguilar-Elguezabal. 2007. "Characterization of the Film of NiB Electroless on Steel S7, Through Optical Microscopy and Scanning Electron Microscopy." *ECS Transactions*, 3:117–122. doi:10.1149/1.2753246.

Dong, D., X. H. Chen, W. T. Xiao, G. B. Yang, and P. Y. Zhang. 2009. "Preparation and Properties of Electroless Ni–P–SiO_2 Composite Coatings." *Applied Surface Science* 255 (15): 7051–7055. doi:10.1016/J.APSUSC.2009.03.039.

Ebrahimi, F., S. S. Yazdi, M. H. Najafabadi, and F. Ashrafizadeh. 2015. "Influence of Nanoporous Aluminum Oxide Interlayer on the Optical Absorptance of Black Electroless Nickel–phosphorus Coating." *Thin Solid Films* 592: 88–93. doi:10.1016/j.tsf.2015.09.004.

Eliaz, N., and E. Gileadi. 2018. *Physical Electrochemistry: Fundamentals, Techniques and Applications*. Wiley-Vch, Weinheim, Germany.

Fan, N., M. Huang, and L. Wang. 2010. "Electroless Nickel-Boron Plating on Magnesium Alloy." In *2010 11th International Conference on Electronic Packaging Technology & High Density Packaging*, pp. 288–292. IEEE. doi:10.1109/ICEPT.2010.5582328.

Fang, X., H. Zhou, and Y. Xue. 2015. "Corrosion Properties of Stainless Steel 316L/Ni-Cu-P Coatings in Warm Acidic Solution." *Transactions of Nonferrous Metals Society of China (English Edition)* 25 (8): 2594–2600. doi:10.1016/S1003-6326(15)63880-8.

Farzaneh, A., M. Sarvari, M. Ehteshamzadeh, and O. Mermer. 2016. "Effect of Zincating Bath Additives on Structural and Electrochemical Properties of Electroless NiP Coating on AA6061." *International Journal of Electrochemical Science* 1111 (10): 9676–9686. doi:10.20964/2016.11.75.

Fayyad, E. M., A. Abdullah, M. Hassan, A Mohamed, C. Wang, G. Jarjoura, Z. Farhat et al. 2018. "Synthesis, Characterization, and Application of Novel NiP-Carbon Nitride Nanocomposites." *Coatings* 8 (1): 37. doi:10.3390/coatings8010037.

Franco, M., W. Sha, S. Malinov, and R. Rajendran. 2013. "Phase Composition, Microstructure and Microhardness of Electroless Nickel Composite Coating Co-Deposited with SiC on Cast Aluminium LM24 Alloy Substrate." *Surface and Coatings Technology* 235: 755–763. doi:10.1016/j.surfcoat.2013.08.063.

Franco, M., W. Sha, V. Tan, and S. Malinov. 2015. "Insight of the Interface of Electroless Ni–P/SiC Composite Coating on Aluminium Alloy, LM24." *Materials & Design* 85: 248–255. doi:10.1016/j.matdes.2015.06.159.

Fukuda, S., K. Shimada, N. Izu, H. Miyazaki, S. Iwakiri, and K. Hirao. 2018. "Effects of Phosphorus Content on Generation and Growth of Cracks in Nickel–phosphorus Platings Owing to Thermal Cycling." *Journal of Materials Science: Materials in Electronics* 29 (14): 11688–11698. doi:10.1007/s10854-018-9267-x.

Gafin, A. H., and S. W. Orchard. 1993. "Catalytic Effects in the Initiation of Autocatalytic Nickel Deposition on Nickel Containing Substrates." *Journal of the Electrochemical Society* 140 (12): 3458. doi:10.1149/1.2221109.

Gao, C., L. Dai, W. Meng, Z. He, and L. Wang. 2017. "Electrochemically Promoted Electroless Nickel-Phosphorus Plating on Titanium Substrate." *Applied Surface Science* 392 (January): 912–919. doi:10.1016/j.apsusc.2016.09.127.

Garcés, Y., H. Sánchez, J. Berríos, A. Pertuz, J. Chitty, H. Hintermann, and E. S. Puchi. 1999. "Fatigue Behavior of a Quenched and Tempered AISI 4340 Steel Coated with an Electroless NiP Deposit." *Thin Solid Films* 355–356 (1): 487–493. doi:10.1016/S0040-6090(99)00673-2.

Garcia, A., T. Berthelot, P. Viel, A. Mesnage, P. Jégou, F. Nekelson, S. Roussel, and S. Palacin. 2010. "ABS Polymer Electroless Plating through a One-Step Poly(Acrylic Acid) Covalent Grafting." *ACS Applied Materials and Interfaces* 2 (4): 1177–1183. doi:10.1021/am1000163.

Ghaderi, M., M. Rezagholizadeh, H. Nasiri-Vatan, and R. Ebrahimi-Kahrizsangi. 2015. "Study of Hot Corrosion Resistance of Electroless Nickel Coating with Different Content of Phosphorus in Molten Salt Deposit Na_2SO_4–NaCl at 650°C." *Surface Engineering and Applied Electrochemistry* 51 (4): 367–373. doi:10.3103/S1068375515040055.

Giampaolo, A. R. Di, J. G. Ordoñez, J. M. Gugliemacci, and J. Lira. 1997. "Electroless Nickel-Boron Coatings on Metal Carbides." *Surface and Coatings Technology* 89 (1–2): 127–131. doi:10.1016/S0257-8972(96)03089-7.

Goettems, F. S., J. Z. Ferreira, B. Agronomia, and P. Alegre. 2017. "Wear Behaviour of Electroless Heat Treated NiP Coatings as Alternative to Electroplated Hard Chromium Deposits." *Materials Research* 20 (5): 1300–1308. doi:10.1590/1980-5373-mr-2017-0347.

Gu, C., J. Lian, G. Li, L. Niu, and Z. Jiang. 2005. "Electroless Ni–P Plating on AZ91D Magnesium Alloy from a Sulfate Solution." *Journal of Alloys and Compounds* 391 (1–2): 104–109. doi:10.1016/j.jallcom.2004.07.083.

Guo, C. Y., J. P. Matinlinna, and A. T. H. Tang. 2012. "A Novel Effect of Sandblasting on Titanium Surface: Static Charge Generation." *Journal of Adhesion Science and Technology* 26 (23): 2603–2613. doi:10.1080/01694243.2012.691007.

Hamid, Z. A., and M. T. A. Elkhair. 2002. "Development of Electroless Nickel–Phosphorus Composite Deposits for Wear Resistance of 6061 Aluminum Alloy." *Materials Letters* 57 (December): 720–726. doi:10.1016/S0167-577X(02)00860-1.

Hamid, Z. A., H. B. Hassan, and A. M. Attyia. 2010. "Influence of Deposition Temperature and Heat Treatment on the Performance of Electroless Ni–B Films." *Surface and Coatings Technology* 205 (7): 2348–2354. doi:10.1016/j.surfcoat.2010.09.025.

Harris, A. F., and A. Beevers. 1999. "The Effects of Grit-Blasting on Surface Properties for Adhesion." *International Journal of Adhesion and Adhesives* 19 (6): 445–452. doi:10.1016/S0143-7496(98)00061-X.

Hashemi, S. H., and A. Ashrafi. 2018. "Characterisations of Low Phosphorus Electroless Ni and Composite Electroless NiP-SiC Coatings on A356 Aluminium Alloy." *Transactions of the IMF* 96 (1): 52–56. doi:10.1080/00202967.2018.1403161.

Hino, M., K. Murakami, M. Hiramatsu, K. Chen, A. Saijo, and T. Kanadani. 2005. "Effect of Zincate Treatment on Adhesion of Electroless NiP Plated Film for 2017 Aluminum Alloy." *Materials Transactions* 46 (10): 2169–2175. doi:10.2320/matertrans.46.2169.

Homjabok, W., S. Permpoon, and G. Lothongkum. 2010. "Pickling Behavior of AISI 304 Stainless Steel in Sulfuric and Hydrochloric Acid Solutions." *Journal of Metals, Materials and Minerals* 20 (2): 1–6.

Homma, T., I. Komatsu, A. Tamaki, H. Nakai, and T. Osaka. 2001. "Molecular Orbital Study on the Reaction Mechanisms of Electroless Deposition Processes." *Electrochimica Acta* 47 (1–2): 47–53. doi:10.1016/S0013-4686(01)00574-6.

Homma, T., M. Tanabe, K. Itakura, and T. Osaka. 1997. "Tapping Mode Atomic Force Microscopy Analysis of the Growth Process of Electroless Nickel-Phosphorus Films an Nonconducting Surfaces." *Journal of the Electrochemical Society* 144 (12): 4123–4127. doi:10.1149/1.1838153.

Hsu, C.-H., S.-C. Chiu, and Y.-H. Shih. 2004. "Effects of Thickness of Electroless NiP Deposit on Corrosion Fatigue Damage of 7075-T6 under Salt Spray Atmosphere." *Materials Transactions* 45 (11): 3201–3208. doi:10.2320/matertrans.45.3201.

Huang, J., and Z. Chen. 2017. "Method for Electroless Nickel Plating on the Surface of $CaCO_3$ Powders." *RSC Advances* 7 (41): 25622–26. doi:10.1039/C7RA03110F.

Huang, P., J. Hu, K. J. Wang, and Y. T. Wang. 2014. "The Influence of Etching Time on the Surface Morphology of Electroless Nickel Plated Acrylic Resins Microsphere." *Advanced Materials Research* 1058 (November): 89–92. doi:10.4028/www.scientific.net/AMR.1058.89.

Huh, S.-H., S.-H. Choi, A.-S. Shin, S.-J. Ham, S. Moon, and H.-J. Lee. 2016. "Nucleation and Growth Behaviors of Pd Catalyst and Electroless Ni Deposition on Cu (111) Surface." *Journal of the Electrochemical Society* 163 (2): D49–D53. doi:10.1149/2.0641602jes.

Kang, E. T., and K. G. Neoh. 2009. "Surface Modification of Polymers." *Encyclopedia of Polymer Science and Technology* 13 (3). doi:10.1002/0471440264.pst358.

Karabulut, A., M. Durmaz, B. Kilinc, U. Sen, and S. Sen. 2017. "Effect of H_3BO_3 on the Corrosion Properties of NiB Based Electroplating Coatings." *Acta Physica Polonica A* 131 (1): 147–149. doi:10.12693/APhysPolA.131.147.

Karmalkar, S. 1997. "A Novel Activation Process for Autocatalytic Electroless Deposition on Silicon Substrates." *Journal of the Electrochemical Society* 144 (5): 1696. doi:10.1149/1.1837662.

Karmalkar, S. 1999. "A Study of Immersion Processes of Activating Polished Crystalline Silicon for Autocatalytic Electroless Deposition of Palladium and Other Metals." *Journal of the Electrochemical Society* 146 (2): 580. doi:10.1149/1.1391647.

Karmalkar, S., and V. P. Kumar. 2004. "Effects of Nickel and Palladium Activations on the Adhesion and I-V Characteristics of As-Plated Electroless Nickel Deposits on Polished Crystalline Silicon." *Journal of the Electrochemical Society* 151 (9): C554. doi:10.1149/1.1773582.

Karrab, S. A., M. A. Doheim, M. S. Aboraia, and S. M. Ahmed. 2013. "Effect of Heat Treatment and Bath Composition of Electroless Nickel-Plating on Cavitation Erosion Resistance." *Journal of Engineering Science Assiut University* 4 (5): 1989–2011.

Kaya, B., T. Gulmez, and M. Demirkol. 2008. "Preparation and Properties of Electroless NiB and NiB Nanocomposite Coatings." In *Proceedings of the World Congress on Engineering and Computer Science 2008*, pp. 22–24.

Kerber, S. J., and J. Tverberg. 2000. "Stainless Steel: Surface Analysis." *Advanced Materials and Processes* 158 (5): 33–36.

Kerr, C., D. Barker, and F. C. Walsh. 1996. "Studies of Porosity in Electroless Nickel Deposits on Ferrous Substrates." *Transactions of the IMF* 74 (6): 214–220. doi:10.1080/00202967.1996.11871129.

Khan, E., C. F. Oduoza, and T. Pearson. 2007. "Surface Characterization of Zincated Aluminium and Selected Alloys at the Early Stage of the Autocatalytic Electroless Nickel Immersion Process." *Journal of Applied Electrochemistry* 37: 1375–1381. doi:10.1007/s10800-007-9397-y.

Khosravipour, M A. 2004. "Electroless Deposition of Ni-Cu-P Alloy on 304 Stainless Steel by Using Thiourea and Gelatin as Additives and Investigation of Some Properties of Deposits." *International Journal of Iron and Steel Society of Iran* 1 (1): 29–34.

Kim, T.-Y., B.-W. Ahn, Y.-S. Kim, J.-H. Ahn, and S.-J. Suh. 2016. "All-Wet Metallization of Chemically Modified Polydimethylsiloxane with Self-Assembled Monolayers." *Journal of Nanoscience and Nanotechnology* 16 (11): 11256–11261. doi:10.1166/jnn.2016.13489.

Kimura, M., H. Yamagiwa, D. Asakawa, M. Noguchi, T. Kurashina, T. Fukawa, and H. Shirai. 2010. "Site-Selective Electroless Nickel Plating on Patterned Thin Films of Macromolecular Metal Complexes." *ACS Applied Materials and Interfaces* 2 (12): 3714–3717. doi:10.1021/am100853t.

Kinast, J., E. Hilpert, N. Lange, A. Gebhardt, R.-R. Rohloff, S. Risse, R. Eberhardt, and A. Tünnermann. 2014. "Minimizing the Bimetallic Bending for Cryogenic Metal Optics Based on Electroless Nickel." In *Advances in Optical and Mechanical Technologies for Telescopes and Instrumentation* edited by R. Navarro, C. R. Cunningham, and A. A. Barto, pp. 915136. doi:10.1117/12.2056271.

Korotin, D. M., S. Bartkowski, E. Z. Kurmaev, M. Meumann, E. B. Yakushina, R. Z. Valiev, and S. O. Cholakh. 2012. "Surface Characterization of Titanium Implants Treated in Hydrofluoric Acid." *Journal of Biomaterials and Nanobiotechnology* 3 (1): 87–91. doi:10.4236/jbnb.2012.31011.

Krishnan, K. H., S. John, K. N. Srinivasan, J. Praveen, M. Ganesan, and P. M. Kavimani. 2006. "An Overall Aspect of Electroless NiP Depositions—A Review Article." *Metallurgical and Materials Transactions A* 37 (6): 1917–1926. doi:10.1007/s11661-006-0134-7.

Kumar, S. M., R. Pramod, M. E. Shashi Kumar, and H. K. Govindaraju. 2014. "Evaluation of Fracture Toughness and Mechanical Properties of Aluminum Alloy 7075, T6 with Nickel Coating." *Procedia Engineering* 97: 178–185. doi:10.1016/j.proeng.2014.12.240.

Kunimoto, M., H. Nakai, and T. Homma. 2011a. "Density Functional Theory Analysis for Orbital Interaction between Hypophosphite Ions and Metal Surfaces." *Journal of the Electrochemical Society* 158 (10): D626. doi:10.1149/1.3623782.

Kunimoto, M., T. Shimada, S. Odagiri, H. Nakai, and T. Homma. 2011b. "Density Functional Theory Analysis of Reaction Mechanism of Hypophosphite Ions on Metal Surfaces." *Journal of the Electrochemical Society* 158 (9): D585. doi:10.1149/1.3609000.

la Fuente, D. de, E. Otero-Huerta, and M. Morcillo. 2007. "Studies of Long-Term Weathering of Aluminium in the Atmosphere." *Corrosion Science* 49 (7): 3134–3148. doi:10.1016/j.corsci.2007.01.006.

Lamolle, S. F., M. Monjo, M. Rubert, H. J. Haugen, S. P. Lyngstadaas, and J. E. Ellingsen. 2009. "The Effect of Hydrofluoric Acid Treatment of Titanium Surface on Nanostructural and Chemical Changes and the Growth of MC3T3-E1 Cells." *Biomaterials* 30 (5): 736–742. doi:10.1016/j.biomaterials.2008.10.052.

Lee, S. B., and Y. M. Kim. 2011. "Signature of Surface Energy Dependence of Partial Dislocation Slip in a Gold Nanometer-Sized Protrusion." *Scripta Materialia* 64 (12): 1125–1128. doi:10.1016/j.scriptamat.2011.03.008.

Lee, W. H., C. H. Huang, Y. T. Sun, H. Chang, and C. Y. Hsu. 2018. "Tribological Properties of NiP Electroless Coatings on Low-Carbon Steel Substrates Using an Environmentally Friendly Pretreatment." *International Journal of Electrochemical Science* 13 (2): 2044–2053. doi:10.20964/2018.02.64.

Li, L., M. An, and G. Wu. 2006. "A New Electroless Nickel Deposition Technique to Metallise SiCp/Al Composites." *Surface and Coatings Technology* 200 (16–17): 5102–5112. doi:10.1016/j.surfcoat.2005.05.031.

Li, L.-F., P. Caenen, M. Daerden, D. Vaes, G. Meers, C. Dhondt, and J.-P. Celis. 2005. "Mechanism of Single and Multiple Step Pickling of 304 Stainless Steel in Acid Electrolytes." *Corrosion Science* 47 (5): 1307–1324. doi:10.1016/j.corsci.2004.06.025.

Li, W., G. Shi, and Y. Lu. 2014. "Copper-Catalyzed Electroless Nickel Coating on Poly(Ethylene Terephthalate) Board for Electromagnetic Application." *International Journal of Materials Research* 105 (8): 797–801. doi:10.3139/146.111081.

Liew, K. W., S. Y. Chia, C. K. Kok, and K. O. Low. 2013. "Evaluation on Tribological Design Coatings of Al_2O_3, NiP-PTFE and Mos2 on Aluminium Alloy 7075 under Oil Lubrication." *Materials & Design* 48 (June): 77–84. doi:10.1016/j.matdes.2012.08.010.

Lin, C. J., and J. L. He. 2005. "Cavitation Erosion Behavior of Electroless Nickel-Plating on AISI 1045 Steel." *Wear* 259 (1–6): 154–159. doi:10.1016/j.wear.2005.02.099.

Lin, J., C. Wang, S. Wang, Y. Chen, W. He, and D. Xiao. 2016. "Initiation Electroless Nickel Plating by Atomic Hydrogen for PCB Final Finishing." *Chemical Engineering Journal* 306 (December): 117–123. doi:10.1016/j.cej.2016.07.033.

Liu, H. Y., X. J. Wang, L. P. Wang, F. Y. Lei, X. F. Wang, and H. J. Ai. 2008. "Effect of Fluoride-Ion Implantation on the Biocompatibility of Titanium for Dental Applications." *Applied Surface Science* 254 (20): 6305–6312. doi:10.1016/j.apsusc.2008.03.075.

Liu, S., X. Bian, J. Liu, C. Yang, X. Zhao, J. Fan, K. Zhang, et al. 2015. "Structure and Properties of Ni–P–graphite (C g)–TiO_2 Composite Coating." *Surface Engineering* 31 (6): 420–426. doi:10.1179/1743294414Y.0000000445.

Liu, X.-K., Z.-L. Liu, P. Liu, Y.-H. Xiang, W.-B. Hu, and W.-J. Ding. 2010. "Properties of Fluoride Film and Its Effect on Electroless Nickel Deposition on Magnesium Alloys." *Transactions of Nonferrous Metals Society of China* 20 (11): 2185–2191. doi:10.1016/S1003-6326(09)60440-4.

Liu, Z. M., and W. Gao. 2006c. "Scratch Adhesion Evaluation of Electroless Nickel Plating on Mg and Mg Alloys." *International Journal of Modern Physics B* 20 (25–27): 4637–4642.

Liu, Z., and W. Gao. 2006a. "The Effect of Substrate on the Electroless Nickel Plating of Mg and Mg Alloys." *Surface and Coatings Technology* 200 (11): 3553–3560. doi:10.1016/j.surfcoat.2004.12.001.

Liu, Z., and W. Gao. 2006b. "Electroless Nickel Plating on AZ91 Mg Alloy Substrate." *Surface and Coatings Technology* 200 (16–17): 5087–5093. doi:10.1016/j.surfcoat.2005.05.023.

Loto, C. A. 2016. "Electroless Nickel Plating–A Review." *Silicon* 8 (2): 177–186. doi:10.1007/s12633-015-9367-7.

Mahmoud, S. S. 2009. "Electroless Deposition of Nickel and Copper on Titanium Substrates: Characterization and Application." *Journal of Alloys and Compounds* 472 (1–2): 595–601. doi:10.1016/j.jallcom.2008.05.079.

Mallory, G. O., and J. B. Hajdu. 1990. *Electroless Plating: Fundamentals and Applications.* Orlando, FL: American Electroplaters and Surface Finishers Society.

Mandich, N. V. 2003. "Surface Preparation of Metals Prior to Plating: Part 1." *Metal Finishing* 101 (9): 8–22. doi:10.1016/S0026-0576(03)90245-5.

Merino, M. C., A. Pardo, R. Arrabal, S. Merino, P. Casajús, and M. Mohedano. 2010. "Influence of Chloride Ion Concentration and Temperature on the Corrosion of Mg–Al Alloys in Salt Fog." *Corrosion Science* 52 (5): 1696–1704. doi:10.1016/j.corsci.2010.01.020.

Mohandas, A., and N. Radhika. 2017. "Studies on Mechanical Behaviour of Aluminium/Nickel Coated Silicon Carbide Reinforced Functionally Graded Composite." *Tribology in Industry* 39 (2): 145–151. doi:10.24874/ti.2017.39.02.01.

Mukhopadhyay, A., T. B. Kumar, and P. Sahoo. 2017. "Effect of Heat Treatment on Microstructure and Corrosion Resistance of NiB-W-Mo Coating Deposited by Electroless Method." *Surface Review and Letters* 25 (8): 1950023. doi:10.1142/S0218625X19500239.

Murakami, K, M. Hino, N. Nagata, and T. Kanadani. 2013b. "Effect of Alloying Elements and Generation of Hydrogen Gas on Zincate Treatment and Electroless Nickel-Phosphorus Plating for Aluminum Alloys." *Nippon Kinzoku Gakkaishi/Journal of the Japan Institute of Metals* 77 (12): 599–603. doi:10.2320/jinstmet.JC201309.

Murakami, K., M. Hino, M. Hiramatsu, K. Osamura, and T. Kanadani. 2006. "Influence of Zincate Treatment on Adhesion Strength of Electroless Nickel-Phosphorus Plated Film for Commercial Pure Aluminium." *Materials Science Forum* 519–521 (10): 759–64. doi:10.4028/www.scientific.net/MSF.519-521.759.

Murakami, K., M. Hino, M. Ushio, D. Yokomizo, and T. Kanadani. 2013a. "Formation of Zincate Films on Binary Aluminum Alloys and Adhesion of Electroless Nickel-Phosphorus Plated Films." *Materials Transactions* 54 (2): 199–206. doi:10.2320/matertrans. L-M2012830.

Nakao, S., D.-H. Kim, K. Obata, S. Inazawa, M. Majima, K. Koyama, and Y. Tani. 2003. "Electroless Pure Nickel Plating Process with Continuous Electrolytic Regeneration System." *Surface and Coatings Technology* 169–170 (June): 132–134. doi:10.1016/S0257-8972(03)00193-2.

Nascimento, M. P., H. J. C. Voorwald, R. C. Souza, and W. L. Pigatin. 2001. "Evaluation of an Electroless Nickel Interlayer on the Fatigue & Corrosion Strength of Chromium-Plated AISI 4340 Steel." *Plating and Surface Finishing* 88 (4): 84–90.

National Physical Laboratory. 1982. *Guides to Good Practice in Corrosion Control: Surface Preparation for Coating.* Middlesex, UK: NPL.

Nikitasari, A., and E. Mabruri. 2016. "Study of Electroless Ni-W-P Alloy Coating on Martensitic Stainless Steel." In *AIP Conference Proceedings*, 1725: 020053. doi:10.1063/1.4945507.

Oda, Y., N. Fukumuro, and S. Yae. 2018. "Intermetallic Compound Growth between Electroless Nickel/Electroless Palladium/Immersion Gold Surface Finish and Sn-3.5Ag or Sn-3.0Ag-0.5Cu Solder." *Journal of Electronic Materials* 47 (4): 2507–2511. doi:10.1007/s11664-018-6067-2.

Olarewaju Ajibolam, O., A. Adebayo, and D. T. Oloruntoba. 2015. "Corrosion of Heat Treated Electroless-Ni Plated Mild Carbon Steels in Dilute H_2SO_4." *International Journal of Materials Science and Applications* 4 (5): 333. doi:10.11648/j.ijmsa.20150405.18.

Oliveira, M. C. L. de, O. V. Correa, B. Ett, I. J. Sayeg, N. B. Lima, and R. A. Antunes. 2017. "Influence of the Tungsten Content on Surface Properties of Electroless Ni-W-P Coatings." *Materials Research* 21 (1): 0. doi:10.1590/1980-5373-mr-2017-0567.

Oskouei, R., M. Barati, and R. Ibrahim. 2016. "Surface Characterizations of Fretting Fatigue Damage in Aluminum Alloy 7075-T6 Clamped Joints: The Beneficial Role of Ni–P Coatings." *Materials* 9 (3): 141. doi:10.3390/ma9030141.

Pan, D., L. Shuansuo, L. Jianming. 2013a. HT200 grey cast iron workpiece chemical nickel-plating treating agent, and production method thereof. CN104711546A, issued 2013.

Pan, J. X., R. J. Chen, and C. L. Wu. 2013b. "Research on Micro-Structure of Electroless NiB Coatings." *Advanced Materials Research* 602–604: 1641–1645.

Pancrecious, J. K., J. P. Deepa, R. Ramya, T. P. D. Rajan, E. Bhoje Gowd, and B. C. Pai. 2015. "Ultrasonic-Assisted Electroless Coating of NiB Alloy and Composites on Aluminum Alloy Substrates." *Materials Science Forum* 830–831 (September): 687–90. doi:10.4028/www.scientific.net/MSF.830-831.687.

Pancrecious, J. K., S. B. Ulaeto, R. Ramya, T. P. D. Rajan, and B. C. Pai. 2018. "Metallic Composite Coatings by Electroless Technique—A Critical Review." *International Materials Reviews* 63 (8): 488–512. doi:10.1080/09506608.2018.1506692.

Park, I.-C., and S.-J. Kim. 2018. "Effect of Stabilizer Concentration on the Cavitation Erosion Resistance Characteristics of the Electroless Nickel Plated Gray Cast Iron in Seawater." *Surface and Coatings Technology*. doi:10.1016/j.surfcoat.2018.08.098.

Paunovic, M., and M. Schlesinger. 2006. *Fundamentals of Electrochemical Deposition*, 2nd ed., Vol. 5. Hoboken, NJ: John Wiley & Sons.

Pázmán, J., V. Mádai, J. Tóth, and Z. Gácsi. 2012a. "Production and Investigation of Al/SiC(Ni)p Composites." *International Journal of Microstructure and Materials Properties* 7 (2–3): 220. doi:10.1504/IJMMP.2012.047501.

Pázmán, J., V. Mádai, Z. Gácsi, and A. Kovács. 2012b. "Arrangement of the Al-Ni Phases in Al/SiC(Ni)p Composites." *International Journal of Microstructure and Materials Properties* 7 (1): 49–63. doi:10.1504/IJMMP.2012.045802.

Petro, R. 2014. "Modern Applications of Novel Electroless Plating Techniques." Electronic theses and dissertations. University of Windsor.

Petro, R., and M. Schlesinger. 2011. "Direct Electroless Deposition of Nickel Boron Alloys and Copper on Aluminum Containing Magnesium Alloys." *Electrochemical and Solid-State Letters* 14 (4): D37. doi:10.1149/1.3537041.

Petukhov, D. I., M. N. Kirikova, A. A. Bessonov, and M. J. A. Bailey. 2014. "Nickel and Copper Conductive Patterns Fabricated by Reactive Inkjet Printing Combined with Electroless Plating." *Materials Letters* 132 (October): 302–306. doi:10.1016/j.matlet.2014.06.109.

Rabiatul Adawiyah, M. A., and O. Saliza Azlina. 2018. "Comparative Study on the Isothermal Aging of Bare Cu and ENImAg Surface Finish for Sn-Ag-Cu Solder Joints." *Journal of Alloys and Compounds* 740 (April): 958–966. doi:10.1016/j.jallcom.2018.01.054.

Rahmat, M. A., R. H. Oskouei, R. N. Ibrahim, and R. K. Singh Raman. 2013. "The Effect of Electroless NiP Coatings on the Fatigue Life of Al 7075-T6 Fastener Holes with Symmetrical Slits." *International Journal of Fatigue* 52: 30–38. doi:10.1016/j.ijfatigue.2013.02.007.

Rajendran, R., W. Sha, and R. Elansezhian. 2010. "Abrasive Wear Resistance of Electroless Ni–P Coated Aluminium after Post Treatment." *Surface and Coatings Technology* 205 (3): 766–772. doi:10.1016/j.surfcoat.2010.07.124.

Ramaseshan, R., S. K. Seshadri, and N. G. Nair. 2001. "Electroless Nickel-Phosphorus Coating on Ti and Al Elemental Powders." *Scripta Materialia* 45 (2): 183–189. doi:10.1016/S1359-6462(01)01013-2.

Rantell, A., and A. Holtzman. 1974. "The Role of Accelerators Prior to Electroless Plating of ABS Plastic." *Transactions of the IMF* 52 (1): 31–38. doi:10.1080/00202967.1974.11870301.

Rao, Q., G. Bi, Q. Lu, H. Wang, and X. Fan. 2005. "Microstructure Evolution of Electroless NiB Film during Its Depositing Process." *Applied Surface Science* 240: 28–33.

Rice, D. W. 1981. "Atmospheric Corrosion of Copper and Silver." *Journal of the Electrochemical Society* 128 (2): 275. doi:10.1149/1.2127403.

Riddle, Y. W., and T. O. Bailer. 2005. "Friction and Wear Reduction via an NiB Electroless Bath Coating for Metal Alloys." *Jom* 40–45. doi:10.1007/s11837-005-0080-7.

Rudnik, E., T. Jucha, L. Burzynska, and K. Ćwięka. 2011. "Electro- and Electroless Deposition of Ni/SiC and Co/SiC Composite Coatings on Aluminum." *Materials Science Forum* 690 (June): 377–380. doi:10.4028/www.scientific.net/MSF.690.377.

Saeedi Heydari, M., H. R. Baharvandi, and S. R. Allahkaram. 2018. "Electroless Nickel-Boron Coating on B_4C-Nano TiB_2 Composite Powders." *International Journal of Refractory Metals and Hard Materials* 76 (November): 58–71. doi:10.1016/j.ijrmhm.2018.05.012.

Sahoo, P., and S. K. Das. 2011. "Tribology of Electroless Nickel Coatings–A Review." *Materials and Design* 32 (4): 1760–1775. doi:10.1016/j.matdes.2010.11.013.

Salahinejad, E., R. Eslami Farsani, and L. Tayebi. 2017. "Synergistic Galvanic-Pitting Corrosion of Copper Electrical Pads Treated with Electroless Nickel-Phosphorus/Immersion Gold Surface Finish." *Engineering Failure Analysis* 77 (July): 138–145. doi:10.1016/j.engfailanal.2017.03.001.

Schilling, M. L., H. E. Katz, F. M. Houlihan, S. M. Stein, R. S. Hutton, and G. N. Taylor. 1996. "Selective Electroless Nickel Deposition on Patterned Phosphonate and Carboxylate Polymer Films." *Journal of the Electrochemical Society* 143 (2): 691. doi:10.1149/1.1836502.

Sevonkaev, I., D. V. Goia, and E. Matijević. 2008. "Formation and Structure of Cubic Particles of Sodium Magnesium Fluoride (Neighborite)." *Journal of Colloid and Interface Science* 317 (1): 130–136. doi:10.1016/j.jcis.2007.09.036.

Sheng, G. Y. 2016. "Study on Growth Mode and Properties of Electroless Nickel Plating on Aluminum Alloy." *Chemical Engineering Transactions* 55: 319–324. doi:10.3303/CET1655054.

Shirmohammadi Yazdi, S., F. Ashrafizadeh, and A. Hakimizad. 2013. "Improving the Grain Structure and Adhesion of NiP Coating to 3004 Aluminum Substrate by Nanostructured Anodic Film Interlayer." *Surface and Coatings Technology* 232 (October): 561–566. doi:10.1016/j.surfcoat.2013.06.028.

Singh, A. K., V. K. Bajpai, and C. S. Solanki. 2014. "Effect of Light on Electroless Nickel Deposition for Solar Cell Applications." *Energy Procedia* 54 (January): 763–770. doi:10.1016/j.egypro.2014.07.318.

Soares, M. E., P. Soares, P. R. Souza, R. M. Souza, and R. D. Torres. 2017. "The Effect of Nitriding on Adhesion and Mechanical Properties of Electroless Ni–P Coating on AISI 4140 Steel." *Surface Engineering* 33 (2): 116–121. doi:10.1080/02670844.2016.1148831.

Song, Y., D. Shan, and E. Han. 2006. "Initial Deposition Mechanism of Electroless Nickel Plating on Az91d Magnesium Alloys." *Canadian Metallurgical Quarterly* 45 (2): 215–222.

Staia, M. H., E. J. Castillo, E. S. Puchi, B. Lewis, and H. E. Hintermann. 1996. "Wear Performance and Mechanism of Electroless NiP Coating." *Surface and Coatings Technology* 86–87: 598–602. doi:10.1016/S0257-8972(96)03086-1.

Staia, M. H., E. S. Puchi, E. Castillo, D. B. Lewis, and M. Jeandin. 1997. "Characterisation of Ni–P Electroless Deposits with Moderate Phosphorus Content." *Surface Engineering* 13 (4): 335–338. doi:10.1179/sur.1997.13.4.335.

Strandjord, A. J. G., S. Popelar, and C. Jauernig. 2002. "Interconnecting to Aluminum- and Copper-Based Semiconductors (Electroless-Nickel/Gold for Solder Bumping and Wire Bonding)." *Microelectronics Reliability* 42 (2): 265–283. doi:10.1016/S0026-2714(01)00236-0.

Straumanis, M. E., and P. C. Chen. 1951. "The Mechanism and Rate of Dissolution of Titanium in Hydrofluoric Acid." *Journal of the Electrochemical Society* 98 (6): 234–240. doi:10.1149/1.2430103.

Stremsdoerfer, G., H. Omidvar, P. Roux, Y. Meas, and R. Ortega-Borges. 2008. "Deposition of Thin Films of Ni–P and Ni–B–P by Dynamic Chemical Plating." *Journal of Alloys and Compounds* 466 (1–2): 391–397. doi:10.1016/j.jallcom.2007.11.052.

Su, Y., B. Zhou, L. Liu, J. Lian, and G. Li. 2015. "Electromagnetic Shielding and Corrosion Resistance of Electroless NiP and NiP-Cu Coatings on Polymer/Carbon Fiber Composites." *Polymer Composites* 36 (5): 923–930. doi:10.1002/pc.23012.

Sudagar, J., J. Lian, and W. Sha. 2013. "Electroless Nickel, Alloy, Composite and Nano Coatings—A Critical Review." *Journal of Alloys and Compounds* 571 (September): 183–204. doi:10.1016/j.jallcom.2013.03.107.

Sudagar, J., K. Venkateswarlu, and J. Lian. 2010. "Dry Sliding Wear Properties of a 7075-T6 Aluminum Alloy Coated with NiP (h) in Different Pretreatment Conditions." *Journal of Materials Engineering and Performance* 19 (6): 810–818. doi:10.1007/s11665-009-9545-0.

Sun, S., J. Liu, C. Yan, and F. Wang. 2008. "A Novel Process for Electroless Nickel Plating on Anodized Magnesium Alloy." *Applied Surface Science* 254 (16): 5016–5022. doi:10.1016/j.apsusc.2008.01.169.

Sviridov, V. V., T. V. Gaevskaya, L. I. Stepanova, and T. N. Vorobyova. 2003. "Electroless Deposition and Electroplating of Metals." *Chemical Problems of the Development of New Materials and Technologies*. BSU, pp. 9–59.

Szirmai, G., J. Tóth, T. I. Török, and N. Hegman. 2010. "An Experimental Study on the Effect of Aqueous Hypophosphite Pre-Treatment Used on an Aluminium Alloy Substrate before Electroless Nickel Plating." *Materials Science Forum* 659 (September): 103–108. doi:10.4028/www.scientific.net/MSF.659.103.

Takács, D., L. Sziráki, T. I. Török, J. Sólyom, Z. Gácsi, and K. Gál-Solymos. 2007. "Effects of Pre-Treatments on the Corrosion Properties of Electroless Ni–P Layers Deposited on $AlMg_2$ Alloy." *Surface and Coatings Technology* 201 (8): 4526–4535. doi:10.1016/j.surfcoat.2006.09.045.

Tang, J., and Y. Zuo. 2008. "Study on Corrosion Resistance of Palladium Films on 316L Stainless Steel by Electroplating and Electroless Plating." *Corrosion Science* 50 (10): 2873–2878. doi:10.1016/j.corsci.2008.07.014.

Tang, Z. Q., S. R. Hu, and Y. C. Chao. 2014. "Electroless Nickel Plating on Copper Foil." *Advanced Materials Research* 926–930 (May): 103–107. doi:10.4028/www.scientific.net/AMR.926-930.103.

Teixeira, L. A. C., and M. C. Santini. 2005. "Surface Conditioning of ABS for Metallization without the Use of Chromium Baths." *Journal of Materials Processing Technology* 170 (1–2): 37–41. doi:10.1016/j.jmatprotec.2005.04.075.

Tian, D., D. Y. Li, F. F. Wang, N. Xiao, R. Q. Liu, N. Li, Q. Li, W. Gao, and G. Wu. 2013. "A Pd-Free Activation Method for Electroless Nickel Deposition on Copper." *Surface and Coatings Technology* 228 (August): 27–33. doi:10.1016/j.surfcoat.2013.03.048.

Tsai, T. K., S. J. Hsueh, J. H. Lee, and J. S. Fang. 2012. "Optical Properties and Durability of Al_2O_3-NiP/Al Solar Absorbers Prepared by Electroless Nickel Composite Plating." *Journal of Electronic Materials* 41 (1): 53–59. doi:10.1007/s11664-011-1746-2.

Uma Rani, R., A. K. Sharma, C. Minu, G. Poornima, and S. Tejaswi. 2010. "Studies on Black Electroless Nickel Coatings on Titanium Alloys for Spacecraft Thermal Control Applications." *Journal of Applied Electrochemistry* 40 (2): 333–339. doi:10.1007/s10800-009-9980-5.

Umasankar, V., S. Karthikeyan, and M. A. Xavior. 2014. "The Influence of Electroless Nickel Coated SiC on the Interface Strength and Microhardness of Aluminium Composites." *Journal of Materials and Environmental Science* 5 (1): 153–158.

Unigovski, Y. B., A. Grinberg, E. Gerafi, and E. M. Gutman. 2013. "Low-Cycle Fatigue of a Multi-Layered Aluminum Sheet Alloy." *Surface and Coatings Technology* 232 (October): 695–702. doi:10.1016/j.surfcoat.2013.06.080.

Vaškelis, A., A. Jagminienė, L. Tamašauskaitė–Tamašiūnaitė, and R. Juškėnas. 2005. "Silver Nanostructured Catalyst for Modification of Dielectrics Surface." *Electrochimica Acta* 50 (23): 4586–4591. doi:10.1016/j.electacta.2004.10.093.

Vijayanand, M., and R. Elansezhian. 2014a. "Effect of Different Pretreatments and Heat Treatment on Wear Properties of Electroless NiB Coatings on 7075-T6 Aluminum Alloy." *Procedia Engineering* 97: 1707–1717. doi:10.1016/j.proeng.2014.12.322.

Vijayanand, M., and R. Elansezhian. 2014b. "Study on Influence of Zwitterionic Surfactant on the Surface Finish and Surface Morphology of Electroless NiB Coatings on Al 7075-T6 Alloy." *Advanced Materials Research* 984–985 (July): 476–481. doi:10.4028/www.scientific.net/AMR.984-985.476.

Vite-Torres, J., M. Vite-Torres, R. Aguilar-Osorio, and J. E. Reyes-Astivia. 2015. "Tribological and Corrosion Properties of Nickel Coatings on Carbon Steel." *FME Transactions* 43 (3): 206–210. doi:10.5937/fmet1503206V.

Vitry, V., A. Sens, A.-F. Kanta, and F. Delaunois. 2012b. "Wear and Corrosion Resistance of Heat Treated and As-Plated Duplex NiP/NiB Coatings on 2024 Aluminum Alloys." *Surface and Coatings Technology* 206 (16): 3421–3427. doi:10.1016/j.surfcoat.2012.01.049.

Vitry, V., A. Sens, A-F. Kanta, and F. Delaunois. 2012c. "Experimental Study on the Formation and Growth of Electroless Nickel-Boron Coatings from Borohydride-Reduced Bath on Mild Steel." *Applied Surface Science* 263 (December): 640–647. doi:10.1016/j.apsusc.2012.09.126.

Vitry, V., A.-F. F. Kanta, and F. Delaunois. 2010. "Initiation and Formation of Electroless Nickel-Boron Coatings on Mild Steel: Effect of Substrate Roughness." *Materials Science and Engineering B: Solid-State Materials for Advanced Technology* 175 (3): 266–273. doi:10.1016/j.mseb.2010.08.003.

Vitry, V., A.-F. Kanta, J. Dille, and F. Delaunois. 2012a. "Structural State of Electroless Nickel-Boron Deposits (5wt.% B): Characterization by XRD and TEM." *Surface and Coatings Technology* 206 (16): 3444–3449. doi:10.1016/J.SURFCOAT.2012.02.003.

Vitry, V., and F. Delaunois. 2015. "Applied Surface Science Formation of Borohydride-Reduced Nickel–Boron Coatings on Various Steel Substrates." *Applied Surface Science* 359: 692–703. doi:10.1016/j.apsusc.2015.10.205.

Vitry, V., F. Delaunois, and C. Dumortier. 2008. "Mechanical Properties and Scratch Test Resistance of Nickel-Boron Coated Aluminium Alloy after Heat Treatments." *Surface and Coatings Technology* 202 (14): 3316–3324. doi:10.1016/j.surfcoat.2007.12.001.

Vitry, V., L. Bonin, and L. Malet. 2017. "Chemical, Morphological and Structural Characterisation of Electroless Duplex NiP/NiB Coatings on Steel." *Surface Engineering* 34 (6): 1–10. doi:10.1080/02670844.2017.1320032.

Volovitch, P., C. Allely, and K. Ogle. 2009. "Understanding Corrosion via Corrosion Product Characterization: I. Case Study of the Role of Mg Alloying in Zn-Mg Coating on Steel." *Corrosion Science* 51 (6): 1251–1262. doi:10.1016/j.corsci.2009.03.005.

Voorwald, H. J. C., R. Q. Padilha, L. W. Pigatin, M. O. H. Cioffi, and M. P. Silva. 2007. "Influence of Electroless Nickel Interlayer Thickness on Fatigue Strength of Chromium-Plated AISI 4340 Steel." In *ESIS-ECF 15 Sweden*, pp. 895–704.

Wang, C., Z. Farhat, G. Jarjoura, M. K. Hassan, and A. M. Abdullah. 2018. "Indentation and Bending Behavior of Electroless NiP-Ti Composite Coatings on Pipeline Steel." *Surface and Coatings Technology* 334 (January): 243–252. doi:10.1016/j.surfcoat.2017.10.074.

Wang, M.-Q., J. Yan, S.-G. Du, and S.-H. Meng. 2013. "Copper Nanoparticles Seeded Functionalized-PVC Plastic Surface for Electroless Nickel Deposition." *Surface and Interface Analysis* 45 (13): 1899–1902. doi:10.1002/sia.5337.

Wang, X. Y., and D. Y. Li. 2003. "Mechanical, Electrochemical and Tribological Properties of Nano-Crystalline Surface of 304 Stainless Steel." *Wear* 255 (7–12): 836–845. doi:10.1016/S0043-1648(03)00055-3.

Wang, Y., X. Shu, S. Wei, C. Liu, W. Gao, R. A. Shakoor, and R. Kahraman. 2015. "Duplex Ni–P–ZrO_2/Ni–P Electroless Coating on Stainless Steel." *Journal of Alloys and Compounds* 630: 189–194. doi:10.1016/j.jallcom.2015.01.064.

Wang, Z. C., F. Jia, L. Yu, Z. B. Qi, Y. Tang, and G.-L. Song. 2012a. "Direct Electroless Nickel–Boron Plating on AZ91D Magnesium Alloy." *Surface and Coatings Technology* 206 (17): 3676–3685. doi:10.1016/j.surfcoat.2012.03.020.

Wang, Z.-C., L. Yu, F. Jia, and G.-L. Song. 2012b. "Effect of Additives and Heat Treatment on the Formation and Performance of Electroless Nickel-Boron Plating on AZ91D Mg Alloy." *Journal of the Electrochemical Society* 159 (7): D406–D412. doi:10.1149/2.012207jes.

Wei, X., and D. K. Roper. 2014. "Tin Sensitization for Electroless Plating Review." *Journal of the Electrochemical Society* 161 (5): D235–D242. doi:10.1149/2.047405jes.

Woo, B. H., M. Sone, A. Shibata, C. Ishiyama, K. Masuda, M. Yamagata, and Y. Higo. 2009. "Effects of Sc-CO_2 Catalyzation in Metallization on Polymer by Electroless Plating." *Surface and Coatings Technology* 203 (14): 1971–1978. doi:10.1016/j.surfcoat.2009.01.031.

Wu, L.-P., J.-J. Zhao, Y.-P. Xie, and Z.-D. Yang. 2010. "Progress of Electroplating and Electroless Plating on Magnesium Alloy." *Transactions of Nonferrous Metals Society of China* 20 (July): s630–s637. doi:10.1016/S1003-6326(10)60552-3.

Xiang, Y., W. Hu, X. Liu, C. Zhao, and W. Ding. 2001. "Initial Deposition Mechanism of Electroless Nickel Plating on Magnesium Alloys." *Transactions of the Institute of Metal Finishing* 79 (1): 30–32. doi:10.1080/00202967.2001.11871356.

Xie, R., H. Zhang, J. Zou, N. Lin, Y. Ma, Z. Wang, W. Tian et al. 2016. "Effect of Adding Lanthanum (La^{3+}) on Surface Performance of Ni-P Electroless Plating Coatings on RB400 Support Anchor Rod Steel." *International Journal of Electrochemical Science* 11 (5): 3269–3284. doi:10.20964/110343.

Xu, S.-A. 2016. "Effect of Deposition Time and Temperature on the Performance of Electroless NiP Coatings." *International Journal of Electrochemical Science* 11 (October): 8817–8826. doi:10.20964/2016.10.55.

Yagi, S., K. Murase, S. Tsukimoto, T. Hirato, and Y. Awakura. 2005. "Electroless Nickel Plating onto Minute Patterns of Copper Using Ti(IV)/Ti(III) Redox Couple." *Journal of the Electrochemical Society* 152 (9): C588. doi:10.1149/1.1973244.

Yamamoto, A., T. Yamada, T. Naganawa, and T. Natori. 1989. "Study on Electroless Nickel Plating on Cast Iron." *The Journal of the Japan Foundrymen'S Society* 61 (3): 177–182. doi:10.11279/imono.61.3_177.

Yee, P. K., W. T. Wai, and Y. F. Khong. 2014. "Palladium-Copper Inter-Diffusion during Copper Activation for Electroless Nickel Plating Process on Copper Power Metal." In *Proceedings of the International Symposium on the Physical and Failure Analysis of Integrated Circuits, IPFA*, pp. 219–222. IEEE. doi:10.1109/IPFA.2014.6898185.

Yunes, L. F., and M. Azadbeh. 2013. "The Effect of Surface Mechanical Activation on Protection Properties of Zinc Phosphate Coating." *World Applied Sciences Journal* 26 (10): 1345–1350. doi:10.5829/idosi.wasj.2013.26.10.1139.

Zahran, R., J. I. Rosales Leal, M. A. Rodríguez Valverde, and M. A. Cabrerizo Vílchez. 2016. "Effect of Hydrofluoric Acid Etching Time on Titanium Topography, Chemistry, Wettability, and Cell Adhesion." *PLoS One* 11 (11): e0165296. doi:10.1371/journal.pone.0165296.

Zangeneh-Madar, K., and A. Jafari. 2012. "Characterisation of Electroless Nickel Plated Titanium Powder." *Surface Engineering* 28 (6): 393–399. doi:10.1179/1743294411Y.0000000067.

Zeb, G., X. T. Duong, N. P. Vu, Q. T. Phan, D. T. Nguyen, V. A. Ly, S. Salimy, and X. T. Le. 2017. "Chemical Metallization of KMPR Photoresist Polymer in Aqueous Solutions." *Applied Surface Science* 407 (June): 518–525. doi:10.1016/j.apsusc.2017.02.231.

Żenkiewicz, M., K. Moraczewski, P. Rytlewski, M. Stepczyńska, and B. Jagodziński. 2015. "Electroless Metallization of Polymers." *Archives of Materials Science and Engineering* 74 (2): 67–76. doi:10.14314/polimery.2017.163.

Zhang, H., S. Wang, G. Yao, and Z. Hua. 2009. "Electroless Ni–P Plating on Mg–10Li–1Zn Alloy." *Journal of Alloys and Compounds* 474 (1–2): 306–310. doi:10.1016/j.jallcom.2008.06.067.

Zhang, W. X., Z. H. Jiang, G. Y. Li, Q. Jiang, and J. S. Lian. 2008. "Electroless NiP/NiB Duplex Coatings for Improving the Hardness and the Corrosion Resistance of AZ91D Magnesium Alloy." *Applied Surface Science* 254 (16): 4949–4955. doi:10.1016/j.apsusc.2008.01.144.

Zheng, H. Z., Y. Mo, and W. B. Liu. 2012. "Fabrication of Nano-Al_2O_3/Ni Composite Particle with Core-Shell Structure by a Modified Electroless Plating Process." *Advanced Materials Research* 455–456 (January): 49–54. doi:10.4028/www.scientific.net/AMR.455-456.49.

Zuleta, A. A. 2008. "Obtención y Evaluación Del Comportamiento a Altas Temperaturas de Recubrimientos NiP Modificados Con Magnetitas Sintéticas Puras y En Presencia de Al o Ce." Universidad de Antioquia.

Zuleta, A. A., E. Correa, C. Villada, M. Sepúlveda, J. G. Castaño, and F. Echeverría. 2011. "Comparative Study of Different Environmentally Friendly (Chromium-Free) Methods for Surface Modification of Pure Magnesium." *Surface and Coatings Technology* 205 (23–24): 5254–5259. doi:10.1016/j.surfcoat.2011.05.048.

Zuleta, A. A., E. Correa, J. G. Castaño, F. Echeverría, A. Baron-Wiecheć, P. Skeldon, and G. E. Thompson. 2017. "Study of the Formation of Alkaline Electroless NiP Coating on Magnesium and AZ31B Magnesium Alloy." *Surface and Coatings Technology* 321: 309–320. doi:10.1016/j.surfcoat.2017.04.059.

Zuleta, A. A., E. Correa, M. Sepúlveda, L. Guerra, J. G. Castaño, F. Echeverría, P. Skeldon, and G. E. Thompson. 2012. "Effect of NH_4HF_2 on Deposition of Alkaline Electroless NiP Coatings as a Chromium-Free Pre-Treatment for Magnesium." *Corrosion Science* 55 (February): 194–200. doi:10.1016/j.corsci.2011.10.028.

Zuleta, A. A., O. A. Galvis, J. G. Castaño, F. Echeverría, F. J. Bolivar, M. P. Hierro, and F. J. Pérez-Trujillo. 2009. "Preparation and Characterization of Electroless Ni–P–Fe_3O_4 Composite Coatings and Evaluation of Its High Temperature Oxidation Behaviour." *Surface and Coatings Technology* 203 (23): 3569–3578. doi:10.1016/J.SURFCOAT.2009.05.025.

3 Electroless Nickel-Phosphorus Deposits

Suman Kalyan Das and Prasanta Sahoo

CONTENTS

3.1 INTRODUCTION

Electroless nickel coatings have become very popular over the time due to their easy process and relatively cheaper and simple setup. Since the discovery of the process around seven decades ago, electroless nickel plating has seen a tremendous development, which has broken the barriers of research and development and caught the attention of the industries. With virtually innumerable combinations of alloys and composites possible to be developed, electroless deposition process has something to offer for everybody. From simple needle to mighty space vehicles, electroless nickel coatings have been a popular means for providing surface protection.

Although a variety of tailor-made electroless alloys have been developed by researchers over the years, electroless nickel-phosphorus (NiP)-based coatings have remained popular, as the same can be developed very easily and can also give rise to a hefty combination of alloys and composites each possessing special characteristics suitable for a particular application. Binary NiP coating is a dense alloy of nickel and phosphorus. Phosphorus is always present in the coatings when reduction is performed by hypophosphite. Electroless coatings formed from hypophosphite baths seemed to have excellent corrosion and abrasion resistance. The coatings can be further strengthened by a suitable heat treatment cycle to make the deposit hard and wear-resistant.

Electroless nickel coatings have been primarily used to prevent wear and corrosion, and also found suitable as an alternative to hard chromium, because chromium baths pose hazards to both human health and the environment. Electroless nickel coatings have garnered widespread attention in the last couple of decades, as the properties of these coatings are superior to conventional coatings, which have improper adherence and poor properties. The several advantages of electroless over the conventional plating techniques include the quality of the deposit, uniformity and excellent tribological properties. Industries have begun adopting this coating, owing to its enhanced wear and corrosion behavior. As already discussed in a previous chapter, electroless nickel coatings can be applied to nonconductors and to components with intricate shapes. In the machinery construction, components usually coated with electroless nickel are lathe beds, shafts, levers, calipers and fasteners. Applications of electroless nickel in chemical and plastic industries include coatings for autoclaves, filters, pipe works, heat exchangers, screw feeders, valves, pump components and so on. The aesthetic attribute and functionality of electroless nickel are used in automobile industry for brake pistons, radiator jackets, clutch bosses, steering gears and so on. Apart from these, electroless nickel coatings are also used in hydraulics, mining, offshore technology, oil and gas, electrical and electronics, aerospace and printing industries (Sahoo and Das 2011). Figure 3.1 illustrates some components plated with electroless NiP.

3.1.1 Types of Electroless Nickel Plating

As already discussed, a variety of electroless nickel coatings are possible to be developed and many of which have been realized. Electroless NiP coatings can be divided into two broad categories.

FIGURE 3.1 Electroless NiP-plated components (a) Machine components, (b) Fittings with nut and bolt, (c) Special screws and pins and (d) Computer hard disk.

3.1.1.1 Alloy and Poly-Alloy Coatings

Electroless nickel coatings have displayed satisfactory performance for a variety of applications, and their performance can be further boosted by adding alloy elements in the NiP coating matrix. Properties, namely chemical, mechanical, physical, magnetic and so on, can be suitably enhanced through this technique. A number of alloys can be readily deposited by combining metals that are independently deposited by electroless method from similar baths; an example being nickel and cobalt from alkaline baths. Additionally, and more importantly, certain metals that cannot themselves be deposited by the autocatalytic mechanism can be induced

TABLE 3.1
Various Alloys of Electroless Nickel-Phosphorus

Types	Deposits
Binary alloys	NiP, Ni-B
Ternary alloys	Ni-P-B, Ni-W-P, Ni-Cu-P, Ni-Co-P, Ni-P-Sn
Quaternary alloys	Ni-W-Cu-P

to co-deposit with an electrolessly depositing metal. Table 3.1 lists the common alloy and poly-alloys of nickel, which are developed through electroless method. Each of the alloy coatings has specific advantages. For instance, upon introduction of copper into the coating, its corrosion resistance increases. Again, Ni-W-P coatings have good thermal resistance and find applications in printing operations. To have the benefit of both the elements, quaternary electroless Ni-W-Cu-P coatings is carried out using nickel sulfate and chloride-based baths. Similarly, many alloys can be steadily deposited by a combination of certain metals that can be individually deposited through electroless method using similar baths. Applications based on tribology have encouraged a number of ternary and poly-alloy coatings such as Ni-Mo-P, Ni-Co-P, Ni-W-Cr-P and so on.

3.1.1.2 Composite Coatings

Composite coatings can very easily be prepared by co-deposition of composite materials in the alloy coating. The composite materials are mainly added to the electroless bath in the form of particles, which get incorporated into the NiP matrix when the bath is agitated. Although electroless nickel alloy coatings can serve many purposes, the quest for improved properties such as higher hardness, lubricity, anti-sticking and anti-wear properties has led to the incorporation of many soft and hard particles in the matrix of the electroless nickel, namely polytetrafluoroethylene (PTFE), ZrO_2, Al_2O_3, SiC and so on. The choice of the particles depends on the specific property that is desired. Various hard and soft composites of electroless nickel are listed in Table 3.2.

When the coating is to be made hard and wear-resistant, composite coatings usually have co-deposited hard particles such as WC, SiC, Al_2O_3, B_4C and diamond. Whereas, when lubrication and low friction are the main criteria, soft particles, namely WS_2, MoS_2, PTFE and graphite are incorporated into the coating.

The preparation of electroless nickel-based composite coatings with excellent comprehensive properties is highly dependent on the stable dispersion of the nanoparticles in plating bath; otherwise, the so-called composite coatings would have non-uniformly distributed particulates and numerous defects, owing to the segregation and agglomeration of the nanoparticles with high surface energy and activity in the plating bath. Fortunately, this can be conveniently realized by capping the nanoparticles with specially selected surface-modifying agents.

TABLE 3.2
Various Hard and Soft Electroless Nickel Composites

Hard Composites (Wear-Resistant)	Soft Composites (Low COF)
NiP-SiC, NiP-Si_3N_4, Ni-P-Al_2O_3, Ni-P-SiO_2, Ni-P-B_4C, Ni-P-Diamond, Ni-P-BN, Ni-P-WC, Ni-P-TiO_2, Ni-P-Cr_2O_3	Ni-P/cenosphere, Ni-P-MoS_2, Ni-P-PTFE, Ni-P-CNT, Ni-P-Gr-SiC, Ni-P-ZnO, Ni-P-$K_2Ti_6O_{13}$ whiskers, Ni-P-ZrO_2, Ni-P-PTFE-SiC, Ni-P-CNT-SiC

Source: Sahoo, P. and Das, S.K., *Mater. Des.*, 32, 1760–1775, 2011.

3.2 DEVELOPMENT AND CHEMISTRY OF ELECTROLESS NiP DEPOSITION

3.2.1 Components of NiP Bath

Electroless NiP coatings are heavily dependent on the bath constituents. The plating speed and the properties of the deposit are affected significantly by the bath composition. Now, electroless plating is an autocatalytic process where the substrate develops a potential when it is dipped in electroless solution, also called bath, that contains a source of metallic ions, reducing agent, complexing agent, stabilizer and other components dissolved in water. Due to the developed potential, both positive and negative ions are attracted toward the substrate surface and release their energy through charge transfer process. While releasing energy, the ions also get deposited onto the substrate surface. The primary components of electroless NiP bath and their functions are listed below (Loto 2016):

1. A source of nickel cations (namely, nickel chloride or nickel sulfate) to supply metallic nickel.
2. Hypophosphite anions (commonly sodium hypophosphite monohydrate), to supply the catalytic active hydrogen atoms for reducing the nickel ions to the metal and to supply the phosphorus of the deposited alloy.
3. An organic chelating agent (organic hydrocarboxylic acids such as hydroxyacetic, hydroxypropionic, citric or malic acid) to complex the nickel ions so that nickel phosphate precipitation is prevented and also to serve as a buffer to prevent a rapid decrease in the pH.
4. Stabilizers to prevent solution decomposition by "masking" active nuclei. The commonly used stabilizers include thiourea, sodium ethylxanthate, lead or tin sulfide.
5. Additives to increase the rate of nickel deposition by activating the hypophosphite anions, and thus to counteract the slowing effect of chelating agents and stabilizers. Chemicals include succinic anions, adipic anions and alkali fluorides.
6. pH buffers to regulate the pH of the solution and keep it constant during continuous operation. These include acids (H^+ ions) such as sulfuric acid and hydrochloric acid and alkalizers (OH^- ions) such as caustic soda and sodium carbonate.
7. Wetting agents to reduce the surface tension and promote wetting by the solution of the parts to be plated. Compounds such as sulfated alcohols, sulfonates of fatty acids or sulfonated oil fractions are used for this purpose.

Table 3.3 summarizes the bath components for electroless NiP deposition (Krishnan et al. 2006).

TABLE 3.3
Bath Components and Their Function

Component	Function	Examples
Metal ions	Source of metal	Nickel chloride, nickel sulfate, nickel acetates, sodium hypophosphite
Hypophosphite ions	Reducing agent	Sodium hypophosphite
Complexants	Form Ni complexes, prevent excess free Ni ion concentration, stabilizing and preventing Ni phosphate precipitation; also act as pH buffers	Monocarboxylic acids, dicarboxylic acids, hydrocarboxylic acids, ammonia, alkanolamines, etc.
Accelerators (exultants)	Active reducing agent and accelerate deposition; mode of action opposes stabilizers and complexants	Anions of some mono- and dicarboxylic acids, fluorides, borates
Stabilizers (inhibitors)	Prevent solution breakdown by shielding catalytically active nuclei	Pb, Sn, As, Mo, Cd or Th ions, thiourea, etc.
Buffers	For long-term pH control	Sodium salt of certain complexants, c depends on pH range used
pH regulators	For subsequent pH adjustment	Sulfuric and hydrochloric acids, soda, caustic soda, ammonia
Wetting agents	Increase wettability of surfaces to be coated	Ionic and nonionic surfactants

Source: Krishnan, K.H. et al., *Metall. Mater. Trans. A*, 37, 1917–1926, 2006.

3.2.2 Choice of Substrate

As already mentioned, electroless nickel coatings can be deposited on any surface, be it metallic or nonmetallic. In fact, it is one of the advantages and adds to the versatility of the deposition process. Electroless nickel coatings are many times deposited over softer alloys of aluminum and magnesium to enhance their surface hardness. Moreover, electroless nickel coatings have been developed over steel substrates to provide both hardness and resistance against corrosion. Table 3.4 lists the commonly used substrates for electroless nickel deposition. It is observed that the substrate material has a strong effect on the electroless nickel coatings (Liu and Gao 2006a). For instance, electroless nickel coatings display a very good deposition rate over some substrates, whereas others require activation to initiate the deposition process. Besides, the substrate material influences the nucleation process, chemical composition and microstructure of the NiP coatings.

TABLE 3.4
Popular Substrates Material for Electroless Nickel Deposit

S. No.	Substrate Material	Reason	Purpose
1.	Steel	Most common engineering material	For overall surface protection
2.	Magnesium and alloys	Lightweight and high specific strength	Enhance anti-wear and anti-corrosion characteristics (Ban et al. 2014)
3.	Aluminum and alloys	Lightweight	Enhance anti-wear and anti-corrosion characteristics
4.	Copper	Electrical conductor, used in circuits	For electrical conductivity and solderability
5.	Plastics	As replacement to conventional metals and alloys, lightweight and corrosion-resistant	Metallization purposes when electrical conductivity is required

3.2.2.1 Activation of Non-catalytic Surface

Although electroless nickel coatings can be developed over a host of surfaces, some of them don't support spontaneous deposition but instead require some sort of pre-treatment before deposition to initiate the deposition process. This step is known as activation, which is carried out for non-catalytic surfaces, namely nonconductors, non-catalytic metals, non-catalytic semiconductors and so on, by generating catalytic nuclei on the surfaces. The activation process can be categorized into two broad types: electrochemical process and photochemical process (Paunovic and Schlesinger 2006).

In case of NiP coatings, the electrochemical activation process is more popular. Elements namely, palladium (Pd), tin (Sn), and so on are used as the catalyst. The commonly used sensitizing solution with palladium chloride is presented in Table 3.5 (Schlesinger 2011). In the electrochemical method, catalytic nuclei of metal M on a non-catalytic surface is generated in an electrochemical oxidation–reduction reaction (Paunovic and Schlesinger 2006),

$$M^{z+} + \text{Red} \rightarrow M + \text{Ox} \tag{3.1}$$

where M^{z+} is the metallic ion and M is the metal catalyst. The preferred catalyst is Pd, and thus the preferred nucleating agent M^{z+} is Pd^{2+} (from $PdCl_2$). The preferred reducing agent (Red) in this case is Sn^{2+} ion (from $SnCl_2$). In this example, the overall reaction of activation, according to a simplified model, is (Paunovic and Schlesinger 2006)

$$Pd^{2+} + Sn^{2+} \rightarrow Pd + Sn^{4+} \tag{3.2}$$

Sn^{2+} can reduce Pd^{2+} ions because the standard oxidation–reduction potential of Sn^{4+}/Sn^{2+} is 0.15 V and that of Pd^{2+}/Pd is 0.987 V. The flow of electrons is from a more electronegative couple (here Sn^{4+}/Sn^{2+}) toward a less electronegative (more positive) couple (here Pd^{2+}/Pd). As the standard potential of Au^{+}/Au is 1.692 V, Sn^{2+} ions can reduce Au^{+} to produce Au catalytic nuclei.

TABLE 3.5
Sensitizing Solution with Palladium

Components	Concentration (at 23°C)
$PdCl_2$	0.1 g/L
HCl (12 M)	0.1 mL/L
H_2O	Balance to 1 L

Source: Schlesinger, M., Electroless deposition of nickel. In *Modern Electroplating*, M. Schlesinger and M. Paunovic (Eds.), John Wiley & Sons, Hoboken, NJ, 2011.

3.2.3 Electroless NiP Deposition Process

Electroless depositions are redox reactions with key components as the oxidizer and reducing agent in the bath solution. Electroless NiP coating is carried out by immersion of the substrate in a bath containing nickel salt and hypophosphite as the reducing agent with other ingredients, namely buffers, stabilizers, exultants, etc. There are two types of NiP baths (Krishnan et al. 2006) based on their pH levels.

3.2.3.1 Acid Baths

Acid baths are the most popular for the development of electroless nickel coatings and have some inherent advantages when compared to the alkaline baths. For example, the deposition rate is higher (20–25 μm/h) and the compositions are more stable because there is no loss of the complexing agent by evaporation. Moreover, coatings obtained from acid solutions are of better quality. The properties of the electroless NiP alloy can be easily controlled by controlling the amount of phosphorus in the deposit. Hence, acid solutions are generally preferred in many applications. Reactions occurring in electroless nickel deposition with hypophosphite ion as the reducing agent may be represented as follows (Hari Krishnan et al. 2006):

$$Ni^{2+} + H_2PO_{2^-} + H_2O \xrightarrow{\text{catalytic active surface}} Ni + 2H^+ + H(HPO_3)^- \quad (3.3)$$

$$HPO_2^- + H_2O \xrightarrow{\text{catalytic active surface}} H\ (HPO_3)^- + 1/2H_2 \quad (3.4)$$

The deposition reaction proceeds forward due to the following factors: (1) reduction in nickel ion concentration, (2) conversion of the hypophosphite to phosphate, (3) increase in hydrogen ion concentration and (4) adsorption of hydrogen gas by the deposit.

3.2.3.2 Alkaline Bath

Instability is the main disadvantage of alkaline baths particularly at elevated temperatures (greater than 90°C). Moreover, at higher temperature, there is a loss of ammonia, which is added to maintain the pH at that temperature. The reduction of nickel in alkaline solutions follows the same pattern as in acid solutions. Here, unlike acid baths, the rate of electroless nickel deposition depends on the hypophosphite concentration and increases with it. However, very high concentrations of hypophosphite make the bath unstable due to homogeneous deposition in the bulk of the solution. Temperature influences the rate of deposition of nickel in the same way as in acid solutions. However, temperatures above 90°C make the control of pH very difficult.

3.2.4 Effect of Bath Composition/Deposition Conditions on Electroless Nickel Properties

3.2.4.1 Effect of Nickel Source Concentration

Nickel source plays an important role in defining the properties of electroless NiP coating. The commonly used nickel salts include nickel sulfate, nickel chloride, nickel acetate and so on. However, nickel sulfate is the most preferred, as chloride anion (from nickel chloride) may be treacherous particularly when plating aluminum. Now, aluminum inherently exhibits high corrosion resistance due to the presence of passive oxide film covering its surface. However, in environments that contain aggressive anions such as chloride, Cl^-, the passive film becomes unstable and degrades locally resulting in film breakdown and pitting corrosion. Moreover, chloride can also pose a problem in case of coating steels for corrosion-based applications. The use of nickel acetate does not yield any significant improvement in bath performance or deposit quality when compared to nickel sulfate (Mallory and Hadju 1991). Moreover, its cost is quite higher compared to that of nickel sulfate. However, according to Mallory and Hadju (1991), the ideal source of nickel ions is the nickel salt of hypophosphorus acid, $Ni(H_2PO_2)_2$. There are two-way benefits of using this particular salt. First the buildup of alkali metal ions viz. sulfates and chlorides in the bath can be eliminated. Second, the hypophosphite ion can replenish the reactants consumed during metal deposition. The dependence of deposition rate of NiP on nickel source concentration is shown in Figure 3.2. As can be observed from the plot, an optimum concentration of nickel concentration is required to obtain the maximum electroless nickel deposition rate in the bath.

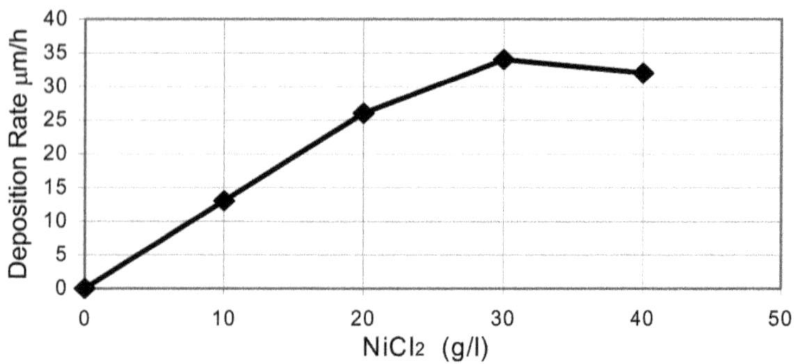

FIGURE 3.2 Effect of nickel chloride concentration on the deposition rate of electroless nickel. (From Taheri, R., Evaluation of electroless nickel-phosphorus (electroless nickel) coatings, PhD thesis, University of Saskatchewan, 2003.)

3.2.4.2 Effect of Reducing Agent Concentration

Variation of hypophosphite concentration affects the rate of reduction considerably. Although the increase in hypophosphite concentrations improves the rate of reduction of nickel, the excess concentration of reducing agents may cause reduction to occur in the bulk of the solution, which finally results in solution decomposition. The following text presents a detailed deliberation (Taheri 2003) about the chemical pathway of the various reactions, which helps in realizing the effect of reducing agent in a better manner.

The deposition rate of electroless nickel coating is dependent on two significant factors, namely Ni concentration and $Ni^{2+}/H_2PO_2^-$ ratio. Riedel (1991) made the following propositions:

1. The concentration of hypophosphite should lie between 0.15 and 0.35 mol/L
2. The optimum $Ni^{2+}/H_2PO_2^-$ ratio should be maintained between 0.25 and 0.60, preferably between 0.30 and 0.45.

If the following equation is considered, three moles of hypophosphite is required to be consumed to reduce one mole of Ni ion.

$$3NaH_2PO_2 + 3H_2O + NiSO_4 \rightarrow 3NaH_2PO_3 + H_2SO_4 + 2H_2 + Ni \qquad (3.5)$$

Thus, from the equation, the typical $Ni^{2+}/H_2PO_2^-$ ratio is 0.33, which is quite close to the measuring range of 0.30–0.45. As illustrated in Figure 3.3, the highest deposition rate can be obtained when the $Ni^{2+}/H_2PO_2^-$ ratio is approximately

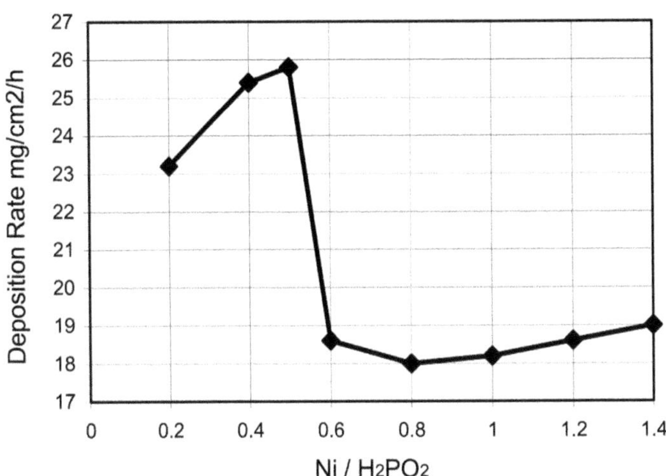

FIGURE 3.3 Dependence of deposition rate in an acetate-containing solution on the Ni^{2+}/H_2PO_2- ratio: $Na_2H_2PO_2$ = 0.224 mol/L; acetate ion = 0.12 mol/L; initial pH = 5.5. (From Taheri, R., Evaluation of electroless nickel-phosphorus (electroless nickel) coatings, PhD thesis, University of Saskatchewan, 2003.)

0.45. If the $Ni^{2+}/H_2PO_2^-$ ratio is too low, the lack of Ni ions in the solution causes a brownish coating on the specimen (Taheri 2003). In addition, as the $Ni^{2+}/H_2PO_2^-$ ratio decreases, the concentration of the hypophosphite ion increases, thus increasing the risk of solution decomposition. When $Ni^{2+}/H_2PO_2^-$ ratio is high, the deposition rate decreases. Besides, the phosphorus content of the deposit also decreases.

The appropriate amount of hypophosphite, however, can be adjusted by observing the bath condition during the course of reaction (Agarwala and Agarwala 2003). Weak hydrogen evolution is an indication of a low concentration of hypophosphite and a vigorous hydrogen evolution indicates excess hypophosphite. Hence, with some experience, one can roughly decide the suitable concentration of reducing agent in the solution for optimum deposition.

Moreover, from the above discussion, it is further established that a correlation exists between the electroless nickel bath composition and the deposition rate. This is why the accurate control of the replenishment process is required. Most of the commercially available electroless nickel solutions come with replenishment instructions. An analytic titration method with ethylenediaminetetraacetic acid (EDTA) is commonly used to estimate the concentration of nickel as well as hypophosphite and orthophosphite in the bath.

3.2.4.3 Effect of Complexing Agent

Organic salts are used as complexing agents in electroless nickel bath. When nickel salts come in contact with reducing agent, often an uncontrolled reaction occurs, which results in bath decomposition. Complexing agent forms metastable complexes (chelates), thus controlling this reaction. Besides, in case of acidic hypophosphite bath, the precipitate interferes with chemical balance of the solution by removal of nickel ions, which has a detrimental effect on the quality of deposit. The primary functions of a typical complexing agent are listed as follows:

- Exerts a buffering action that prevents rapid pH change.
- Retards the tendency for the precipitation of nickel salts, e.g., basic salts or phosphites.
- Forms nickel complex and reduces the concentration of free nickel ions to avoid spontaneous decomposition of electroless nickel solutions.
- Controls the reaction so that it occurs only on the catalytic surface; when complexing agents are added, the nickel complex controls the number of free ions that is available to participate in the deposition reaction. Therefore, the larger the number of free nickel ions adsorbed on the catalytic surface, the greater is the deposition rate.

Complexing agent affects the deposition reaction and also the deposit characteristics, namely hardness and corrosion resistance. Common complexing agents for electroless nickel deposition are given in Table 3.6 (Schlesinger 2011). The plating rate is inversely proportional to the complexing ion's stability constant, as can be seen from the table.

TABLE 3.6
Commonly Used Complexing Agents for Electroless Nickel Deposition

Anion	Acid	Stability, pK = −logK
Acetate	CH_3COOH	1.5
Succinate	$HOOCCH_2CH_2COOH$	2.2
Aminoacetate	NH_2CH_2COOH	6.1
Malonate	$HOOCHCH_2COOH$	4.2
Pyrophosphate	$H_2O_3POPO_3H_2$	5.3
Malate	$HOOCCH_2(OH)COOH$	3.4
Citrate	$HOOCCH_2(OH)C(COOH)_2$	6.9

Source: Schlesinger, M., Electroless deposition of nickel. In *Modern Electroplating*, M. Schlesinger and M. Paunovic (Eds.), John Wiley & Sons, Hoboken, NJ, 2011.

3.2.4.4 Effect of Stabilizer

Although complexing agent forms metastable complexes with nickel ions, the bath may decompose any time especially when the deposition period is needed to be kept longer. Bath decomposition is indicated by the formation of finely divided black particles of nickel and nickel phosphide throughout the bulk of the solution. Decomposition is usually initiated by the presence of colloidal, solid particles in the solution. These particles may be the result of the presence of foreign matter (such as dust or blasting media) or may be generated in the bath as the concentration of orthophosphite exceeds its solubility limit. Whatever the source, the large surface area of the particles catalyzes reduction, leading to a self-accelerating chain reaction and decomposition. This is usually preceded by increased hydrogen evolution and the appearance of a finely divided black precipitate throughout the solution.

The decomposition of bath results in the wastage of chemicals and associated resources. Hence, the reduction reaction in the electroless nickel bath must be controlled so that deposition occurs at a predictable rate and only on the substrate to be plated. In order to accomplish this, inhibitors, also known as stabilizers, are added to the bath. These inhibitors are absorbed on any colloidal particles present in the solution and prevent the reduction of nickel on their surface. The addition of inhibitors can have harmful and beneficial effects on the plating bath and its deposit. The amount of inhibitor used is critical. When, added in small amounts, some inhibitors increase the rate of deposition and/or the brightness of the deposit, others, especially metals or sulfur compounds, increase internal stress and porosity and reduce ductility, thus reducing the ability of the coating to resist corrosion and wear.

To use stabilizers effectively, the compatibility of the stabilizer with the process being used must be checked to avoid any adverse loss in catalytic activity due to a synergistic action with any other additive present in the bath. When two or more stabilizers are used, it is important to make sure that the action of one stabilizer does not inhibit or lower the effectiveness of the other. Finally, the stabilizers must be selected on the basis that they only affect the plating process in a manner that the resultant deposit will be able to meet any required performance criteria.

The popular stabilizers used for stabilizing electroless nickel baths are heavy metal cations, namely Lead, Tin, Thallium and so on. Besides, chemicals such as thiourea are effectively used as stabilizers in electroless nickel baths. The stabilizer concentration in the plating bath is very important. As stabilizer acts as a poison for the deposition reaction, too much of it in the solution may stop the reaction completely. The concentration of stabilizer at which this happens is known as the critical concentration of the stabilizer for the electroless nickel solution. Obviously, the critical concentration is dependent on the other ingredient in the bath and their concentrations. Figure 3.4 shows the general trend of effect of stabilizer concentration on the electroless nickel deposition rate. Hence, it can be observed that the deposition rate is maximum for a particular optimum stabilizer concentration. Table 3.7 lists the common stabilizers for electroless nickel deposition along with their usual concentrations.

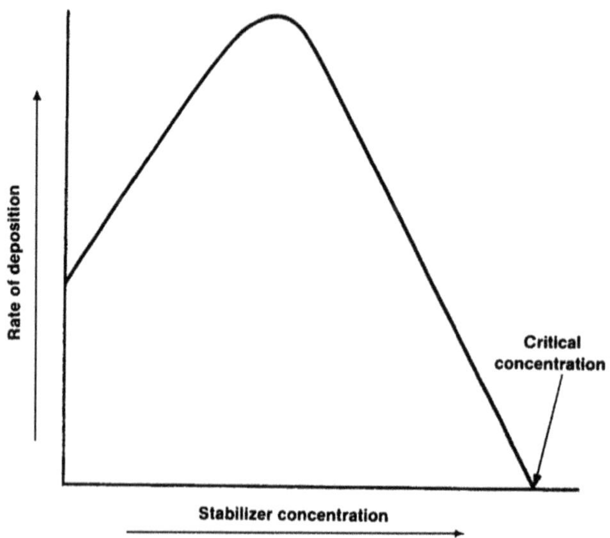

FIGURE 3.4 Effect of concentration of stabilizer on NiP deposition rate. (From Mallory, G.O. and Hadju, J.B., *Electroless Plating: Fundamentals and Applications*, AESF, Orlando, FL, 1991.)

TABLE 3.7
Common Stabilizers for Electroless Nickel Deposition

Stabilizer	Concentration
Thiourea	0.4–1.0 ppm
Lead acetate	0.5–2.0 ppm
Lead nitrate	0.5–2.0 ppm

Stabilizers can also affect the properties of electroless nickel deposit in the following ways (Baskaran et al. 2006):

- The element contained in the stabilizer may co-deposit in the coating, namely sulfur and lead co-deposition in case of mercaptobenzothiazole and lead acetate.
- They can act as brighteners and/or leveling agents.
- They might influence the phosphorus content of the electroless nickel coatings.
- They can increase the porosity of the electroless nickel coatings.

3.2.4.5 Effect of Bath pH and Need for Buffers

The pH of the solution has significant influence on the reactions involved in the deposition of electroless nickel coatings. An increase in the pH of the solution leads to the acceleration of the nickel reduction reaction but retards the self-reduction of phosphorus from hypophosphite, as hydroxide ions are generated in the case of latter. Thus, deposits developed from high pH baths generally have low phosphorus content. However, because the nickel reduction reaction predominantly controls the deposition rate, increasing the pH of the solution increases the deposition rate. Figure 3.5 illustrates clearly the effect of pH of the solution on

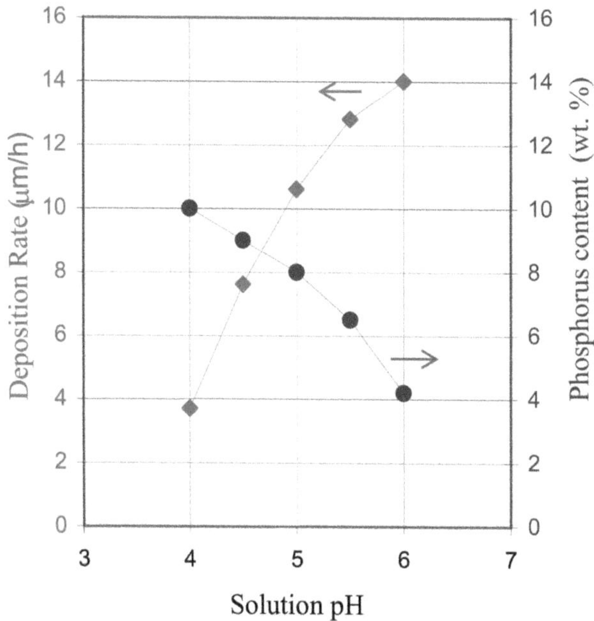

FIGURE 3.5 Effect on solution pH on deposition rate and phosphorus content of the coating. (From Fields, W. and Zickearff, J.R., Electroless, Publication of ASM committee on electroless nickel plating, 1984.)

both deposition rate and phosphorus content (Fields and Zickearff 1984). Some more details about the effect of pH on electroless nickel deposition have been reported by Riedel (1997).

- Raising the bath pH has the following effects:
  1. Increasing the deposition rate in an almost linear manner.
  2. Modification of the hypophosphite reaction from catalytic to homogeneous. A consequence of this can be spontaneous decomposition of the solution with nickel deposition.
  3. Lowering the solubility of the nickel phosphate. The deposition of this unwanted component may initiate decomposition and often leads to rough deposits.
  4. Reducing the phosphorus content of the deposit.
- Lowering the pH can lead to:
  1. Prevention of the precipitation of basic salts and hydroxides.
  2. Lowering of the reducing power of the hypophosphite.
  3. More effective buffering action of species in the bath.

As the deposition progresses, pH of the bath decreases with the increase in H^+ concentration, as nickel ions reduce to form metal nickel. If the pH value drops below 4.0, the deposition almost stops (Taheri 2003). Thus, in order to regulate the level of pH, the electroless bath uses another additive (buffer) during its operation, where there is a continuous supply of OH^- ions to balance the excess H^+ in the bath. Some of the common buffers are organic acids (Schlesinger 2011), namely acetic acid, propionic acid, glutaric acid, succinic acid and adipic acid. The content of phosphorus in the coating increases with decreasing level of pH. Subsequently, the decrease in pH level may prevent deposition and lower the reducing power of the reducing agent. At a pH level of 3, the deposition rate seems to retard and the deposit formed over the substrate is attacked by the solution itself (Taheri 2003). Both nano- and micro-sized particles and the dispersion medium can affect the stability of dispersions, structure and properties within a pH range of 4–6.5. On the other hand, increasing the pH may tend to increase the deposition rate and lower the solubility of nickel salts. The bath pH should be controlled at an optimum value, to attain the desired characteristics of the deposit. According to the difference in pH values of plating bath, the techniques can be classified as acid electroless plating and alkali electroless plating (Table 3.8).

3.2.4.6 Effect of Bath Temperature

Bath temperature is one of the significant parameters that influence the rate of electroless nickel deposition. The reduction and oxidation processes that occur in the entire reaction require an external energy source in the form of heat, which holds good for all types of bath. Most of the reactions involved in the deposition process are endothermic. As a result, by increasing the temperature, the deposition rate increases. These reactions of the plating using hypophosphite baths operate between 60°C and 95°C. There is an exponential relationship between the temperature and

TABLE 3.8
pH Values of Electroless NiP-Based Solutions for Coatings on Various Substrates

Coatings	Substrate	Solution Properties (pH)	References
NiP	Steel ball bearings	4.8	(Zhang et al. 2014)
	St-37 steel, Ck45 steel	4.8	
	Mild steel sheets	4.8	
	Mild steel sheets	9.5	
	20#steel	4.0–5.0	
	Low carbon steel	4.7	
	Foam samples	9.5	
	Q235 carbon steel	4.5–5.0	
	Magnesium-Aluminum-Zinc (AZ91) alloy	5.5	(Anik and Körpe 2007)
	AISI 1045	8	(Berríos et al. 1998)
	AISI 1045	4.6–4.8	(Contreras et al. 1999)
	Aluminum alloy	4.8–5.3	(Gordani et al. 2008)

Note: Shaded block indicates acidic solution.

the deposition rate of the coatings, which is independent of the acidity or alkalinity of the bath (shown in Figure 3.6). As shown in the figure, the deposition rate increases with increasing bath temperature. Most of the acidic baths are operated at 80°C–90°C, while the alkaline baths can be operated at lower temperatures (as low as 40°C). This is why alkaline baths are used for coating plastic substrates.

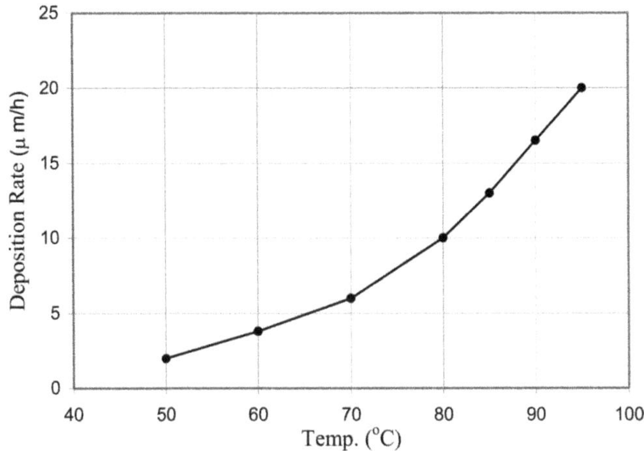

FIGURE 3.6 Effect of solution temperature on the deposition rate of electroless nickel coating in acidic bath. (From Taheri, R., Evaluation of electroless nickel-phosphorus (electroless nickel) coatings, PhD thesis, University of Saskatchewan, 2003.)

Practically useful deposition rates are only obtained on metals above 80°C corresponding to the level of pH. Although there is a higher deposition rate at temperatures beyond 90°C, there is a fear of bath instability or decomposition. Deposition is completely possible in weak alkaline electrolytes at a temperature of approximately 60°C but suitable only for metallizing nonconductors with poor resistance to melting at high temperatures. Therefore, there is a necessity for accurate controlling of temperature of electroless NiP baths. Operating at a temperature above 87°C promotes the formation of amorphous phase and considerable improvement in the coating adherence and efficiency. In addition to deposition rate, temperature also affects the amount of phosphorus content in the coating matrix. Besides, temperature also has influence over the aesthetics of the developed coating as indicated in Table 3.9. Table 3.10 presents typical bath compositions for acidic and alkaline electroless nickel baths along with the deposition conditions (Schlesinger 2011).

3.2.4.7 Effect of Surfactant

Surfactants are wetting agents that lower the surface tension of a liquid, allowing easier spreading, and lower the interfacial tension between two liquids or a liquid and solid surface. Surfactants reduce the surface tension of water by adsorbing at the liquid–gas interface. In an electroless nickel bath, the presence of surfactant promotes the coating deposition reaction between the bath solution and the immersed substrate surface (Elansezhian et al. 2009).

Many surfactants assemble in the bulk solution into aggregates known as micelles. The concentration at which surfactants begin to form micelles is known as the critical micelle concentration or CMC. Hence, the surfactant added into electroless bath should be below CMC for higher effectiveness.

The addition of surfactant affects the surface roughness of the deposit along with some of the physical properties such as hardness. Moreover, the surfactant is also found to have impact on the phosphorus content of the deposit and also its corrosion-resistant properties.

TABLE 3.9
Effect of Bath Temperature on the Appearance of Electroless Nickel Deposits

Temperature (°C)	NiP Deposits
70	Dull bright
75	Bright
80	Very bright
85	Black
90	Black
95	Black burnt
100	Black burnt

Source: Osifuye, C.O. et al., *Int. J. Electrochem. Sci.*, 9, 14, 2014.

TABLE 3.10
Common Baths Constituents and Deposition Conditions for Electroless Nickel Coatings

	Acid Baths				Alkaline Baths			
Bath Constituents (gL^{-1})	**1**	**2**	**3**	**4**	**5**	**6**	**7**	**8**
Nickel chloride, $NiCl_2$, $6H_2O$	30	30	—	21	26	30	20	—
Nickel sulfate, $NiSO_4$, $6H_2O$	—	—	25	—	—	—	—	25
Sodium hypophosphate, NaH_2PO_3, H_2O	10	10	23	24	24	10	20	25
Hydroxyacetic acid, $HOCH_2COOH$	35	—	—	—	—	—	—	—
Sodium citrate, $Na_3C_6H_5O_7$, H_2O	—	12.6	—	—	—	84	10	—
Sodium acetate, $NaC_2H_3O_2$	—	5	9	—	—	—	—	—
Succinic acid, $C_4H_6O_4$	—	—	—	7	—	—	—	—
Sodium fluoride, NaF	—	—	—	5	—	—	—	
Lactic acid, $C_3H_6O_3$	—	—	—	—	27	—	—	—
Propionic acid, $C_3H_6O_2$	—	—	—	—	2.2	—	—	—
Ammonium chloride, NH_4Cl	—	—	—	—	—	50	35	—
Sodium pyrophosphate, $Na_4P_2O_7$	—	—	—	—	—	—	—	50
Lead ion, Pb^{2+}	—	—	0.001	—	0.002	—	—	—
pH	4–6	4–6	4–8	6	4–6	8–10	9–10	10–11
Temperature (°C)	100	100	85	100	100	95	85	70

Source: Schlesinger, M., Electroless deposition of nickel. In *Modern Electroplating*, M. Schlesinger and M. Paunovic (Eds.), John Wiley & Sons, Hoboken, NJ, 2011.

When surfactant concentration is below CMC, the surfactant molecules are loosely integrated into the water structure and they are termed as monomer. At CMC region, the structure of surfactant-water gets modified such that the surfactant molecules start building up own micelles structure. The CMC value of surfactants may be significantly modified with ionic liquids at appropriate room temperature, and it varies for specific applications. A surfactant can be classified by the presence of formally charged groups in its head:

- Anionic surfactant hydrophilic group bears a negative charge, for example, sulfonate salts, alcohol, sulfates, alkyl benzenes, phosphoric esters and carboxylic acid salts.
- Cationic surfactant hydrophilic group bears a positive charge, for example, polyamides and their salts, quaternary ammonium salts and amine oxides, salt of long-chain amine.
- Nonionic surfactant hydrophilic group has no charge, for example, polyethylenated alkyl phenols, alcohol ethoxylates and alkanolamides.
- Zwitterionic surfactant both positive and negative charges may be present in the hydrophilic, for example, long chain amino acid, betaines and sulfobetaines.

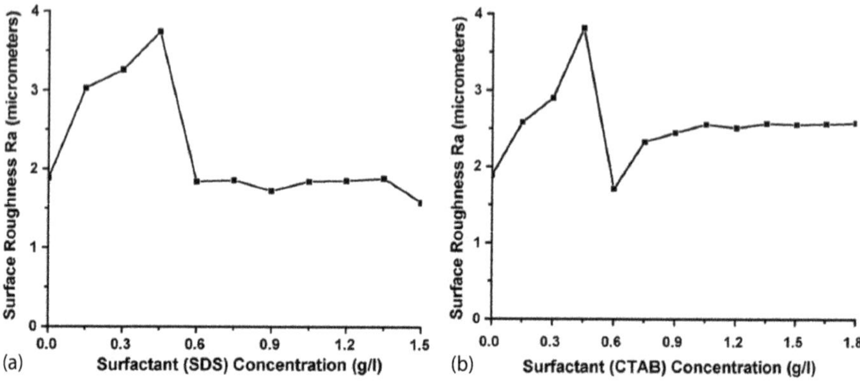

FIGURE 3.7 Average surface roughness of electroless nickel deposits vs. concentration of (a) SDS and (b) CTAB. (From Elansezhian, R. et al., *Surf. Coat. Technol.*, 203, 709–712, 2008.)

The addition of surfactant is found to improve the surface finish of electroless nickel deposits. Elansezhian et al. (2009) studied the effect of two surfactants, namely sodium dodecyl sulfate (SDS) and cetyl trimethyl ammonium bromide (CTAB) on the surface topography of electroless NiP coating. It was found that surface finish of the coated layer significantly improved when the concentration of the surfactant exceeded approximately 0.6 g/L. However, at the lower levels of concentration, the surface finish was found to be poor (Figure 3.7). The hardness of deposits also increased with the addition of the surfactants. Moreover, by adding the above two surfactants during deposition of NiP, the phosphorus content had increased, resulting in improved quality of the deposits. Particularly, it improved the corrosion resistance of the coating.

3.2.4.8 Effect of Bath Loading

Bath loading is a term used to define the ratio of the surface of the working piece to the volume of solution in the tank. Commercial baths are operated in a bath loading range of 0.1–1.0 dm^2/L depending on the bath solution (Riedel 1997). Figures 3.8 and 3.9 show the effect of bath load on the deposition rate. As shown in the figure, the deposition rate decreases with increasing bath load. Therefore, an optimum bath load is required to provide the acceptable deposition rate and bath efficiency.

Different coating characteristics can be obtained depending upon the loading of the bath. Optimum loading conditions will depend on the type of bath or proprietary formulations. For low levels of bath loading, the potential for the stabilizer's adsorption is greater, especially on intricate geometries. The poor coverage can result in inconsistent plate thickness and increased porosity with a decrease in phosphorus content in the deposit.

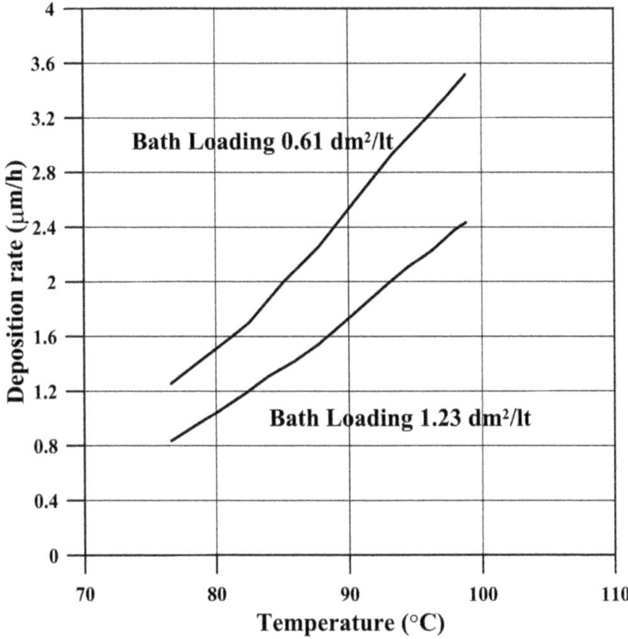

FIGURE 3.8 Effect of bath load and temperature on deposition rate. (From Taheri, R., Evaluation of electroless nickel-phosphorus (electroless nickel) coatings, PhD thesis, University of Saskatchewan, 2003.)

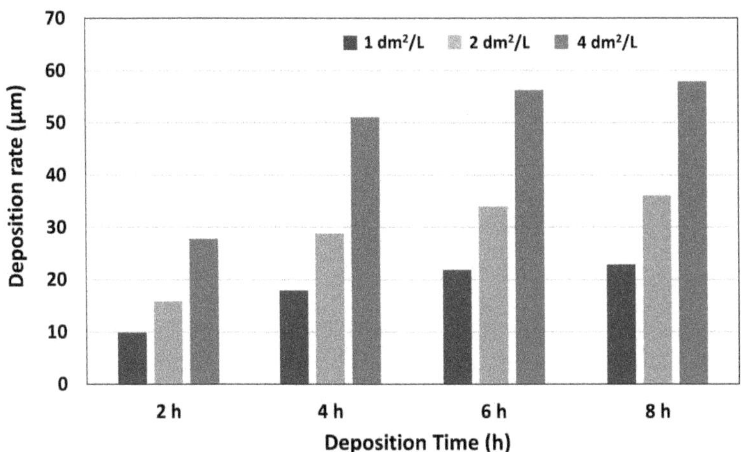

FIGURE 3.9 Effect of bath load on deposition rate. (From Taheri, R., Evaluation of electroless Nickel-phosphorus (electroless nickel) coatings, PhD thesis, University of Saskatchewan, 2003.)

3.2.5 Effect of Agitation on NiP Deposition

Solution stirring or agitation is generally recommended in the case of electroless nickel plating. One of the reasons is that the higher the rates of diffusion and convection, the better the reacting ions approach the coating workpiece and the reaction products removed. Besides, it is also true for workpieces having bores or holes (namely inside of tubes or pipes). The lack of agitation of the bath solution during deposition may also cause problems. Solution stratification can occur, resulting in gas pitting, patterns and/or streaking of the deposit. Proper agitation enables uniform distribution of plating chemicals and helps in avoiding localized overheating, which may become potential zones for the initiation of bath decomposition.

In electroless plating, the agitation of the solution is normally achieved by magnetic stirring, aeration (bubbling of clean air) or continuous pumping. Agitation of small substrates is possible by rotation but the same for large or heavy substrates may be expensive or infeasible. Agitation of the solution provides a fresh supply of solution to the parts and enhances the removal of hydrogen gas produced during deposition. It is found that agitation improves the rate of nickel deposition and affects to some extent the composition (phosphorus content) of the electroless nickel deposit.

Apart from the conventional agitation techniques, ultrasound assistance is used as an alternate method of agitation in electroless nickel coatings. In the presence of ultrasound, a pseudo "ultrasonic stirring" results in agitation of the solution both by means of acoustic streaming and cavitation. As an ultrasound wave propagates through a liquid medium, the liquid is agitated, leading to enhanced mass transfer. Therefore, the imposition of ultrasound accelerates the diffusion of the ions from the bulk solution to the substrate surface, resulting in an increase in the electroless deposition rate and benefiting the formation of thick coating for the same deposition time as electroplating in static electrolyte. Moreover, due to the cavitation resulting from ultrasonic irradiation, agitation in electroless nickel bath is more agitated than the conventional stirring techniques. Under ultrasonic agitation, electroless nickel coating becomes smooth and compact, has refined grains and is free of cracks and pores, leading to a significant improvement in the coating corrosion resistance. The crystallinity of the coating is also improved by ultrasonic irradiation (UI), transforming from amorphous state to a mixture of amorphous and nanocrystalline state (Ban et al. 2014). Moreover, the coatings display a significant reduction in the porosity, as hydrogen bubbles during coating deposition is continuously removed by agitation.

3.2.6 Effect of Bath Age

Most electroplating solutions maintain metal by dissolving anodes. However, as the conventional electrodes are absent, electroless nickel requires addition of primary bath chemicals to maintain correct balance and nickel metal in the correct range. Electroless nickel plating solutions require frequent, or better still, continuous additions of all the critical components for replenishment. Nickel and sodium hypophosphite are continuously consumed, and the pH changes unless additions are made. If replenishment is not made, the pH changes together with the amount of critical materials that affect the plating rate.

Now, electroless nickel baths have a finite life. Bath age is defined in terms of the number of times the entire nickel ion content (g/L) is consumed and replenished. Each such replenishment is called a turnover (Taheri 2003). Typically, a bath contains 6 g/L Ni ion. By the time the amount of Ni ions replaced reaches 30–80 g/L, after turnover reaches 5–13, the whole bath has to be discarded. By increasing the amount of Ni replaced, the deposition rate and, consequently, the efficiency of the bath decrease. Bath age has a profound effect on the electroless nickel deposit composition and other electroless nickel properties such as internal stress, ductility, corrosion resistance and fatigue resistance. Riedel (1997) has studied the effect of bath age on the properties of electroless nickel coatings. The study shows that after 4–5 turnovers, the deposition rate is reduced. Figure 3.10 shows that, as the electroless nickel bath gets older, the deposition rate decreases. On the other hand, the phosphorus content of the coating increases with bath age. The effect of bath age on the properties of electroless nickel deposits is low when up to 5–6 turnovers. As the bath solution exceeds its sixth turnover, the properties of the deposit are deleteriously changed. Some studies have shown that even the phosphorus content stays constant for up to six turnovers (Riedel 1997). It is also observed that after five turnovers, the cyclic corrosion resistance of the electroless nickel deposit drops significantly. Thus, as a rule of thumb in most of the commercial plating shops, the bath content is replaced with a new solution after the fifth turnover. The bath age also has a pronounced effect on the deposition rate. As shown in Figure 3.11, as the bath gets older, the plating rate decreases. This is due to the decomposition of chemicals in the bath including the complexants and accelerators.

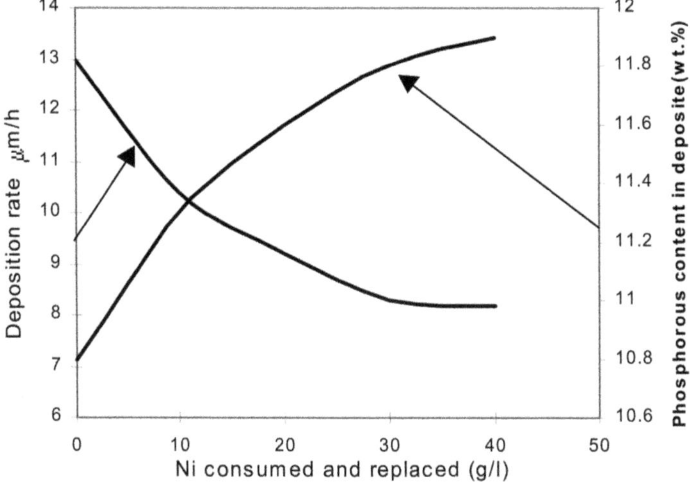

FIGURE 3.10 Dependence of deposition rate and phosphorus content of the electroless nickel coating on the bath age. (From Linka, G. and Riedel, W., *Galvanotechnik*, 6, 1986.)

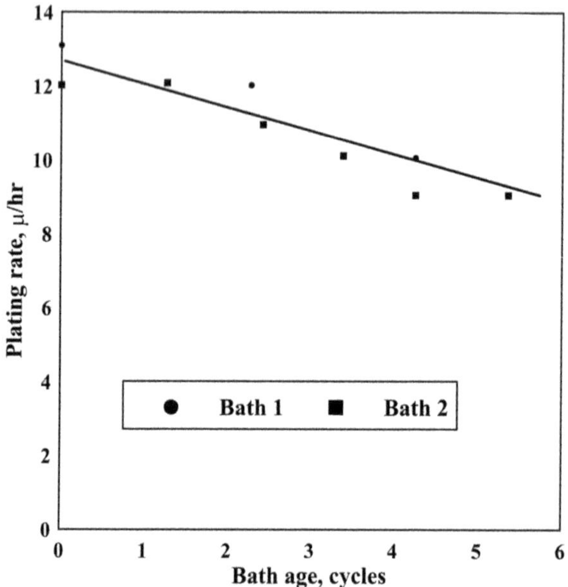

FIGURE 3.11 Effect on deposition rate due to bath age. (From Duncan, R.N., *Surf. Finish.*, 10, 5, 1983.)

3.2.7 Reasons for Spontaneous Decomposition of Electroless Nickel Bath

Electroless nickel baths are deliberately kept at the verge of instability for the deposition occurs. A highly stable bath will not produce deposition at all or may be at an unacceptably poor rate. Thus, there is a high chance of the bath getting decomposed, or in other words, spontaneous reaction starts to occur throughout the solution with the formation of nickel phosphides. As a result of this, the bath turns black in color with heavy hydrogen evolution. Once the bath decomposes, it is unsuitable for carrying out further deposition. This is a loss as the nickel can't be recovered from such a solution state and the chemicals are wasted.

One of the primary factors of bath decomposition is the temperature of the bath. It has been observed that the deposition is highly influenced by the temperature of the solution and increases exponentially with the increase in temperature. Hence, too high temperature may render the bath unstable, resulting in its spontaneous decomposition. Besides, improper bath composition and rapid change in pH of the solution may also make the bath decompose. The common causes of electroless nickel bath decomposition are listed below (Krishnan et al. 2006):

1. Local overheating;
2. Rapid addition of sodium hypophosphite;
3. High phosphite concentration;
4. Rapid addition of an alkali that may change the pH of the solution drastically;

5. Excessive deposits of nickel on tank walls or on heating coils;
6. High pH during bath preparation, causing the precipitation of nickel compounds;
7. Incomplete removal of palladium after the use of palladium activation;
8. Instability of fresh baths;
9. Low bath loadings;
10. Overuse of baths, leading to the precipitation of phosphates.

To obtain a quality electroless nickel deposit, many trials have to be conducted. Otherwise, bath composition and deposition conditions can be selected from an established process.

3.3 MICROSTRUCTURAL ASPECTS OF ELECTROLESS NICKEL COATING

3.3.1 Surface Morphology and Microstructure

The electroless nickel coatings can be amorphous or crystalline; the microstructure transition from nanocrystalline to amorphous occurs in a progressive manner with the P concentration, because these atoms disorder the regular disposition of nickel atoms, the phenomenon in which the electroless NiP coatings are classified according to the phosphorus content into three levels: high, medium and low (Acuña and Echeverría 2015). Although there is no consensus on the actual microstructure of each of these levels, it is commonly accepted that coatings with low phosphorus concentration correspond to pure nanocrystalline nickel face-centered cubic (FCC) grains (particle size between 5 and 30 nm) surrounded by grains rich in phosphorus; coatings with high phosphorus concentration show a totally amorphous structure (however, it has been observed that these coatings may contain small microcrystalline regions, probably as a consequence of local inhomogeneities in the P distribution) and coatings with average P concentration exhibit a mixture of these structures, which can be described as nanocrystals "floating" in an amorphous phase of Ni and P.

The properties of electroless nickel coatings are heavily dependent on their microstructural characteristics. The microstructure of the deposits is again controlled by its phosphorus content. Based on the bath ingredients and their concentrations, the phosphorus content can range typically from 2% to 12% by weight. Now, the properties of electroless NiP deposits are largely dependent on the amount of phosphorus present in the deposit. Hence, traditionally, the deposits are categorized according to their phosphorus content, as given in Table 3.11.

TABLE 3.11
Electroless Nickel Coating Categorization Based on P Content

Low phosphorus	1–5% P
Medium phosphorus	6–8% P
High phosphorus	9–12% P

TABLE 3.12
Structure of As-Deposited Electroless Nickel Coatings

As-Deposited Electroless Nickel Coating	Type of Structure
Low phosphorus	Crystalline
Medium phosphorus	Mixed crystalline and amorphous
High phosphorus	Amorphous

The as-deposited electroless nickel coatings in general have a tendency to show an amorphous structure but sometimes display a crystalline structure and even a mixture of both. Crystalline materials are arranged in a highly ordered microscopic structure and display a long-range periodicity that lacks in the case of amorphous materials. The established trend for understanding the structure of electroless nickel deposits is shown in Table 3.12. This is due to large phosphorus segregation (in high P coating), which prevents the nucleation of FCC Ni phase resulting in an amorphous structure in the as-deposited coatings. By increasing the phosphorus percentage, lattice disorder of crystalline phase is also increased. Crystallite size of nickel is increased from 50 to 600 Å, from as-deposited condition compared to coatings heat-treated at 400°C (Sampath Kumar and Kesavan Nair 1996). It is also observed that, at higher heat-treatment temperature (i.e., approximately 600°C), crystalline nickel phosphide (Ni_3P) phase is formed with coating, displaying the least quantity of phosphorus.

The amorphous electroless nickel deposits crystallize upon heat treatment over a certain temperature, thus changing the properties of the deposit. The degree of crystallinity, however, is a complex function of a number of factors, namely phosphorus content, heat-treatment temperature, heating rate and period of heat treatment. Besides, previous thermal history of the sample may also affect its crystallinity.

The amorphous structure of electroless nickel coatings (P content greater than 9 wt% (19 at.%)) is characterized by surface morphology of nodules in a "cauliflower" shape (see Figure 3.12a); these are the result of the high concentration of phosphorus, it distorts the crystalline structure of nickel, deforming it to an amorphous structure (Acuña and Echeverría 2015). The lateral growth of the coatings is limited by the presence of stabilizers leading to a columnar growth (see Figure 3.12b) rather than a stratified growth in layers. However, it was observed that the nodular structure reduction occurs with the phosphorus content. Amorphous electroless nickel coatings exhibit higher corrosion resistance than crystalline coatings; these are characterized by a large amount of grain boundaries that act as preferably corrosion routes due to the formation of a large number of micro-chemical cells with the substrate.

3.3.1.1 Effect of Phosphorus Content

One of the yields of studying the microstructure of electroless nickel coatings is to have a better understanding about the properties of the deposit. The micro-level events can many times be correlated with macro-level happenings and are useful in predicting the physical characteristics of the material. As the deposit is an alloy of nickel and phosphorus, many of the microstructural transformations follow the NiP phase

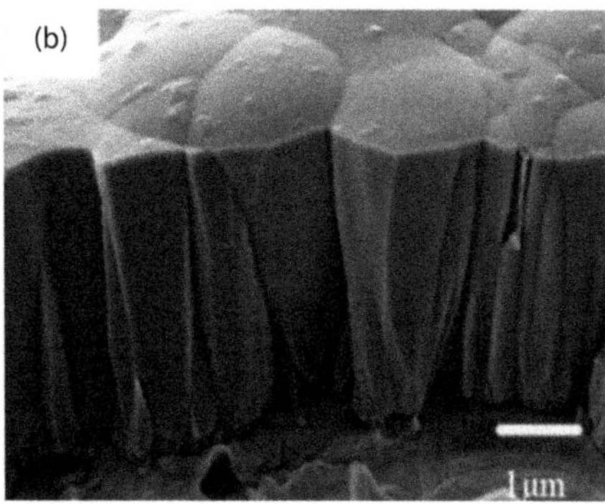

FIGURE 3.12 Electroless nickel coatings. (a) Nodular surface morphology. (From Biswas, A. et al., *Surf. Coat. Technol.*, 328, 102–114, 2017.) (b) Coating section showing columnar growth. (From Wang, W. et al., *Appl. Surf. Sci.*, 367, 528–532, 2016.)

diagram. Now, the NiP phase diagram was originally developed by Konstantinov in 1908, long before the discovery of the electroless nickel process (Taheri 2003). Below the melting point, the conventional diagram shows only two phases for the alloy: Ni_3P, an intermetallic compound containing 15 wt% phosphorus, and α phase, which is a solid solution of 0.17 wt% phosphorus in nickel. The region between these two phases consists of a mixture of α and Ni_3P. The conventional NiP phase diagram can be used to describe the microstructure of the alloys in their equilibrium condition (e.g., the solidified alloys after melting or the electroless nickel coating after the heat treatment). Mistakenly, many authors have used the conventional NiP phase diagram to explain the behavior of the as-deposited electroless nickel coating (Riedel 1991), although the as-deposited electroless nickel is not in its equilibrium state. Therefore, the prediction of the microstructure of the as-deposited electroless nickel from the conventional phase diagram (Figure 3.13a) is not accurate. As the nonequilibrium NiP phase diagram shows, there are two additional phases not contained in the equilibrium phase diagram. Phase β is a crystalline solution of phosphorus in nickel (same as α but with 4.5 wt% P). The second phase is γ, which is a totally amorphous phase with 11–15 wt% phosphorus. Between these phases, region 4.4–11 wt%, β and γ coexist.

Zhang and Yao (1999) conducted a comprehensive study on the microstructure of electroless nickel coatings. They also studied the effect of heat treatment on the transformation of amorphous to crystalline microstructure of electroless nickel coatings. Figure 3.14 shows the TEM images and electron diffraction patterns of electroless Ni-1.5 wt% P deposits under different heat-treatment conditions. The diffraction pattern shows continuous well-defined rings, which are a characteristic of a microcrystalline structure for deposits after 1 h annealing at 200°C. All rings were determined to be from FCC nickel, and there was no evidence of any nickel phosphide. The grain size was approximately 10 nm. Annealing at 400°C for 1 h gave rise to the precipitation of b.c.t. Ni_3P, which was determined from X-ray diffraction (XRD). Due to the fine size of Ni and Ni_3P, the diffraction pattern appears to be several discontinuous rings (Figure 3.14d). Increasing the annealing temperature to 600°C resulted in coarsening of microstructure (Figure 3.14e).

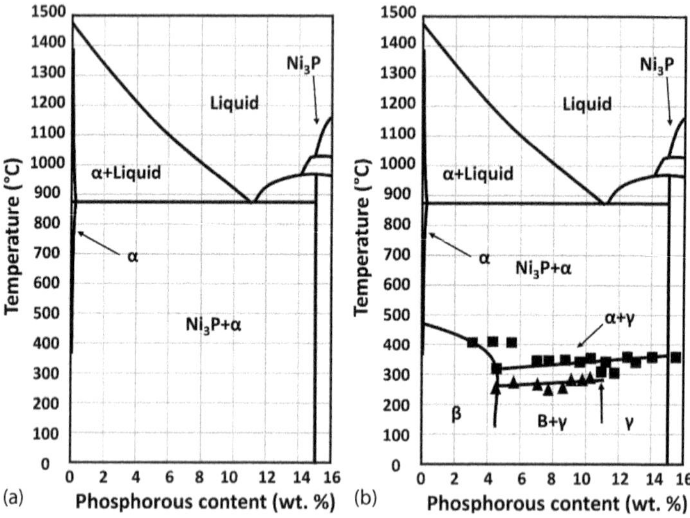

FIGURE 3.13 NiP binary phase diagram: (a) NiP conventional phase diagram; (b) Nonequilibrium NiP phase diagram. (From Taheri, R., Evaluation of electroless nickel-phosphorus (electroless nickel) coatings, PhD thesis, University of Saskatchewan, 2003.)

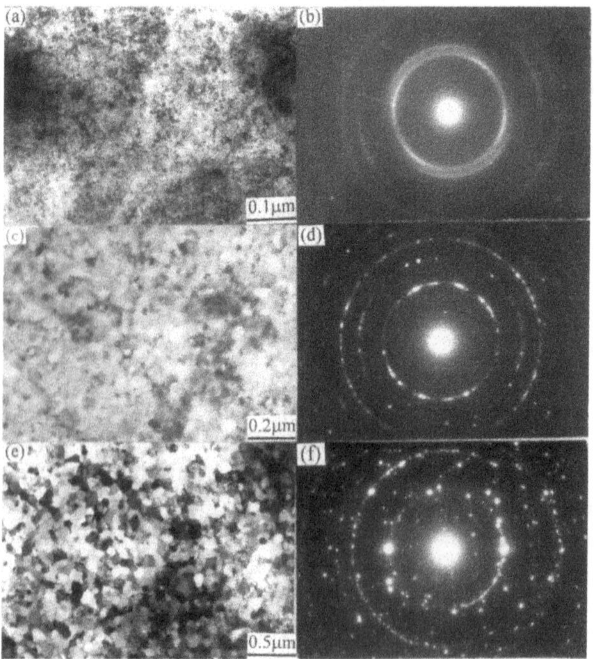

FIGURE 3.14 TEM results of low-phosphorus electroless nickel coating (1.5 wt% P) after various heat treatment processes: (a) micrograph and (b) diffraction pattern of 200°C for 1 h, (c) micrograph and (d) diffraction pattern of 400°C for 1 h and (e) micrograph and (f) diffraction pattern of 600°C for 1 h. (From Zhang, Y.Z. and Yao, M., *Trans. IMF*, 77, 78–83, 1999.)

3.3.1.2 Influence of Heat Treatment on Coating Structure

Heat treatment is an important factor that affects the thickness, hardness, structure and morphology of deposit. Generally, acknowledged optimal heat-treatment regime is 400°C for 1 h, as it results in maximal hardness of electroless NiP coatings. The hardness increase is attributed to the crystallization of nickel and to the precipitation of fine particles of Ni_3P phase. The use of higher heat-treatment temperatures and longer times leads to the progressive hardness decrease, which can be attributed to the nickel grain growth and to the phosphides coarsening. Heat treatment also has a profound effect on the corrosion resistance of electroless nickel coatings. It is invariably found that as-deposited coatings exhibit the best corrosion-resistant properties due to their amorphous structure. However, with the heat treatment, the corrosion resistance of the coating gradually decreases. This is attributed to the advent of crystallinity in the coating due to the heat treatment. Crystallinity increases the grain boundaries, which form active sites for corrosion attack.

3.3.1.2.1 Phase Transformation Behavior of Electroless Nickel Coating

Amorphous alloys are highly disordered in structure and are metastable (Masoumi et al. 2012). When they are heated at a relatively high temperature, crystallization reaction will occur, leading to changes in the alloy properties. This transformation is the main reason for increasing the hardness of as-deposited specimens with increasing wear test temperature. It is known from the theory of hardening of materials that when a fine precipitate is formed in a solid solution, an additional barrier is formed for the movement of dislocation to propagate. It is generally believed that these dislocations cut through the precipitate particles until they are coherent to the matrix. The degree of crystallinity affects resultant properties and is a complex function of a number of factors, namely phosphorus content, heating rate, heat-treatment temperature and time at heat-treatment temperature and previous thermal history.

However, as the heat-treatment temperature increases above 400°C, in wear test conditions such as room and 550°C temperatures, the finely dispersed crystallites of Ni_3P and Ni agglomerate (grain coarsening), and they become incoherent to the matrix. Hence, it will be easier for a dislocation to loop between these coarse particles rather than to shear them (Orowan mechanism). It can be the reason for the decrease in hardness of coatings after 400°C, which is in good agreement with similar results as described in the previous literatures.

In case of electroless nickel coatings, phase transition is not detected below 300°C, as the transformation occurs somewhere in between 325°C and 375°C. However, the rate of heating is found to have an influence over phase transition; increasing the heating rate results in shifting (increase) of the crystalline temperature as shown in Figure 3.15. The presence of exothermic peaks in the figure is indicative of the transformation of nanocrystalline/amorphous deposit into crystalline deposit. It can also be observed that the higher scanning rates improve peak sensitivity. The thermogram obtained at 50°C/min scanning rate for electroless NiP deposit shows a single, sharp exothermic peak in the temperature range of 350°C–370°C. The crystallization temperature for the deposit is 372°C and the corresponding enthalpy (ΔH) value is −63.5 J/g. In addition, due to the high temperature, a much weaker and broader peak at approximately 448°C is observed.

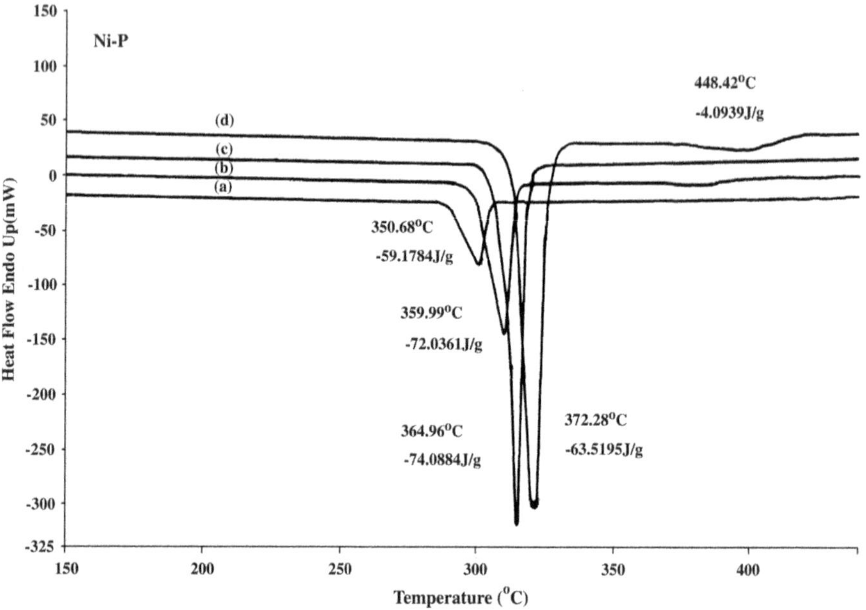

FIGURE 3.15 Differential Scanning Calorimetry (DSC) thermograms obtained for NiP at different scanning rates: (a) 10°C/min, (b) 20°C/min, (c) 30°C/min and (d) 50°C/min. (From Balaraju, J.N. et al., *Surf. Coat. Technol.*, 201, 507–512, 2006a.)

To assess the effect of heat-treatment temperature and period of heat treatment, the XRD plots (Figure 3.16) for high-phosphorus electroless nickel coating heat-treated for 1, 2 and 4 h are studied. It is found that below phase transition temperature (PTT) (338°C), the coating displays a broad peak indicating amorphous structure. However, for longer durations of heating (2 and 4 h) at 200°C, Ni (111) peak is detected. Upon heat treatment at 338°C for 1 h, crystalline peaks of Ni and Ni_3P are manifested on the XRD plot. With further increase in the heat-treatment temperature (i.e., 400°C–800°C), more crystalline phases appear in the XRD plots. It is found that heating of electroless nickel deposit at higher temperature produces better reflection peaks in XRD results that imply a better crystalline texture in the coating. Again, it is observed that increase in the heat-treatment temperature does not always ensure about the appearance of new peaks. Up to 600°C, the peak refinement is observed particularly in between 2θ range of 35°–55°. It is shown that Ni_3P is the major nickel phosphide formed due to heat treatment. Ni_3P is a stable phosphide phase and being hard is believed to impart hardness to the coating. Apart from Ni_3P, other phosphides, namely $Ni_{12}P_5$ is detected, which is, however, considered to be a metastable phosphide state. However, not many metastable phases are detected. When the composition of the coating is near eutectic, the chances of the formation of metastable phases is minimum and Ni_3P phase is directly attained.

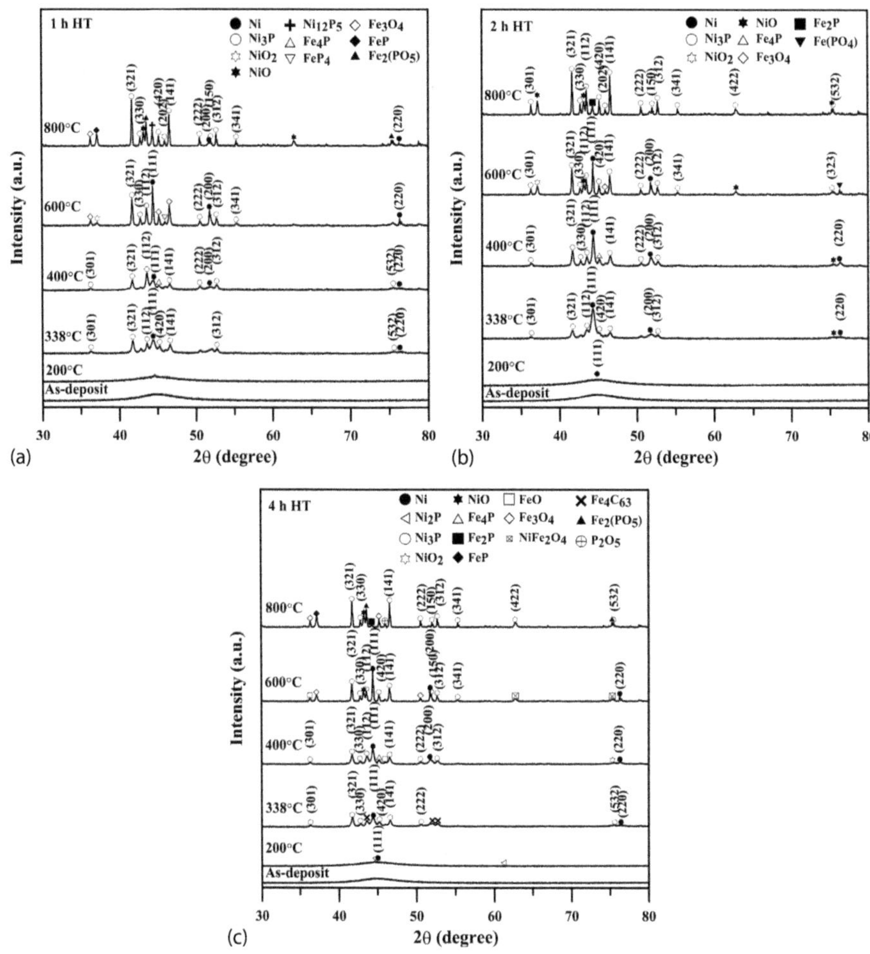

FIGURE 3.16 XRD plots of electroless nickel coatings at various heat treatment temperatures for (a) 1 h, (b) 2 h and (c) 4 h duration (From Biswas, A. et al., *Surf. Coat. Technol.*, 328, 102–114, 2017.)

Crystalline nickel is also detected in some of the XRD plots (Figure 3.16a). The nickel and nickel phosphide-based peaks are found to be small and wide at low-temperature heat treatment. This is indicative of a fine grain size and presence of residual stress (Keong et al. 2003). With an increase in the heat-treatment temperature, the peaks become finer and sharper, thus indicating a coarser grain size and the absence of internal stresses. The major Ni peak diffracted corresponds to Ni (111) plane whose intensity in the diffracted plot increases continuously up to a temperature of 600°C. This is indicative of a preferred orientation in Ni (111) plane. However, for heat treatment at 800°C, the Ni (111) peak vanishes suddenly. This may mean that at higher temperature, free nickel tends to bond with other elements to form newer compounds. However, the coating is not totally devoid of nickel which is present as very small peaks of (200) and (220). The nickel phosphide peaks are also found to increase in intensity with an increase in heat-treatment temperature. The increasing intensity with heat-treatment temperature is observed for Ni_3P (321) peak. Other than (321), other notable peaks of Ni_3P are (312), (112) and (141).

By observing the formation of intermetallic compounds of iron and phosphorus and their oxides, it can be concluded that inter-diffusion of iron occurs into the coating from the steel substrate surface while phosphorus diffuses into the substrate from the electroless nickel coating. Now, Fe_3P is known to be the most stable iron phosphide. However, the phosphides detected are believed to be some intermediate compounds formed. Above 600°C heat treatment, oxide formations are also observed in the coating. This is due to the reaction between the coating elements and oxygen from atmospheric air present in the furnace in which the heat treatment operations are carried out. Nickel oxide peaks, namely NiO and NiO_2 are observed in the XRD plots. Apart from these, oxides of iron peaks, namely Fe_3O_4 and $Fe_2(PO)_5$ are also observed. It is thus confirmed that during oxidation of electroless nickel coating on ferrous substrates, iron is diffused in the coating and reaches very near to the coating surface, thus forming oxides. However, another proposition is that with an increase in heat-treatment temperature, iron concentration increases, resulting in the substitution of nickel in the NiO lattice. It can thus be concluded from these observations that diffusion mechanism of Fe atoms is augmented by the heat-treatment temperature. In reality, the movement of atoms is based on their vibration energy that can be supplied by the heat treatment. Under low-temperature heat treatment (generally below 400°C), atoms can move in short ranges, which is given the name of structural relaxation (to obtain the ground state relaxed geometry of the system under consideration), namely annihilation of point defects and dislocations within grains and grain boundary zones (Ashassi-Sorkhabi and Rafizadeh 2004). With increase in temperature, greater atomic vibration energy is available, which is favorable for long-range mobility of atoms. Moreover, with an increase in temperature of the metal, more vacancies are present together with higher availability of thermal energy. All these conditions are favorable for higher diffusion rate of atoms at higher heat-treatment temperature (Rabizadeh et al. 2010). Besides, the formation of Fe-based oxidation products observed for electroless nickel coating is governed mainly by the diffusion and inter-diffusion mechanisms that occur at the coating–substrate interface rather than by the oxidation properties of the coating (Eraslan and Ürgen 2015).

The XRD plots of electroless nickel coatings heat-treated for 2–4 h durations are shown in Figure 3.16b and c, respectively. The same plot with a broad peak as in the case of 1 h heat treatment is observed for samples heat-treated at 200°C for both 2 and 4 h durations. However, a low intensity (111) Ni peak at approximately 44.5° is detected for both 2 and 4 h duration heat-treatments. Furthermore, for 4 h heat treatment, peaks of metastable phosphide Ni_2P are observed. This indicates that crystallinity is initiated in the amorphous coating with an increase in the duration of heat treatment even below the PTT (338°C for the present case).

At the PTT, apart from the Ni and Ni_3P phases, some other phases such as NiO oxide phase are observed for 2 h heat-treated coatings. Phosphorus oxide (P_2O_5) is also observed at a temperature of 800°C, which is formed due to the migration of phosphorus atoms to the surface of the coating as is also reported by Tomlinson and Wilson (1986). Thus, prolonged exposure to oxygen in air at high temperature results in the formation of oxides. Increasing the heat-treatment time results in the increase of iron concentration in the coating, which is detected in various forms such as carbide (Fe_4C_{63}) and oxide ($NiFe_2O_4$) as observed for 4 h heat-treatment plot. Notably, there is an increase in the concentration of ferrous compounds (both phosphides and oxides) with an increase in heat-treatment duration. Thus, it can be concluded from the XRD results that higher heat-treatment temperature results in the increased diffusion of iron atoms into the coating, which is synergistically augmented by the increased duration of heat treatment. Both the factors together result in an enhanced iron intrusion in the coating that substitutes nickel in various lattices of nickel compounds. With an increase in the heat-treatment temperature and time, iron concentration increase, resulting in total elimination of nickel (Eraslan and Ürgen 2015). It is consistently observed that, for all the heat-treatment durations, there is an indication of preferred orientation toward Ni (111) plane that is disrupted at 800°C heat-treatment temperature, which may partially be due the increased diffusion of iron atoms into the coating and also the vulnerability of the coating to oxidation. In connection with this, another interesting thing to note is that the average intensity of peaks reduces at heat treatment for 4 h compared to the same for 1 h.

From the compositional analysis, iron is detected as an element particularly at higher temperatures supporting its diffusion into the coating system. Higher weight percentages are observed for longer duration of heat treatment. Another major element to be detected in the coating is oxygen. Figure 3.17a shows that weight percentage of oxygen in the coating exhibits an increasing trend (implies formation of oxidation products) with heat-treatment temperature after displaying an initial fall at a temperature of 400°C. Furthermore, this trend is observed for all the heat-treatment durations except for higher average oxygen content for longer durations. The percentage increase in the oxygen content in the coating from a temperature of 200°C to 800°C is above 80%, which is tremendous and implies high level of oxidation. Now, this elevated level of oxidation suggests the presence of some other mechanism other than simple surface oxidation. Oxygen may be entering inside the coating by diffusion through micro-cracks induced by growth stresses (Tomlinson and Wilson 1986). With the presence of oxygen inside the coating, the overall chances of the formation of oxides also increase. From Figure 3.17b, it can also be noted that the phosphorus content of the coating decreases with the increase in the heat-treatment temperature. These apparent losses in nickel and phosphorus weight percentage are due to the heavy incorporation of oxygen into the coating and the movement of nickel and phosphorus into the substrate due to inter-diffusion phenomenon, which has already been discussed. Moreover, electroless nickel coatings start to oxidize above 600°C, and above 800°C, decomposition of phosphide and evaporation of phosphorus from the coating occur (Tomlinson and Wilson 1986).

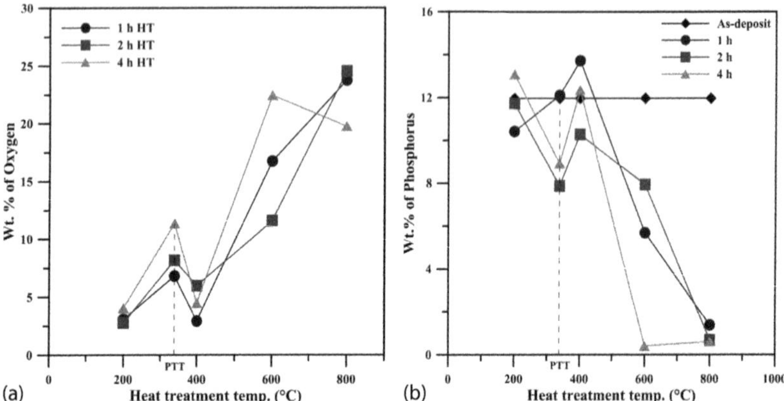

FIGURE 3.17 (a) Oxygen content vs. heat treatment temperature and (b) phosphorus content vs. heat treatment temperature. (From Biswas, A. et al., *Surf. Coat. Technol.*, 328, 102–114, 2017.)

The microstructure of NiP coatings depended on the phosphorus content and heat-treatment temperature. The relative strength of Ni (111) peak becomes weak and broad as the phosphorus content is increasing. The diffraction peaks of NiP deposits were almost unaffected when being heat-treated at 200°C except for the main peak becoming strengthening (Cheng et al. 2011). In addition, the half-width ratio of diffraction peak is larger at the higher phosphorus content. It implied that the nickel crystalline lattice sizes decrease with increasing phosphorus content. These results might be attributed to the increasing of the number of phosphorus atoms within the FCC Ni lattice. It causes the increase of lattice strain, which will further reduce the growth rate of crystallite. Consequently, it leads to a smaller crystalline size (Balaraju et al. 2006b).

3.4 SPECIFICATIONS AND PROPERTIES OF NiP COATING (AS-PLATED AND HEAT-TREATED)

3.4.1 Physical Properties

3.4.1.1 Density

The density of pure nickel is 8900 kg/m^3 whereas the same for phosphorus is 1820 kg/m^3. Hence, electroless NiP being an alloy of both the abovementioned elements has a density close to pure nickel but somewhat lower than that. The density depends on the percentage of phosphorus in the deposit and decreases appreciably with increasing phosphorus content. For instance, a deposit containing 3% phosphorus has a density of 8520 kg/m^3 while at 11%, the density is only 7750 kg/m^3 (Parkinson 1997). A plot depicting the relationship between phosphorus content and density is shown in Figure 3.18. It is worthwhile to mention here that the density of the deposit is not consistent throughout the coating thickness, as the concentration of potential nickel and phosphate ions for deposition changes due to a continuous change in the concentration of the bath (due to coating deposition).

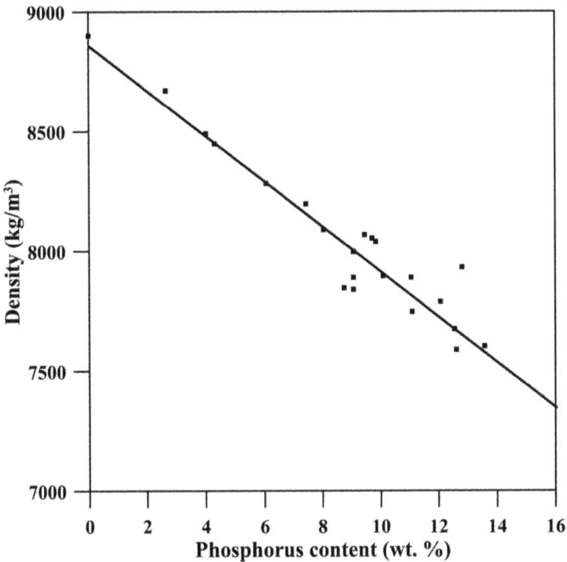

FIGURE 3.18 Effect of composition on deposit density. (From Parkinson, R., Properties and applications of electroless nickel, Nickel Development Institute, NiDL Technical Series No. 10081, 1997.)

3.4.1.2 Melting Point

Melting point of a solid is the first indication about its thermal stability. Electroless NiP deposit cannot match the high-temperature performance of pure nickel, namely high-temperature oxidation resistance. The melting point of pure nickel is 1455°C, but the inclusion of phosphorus (11 wt%) brings down the melting point to approximately 880°C. This is the lowest melting point (eutectic temperature) for the NiP system. Figure 3.19 depicts the actual effect of phosphorus content on the melting point of the electroless NiP deposit (Parkinson 1997). Electroless NiP deposits containing 7.9 wt% P have a melting point of 890°C (Krishnan et al. 2006).

3.4.1.3 Hardness of Electroless Nickel Deposit

Electroless nickel coatings are known for their hardness, which further increases after heat treatment. Now, hardness is simply determined and the most widely accepted parameter used to characterize the surface contact response of diverse engineering materials (Keong et al. 2003). Like many other materials, hardness is the parameter that has been studied most comprehensively in electroless nickel deposits. One of the primary reasons for this could be that the parameter has a direct relationship with the wear and abrasion resistance that is a vital property of the deposits. The factors on which the hardness of electroless nickel coatings depends are:

1. Phosphorus content;
2. Heat-treatment temperature;
3. Period of heat treatment.

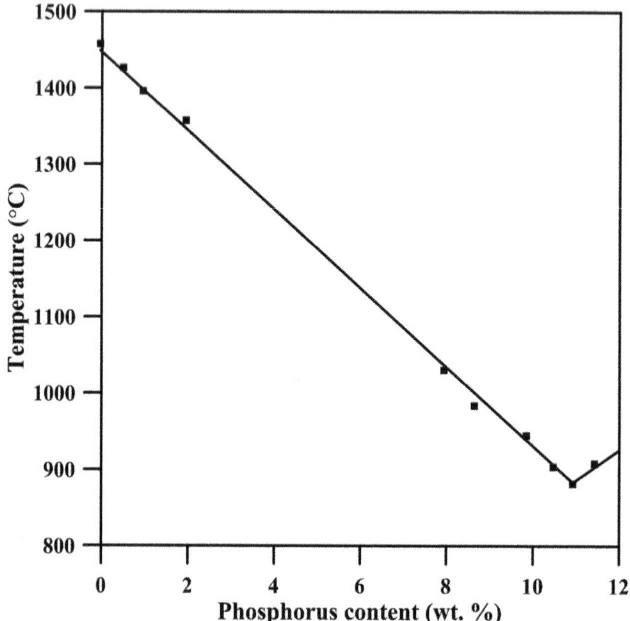

FIGURE 3.19 Effect of phosphorus percentage on the melting point of deposit. (From Parkinson, R., Properties and applications of electroless nickel, Nickel Development Institute, NiDL Technical Series No. 10081, 1997.)

3.4.1.3.1 Effect of Phosphorus Content on the Deposit Hardness

The hardness of electroless nickel deposits, as with many other properties, is directly affected by the phosphorus content. The hardness value may vary from 500 to 700 $HV_{0.1}$ depending on the phosphorus content (Keong et al. 2003). As shown in Figure 3.20, increasing the phosphorus content of the deposit lowers the hardness of the coating. The maximum hardness is obtained at approximately 4 wt% phosphorus where the microstructure consists of the single β phase (see Figure 3.13). As the phosphorus content increases, the amount of β phase decreases while the amount of the γ phase increases. This causes a reduction in the hardness because the γ phase is a softer phase compared to the β phase (Taheri 2003). The minimum hardness value is obtained when the microstructure consists of the single γ phase at 11 wt% phosphorus.

Another probable reason for higher hardness value in the deposits with lower phosphorus level is because of the higher residual stresses (Keong et al. 2003).

Moreover, it is also reported that as-deposited low P electroless nickel coatings are microcrystalline in nature whereas the high P coatings are amorphous. Now, in case of amorphous structure, due to the absence of proper grain boundary, any slip or dislocation can travel longer and relatively easier. However, in case of microcrystalline structure, the slip or dislocation are obstructed at the grain boundaries and thus face resistance. Now, commonly, hardness is a parameter characterizing

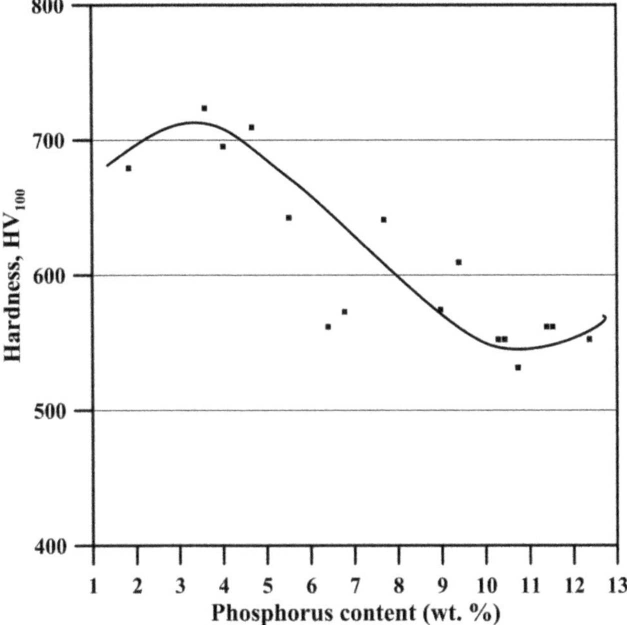

FIGURE 3.20 Effect of P content on as-deposit hardness. (Duncan 1996a).

material local resistance against plastic deformation by the penetration of a hard body (indenter). Furthermore, plastic deformation is characterized by a collection of slips and dislocations. Thus, any resistance to these dislocations would surely lead to a higher hardness for the material. This is another explanation of why low P electroless nickel coatings exhibit higher hardness than high P coatings. Due to this, there is a considerable demand for low P coatings, which has been discussed in one of the following sections.

3.4.1.3.2 Effect of Heat Treatment on Hardness of Electroless Nickel Deposits

NiP, which is a supersaturated alloy in as-deposited state, can be strengthened by precipitation of nickel phosphide crystallites with suitable heat treatments. The phosphides act as barriers for dislocation movement, thereby increasing the hardness further. However, the hardness of NiP films degrades with excessive annealing due to grain coarsening (Palaniappa and Seshadri 2008), leading to surface brittleness and enhanced dislocation propagation. Heat treatment for electroless nickel coatings is carried out keeping in mind the following objectives (Biswas et al. 2017):

1. Increasing hardness of deposit together with its abrasion resistance,
2. Increasing coating adhesion in the case of certain substrates,
3. Elimination of hydrogen embrittlement if any, and
4. Increasing temporarily corrosion resistance (subject to the type of deposit).

Post-heat treatment has a significant effect on the hardness of the electroless nickel deposit. As a matter of fact, one of the outstanding characteristics of electroless nickel coatings is the possibility of obtaining very high hardness values (980–1050 VHN) through an appropriate heat treatment. This provides a unique wear and erosion resistance. The parameters of the post-heat treatment (time and temperature) are defined by the phosphorus content. If the phosphorus content stays in one phase region, i.e., below 4.5 or above 11, there would be only one reaction in which the β or γ phase is transformed into α and Ni_3P (Taheri 2003). This transformation occurs at 400°C for low phosphorus and 330°C–360°C for high-phosphorus content alloys (Duncan 1996a). However, in the region between 4.5 and 11 wt% phosphorus, there is another transformation in which β is transformed into α nickel. This transformation occurs at 250°C–290°C and causes the precipitation of fine particles throughout the coating. The summary of the hardness behavior of electroless nickel coating in terms of the hardness change, relating to phosphorus content and heating temperature, is presented in Figure 3.21. Figure 3.21 also shows that maximum hardness is obtained after a heat treatment at 400°C for 1 h (Baudrand 1978, Riedel 1997). Table 3.13 compares the Vickers hardness values of both as-deposited and heat-treated electroless nickel coatings of various P contents.

3.4.1.3.3 Other Factors Affecting Hardness

Electroless nickel deposits with high hardness and wear resistance can also be developed without the necessity of subsequent thermal treatment. Various ratios of lactic acid and sodium acetate were adopted as complexing agents in acidic electroless nickel baths, with a premise that the ions of nickel were just complexed by the

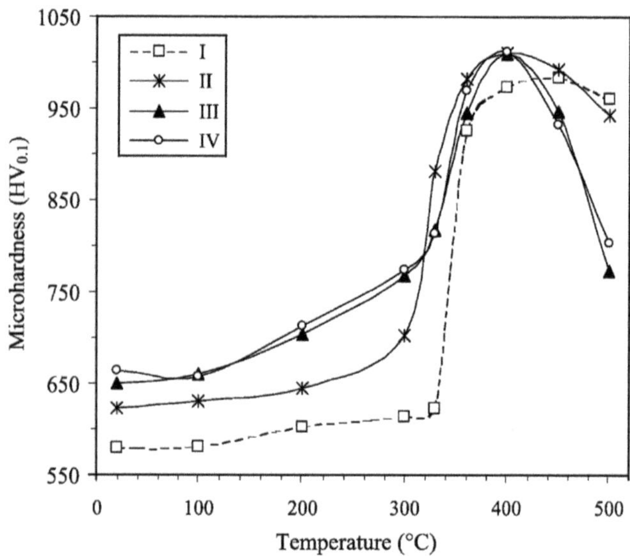

FIGURE 3.21 Vickers hardness measured at cross sections for Samples I–IV after heating at different temperatures for 1 h: Sample composition (wt% P) I: 12%; II: 8.8%; III: 6.7%; IV: 6.3%. (From Keong, K.G. et al., *Surf. Coat. Technol.*, 168, 263–274, 2003.)

TABLE 3.13
Hardness of As-Deposited and Heat-Treated Electroless Nickel Coating for Various P Content

Phosphorus Content	Hardness (As-Deposited), HK_{100}	Hardness (Heat-Treated), HK_{100}
2%–3%	700	1000
6%–9%	550	920
10%–12%	510	880

Source: Agarwala, R.C. and Agarwala, V., *Sadhana*, 28, 475–493, 2003.

complexing agent. Results indicated that the deposition rate first increased and then decreased with increasing atom ratio of lactic acid. When the atom ratio of lactic acid to sodium acetate was 4:6, the deposition rate reached the maximum, and the coatings obtained from such bath had high microhardness (820 HV) and wear resistance (Sahoo and Das 2011).

The addition of the surfactants, namely SDS and CTAB resulted in an increase of more than 50% in as-deposited hardness of the NiP coating. This is due to the fact that the addition of the surfactants results in a mixture of nanocrystalline and amorphous structure (Elansezhian et al. 2008). Table 3.14 presents a summary of the physical and mechanical properties of electroless nickel coatings.

TABLE 3.14
Summary of Physical and Mechanical Properties of Electroless Nickel Coatings

Property	Ni-3% P	Ni-8% P	Ni-11% P
Composition range: balance nickel	3–4% P	6–9% P	11–12% P
Structure[a]	m-c	m-c-a	a
Internal stress (MPa)	−10	+40	−20
Final melting point (°C)	1275	1000	880
Density (gm/cm^3)	8–6	8–1	7–8
Coefficient of thermal expansion (mm/m-°C)	12–4	13–0	12–0
Electrical resistivity (mW-cm)	30	75	100
Thermal conductivity (W/cm-K)	0–6	0–0.05	0–0.08
Specific heat (J/kg-K)	1000	ND	460
Magnetic coercivity (A/m)	10,000	110	0
Tensile strength (MPa)	300	900	800
Ductility (%)	0–7	0–7	1–5
Modulus of elasticity (GPa)	130	100–120	170
Hardness of as-deposited (HV_{100})	700	600	530
Hardness of heat-treated (HV_{100})	960	1000	1050

Source: Agarwala, R.C. and Agarwala, V., *Sadhana*, 28, 475–493, 2003.

[a] a: amorphous; m-c: microcrystalline; m-c-a: mixed crystalline and amorphous; c: crystalline; ND: not determined.

3.4.2 Mechanical Properties

3.4.2.1 Tensile Strength

The tensile strength of electroless nickel deposits is related to their phosphorus content and, consequently, the microstructure (Taheri 2003). Generally, an amorphous microstructure has high tensile strength and strain due to the absence of defects and high-stress areas normally located in the grain boundaries. As a result, by increasing the phosphorus content, a high tensile strength and strain can be achieved. This is illustrated in Figure 3.22 (Baudrand 1978, Duncan 1996a). For deposits (in the as-deposited condition) with relatively low phosphorus content, the values of tensile strength were found to be between 450 and 550 MPa. After heat treatment between 300°C and 600°C, these values fall to 200–320 N/mm^2.

Figure 3.23 shows the effect of phosphorus content on the elongation properties of an electroless nickel deposit. As illustrated, between 4.5 and 11 wt% phosphorus (when the microstructure consists of both β and γ), the ductility is at its lowest value. As the phosphorus content falls below 4% or rises above 11 wt%, the electroless nickel deposit becomes more ductile.

3.4.2.2 Ductility

The ductility of electroless and electrodeposited nickel is arguably lesser (less than 2%) than those analogously formed from the molten state (3%–30%) (Krishnan et al. 2006). This is unfortunate, because good elastic as well as plastic properties are an important feature in the selection of metallic materials. The preferred technique for quantitative measurements of ductility of the deposits is the micromechanical bulge test, a development of the long-established Ericsson technique.

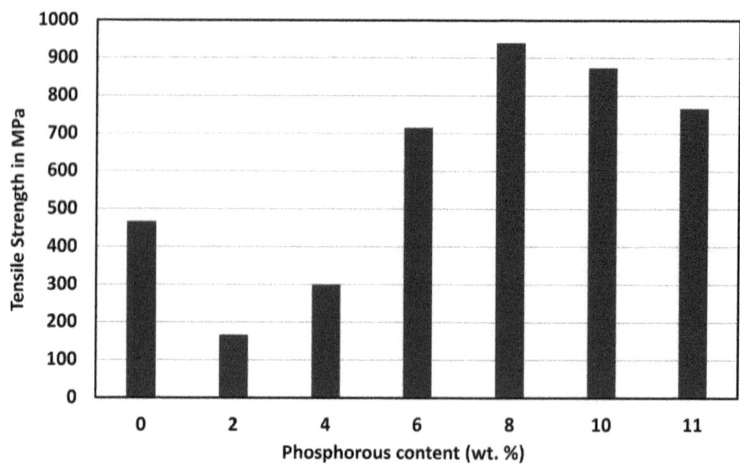

FIGURE 3.22 Effect of P content on tensile strength of electroless nickel deposit. (Duncan 1996a).

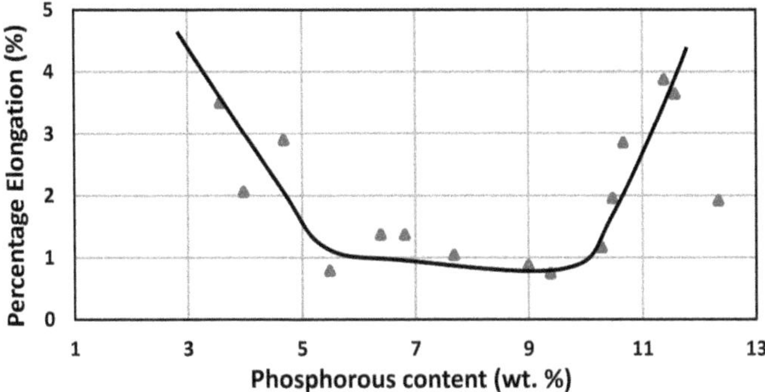

FIGURE 3.23 Dependence of elongation on phosphorus content of electroless nickel deposit. (Duncan 1996a).

3.4.2.3 Fatigue Behavior

Fatigue and corrosion fatigue properties of electroless nickel coatings have been studied by many researchers (Puchi et al. 1996, Taheri 2003). However, in many cases, the results obtained are controversial. The general understanding concerning the fatigue behavior is that due to their tendency to crack under cyclic loads, electroless nickel coatings can cause a significant reduction in the fatigue properties of steel substrates, although the magnitude of reduction in fatigue strength would depend upon composition, heat treatment and coating thickness. The magnitude of reduction of fatigue strength depends on the composition, heat treatment and thickness of the coating, as well as original fatigue strength of the steel. Several investigations have shown that the use of electroless nickel coatings causes a 10%–50% reduction in fatigue strength and endurance limit of steel substrate (Krishnan et al. 2006). It is found that electroless nickel coatings had a detrimental effect on corrosion fatigue properties of AISI 1045 substrate in 3% sodium chloride solution (Chitty et al. 1999). Heat treatment of the coating doesn't help as well. Moreover, based on other studies (Taheri 2003), the reduction in fatigue strength could range between 10% and 50%. In addition, a study (Yamasaki et al. 1981) shows that electroless nickel coatings deteriorate the cyclic stress properties of the substrate. The effect is dependent on the pH of the electroless nickel solution and phosphorus content of the coating. As shown in Figure 3.24, a coating with a higher phosphorus level (amorphous at pH below 4.7) has better fatigue resistance than that of a low-phosphorus (crystalline at pH above 4.7) electroless nickel coating. This is directly related to the brittleness of the electroless nickel coatings with lower phosphorus content. Figure 3.25 shows the variation of fracture strain of NiP, 7 wt% P, after cyclic stressing at $\sigma_a = 275$ MPa. This figure shows a typical fatigue behavior of a metal. Puchi et al. (1996) also studied the effect of electroless nickel coatings on the fatigue properties of their substrates. Their results indicated that electroless nickel coating, 10 wt% P, improves the fatigue properties of AISA 1010 and 1045 substrates.

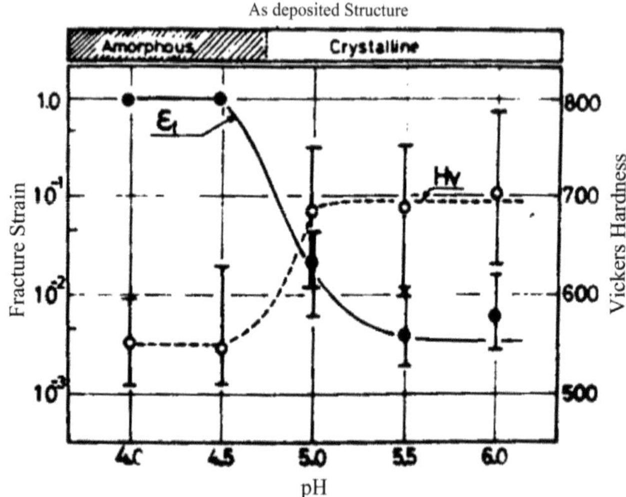

FIGURE 3.24 Effect of pH on the fracture strain and Vickers hardness of electroless nickel coatings. (From Taheri, R., Evaluation of electroless nickel-phosphorus (electroless nickel) coatings, PhD thesis, University of Saskatchewan, 2003.)

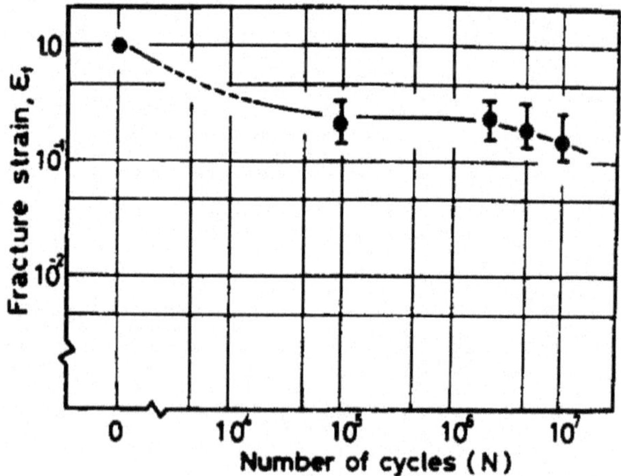

FIGURE 3.25 Variation of fracture strain of NiP, 7 wt% P, after cyclic stressing at $\sigma_a = 275$ MPa. (From Puchi, E.S. et al., *Thin Solid Films*, 290–291, 370–375, 1996.)

Furthermore, the effect of the agitation factor on the fatigue properties of electroless nickel coating has also been studied (Prasad et al. 1994). They found that using an ultrasonic agitation system was significantly beneficial in improving the fatigue properties of the coating (Figure 3.26). They correlated the improving effect of an ultrasonic agitation bath on the fatigue properties of electroless nickel coatings with the superior surface properties of electroless nickel coatings deposited in an ultrasonic bath. Based on their results, electroless nickel coating deposited in a bath with an ultrasonic agitation system had lower roughness values (R_a) that that deposited in a still bath.

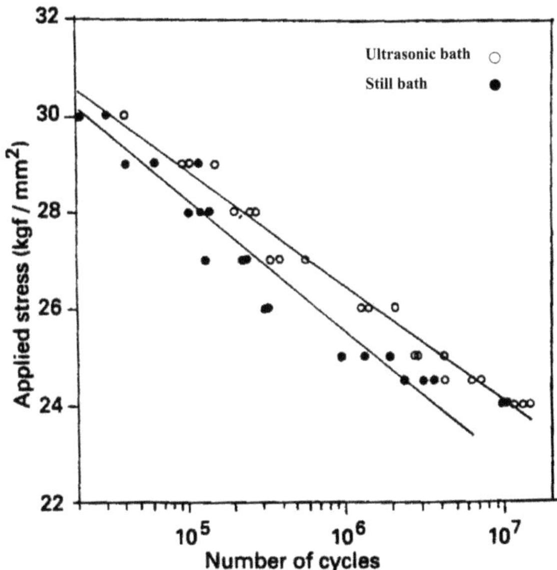

FIGURE 3.26 Fatigue life of nickel coating at different stresses. (From Prasad, P.B.S.N.V. et al., *J. Mater. Sci. Lett.*, 13, 2, 1994.)

3.4.2.4 Adhesion Properties

Coatings with poor adhesion do not provide good protection. Most of the test methods applied such as pull-off test, ring-shear test, peel test and three-point bend test give semiquantitative or qualitative results only. In general, electroless nickel coatings have superior adhesion compared with the nickel electrodeposition method. This is due to the existence of a stronger metal-to-metal bonding during the electroless deposition. However, in case of coating non-ferrous alloys such as magnesium and aluminum, the adherence of the electroless nickel coatings to the substrate is sometimes poor. Hence, the study of adhesion is mainly concentrated in studies using one of these alloys as substrates.

Adhesion of electroless nickel deposits to iron, copper, nickel and alloys of these metals are better than electrodeposited nickel and are further increased by heat treatment at 200°C–400°C for 1 h. Heat treatment is required for improving the adhesion of electroless nickel on aluminum alloys, stainless steel, chrome steels and high carbon steels. Heat treatment is found to enhance the adhesion between NiP coating and 7075 aluminum alloy up to a certain temperature (300°C) above which adhesion decreased (Arabani and Vaghefi 2006). Liu and Gao (2006b) investigated the adhesion of NiP coatings in different variants of magnesium alloys by scratch test and found that adhesion strength of the coatings on AZ31 and AZ91 magnesium alloys is higher than that on pure magnesium. The critical load (a measure of adhesion strength) in case of AZ31 alloy reached 13.1 N. Zincating is found to increase the adhesion electroless nickel and the substrate alloy (Hino et al. 2009).

3.4.2.5 Thermal Characteristics

The coefficient of thermal expansion is one of the important characteristics of coatings. If there is a significant difference between the thermal expansion coefficients of the coating and the substrate, the coating might fail due to poor adhesion or cracking caused by residual thermal stresses at higher working temperatures (Taheri 2003). The residual thermal stresses created at the substrate-coating interface cause initiation and propagation of cracks that may lead to failure. However, unlike many conventional coating methods, electroless nickel coatings have a wide range of thermal expansion coefficients depending on the phosphorus content. Thermal expansion measurements are performed with a dilatometer instrument (Krishnan et al. 2006). The coefficient of thermal expansion of electroless nickel coatings varies from 22.3 μm/m/°C at 3 wt% phosphorus to 11.1 μm/m/°C at 11 wt% phosphorus (Taheri 2003). Thus, whenever thermal expansion is the main concern, it is possible to minimize the residual thermal stresses by controlling the phosphorus content. Figure 3.27 shows the dependence of the thermal expansion coefficient of electroless nickel deposit on the phosphorus content (Baudrand 1978). It is observed that after heat treatment, the coefficient of thermal expansion reduces (Krishnan et al. 2006). The specific thermal conductivity values for NiP deposits with 8.9 wt% P are 0.0105–0.0135 cal/cm/s/deg.

3.4.3 Roughness of NiP Coatings

Roughness is generally an undesirable property, as it may cause friction, wear, drag and fatigue, but it is sometimes beneficial, as it allows surfaces to trap lubricants and prevents them from welding together. Electroless nickel deposition has become commercially important for finishing steel, aluminum, copper, plastics and many other materials (Mallory and Hadju 1991). Electroless nickel coatings are very uniform and they follow the surface profile of the substrate rather than just fill the spaces between surface asperities. Thus, in general, the roughness of the coating does

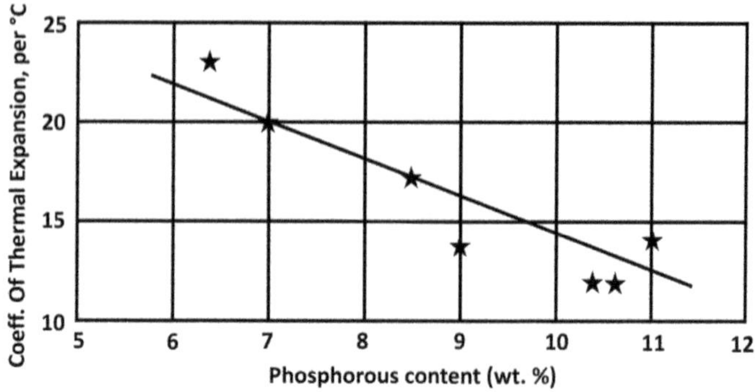

FIGURE 3.27 Effect of phosphorus content on the thermal expansion coefficient. (From Baudrand, D.W., *Metals Handbook*, American Society for Metals, Ohio, 1978.)

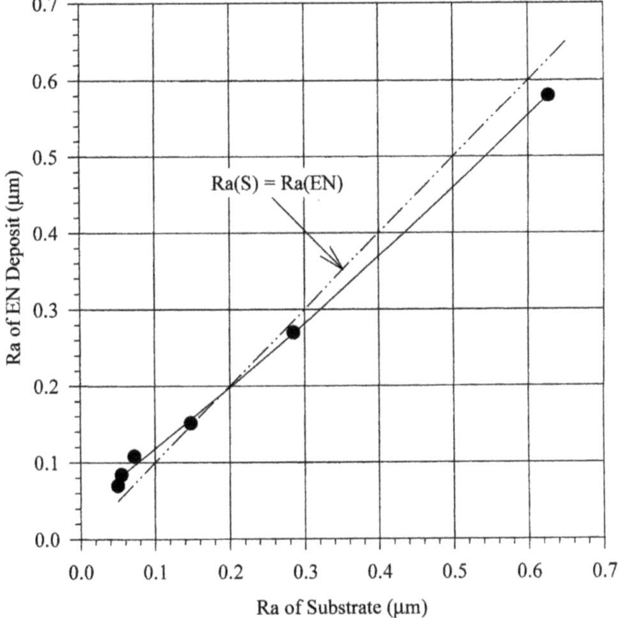

FIGURE 3.28 Relationship between substrate and electroless nickel coating R_a parameter for 1018 steel. (From Taheri, R. et al., *Wear*, 249, 389–396, 2001.)

not vary much from the roughness of the substrate. However, electroless coatings are suspected to have some smoothening effect above a critical substrate roughness. A plot of R_a (S) = R_a(electroless nickel) (where R_a(S) refers to the R_a of the substrate and R_a(electroless nickel) refers to the R_a of the electroless nickel coating) cuts the experimental plot at a single point (Figure 3.28), thereby confirming the existence of a transition or critical substrate roughness for electroless nickel coating process. From Figure 3.28, it can be seen that the transition surface roughness is approximately 0.19 μm for 1018 steel (Taheri et al. 2001). Thus, it can be concluded that on very smooth substrate surfaces, the R_a values increase, whereas, on rough surfaces, the application of electroless nickel coatings tends to decrease the R_a values. Similar results were obtained for the R_z parameter. The effect of coating thickness on the substrate roughness is illustrated in Figure 3.29 (Taheri et al. 2001). A close examination of the plot reveals no dramatic result. At a given substrate R_a, coating thickness does not affect the resulting electroless nickel R_a parameter substantially. The R_a(S) = R_a(electroless nickel) plot that passes through points "A" and "B" is essentially coincident with the straight line fitted to the data in Figure 3.29. This indicates that the substrate R_a parameter remains practically unchanged irrespective of the electroless nickel coating thickness. These results suggest that electroless nickel coating does not necessarily seal off the substrate asperities. Rather, it follows the surface morphology of the substrate material. In general, the roughness of heat treatments of the coatings caused a further decrease in average and maximal roughness

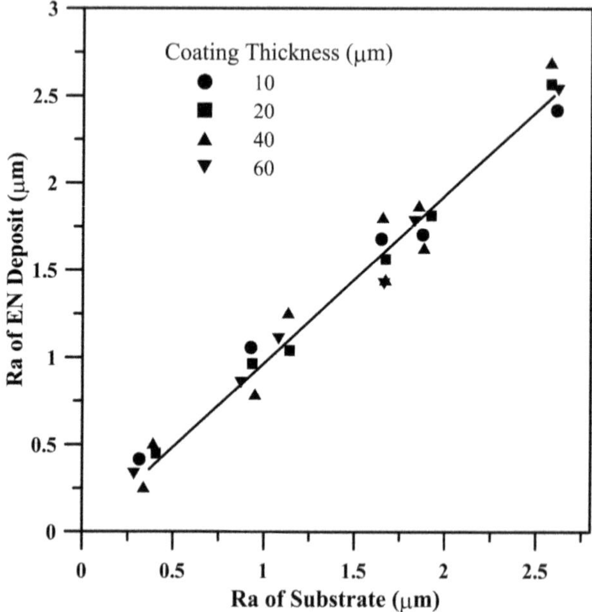

FIGURE 3.29 The effect of high-phosphorus electroless nickel coating thickness on the surface morphology of the substrate. (From Taheri, R. et al., *Wear*, 249, 389–396, 2001.)

(Palaniappa and Seshadri 2008). However, there are also instances of increase in roughness of the sample after coating (Kanta et al. 2009).

The use of a NiP pre-coating is believed to have a leveling effect on this etched surface and produces a smoother top deposit in the end. The presence of surfactants is also found to affect the surface roughness of electroless nickel coatings. The average roughness value of NiP coatings is found to reduce at higher concentration (>0.6 g/L) of surfactants. This is due to the fact that the amount of nickel particles deposited on the coating surface is increased at higher concentration of SDS and CTAB, as the contact angle is reduced, and this leads to the better wettability of NiP deposit (Elansezhian et al. 2009).

Now, a surface generated by machining is composed of a large number of length scales of superimposed roughness and normally characterized by three different types of parameters, namely amplitude parameters (center line average roughness, root-mean-square roughness, skewness, kurtosis, peak-to-valley height, etc.), spacing parameters (mean line peak spacing, high spot count, peak count, etc.) and hybrid parameters (Sahoo 2008). The amplitude parameters are a measure of the vertical characteristics of the surface deviations, whereas spacing parameters are the measures of the horizontal characteristics of the surface deviations. Hybrid parameters on the other hand combine both the vertical and horizontal characteristics of surface deviations. Thus, consideration of only one parameter, namely center line average roughness is not sufficient to describe the surface quality, although it is the most popularly used roughness parameter. This is the justification for the evaluation

TABLE 3.15
Multi-Roughness Values of Substrates Before and After Coating

Roughness Parameters	Substrate Roughness	After-coating Roughness: Nickel Salt—40 g/L Sodium Hypophosphite—17 g/L Sodium Succinate—12 g/L pH = 4.5 Deposition Temperature = 85°C	Optimal coating condition: Nickel Salt—30 g/L Sodium Hypophosphite—17 g/L Sodium Succinate—12 g/L pH = 4.5 Deposition Temperature = 80°C
R_a (μm)	0.614	0.419	0.309
R_q (μm)	0.721	0.589	0.349
R_{sk}	1.45	1.376	0.879
R_{ku}	6.85	6.360	5.350
R_{sm} (mm)	0.153	0.112	0.710

of multi-roughness characteristics of electroless nickel coatings. Sahoo (2008) has used five different roughness parameters, namely center line average roughness (R_a), root-mean-square roughness (R_q), skewness (R_{sk}), kurtosis (R_{ku}) and mean line peak spacing (R_{sm}) to study the surface texture generated in electroless NiP coating and minimize the same by optimizing the electroless bath parameters. It was found that the concentration of the reducing agent and its interaction with concentration of the nickel source solution have a significant influence in controlling the roughness characteristics of electroless NiP coating. Table 3.15 presents the multi-roughness values of substrates before and after coating. The roughness values of the optimized coating conditions are also present in the table.

3.4.4 Tribological Characteristics of the Coating

3.4.4.1 Test Methods

To evaluate the tribological characteristics especially friction and wear behavior of electroless NiP deposit, tribotesters of different contact configurations have been utilized. Pin-on-disc is the most popular configuration that has been used worldwide for the evaluation of friction and wear characteristics. However, other configurations, namely block-on-disc, roller-on-plate, ball-on-plate, Taber apparatus and so on have been used. Figure 3.30 illustrates some of the tribo-test configurations. Apart from tests under ambient conditions, high-temperature tests are also conducted keeping in mind today's demanding situations. In most of the cases, the pin/ball/roller are coated with electroless nickel and tested against a harder counter face so that the actual wear of the coating can be captured. Various tribo-test parameters, namely applied load, sliding velocity, sliding distance and test duration are varied to evaluate the dependency of the coating performance on the tribo-test parameters. Tribological tests have also been conducted under both dry and lubricating conditions. Various lubricants have been used to assess the suitability of the lubricant with respect to electroless nickel deposit.

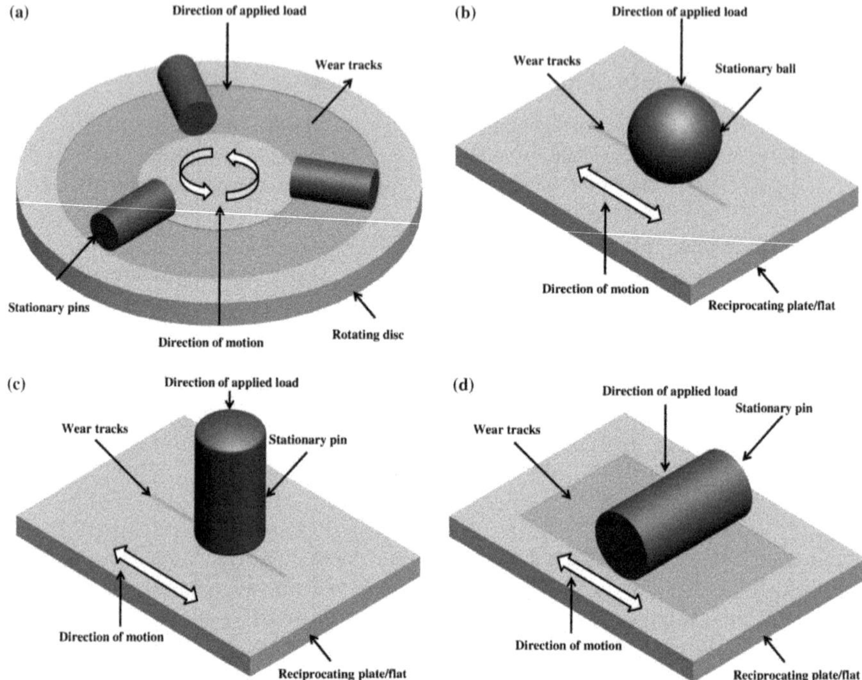

FIGURE 3.30 Common tribo-testing configurations (a) pin-on-disc, (b) ball-on-plate, (c) pin-on-plate and (d) roller-on-plate. (From Zahid, R. et al., *Tribol. Lett.*, 58, 32, 2015.)

3.4.4.2 Friction Behavior

Electroless nickel coatings are in general very smooth and lubricious in nature because of their unique nodular microstructure resembling that of a cauliflower. Electroless deposit is found to reduce the roughness of the blank steel specimen. Figure 3.31 shows the effect of as-plated electroless nickel coatings on the Coefficient of Friction (COF) of substrate surfaces. It is confirmed that electroless nickel coatings generally reduce the kinetic COF of the bare substrate. As expected, the COF of electroless nickel coatings increases with increasing R_a parameter of the test surface. For example, the COF of as-plated high-phosphorus electroless nickel coating is 0.764 for $R_a = 0.254$ µm, whereas it is 0.828 for $R_a = 13.157$ µm. A similar trend is obtained for other classes of electroless nickel coatings.

Good frictional properties are produced by the phosphorus content, which provides natural lubricity, helps minimize heat buildup and reduces scoring and galling (Osifuye et al. 2014). The friction characteristics are also dependent on the condition of the tribological testing and the counter-face material used. Heat treatment in general results in a reduction of the friction coefficient of various electroless coatings compared to the as-deposited coatings. Studies on NiP coating showed that alloys with phosphorus content of 9–13 wt% that are annealed up to 600°C exhibit the best wear resistance in the wear test at room temperature, but it does not coincide with the highest hardness (Masoumi et al. 2012). The amplitude of the friction coefficient

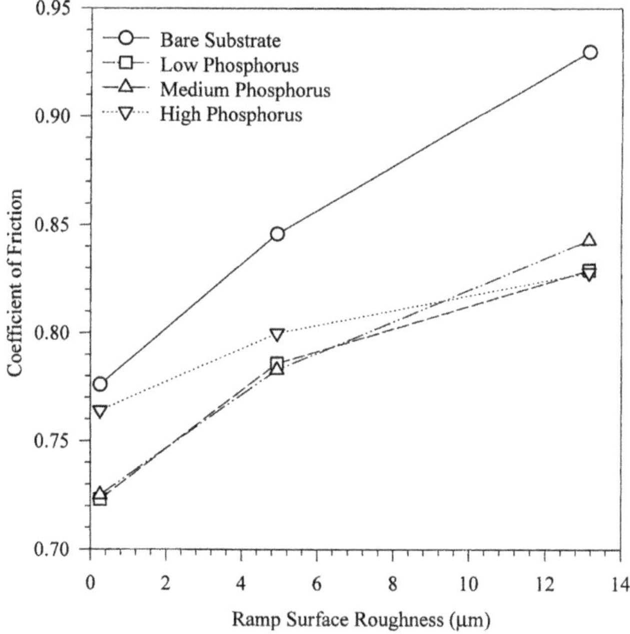

FIGURE 3.31 Effect of P content of as-plated electroless nickel coatings on the COF of various surfaces. (From Taheri, R. et al., *Wear*, 249, 389–396, 2001.)

oscillation of NiP coating after tempering is found to be lower than that of NiP coating (Li et al. 2008). Tsujikawa et al. (2005). The laser-irradiated surface showed lower friction coefficient than that of the furnace-annealed surface with similar hardness (Tsujikawa et al. 2005). The friction coefficient is found to be approximately 0.20–0.30, which is around that of the as-deposited amorphous alloy. The furnace treated samples exhibited a friction value of approximately 0.40.

It is, in general, observed that the friction coefficient of electroless nickel coating decreases with increase in load. The friction coefficient at room temperature lies within the range of 0.15–0.35 when tested under the 15–60 N loading conditions (Staia et al. 1996). The friction study of electroless NiP coating with a ramp apparatus concluded that coatings with high phosphorus content have higher friction coefficient than that with medium- or low-phosphorus electroless coatings (Taheri et al. 2001). The friction coefficient of electroless coating when tested under low loads was found to be as high as 0.7 (Staia et al. 1996).

3.4.4.3 Wear Characteristics

3.4.4.3.1 Wear Measurement

Wear as defined by Rabinowicz (1995) is the interaction between surfaces and more specifically the removal and deformation of material on a surface as a result of mechanical action of the mating surface. Corrosion may be included in wear phenomenon, but the damage is amplified and performed by chemical reactions. Wear is basically an erosion of metal from the contacting surfaces when both the surfaces are in sliding motion. Based on the mechanism, wear may be classified as adhesive wear, abrasive wear, corrosive wear and fatigue wear.

When two flat surfaces come in sliding contact, then adhesive wear occurs. In microscopic view, most of the surfaces consist of asperities or peaks and valleys. A temporary bonding occurs when this type of surfaces comes in contact. During sliding of one surface over another, a fragment of softer surface gets sheared off and gets detached from its base material due to this adhesive bonding. Abrasive wear occurs due to entrapped wear particles within the sliding surfaces. There are generally two modes of abrasive wear: two-body abrasion and three-body abrasion. When hard surface slides over a soft surface, forming a groove in the soft material is called two-body abrasion. In case of three-body abrasion, the hard worn out particles entrapped between two sliding surfaces cause additional grooves in the softer material due to ploughing effect. If wear occurs in corrosive medium, it is called corrosive wear. In case of corrosive wear, both the sliding and corrosion occur simultaneously. Corrosive wear is important in many industries such as mining, mineral processing, chemical processing and so on. Fatigue wear of a material is caused by a cycling loading during sliding and rolling. Surface cracks are produced due to repeated loading and unloading. These cracks come out in the form of large fragment after a critical number of cycles. Besides these, there are some minor forms of wear such as fretting, erosion and percussion.

The traditional means of measuring wear is by the volume or weight loss from a certain specimen after its interaction with another body. The difference in weight before and after the interaction is normally reported as the wear encountered. However, modern tribometers are equipped to report wear undergone as depth of wear. Instruments such as pin-on-disc tribometers, multitribotesters and so on make the specimen slide against a much harder counterface. The wear depth is captured by suitable displacement transducers and plotted real-time in a PC attached to the tribometer. Users can change the sliding speed, applied normal load and the duration of sliding and hence simulate the actual service conditions or any desired condition at which the test is planned.

3.4.4.3.2 Wear Characteristics of Electroless Nickel Coating

One of the unique characteristics of electroless nickel deposition is the superior wear resistance of the coatings. Theoretically, there is a correlation between wear resistance and hardness of a surface. However, the wear properties of a surface are affected by numerous other parameters such as the nature of the applied stress and the surface morphology. The wear resistance of electroless nickel deposits depends on both phosphorus content and the type of post-heat treatment applied. In general, heat treatment increases hardness and hence the wear resistance of the coating but the grain coarsening at higher heat-treatment temperatures negatively affect the wear resistance of the coating. The wear characteristics of as-deposited and heat-treated electroless NiP coating are compared (Balaraju and Seshadri 1999). The coating containing 10%–12% phosphorus and the heat treatment is carried out for 1 h at a constant temperature of 400°C. Wear test is carried out with a disc-on-disc configuration where the counter disc was high carbon, high aluminum steel having a hardness of 60 HRC. The test is carried out with a constant speed of 1000 rpm and loads were varied within the range of 20–60 N. It is revealed that heat-treated NiP samples have low specific wear rate compared to as-deposited coatings. This is believed to be due to very low mutual solubility of nickel phosphide and iron, thus presenting a relatively incompatible surface.

The wear mechanism in electroless nickel coating is primarily found to be either adhesive or abrasive or a combination of both. In general, heat treatment increases the hardness of electroless nickel coating and hence its wear resistance. The optimal heat-treatment temperature for maximum hardness is found to lie around 350°C–400°C. However, at higher heat-treatment temperature, grain coarsening occurs, resulting in softening of the coating and hence declined wear performance. The lowest coating wear rate is achieved for samples annealed at approximately 400°C for 1 h. This can be attributed to the crystallization of nickel and precipitation of phosphides. Further increase of annealing temperature leads to a lower coating wear resistance, which is due to the fact that the nickel and phosphide crystallites progressively coarsen with increasing annealing temperature and time. A similar result was obtained previously by Alirezaei et al. (2007) as shown in Figure 3.32. Figure 3.33 lists the pin-on-disc results of electroless nickel coatings at various heat-treatment conditions. Table 3.16 compares the abrasion resistance of NiP coatings. Laser-irradiated coatings showed lower wear amount than that of as-deposited amorphous alloy and furnace-annealed coatings (Tsujikawa et al. 2005). The wear volume of laser-irradiated electroless nickel samples was approximately one-sixth of that of furnace-treated sample. This splendid wear resistance exhibited by laser irradiation may be attributed to the fine dispersion of Ni_3P compound by rapid heating and cooling phenomena.

Researchers have in general found that annealing at temperatures higher than 400°C adversely affects the wear performance of electroless nickel coatings. Novák et al. (2010) analyzed the wear tracks (Figure 3.34a) and found that abrasion is the

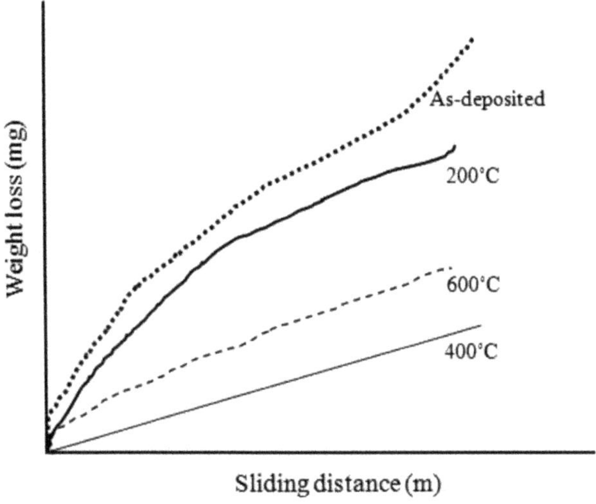

FIGURE 3.32 Variation of wear of NiP heat-treated at various temperatures. (From Alirezaei, S. et al., *Wear*, 262, 8, 2007.)

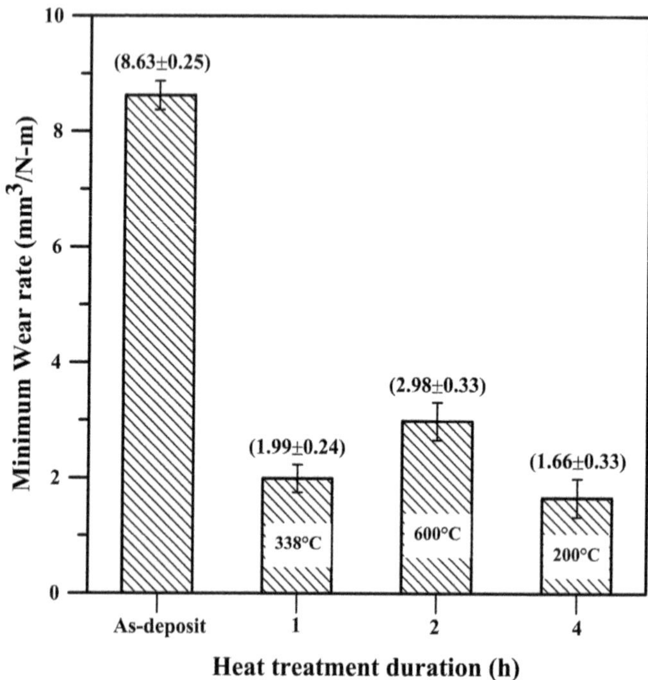

FIGURE 3.33 Minimum wear rate values of electroless nickel coatings (11 wt% P) for various heat treatment temperature and durations (sliding speed, −0.157 m/s; sliding distance, 94.2 m). (From Biswas, A. et al., *Surf. Coat. Technol.*, 328, 102–114, 2017.)

TABLE 3.16
Comparison of Abrasion Resistance of the Electroless Nickel Coatings (5–6 wt% P)

Types of Coatings	Heat Treatment for 1 h	Taber Wear Index (mg/1000) Cycles
Watts nickel	None	25
Electroless NiP	None	17
Electroless NiP	300°C	10
Electroless NiP	400°C	3
Electroless NiP	500°C	6
Electroless NiP	600°C	9
Hard chromium	None	2

major wear mechanism for as-deposited coatings on Al-Si alloy. However, due to the formed intermetallic sublayers, partial coating delamination may occur during tribological testing on the coatings annealed at 450°C and above. Thus, adhesive mechanism is the more dominant form of wear in case of heat-treated electroless nickel coatings (Figure 3.34b).

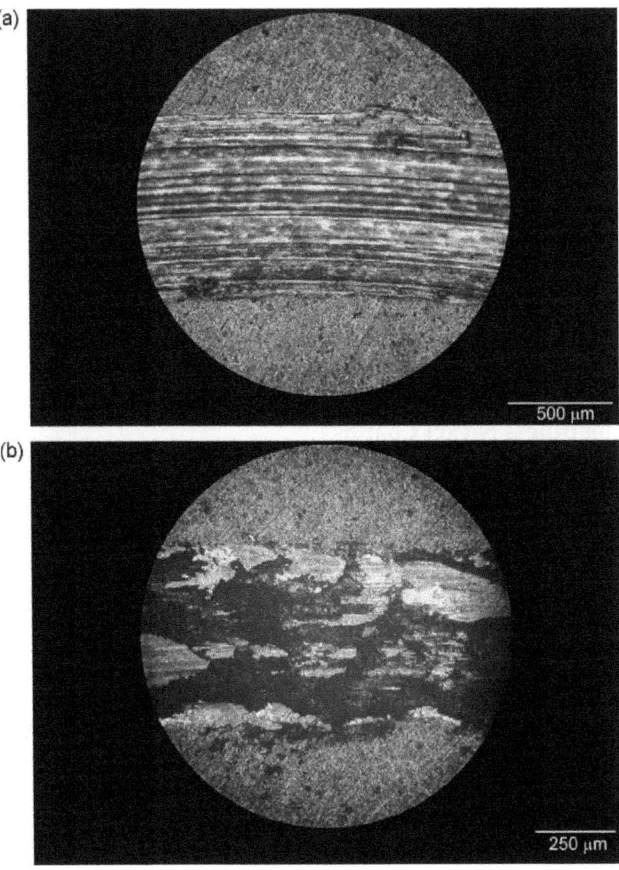

FIGURE 3.34 Wear tracks observed after wear tests on electroless nickel coatings: (a) as-deposited sample and (b) heat-treated sample at 450°C/8 h. (From Novák, M. et al., *Appl. Surf. Sci.*, 256, 2956–2960, 2010.)

Some enthusiasts have also studied the impact of various tribological testing conditions on the wear behavior of electroless nickel coatings. It is found that the rate of wear is strongly dependent on the applied load and it increases along with increasing load (Krishnaveni et al. 2005) irrespective of whether the specimen is NiP-coated or NiP-coated and tempered. But then, load is found to have stronger impact on the wear rate for NiP coating, as Li et al. (2008) have observed that an increase of load from 1 to 1.5 N magnifies the wear rate by almost 10 times for as-deposited NiP, whereas the same magnifies the wear rate by only 1.3 times in case of tempered NiP coatings. The optimization of the coating parameters is also optimized in a study (Sahoo 2009), which reported that annealing temperature and bath temperature have the most significant influence in controlling wear characteristics of electroless NiP coating. Wear is measured in terms of weight loss or mass loss in most of the papers. Some have used depth of wear to represent the magnitude of wear encountered. For their quality resistance against wear, electroless nickel alloy and composite coatings were checked for their potential application as lubricious and anti-wear coatings in small arms weapon actions (Shaffer 2008).

The friction and wear of electroless nickel coatings were investigated under lubricated conditions, particularly with bio-lubricants. Stribeck curves were used to compare the performance of different lubricants. The influence of the different lubricants under study on the wear amount was investigated by measuring the wear scar on the end of each test. The mineral oil was the best lubricant tested with the smallest specific wear rate. It is observed that in boundary lubrication conditions, the wear of the NiP coating occurs by mild abrasion with typical parallel wear scars along the direction of the moving body (Ramalho and Miranda 2007).

3.4.4.4 High-Temperature Friction and Wear Characteristics of Electroless Nickel Coatings

Even though a material may show stupendous mechanical properties in ambient conditions, the same may not be remarkable at high temperature. At elevated operating temperatures, erosion, abrasion and impact are the dominant wear mechanisms hampering the life of costly machine components such as crushers, hammer bars or cutting edges. Hence, the thermal sustainability of a particular material is very important. It is important to know whether the material remains stable and has consistent and reliable performance characteristics under high-temperature conditions and which is also not too much degraded compared to ambient temperature performance. Figure 3.35 shows a high-temperature pin-on-disc setup. Compared to a regular pin-on-disc tribo-tester, this setup has the facility to heat the pin and the counterface whose temperature can be controlled. To minimize heat dissipation, an insulated chamber is provided, which covers the test area. Friction and wear are measured using the usual sensors, which can be found in any pin-disc setup.

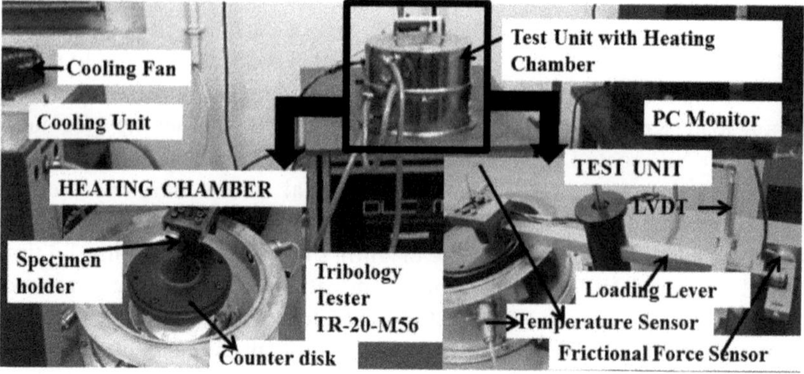

FIGURE 3.35 Pictorial view of wear and friction setup with test unit and heating chamber (inset). (From Kundu, S. et al., *Silicon*, 10, 329–342, 2018.)

Electroless nickel deposits being a category of surface finish coatings have also been investigated for its high-temperature performance, especially its friction and wear behavior. Some key features observed in the case of electroless nickel coatings under high-temperature tests are:

1. Formation of oxide films that are many times responsible for lowering the friction of the coating.
2. Phase transformation sometimes occurs even when the service temperature is below the PTT, which may be due to the flash temperature generated at the contact interface due to sliding.
3. Under high-temperature field, softening of the material occurs, resulting in the deterioration in the wear resistant activity.
4. As-deposited coatings experience an increase in microhardness, which may be because of the partial heat treatment received due to exposure to high temperature during the test.

It is found that the COF of the as-deposited coating mostly increases with the increase in load (for all temperatures) and decreases with the increase in temperatures. The results of elevated temperature tests particularly at 500°C exhibit comparatively lower COF (Figure 3.36) than the others (Kundu et al. 2018). This may be because of the in situ heat treatment due to the exposure to the elevated test temperature, which results in improved hardness of the coating. The improvement in hardness is more for as-deposited coatings compared to heat-treated samples as evident

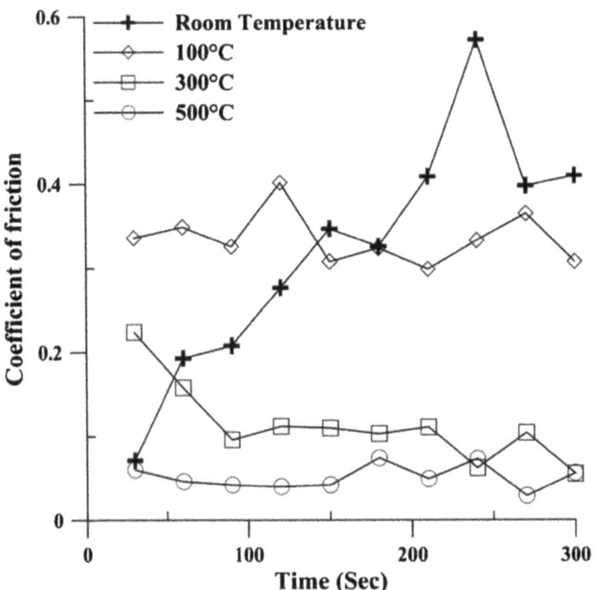

FIGURE 3.36 COF vs. time for as-deposited electroless nickel coatings. (From Kundu, S. et al., *Silicon*, 10, 329–342, 2018.)

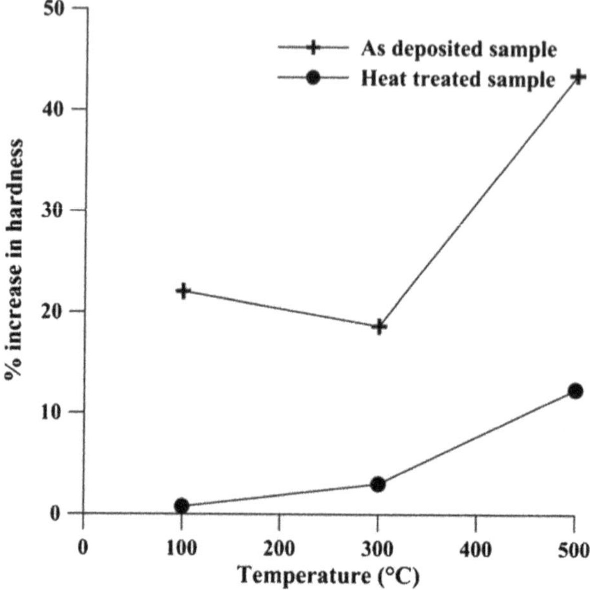

FIGURE 3.37 Percentage increase in hardness after high-temperature tests. (From Kundu, S. et al., *Silicon*, 10, 329–342, 2018.)

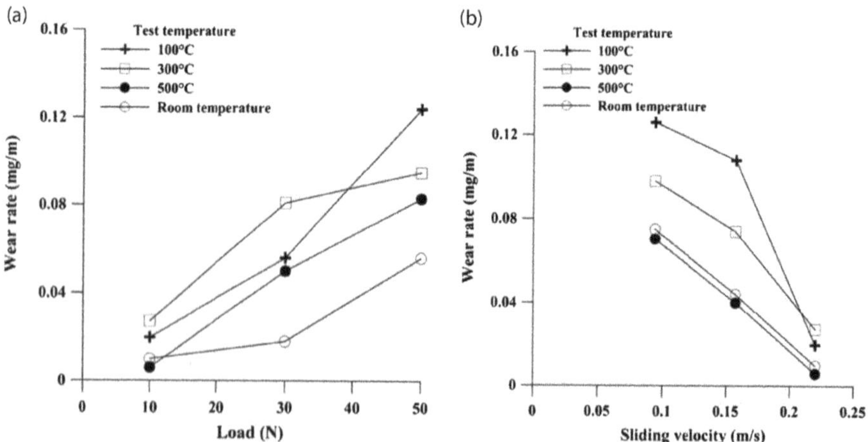

FIGURE 3.38 Wear behavior of as-deposited NiP coatings against (a) applied load and (b) sliding velocity at different test temperatures. (From Kundu, S. et al., *Silicon*, 10, 329–342, 2018.)

from Figure 3.37. It is found that for as-deposited samples, the maximum increase in hardness (>40%) is for the samples subjected to the 500°C test. A similar trend is observed for the wear results. Furthermore, for also high-temperature tests, wear is found to increase with an increase in the applied load (Figure 3.38a) and decreases with n increase in sliding velocity (Figure 3.38b).

3.4.4.4.1 Wear Mechanism of Electroless Nickel Coatings Under High Temperature

The wear mechanism of high P electroless nickel coatings (11 wt% P) depends to a large extent on the attractive force that operates between the atoms of nickel from the coating and iron from the counterface surface (hardened steel). To further understand the role of elevated temperature on the wear mechanism, the worn out surface of the samples is observed by scanning electron microscopy (SEM) (Figure 3.39). Plastic deformation is observed in almost all the samples with the appearance of

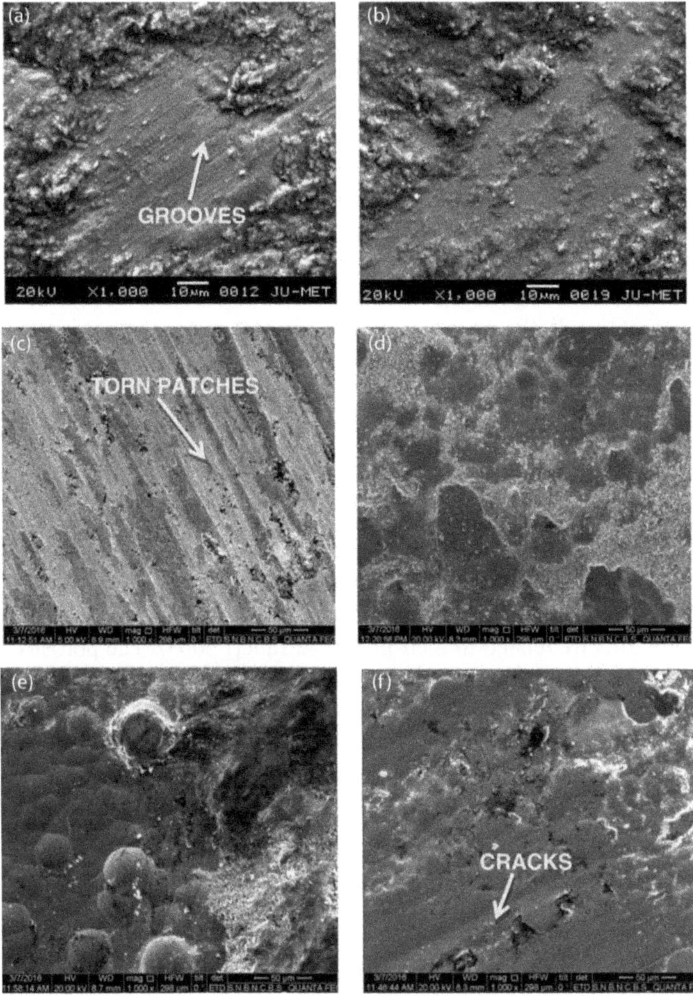

FIGURE 3.39 SEM images after wear test of electroless nickel coatings (11 wt% P): (a) as-deposited tested at 100°C, (b) heat-treated tested at 100°C, (c) as-deposited tested at 300°C, (d) heat-treated tested at 300°C, (e) as-deposited tested at 500°C and (f) heat-treated tested at 500°C. (From Kundu, S. et al., *Silicon*, 10, 329–342, 2018.)

blackish spots especially in samples tested at 300°C and beyond. These black areas may be due to the formation of oxides at high temperature as the test enclosure contains air. Compositional analysis of the tested samples also detects elemental oxygen, which is again an indication of the formation of oxides. Hence, oxidation, a major phenomenon encountered during high-temperature tests.

As-deposited electroless nickel coatings exhibit clear wear tracks especially at lower testing temperatures, which are indicative of the ploughing effect by the counterface and by the wear debris formed, which are strain-hardened. The ploughing effect is the representative of the abrasive wear phenomenon. However, it is not the sole phenomenon governing the wear mechanism of the coatings. The SEM images of the as-deposited coatings tested at 300°C (Figure 3.39c) clearly show the presence of torn patches and in some places even detachment of the coating. The presence of micro-cracks is observed on the worn surface (Figure 3.39f) as a result of the shear stress transmitted to the subsurface layer and the variation of the load experienced at the area of contact (León et al. 2003). This type of morphological feature, commonly known as "prows," is reported for adhesive wear failure of electroless nickel coatings. All this evidence points toward the occurrence of adhesive wear phenomenon in the coatings. The compositional analysis of the wear-tested specimen also shows the presence of iron (Fe) peaks that have definitely come from the steel counterface. This indicates the high mutual solubility of nickel and iron atoms. Overall, the wear mechanism encountered in the test is a mixture of adhesive and abrasive wear phenomena. However, the closer examination of SEM pictures shows that as-deposited coatings exhibit dominant abrasive wear due to the clear presence of wear tracks, whereas heat-treated coatings exhibit a dominant adhesive wear mechanism.

3.4.5 Corrosion Characteristics

3.4.5.1 Porosity

Many of the electroless nickel coatings properties, such as adhesion and corrosion resistance, are directly related to their porosity (Taheri 2003). The porosity of electroless nickel coating is related to many parameters. These include substrate surface characteristics such as roughness and morphology, substrate pretreatment, coating thickness, agitation and filtration. There are many methods for measuring and evaluating the porosity of a coated surface. The following are some of the common porosity measurement methods (Taheri 2003):

- Ferroxy test,
- Neutral salt spray test,
- Acetic acid spray test,
- Copper acceleration acetic acid salt spray test, and
- Saline droplets corrosion test.

Figure 3.40 shows different types of pores that may be created during the electroless nickel process (Leisner and Benzon 1997). The porosity of electroless nickel-coated surfaces is an important criterion defining the quality of the coating, especially when

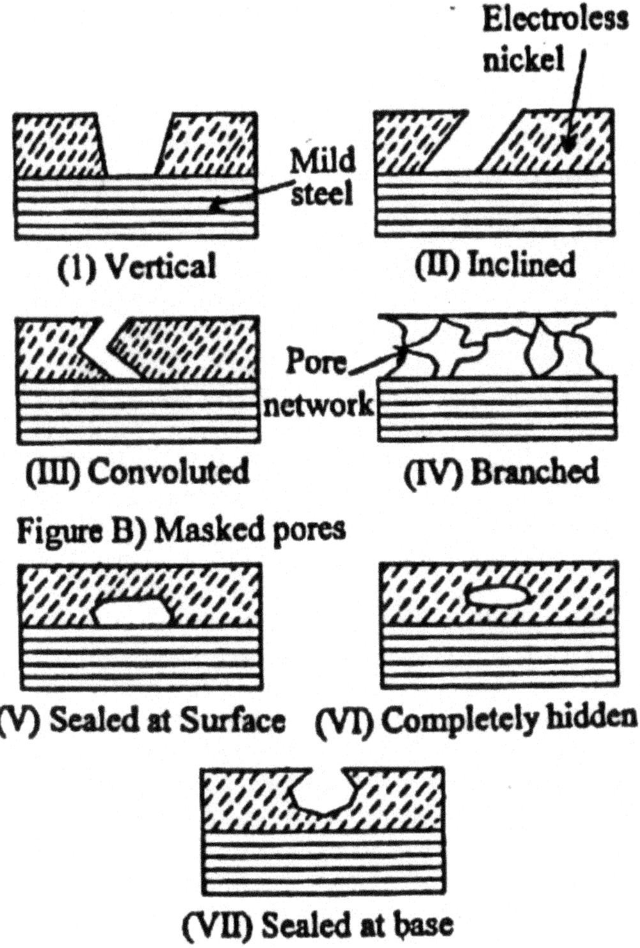

FIGURE 3.40 Different types of pores in electroless nickel deposit. (From Leisner, P. and Benzon, M.E., *T. I. Met. Finish.*, 75, 5, 1997.)

the thickness of the coating is below 10 μm. Some studies show that the open pores on electroless nickel-coated surfaces participate in a galvanic reaction in which the coating surface is the cathode and the small pore area acts as the anode. As a result of the severe galvanic corrosion, the uncoated area of the pore corrodes drastically. On the other hand, based on some other studies conducted in the presence of carbon dioxide and hydrogen sulfide, the existence of man-made pits did not cause any acceleration of the corrosion rate. As the electroless nickel deposition progresses, the porosity of the coating is reduced. This means that the pores with less depth are filled with the NiP alloy. Therefore, in many cases, a minimum thickness of 25 μm is recommended for obtaining a coating with the least amount of pores and the highest corrosion and wear resistance (Deng and Hong 1993).

3.4.5.2 Corrosion Behavior of Electroless Nickel Coating

Electroless NiP coatings are also widely used for corrosion protection application in a variety of environments. They act as barrier coatings, protecting the substrate by sealing it off from the corrosive environments, rather than by sacrificial action. The electroless nickel coating shows superior corrosion resistance compared with electroplated nickel. In this respect, it can be said that the phosphorus content governs the corrosion resistance of the coating. The general consensus is that Ni-high P coating is effective in offering excellent protection, whereas electroless Ni-low P and Ni-medium P coatings are not recommended for severe environments. One of the reasons for this observation is that Ni-high P coating is amorphous in nature without any grain boundaries, which is normally the entry point of the corrosive media. However, the corrosion behavior of electroless nickel coatings is complex and governed by several other factors, namely coating thickness, porosity, type of heat treatment and induced tensile stress of substrates caused by machining or surface-finishing operations (Taheri 2003).

In general, the corrosion resistance of any alloy depends on the ability to form a surface protective film. Phosphorus can make the corrosion potential increase and the corrosion current decrease, and it promotes the anodic and cathodic reactions during the corrosion process, thereby increasing the anodic dissolution of nickel. Accelerated corrosion of nickel provides prerequisites for concentrating P and thereby for the formation of Ni_3P and Ni_xP_y stable intermediate compounds on the surface, which acts as barrier passive film. It is evident from the literature reports on NiP coatings that preferential dissolution of nickel occurs, leading to the enrichment of phosphorus on the surface layer. This enriched phosphorus reacts with water to form a layer of adsorbed hypophosphite anions ($H_2PO_2^-$). This layer in turn will block the supply of water to the electrode surface, thereby preventing the hydration of nickel, which is considered to be the first step to form either soluble Ni^{2+} species or a passive nickel film. Hence, the better corrosion resistance obtained for electroless NiP and poly-alloy coatings is due to the enrichment of phosphorus on the electrode surface (Balaraju et al. 2006c, Sankara Narayanan et al. 2006).

Figure 3.41 illustrates the effect of phosphorus content on the corrosion resistance of an electroless nickel deposit in 10% HCl. The highest corrosion rate is obtained when the microstructure consists of both the β and γ phases. Similar results have been obtained by measuring the time to failure in nitric acid.

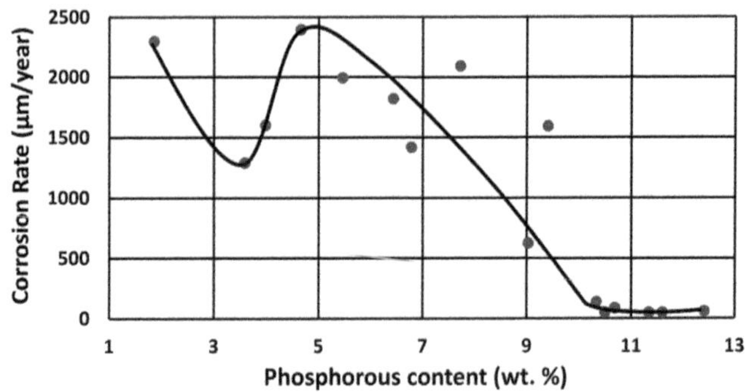

FIGURE 3.41 Effect of phosphorus content on corrosion rate of electroless nickel deposition in 10% HCl. (Duncan 1996a).

Heat treatment is found to decrease the corrosion resistance of electroless coatings invariably. This is attributed to the change of microstructure of the coatings with heat treatment. Electroless coatings in as-deposited condition generally exhibit an amorphous structure that imparts higher corrosion resistance. However, heat treatment induces crystallinity into the deposits, which in turn increases the grain boundaries, that form active sites for corrosion attack (Srinivasan et al. 2010). This is quite evident from Figure 3.42, which shows the effect of phosphorus content on the corrosion potential of the powder metallurgy specimens in 1.0 M HCl solution. A few confirmations can be made from the figure. First, reducing the porosity of the coating results in lower corrosion potentials. Second, heat treatment has a deleterious effect, decreasing the corrosion resistance of the coating. In addition, the corrosion resistance of electroless nickel deposits is directly related to their thickness. It was explained previously that the porosity of an electroless nickel coating is a function of many parameters of which thickness is one. As the coating thickness increases, the density of pores also increases resulting in reduced corrosion resistance.

Figure 3.43 shows the effect of coating thickness on the corrosion resistance of electroless nickel deposits in 0.5 M H_2SO_4. Figure 2.45 shows that the presence of 3 μm electroless nickel coating reduces the I_{corr} = 243 μAcm^{-2} for pure iron to 73 μAcm^{-2}. Furthermore, the I_{corr} drops to 40 μAcm^{-2} when the electroless nickel coating thickness exceeds 5 μm.

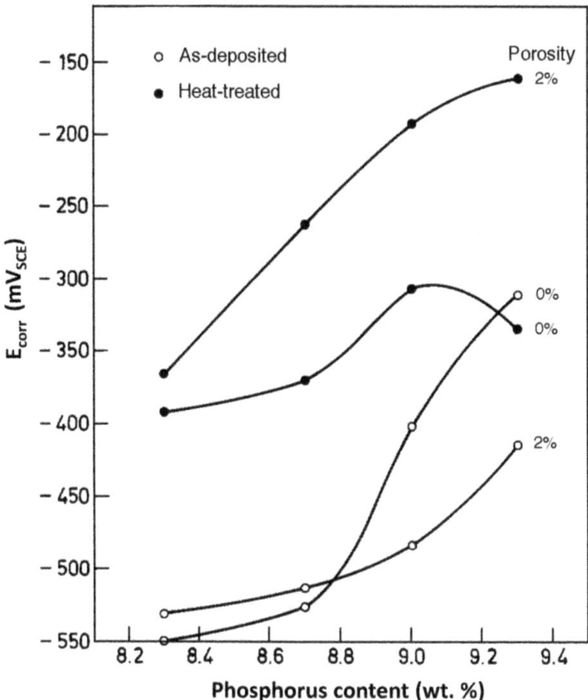

FIGURE 3.42 Effect of phosphorus content on the corrosion potential. (From Singh, D. et al., *Corrosion*, 51, 581–585, 1995.)

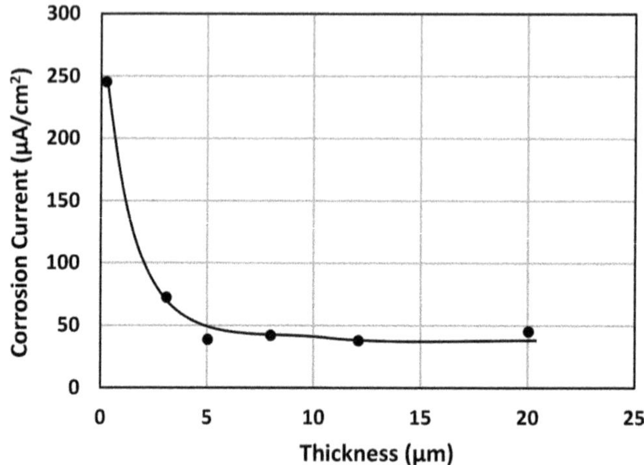

FIGURE 3.43 Dependence of corrosion current on coating thickness (9 wt% P). (From Doong, J.C. et al., *Surf. Coat. Technol.*, 27, 1986.)

Although electroless nickel coatings with high phosphorus show better corrosion resistance than those with low phosphorus content due to their amorphous microstructure. However, in hot alkaline environments, electroless nickel coatings show a reverse corrosion behavior due to the formation of passive layers on the coated surface (Taheri 2003). The top passive layers on low- and medium-phosphorus electroless nickel coatings are NiO and NiO_2. The multilayered passive film shows good stability that leads to high corrosion resistance.

Table 3.17 shows the corrosion rates calculated for different types of electroless nickel coating. As shown in Table 3.18, there is a direct correlation between the coating thickness and its corrosion resistance. The thicker the coating is, the higher the E_{corr} would be. However, based on this study, there is a significant increase in the corrosion resistance of high-phosphorus electroless nickel coating when the thickness exceeds 18 μm.

TABLE 3.17
Electrochemical Results of Electroless Nickel Coatings

System Studied	E_{corr} (mV) vs. Saturated Calomel Electrode (SCE)	i_{corr} (μA/cm²)	R_{ct} (kΩ cm²)	C_{dl} (μF)
EL Ni-Low P coating (L)	−536	4.22	6.90	289
EL Ni-Medium P coating (M)	−434	1.17	24.86	55.60
EL Ni-High P coating (H)	−411	0.60	37.45	49.10
EL NiP-graded coating (LMH)	−403	0.41	38.22	36.70
EL NiP-graded coating (HML)	−481	1.70	14.77	103

Source: Sankara Narayanan, T.S.N. et al., *Surf. Coat. Technol.*, 200, 3438–3445, 2006.
L: Low; M: Medium; H: High.

TABLE 3.18
Tafel Analysis of Electroless Nickel Deposits of Various Thickness (Approximately 4 wt%P)

Deposit Thickness (μm)	Corrosion Potential E_{corr} vs. SCE (v)	Corrosion Current i_{corr} (μA cm^{-2})
0	−0.539	61
1	−0.447	30
3	−0.425	30
6	−0.423	30
12	−0.388	9
18	−0.230	5
24	−0.205	7

Source: Kerr, C. et al., Physical and electrochemical characteristics of electroless nickel on carbon steel, *International Conference on Computer Methods and Experimental Measurements for Surface Treatment Effects*, 1997.

The potentiality of high-phosphorus (12%–14%) NiP coating in marine applications has been investigated by Gao et al. (2007). The corrosion investigations were conducted by immersing the samples in 3.5% NaCl solutions and by standard salt spray test. The tests were conducted for a variety of time periods ranging from 0.5 h to 29 days with intermediate periods of 2, 6 and 13 days. At the beginning of immersion, a passivation film was found to form on the deposit, but it was not integrated. The passivation phenomenon became increasingly obvious as the immersion time increased. The two time constants obtained from the electrochemical impedance spectroscopy (EIS) spectra also confirmed the formation of the passivation film. The scanning electron micrographs showed that the prepared deposits were amorphous. However, after a 15 days standard salt spray test, a few pinholes appeared on the deposit (Figure 3.44) but the corrosion is found to be uniform. However, the weight content of phosphorus on the surface of the deposit was found to be higher, which was beneficial to the formation of the passivation film, than it was before the standard salt spray test, while the nickel content was lower because the dissolved weight of nickel was greater than that of phosphorus. The results from potentiodynamic scan and EIS test showed that passivation film formed on the NiP deposit after immersion in 3.5% NaCl solutions and decreased the corrosion rate of NiP samples.

Balaraju et al. (2006c) performed a special study of the corrosion behavior of electroless NiP coatings in non-deaerated and deaerated conditions, and it was observed that no passivation phenomenon occurs in non-deaerated conditions. Corrosion of NiP coatings was also studied in special environments such as in the presence of carbon dioxide and it minimized the corrosion of N80 steel (Ye et al. 2005). Electroless nickel coating has also been used to protect steel reinforcements embedded in concrete specimen exposed to marine environment against corrosion (Lee et al. 2009).

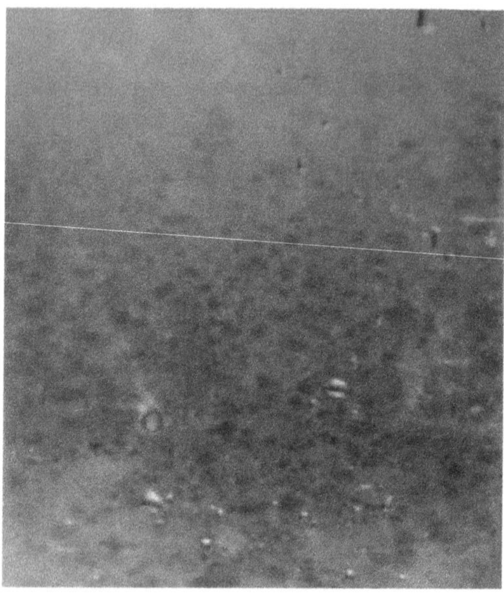

FIGURE 3.44 Image of electroless nickel coating after salt spray test. (From Gao, R. et al., *J. Ocean U China.*, 6, 349–354, 2007.)

3.4.6 Corrosive Wear Behavior of Electroless Nickel

Corrosion and wear are largely found to be interdependent in many applications. Many modern types of machinery are subjected to extreme corrosive environments, namely marine applications, chemical plants and so on. Due to the dynamics involved, many components of these machineries would require protection against both wear and corrosion simultaneously, as they undergo tribo-chemical interactions. Electroless nickel coatings are found to possess resistance against combined corrosion and wear. These coatings are favorably applied to the surface of glass fiber-reinforced plastic (GFRP) substrate for making blades of offshore wind turbine installations (Lee 2008). The deposition of NiP coatings on a carbon fiber-reinforced plastic (CFRP) substrate also increases its wear-corrosion resistance. This wear-corrosion resistance increased as coating thickness and phosphorus content increased and the polishing condition decreased (Lee 2009).

3.4.7 Internal Stress

Electroless nickel deposits contain internal stresses. These stresses are caused by atomic and crystallographic defects such as dislocations (Taheri 2003). There are numerous theories offered to explain the internal stresses in electroless nickel deposits. The mechanical and corrosion resistance properties of electroless nickel coatings are correlated with the state of the internal stress, which has been studied by many researchers. Although the results obtained are in conflict in some cases, they all agree on the significant effect of internal stress on the mechanical properties of electroless nickel deposits. Internal stresses in electroless nickel plating primarily depend on the coating composition, especially the amount of phosphorus in the deposits. Neutral, compressive stresses are developed when steel is coated more than 10 wt% P (Krishnan et al. 2006). At the same time, phosphorus content below this value produces tensile stresses due to the difference in thermal expansion between the deposits and substrate. The high levels of stresses in these coatings promote cracking and porosity. Structural changes during heat treatment at temperatures above 220°C cause volumetric shrinkage of electroless nickel deposits 4–6 wt%. This volumetric shrinkage increases tensile stresses and reduces compressive stresses.

Duncan (1981) has also correlated the effect of phosphorus content on internal stress. It is shown that as the phosphorus content increases, the nature of the internal stress is changed from tensile to compressive. At approximately 11 wt% phosphorus, the coating obtained seems to be free from stress. Additionally, in a recent work, Duncan has suggested a regime for the internal stress in the electroless nickel deposit (Duncan 1996a). According to this proposed theory, the existing internal stress is directly related to the microstructure. As explained earlier, at the phosphorus contents between 4 and 7.5 wt%, the microstructure of an electroless nickel deposit consists of two phases, β and γ. Due to the existence of these two phases, the microstructure is in its most incoherent state and, therefore, very brittle. Thus, the existing internal stress has its maximum tensile value. Below 4 or above 11 wt% phosphorus, the microstructure is either β or γ. Either of these two microstructures is more coherent than the microstructure containing both β and γ, Figure 3.45. As a result, the internal stress of either β or γ when present alone is compressive. Duncan's new finding (Duncan 1996a) consistently explains many of the characteristics of

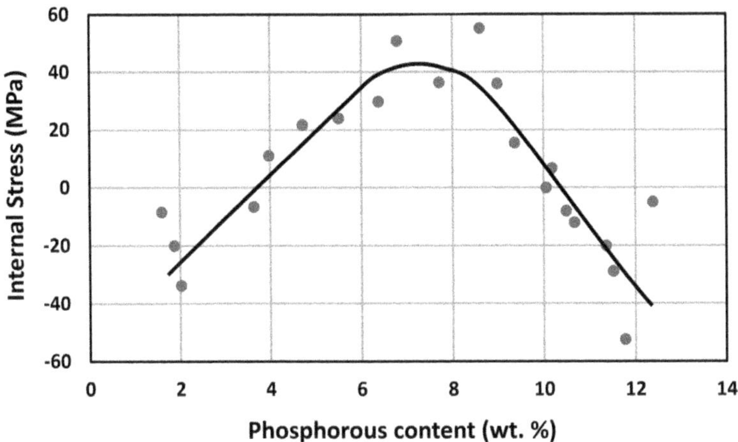

FIGURE 3.45 Dependence of internal stress of electroless nickel deposit on phosphorus content (Duncan 1996a).

electroless nickel coatings, such as wear resistance, tensile strength and ductility. The internal stress of the electroless nickel deposit is also a function of the bath age (Duncan 1996b). Many studies have shown that as the electroless nickel bath gets older, the internal stress tends to become more tensile (Riedel 1997), explaining the fact that the solution needs to be discarded after 5 to 6 turnovers. Riedel reported that after the fifth turnover, the nature of the internal stress significantly changes from compressive to tensile. Based on this, the acceptable life of commercial electroless nickel baths has been limited to previously stated six turnovers.

3.5 SOLDERABILITY OF ELECTROLESS NICKEL

Electroless nickel coatings are solderable but generally under a strong acid flux. Now, such strong flux may be detrimental to other components in the assembly particularly when dealing with electronic devices. Now, nickel and electroless nickel are good conductors compared with thick films used to metalize ceramic devices. Krishnan et al. (2006) have also reported the use of electroless nickel coatings to facilitate soldering of light metals, namely aluminum.

The other advantages of using electroless nickel coatings include the formation of intermetallic compounds that provide strong adhesion. Compounds such as Ni_3Sn, Ni_3Sn_2 and Ni_3Sn_4 are formed, which provide excellent adhesion. In addition, nickel-gold, nickel-silver and nickel-copper compounds can form, although slowly. However, phosphorus and oxides at the surface of electroless nickel deposit interfere with adhesion. To form a good bond to electroless nickel, oxides and surface phosphorus must be removed.

3.5.1 Solder Fluxes

Fluxes are used to remove surface oxides and sulfides, reduce the surface tension of molten solder and prevent oxidation during the heating cycle. There are many solder fluxes available, and they are classified by the degree of activity. Inorganic fluxes are the strongest where the resin-based fluxes are the weakest. For most of the components, rosin mildly activated flux is specified along with conventions Sn-Pb solder. Soluble organic acids and inorganic acids are also used as flux. Preheating the component to 100°C–110°C improves the ease and speed of joints with moderately oxidized surfaces, such as those resulting from steam aging, activated rosin flux or organic acid (which are usually required to obtain wetting of coatings).

3.6 GRADED AND DUPLEX COATINGS

A graded coating composition or structure improves the load carrying capacity by offering smoother transitions in mechanical properties from those of the hard and stiff coating to those of the softer and more flexible substrate. In this way, the contact load can be distributed over larger areas, which reduces the maximum contact stresses and the stress at the coating/substrate interface.

Graded coatings have also been developed for NiP coatings that offer better corrosion resistance than the nongraded coatings ones. The grading in NiP coatings is developed based on the percentage composition of nickel and phosphorus.

Various electroless nickel coatings are available, each having its specific property. Electroless NiP coatings are known specifically for their corrosion-resistant properties, whereas NiB coatings being hard are known for their wear-resistant properties. The complementarity of the properties, with NiP having higher corrosion resistance and NiB having better mechanical behavior, has led to interest in duplex and multilayer coatings (Vitry et al. 2018). A detailed discussion regarding this will be provided in Chapter 7 of this book.

3.7 STANDARDS PERTAINING TO DEVELOPMENT OF ELECTROLESS NICKEL COATINGS

As electroless nickel coatings are getting accepted and started to be applied increasingly, various standards have been framed. Electroless nickel coatings deposited by complying these standards are expected to be of high quality and hence higher reliability. Some discussions of the popular standards for electroless nickel development are as follows.

3.7.1 ASTM B733-15: Standard Specification for Autocatalytic (Electroless) Nickel-Phosphorus Coatings on Metal

This specification establishes the requirements for autocatalytic (electroless) NiP coatings applied from acidic aqueous solutions to metallic products for use in engineering functions operating at elevated temperatures. The coatings covered here are alloys of nickel and phosphorus produced by self-sustaining autocatalytic chemical reduction with hypophosphite. The coatings are grouped into the following classification systems: types, which are based on the general composition with respect to phosphorus; service condition numbers, which are based on the severity of exposure to which the coating is intended to perform and the corresponding minimum thickness that will provide satisfactory performance; and post-heat treatment class, which are based on post-plating heat-treatment temperature and time to produce the desired adhesion and hardness improvements. Prior to plating, substrates should be pretreated by stress relief for reducing risks of hydrogen embrittlement, peening and racking. The coatings shall be sampled and tested accordingly to evaluate both acceptance (appearance, thickness, adhesion and porosity) and qualification requirements (composition, microhardness and hydrogen embrittlement). Thickness shall be assessed either by microscopical method, a magnetic induction instrument, beta backscatter method, a micrometer, weigh-plate-weigh method, coulometric method or X-ray spectrometry. Adhesion shall be examined either by bend, impact or thermal shock tests. And porosity shall be inspected either by ferroxyl test, boiling water test, aerated water test or alizarin test.

The summary of the standard is given as follows (ASTM 2015):

1. This specification covers requirements for autocatalytic (electroless) NiP coatings applied from aqueous solutions to metallic products for engineering (functional) uses.
2. The coatings are alloys of nickel and phosphorus produced by autocatalytic chemical reduction with hypophosphite. Because the deposited nickel alloy is a catalyst for the reaction, the process is self-sustaining. The chemical and physical properties of the deposit vary primarily with its phosphorus content and subsequent heat treatment. The chemical makeup of the plating solution and the use of the solution can affect the porosity and corrosion resistance of the deposit.
3. The coatings are generally deposited from acidic solutions operating at elevated temperatures.
4. The process produces coatings of uniform thickness on irregularly shaped parts, provided the plating solution circulates freely over their surfaces.
5. The coatings have multifunctional properties, such as hardness, heat hardenability, abrasion, wear and corrosion resistance, magnetics, electrical conductivity to provide diffusion barrier and solderability. They are also used for the salvage of worn or mis-machined parts.
6. The low-phosphorus (2%–4% P) coatings are microcrystalline and possess high as-plated hardness (620–750 HK_{100}). These coatings are used in applications requiring abrasion and wear resistance.
7. Lower phosphorus deposits in the range between 1% and 3% phosphorus are also microcrystalline. These coatings are used in electronic applications providing solderability, bondability, increased electrical conductivity and resistance to strong alkali solutions.
8. The medium-phosphorus coatings (5%–9% P) are most widely used to meet the general purpose requirements of wear and corrosion resistance.
9. The high-phosphorus (more than 10% P) coatings have superior salt-spray and acid resistance in a wide range of applications. They are used on beryllium and titanium parts for low-stress properties. Coatings with phosphorus contents greater than 11.2% P are not considered to be ferromagnetic.
10. The values stated in SI units are to be regarded as standard.
11. The following precautionary statement pertains only to the test method portion: "This standard does not purport to address all of the safety concerns, if any, associated with its use. It is the responsibility of the user of this standard to establish appropriate safety and health practices and determine the applicability of regulatory limitations prior to use."

The standard categorizes electroless nickel coating based on phosphorus content as given in Table 3.19. Moreover, the standard also gives minimum thickness of coating based on service condition (shown in Table 3.20).

TABLE 3.19
Phosphorus-Based Categorization of Electroless Nickel Coatings

Deposit Alloy Types	Phosphorus Content (wt%)
Type I	No requirement of phosphorus
Type II	1%–3% (low phosphorus)
Type III	2%–4% (low phosphorus)
Type IV	5%–9% (medium phosphorus)
Type V	10% or more (high phosphorus)

Source: ASTM, Standard Specification for Autocatalytic (Electroless) Nickel-Phosphorus Coatings on Metal, ASTM B733-15, West Conshohocken, ASTM International, 2015.

TABLE 3.20
Minimum Coating Thickness of Electroless Nickel Coating Based on Service Condition

Service Condition	Minimum Coating Thickness Specification	Minimum Thickness (μm)
SC0	Minimum thickness	0.1
SC1	Light service	5.0
SC2	Mild service	13
SC3	Moderate service	25
SC4	Severe service	75

Source: ASTM, Standard Specification for Autocatalytic (Electroless) Nickel-Phosphorus Coatings on Metal, ASTM B733-15, West Conshohocken, ASTM International, 2015.

ASTM B733-15 also defines the heat treatment conditions and divides the coating into 6 classes where the lower class is associated with the most weak coating and higher class with strong coatings. The categorization is provided in Table 3.21.

3.7.2 Electroless Nickel Plating MIL-C-26074

This specification covers the requirements for electroless deposition of nickel or nickel coatings on metal surfaces. The metals specified are iron, copper, aluminum, nickel, cobalt, beryllium and titanium-based alloys. According to the standard, the nickel coatings shall be classified in accordance with the thermal treatment applied subsequent to plating as given in Table 3.22 (MIL-C-26074 1998):

According to the standard, the coatings are divided into three grades as follows:

- Grade A—0.0010-inch minimum deposit thickness,
- Grade B—0.0005-inch minimum deposit thickness, and
- Grade C—0.0015-inch minimum deposit thickness.

TABLE 3.21
Heat Treatment Conditions Associated with Class of Coating

Class of Coating	Heat Treatment Conditions
Class 1	As-deposited, no heat treatment
Class 2	Heat treatment at 260°C–400°C (500°F–752°F) to produce a minimum hardness of 850 HK_{100}
Class 3	Heat treatment at 180°C–200°C (356°F–392°F) for 2–4-h to improve coating adhesion on steel and to provide for hydrogen embrittlement relief
Class 4	Heat treatment at 120°C–130°C (248°F–266°F) for at least 1 h to increase adhesion of heat-treatable (age-hardened) aluminum alloys (such as 7075) and carburized steel
Class 5	Heat treatment at 140°C–150°C (284°F–302°F) for at least 1 h to improve coating adhesion for aluminum, non-age-hardened aluminum alloys, copper, copper alloys and beryllium
Class 6	Heat treatment at 300°C–320°C (572°F–608°F) for at least 1 h to improve coating adhesion for titanium alloys

Source: ASTM, Standard Specification for Autocatalytic (Electroless) Nickel-Phosphorus Coatings on Metal, ASTM B733-15, West Conshohocken, ASTM International, 2015.

TABLE 3.22
Classification of Coating According to Thermal Requirement

Class	Thermal Requirement
Class 1	As-plated, no subsequent heat treatment (a bake for hydrogen embrittlement relief is not considered for heat treatment).
Class 2	Heat-treated to obtain required hardness. May be used on all metals not affected by heating to 500°F and above. Required hardness can be obtained per the following schedule: Temperature, F (time, h) 500 F (4 h or more) 550 F (2 h or more) 650 F (1–1.5 h) 750 F (0.5–1 h)
Class 3	Aluminum alloys, non-heat-treatable, and beryllium alloys processed to improve adhesion of the nickel deposit. Coated non-heat-treatable aluminum parts shall be heated for 1–1.5 h at 375°F±15°F to improve adhesion of nickel deposit.
Class 4	Aluminum alloys, heat-treatable, processed to improve adhesion of the nickel deposit. Coated heat-treatable aluminum alloys shall be heated between 240°F and 260°F for 1–1.5 h to improve adhesion of nickel deposit.

3.7.3 Electroless Nickel Plating AMS 2404

AMS 2404 supersedes AMS-MIL-C-26074, which uses Grade designations to convey thickness requirements. The classification-based thermal requirements are given in Table 3.23.

TABLE 3.23
Classification of Coating Based on Thermal Requirements

Class	Thermal Treatment
Class 1	Except for hydrogen embrittlement relief, no post-plating thermal treatment
Class 2	Thermal treatment at 450°F (232°C) or above to harden the deposit
Class 3	Thermal treatment at 375°F (191°C) to improve adhesion for non-heat-treatable aluminum alloys and beryllium alloys
Class 4	Thermal treatment at 250°F (121°C) to improve adhesion for heat-treatable aluminum alloys

Source: 2404J, A., Plating, Electroless Nickel, AMS, SAE International, 2018.

Some other important points (with serial) from the standard are as follows:

1.3.1 Unless a specific class is specified, Class 1 shall be supplied.
3.1.1 Stress Relief Treatment — Steel parts having a hardness of 40 HRC or higher and which have been ground after heat treatment shall be cleaned to remove surface contamination and suitable stress relieved before preparation for plating. Unless otherwise specified, the stress relief temperature shall not be less than 275°F (135°C) for not less than 5 h for parts having hardness of 55 HRC or higher or not less than 375°F (191°C) for not less than 4 h for other parts.
3.3.2.1 When Class 2 is specified, parts shall be heated to a selected temperature within the range of 450 F–800 F and held for sufficient time to increase the hardness of the deposit. See 3.4.4 and 8.6. Hydrogen embrittlement relief (3.3.1) may be omitted if Class 2 hardening is accomplished within 4 h after plating.
3.3.2.2 When Class 3 is specified, parts shall be heated to 375°F ± 15°F for 1–1.5 h.
3.3.2.3 When Class 4 is specified, parts shall be heated to 250°F ± 10°F for 1–1.5 h.
3.4.1 *Thickness*
Unless otherwise specified, the minimum thickness of the nickel coating shall be 0.0010 inch for aluminum-based alloys, 0.0005 inch for copper, nickel, cobalt, titanium and beryllium alloys and 0.0015 inch for iron-based alloys.
3.4.4 *Hardness*
Class 2 plating shall be not lower than 800 HK_{100} or equivalent determined in accordance with ASTM E384.
3.4.7 *Composition*
The cognizant engineering organization may specify a phosphorus content range of the deposit. When specified, the composition of the deposit shall be determined by a method acceptable to the cognizant engineering organization.

3.8 APPLICATIONS OF ELECTROLESS NICKEL COATINGS

Due to inherent advantages of the deposition process, namely uniform thickness, hardness and corrosion resistance, electroless nickel coatings are also used for deposition on complex shapes and large interior surfaces, for replacing expensive stainless steel vessels in some processing industries, for repairing or salvaging nickel-plated machine parts for improving the adhesion of enamels on steel and application in electronic industries as it is easily solderable.

The applications of NiP coatings may be broadly classified as:

1. For general surface protection (mainly against corrosion);
2. Achieving smooth surface and bright coatings for aesthetics;
3. From tribological aspect (achieving low friction and high wear resistance);
4. Miscellaneous applications, namely printed circuit boards.

3.8.1 Common Industrial Applications

Electroless plating can be used to produce homogeneous consistent coats for several industrial applications. The rapid rate at which deposition occurs and its capability of providing the necessary product quality in a satisfactory process period at reasonably low principal and operating overheads are parts of the significant reasons for the application of this technique. Some of the common industrial applications of electroless nickel coatings are given in Table 3.24.

TABLE 3.24
Some of the Common Industrial Applications of Electroless Nickel Coatings

S. No.	Applications	Property Under Focus
1.	Inner surfaces of pumps, driers, tubes, gasoline containers and tanks, transport cars carrying various chemicals, storage tanks, valves, screws, nut fasteners, etc.	Corrosion resistance and uniformity of deposition
2.	Cylinders for hydraulic pumps, piston rings, piston cylinders, cranks, bearing surfaces, rotating shafts, printed press parts, motor blades, etc.	Wear and abrasion resistance (dry and lubricating conditions)
3.	Aluminum and aluminum parts used in aircraft or spacecraft	Wear resistance and corrosion resistance
4.	Nuclear reactor parts, light alloy dies, radar waveguides and printed circuit boards	Wear resistance and corrosion resistance
5.	Rollers and crimpling tracks in textile industry	Uniformity, wear resistance and corrosion resistance
6.	Brake assemblies, gas turbines, heat exchangers and pump housing in oil and gas industry	Uniformity and hardness
7.	Automotive components such as carburetors, engine bearings, pistons and gears	High-temperature performance, wear resistance
8.	Oil drilling and coal mining and other applications	Corrosion and wear resistance
9.	Connectors and circuit boards	Electrical conductivity, solderability, corrosion resistance
10.	Magnetic discs/hard drive components	Nonmagnetic, wear and corrosion resistance
11.	Food and beverage handling components	Non-toxicity, corrosion resistance and resistance to bacterial growth

Source: Krishnan, K.H. et al., *Metall. Mater. Trans. A*, 37, 1917–1926, 2006.

Electroless nickel, as an engineering coating, is used in many industrial applications in aerospace automotive computers, electronics, food processing, hydraulics machinery, nuclear engineering, oil petrochemicals, plastics, power transmission, printing, pump valves, textiles and so on (Krishnan et al. 2006). A list of items that are electroless nickel-plated is available in a number of publications. Table 3.25 summarizes the industrial applications of electroless nickel plating and its thickness requirement.

TABLE 3.25
Industrial Applications of Electroless Nickel Plating Based on Thickness Requirement

Solution	Type of Work	Thickness (in inches)	Type of Work with Detailed Reason for Use
1.	Chemical process equipment	0.002–0.005	Protects filters, heat exchanger pumps, tanks and pipe fittings
2.	Hydraulic parts, oil-field drilling equipment	0.0005–0.002	Protects recessed areas such as O-ring grooves and other internal parts
3.	Molds for zinc die casting	0.002–0.003	Resists chemical corrosion and provides hardness similar to chromium
4.	Printing press bed and rolls	0.001–0.003	Offers good release characteristics
5.	Plastics extrude dies	0.0008–0.00015	Improve release characteristics and reduction in mold cleaning requirements
6.	Gyro parts	0.001–0.003	Protects against corrosion by ink
7.	Gear and gear assemblies	0.0005–0.001	Provides hardness and wear resistances as well as high corrosion resistances similar to chromium
8.	Spray nozzles	0.0004–0.001	Provides low friction and wear resistances
9.	From complex parts, usually small electrical contacts	0.0003–0.0008	Makes possible accurate control of dimensions and uniform coverage on all areas
10.	Complex stainless steel parts	0.001–0.004	Produces coating uniformity, which preserves the balance of the turbine
11.	Aluminum and high-temperature plastic "Black Boxes" and other electronic junction fittings	0.0003–0.0005	Make it possible to electroform over aluminum mandrels and dissolve out aluminum
12.	Printed circuits	0.0003–0.0005	Provides ease of solderability and allows welding of nickel alloy components to printed circuit boards
13.	Turbine parts	0.0002–0.0005	Produces coating uniformity, which preserves the balance of the turbine

Source: Krishnan, K.H. et al., *Metall. Mater. Trans. A*, 37, 1917–1926, 2006.

TABLE 3.26
Electroless NiP Usage in Industries

Industry	Estimated Usage (%)
Aerospace	9.4
Automotive	9.1
Computers	6.2
Electronics	14.2
Food	3.7
Hydraulics	2.2
Machinery	7.8
Nuclear	5.1
Oil tool	3.2
Petrochemical	2.3
Plastics	4.0
Power transmission	3.1
Printing	3.4
Pumps	4.3
Textile	4.7
Valves	13.4
Other	3.9

Electroless NiP plating has emerged from little more than a laboratory curiosity in the early 1950s to a process utilized today in approximately 1000 installations in the U.S. alone. One of the largest electroless plating facility is in France, where a 100,000 gal tank is used to coat 20-ft-long tube bundle (Shipley 2018). Table 3.26 illustrates the estimated share of the use of electroless nickel coatings for various industries

3.8.2 From Tribological Aspect (Achieving Low Friction and High Wear Resistance)

From the tribological viewpoint, electroless nickel coatings are mainly applied due to their hardness and wear resistance, lubricity (low friction coefficient) and corrosion-resistant properties. Electroless NiP coatings have been successfully applied on GFRP substrate for making blades of offshore wind turbine installations (Lee 2008). An offshore wind turbine should be designed so as to be able to withstand fatigue loads and a hostile environment at all times. Depositing NiP coatings on GFRP substrate increase its wear corrosion resistance, besides imparting conductivity to resist against damage by lightning strikes. Electroless nickel coating has also been used to improve the wear and wear corrosion properties of CFRP materials, which are frequently utilized in the aerospace, aviation and automobile industries due to their light weight (Lee 2009). The potential of NiP coatings for marine applications has been investigated by various researchers (Gao et al. 2007, Wang et al. 2009). Ultra High Molecular Weight Polyethylene (UHMWPE), an advanced engineering plastic used in making seawater hydraulic drive is found to produce the lowest friction coefficient when sliding against electroless NiP coating under seawater lubrication (Wang et al. 2009). The suitability of electroless nickel coatings (NiCoTef®) was checked for applying on to the action components of hand-held automatic weapons (i.e., those other than the barrel and stock) that are subjected to rapid repeated sliding contact and shock loading conditions. The affinity of dirt, dust and fine sand to the residual liquid lubricant films on the surface of parts can lead to malfunction and jamming of weapons when operated in desert or other dusty environments. To avoid this problem, new generations of weapons must make use of "dry" coatings, which can provide low friction between sliding components, corrosion protection, and can last many cycles. It was found that components for which the wear life is a higher priority than having the lowest COF, an electroless nickel coating is a fairly good choice.

Composite coatings of electroless NiP-PTFE were considered for improving the performances of aluminum alloys in the bearings regions of lightweight external gear pumps (Veronesi et al. 2008). The coatings exhibited satisfactory friction and wear behavior when slided against steel. However, the presence of PTFE spheres and the hardness mismatch with the substrate tends to favor tensile cracking of the substrate. Das et al. (2007) reported the application of 4 µm thick electroless NiP-Si_3N_4 coating on the surface of ferrous based bearings that can be successfully used for water-lubricated applications in pH 10 water for 9 years under the actual application environment. The excellent wear resistance plus low friction coefficient of the Ni-P-$C_{graphite}$-SiC coating is found suitable for applications in moulds, automobile parts and other fields (Wu et al. 2006b).

Electroplated hexavalent chromium coatings are applied widely due to their excellent tribological properties. However, chromium coatings being toxic, due to environmental concerns, an alternative to it is always sought. Klingenberg et al. (2005) compared the performances of both types of coatings and suggested that all of the electroless NiP, Ni-Co-P and Co-P processes with occluded diamond particles have the potential to impart the required adhesion, hardness and tribological properties, while reducing the environmental impact of chromium plating processes (Klingenberg et al. 2005). The suitability of electroless nickel coating for application in a commercial brake valve assembly was compared with that of hard chromium coatings (Deshmukh et al. 2005). It was observed that electroless coating performed similar to chromium coating, displaying moderate friction and insignificant wear over the sliding distance under examination. Thus, electroless coating could replace chromium coatings that are much toxic to the environment.

Because of high electrical and thermal conductivity properties, combined with the in situ lubricating ability of these graphite-containing composites, which mix further with ceramic SiC particles, make them potential candidates for such applications as advanced high speed–high load bearing and high speed–high current electrical brushes (Wu et al. 2006a).

Among various metallization processes, electroless metal plating is a preferred way to produce metal-coated fabrics due to the attractive advantages such as uniformity of coverage, excellent conductivity, possibility of metallizing nonconductors and flexibility. NiP plating is applied in order to render abrasion resistance to Electro Magnetic Interference (EMI) shielding fabrics (Hui and Rongli 2008). The abrasion resistance is found to increase with the incorporation of SiO_2 particles in the coating without much change in shielding effectiveness of the fabric.

3.9 FUTURE OF ELECTROLESS NICKEL COATING

The development of electroless plating has largely tackled the problem of corrosion due to its superior, performance enhancing and corrosion-resistant capabilities. Due to the low cost and reliability of equipment, electroless plating is being considered for more and more uses, greatly enhancing its business value.

The electroless plating of metals and alloys serves many useful functions in the electronic devices, such as corrosion protection, conductive circuit elements, via-hole filling for semiconductor integrated circuits, diffusion barriers, through-hole connections for printed wiring boards and flexible circuits. One of the main advantages of electroless nickel plating over regular electroplating is that it is free from flux density and power supply issues. This helps in achieving a uniform deposit regardless of the geometry of the workpiece.

3.9.1 Demand for Low Phosphorus Deposit

The electroless nickel technology has matured over the period since its discovery and its industrial acceptability continues to grow. However, it is quite natural to speculate its future growth. Whether electroless nickel plating would emerge as a major deposition process in the next couple of decades. These along with some other questions will be discussed in this section along with factors that could change the use of electroless nickel plating in the future.

Electroless nickel plating has become a practical and popular finish since its beginnings in the Bureau of Standards laboratory in 1944. There, Abner Brenner and Grace Riddel discovered nickel plating without external electrical current. They were astute enough to recognize the potential and thus developed and patented the first electroless plating processes in 1946. The advantages of electroless nickel coatings have been gradually realized over time and its acceptability has increased. The impressive as-deposited hardness, uniformity, high corrosion and wear resistance and so on make electroless nickel coating attractive. The fact that the properties of the deposit can be further enhanced by suitable thermal treatment is also a striking feature. Properties can also be suitably modified by introducing additional elements and particles into the deposit indicating that the electroless nickel coatings can be tailor-made based on the need of the application. Besides, the ease of development with minimal setup makes acceptable to a majority of concerned people. All these factors make the future of electroless nickel coatings appear bright.

It is already discussed that properties of electroless nickel depend upon the phosphorus content, and it is found that low phosphorus deposit can be heat-treated to a very high hardness number compared to high phosphorus deposits. Maximum hardness is obtained at somewhat different temperatures depending on the phosphorus content.

Low-phosphorus plating solutions are a recent development. As suppliers improve their products, more and more applications will be found. It is estimated that low-phosphorus alloys will be used in an increasing number of applications (Don 2018). Low-phosphorus alloy plating formulations that do not contain sulfur, or other materials that would detract from chemical and corrosion resistance will find the largest market in the future. Furthermore, it is projected that the general use of electroless nickel coatings will increase. Education of design and application engineers in the virtues of electroless nickel is one of the keys to the growth of electroless nickel. In this way, awareness can be created about the product and its promotion would be automatic.

It would be nice to have these variables and the consequences spelled out in the form of specifications corresponding to the various characteristics of the electroless nickel deposits. Porosity would have to be specified for corrosion protection applications. Resistance to chemical attack should be separated from corrosion protection of coatings. Corrosion protection is dependent on a continuous coating with no porosity.

3.9.2 Demand for Bright Electroless Nickel Deposit

There is always a demand for coating solutions that serve their purpose and at the same time are aesthetic. Hence, there is a recent demand for the electroless nickel deposits that are bright. Now, most electroless nickel solutions will deposit semi-bright to bright naturally. Matt or dull deposits are most often the result of something bad in the solution. Many impurities, metallic or organic, can cause dull deposits. Stabilizers used in most formulations will also brighten the deposit as well as stabilize the solution to prevent extraneous plating on particles, tanks, heaters and will prevent spontaneous decomposition of the plating solution. Sulfur-containing stabilizers will lower the chemical resistance and corrosion resistance and increase porosity in the deposit. Many other stabilizers will brighten and stabilize without sacrificing corrosion protection. It should be noted that NiP is not sacrificial to most metals onto which it is deposited. Therefore, for electroless NiP to protect, it must be pore-free. It should also be noted that most of the porosity in electroless NiP deposits have their source in the basis metal and not necessarily a characteristic of the electroless nickel deposit.

A plating solution made of the purest constituents available (Analytical Rated (AR) grade or better) will be semi-bright without stabilizers, but will not be practical due to its instability. (Experimentally, electroless nickel solutions filtered continuously through submicron filters and at a high rate of filtration remain stable without stabilizers). The solutions for memory disc production have only a fraction of the stabilizers that solutions for general use have (Don 2018). It is predicted that there will be a growing use of bright electroless NiP plating in the years to come.

3.9.3 Recommendation for Growth of Electroless Nickel

If platers make their customers aware of the numerous characteristics and virtues of the various electroless nickel plating processes, the use of electroless nickel in all its forms would increase. Educating engineers, designers and purchasing people would move the use of electroless nickel forward. Specifications and standards that include the many variations of electroless NiP and electroless nickel alloys would further help the use of electroless nickel. Delineating the characteristics of the various deposits and how to achieve these characteristics would be helpful. The characteristics such as hardness, wear resistance, as tested by various means, corrosion resistance and protection, magnetics (are different for various P contents) electrical resistivity or conductivity data for each type of alloy, internal stress resulting from the deposit alone, thermal coefficient of expansion for different alloy deposits, tensile strength, elongation, yield, thermal conductivity, modulus of elasticity and perhaps other useful engineering data would help to carry the message. There is much of this in the literature, but not all alloy groups and variations are published. There is much to do. Each plater must think in terms of marketing, and service to help increase the use of plated products.

The global electroless plating market consists of various international, regional and local vendors. The market competition is foreseen to grow at a higher rate with the advances in technological innovation along with the merger and acquisition deals in the future. Based on the quality, reliability and innovations offered by the international vendors in the technology, new vendors are finding it hard to compete with them. Companies providing electroless plating services can use smart control systems to regulate the plating and processing cycles, thereby lowering the environment degrading emissions and saving energy during the idle times.

Varied specific application products are being provided by end users industries such as the chemical industry, oil industry, automotive industry, electronics industry, aerospace industry and machinery industry. One of the major drivers for this market is the continuous requirement for protection of varied assets against corrosion, currently costing companies billions of dollars. In addition, the investments in the research and development facilities by manufacturers such as DOW Chemicals and Atotech Deutschland GmbH to help produce efficient, cost-effective and environment-friendly plating technologies will help the industry grow to its full potential in 4–5 years. The restraining factor in the growth of the electroless plating market remains the volatility of the prices of raw metals. Despite levying an impact on the growth of the global electroless plating market, consumers still prefer products coated with a metallic lustre of their choice.

Electroless plating is a mature industry and has stayed in the market for nearly a decade. This market has reached the stage of maturity in North America and Western Europe. There are approximately 1,000 electroless nickel (electroless nickel) plating facilities in North America. Approximately 66.67% of them are job shops and the remaining 33.33% are the captive facilities of manufacturing plants (Sullivan 2017). The Asia Pacific is an emerging market, as it is anticipated to witness an increase in the automotive manufacturing industry. China, India, Indonesia and Thailand are major market revenue contributors to the automotive industry in the Asia-Pacific region.

3.9.4 Challenges

One of the major challenges the electroless plating industry faces is the need for substantial information on the uses of various substances and exposure. The high probability of requiring an authorization or restriction of the uses also holds back the industry in the European region. However, with electroless plating playing an important role in the electronics industry, micro and nanodevices may be realized using this in near future. Thus, it has high potential to realize high-performance electronic devices with continued movement toward cost reduction, technical innovation and distribution efficiency.

3.10 GREEN ELECTROLESS NICKEL COATING

The ELV/RoHS/WEEE directives have imposed restrictions on certain hazardous substances used in manufactured goods, which can pose harm to consumers and potentially cause a pollution risk to the environment after end-of-life. This has initiated a revolution in all spheres of human life such that nature remains unharmed to the best extent possible. These eco-friendly activities are often termed as "green" activities as they help in maintaining the balance in the ecosystem. Although electroless nickel deposition process is inherently eco-friendly as opposed to other electrolytic coatings, does not require current or anodes. The coating is carried out in a watery solution, containing metallic ions, complexes and stabilizers. Hence, this process consumes relatively lesser energy compared to electrolytic deposition. However, electricity is not completely eliminated as heating of the electroless bath is necessary for the deposition to occur. Hence, non-conventional energy sources could be thought of supplying heat energy to the bath.

Moreover, electroless nickel coating bath has many toxic chemicals as ingredients. Major constituents of the bath, namely nickel ions, phosphorus compounds and organics are highly toxic and hence their release into the environment in the form of spent solution proves detrimental to the nature. Hence, many efforts have been made to increase the metal turnover (MTO) rate, which also serves as a parameter to measure the solution life. One metal turnover is achieved when the quantity of nickel contained in a new bath is totally converted into NiP deposit. In other words, as the metal in electroless nickel bath is consumed, it has to be replaced, of course. When the total additions of replacement metal that has been added are equal to the total amount of metal originally in the bath, that's 1 MTO. The importance of this term is that in addition to consuming metal, the electroless deposition process generates spent reducer and other contaminants that cannot be practically removed, and eventually foul the bath so badly that the same cannot be continued to use further and must be discarded. One measure of the value of an electroless nickel process is how many turnovers can be done before replacement is necessary.

From the viewpoint of resourcing by waste recycling, researchers have developed many approaches to convert the discarded electroless nickel plating solution into valuable compounds. One of the methods that has been traditionally used to reduce the wastage of chemicals was by increasing the MTO. Other methods used for nickel recovery included solvent extraction, electrolysis, electrodialysis, activated carbon adsorption, cathodic deposition-alkaline precipitation and catalytic reduction (Liu et al. 2018). Attempts have been made to recover other elements in the bath solution, namely phosphorus compounds were converted to ferric phosphate (with the UV-Fenton process) (Liu et al. 2014), nickel-phosphorus-sulfur was transformed to graphene/NiAl-layered double-hydroxide composite as a catalyst for methanol fuel cell with an ultrasound-assisted co-precipitation method (Zhu et al. 2015).

The best way may be to develop an electroless nickel plating bath without any emission. That is to say, the spent plating solution can be used directly as a raw material to prepare other material with a high added value. Wastage of chemicals can be reduced by increasing the MTO. This also aids in preventing pollution, as the harmful ions are removed from the final solution, which are discarded and released into the atmosphere.

Continuous effort is also underway to develop alternate electroless nickel baths, which is free from restricted substances, namely lead, thallium and cadmium. Lead and thallium are the popularly used stabilizing agents in electroless nickel bath that prevents spontaneous decomposition of the metastable plating bath. On the other hand, cadmium has been traditionally used as an additive in the medium-phosphorus electroless nickel baths to enhance the brightness and promote leveling of the deposit. According to ELV/RoHS/WEEE directives, the content of lead and cadmium in the coating (by weight) have been restricted to 0.1% and 0.01%, respectively. The Occupational Safety and Health Administration has set the legal limit (permissible exposure limit) for thallium exposure in the workplace as 0.1 mg/m^3 skin exposure over an 8-h workday.

3.11 CLOSURE

From the present discussions, it can be seen that electroless nickel coatings have emerged as suitable coatings that can serve as viable replacements to the conventional electroplating in suitable situations. Their properties such as high hardness, low friction, wear resistance and corrosion resistance have led to their usage in tribological-based applications. Besides, the uniform deposition and the ability to coat any materials have served as an added advantage to their application in various areas. The advantages of modifying the properties of electroless nickel coatings by suitable surface treatments (heat treatment, laser treatment, etc.) and the incorporation of various elements (copper, tungsten, etc.) and particles (SiC, TiO_2, Si_3N_4, etc.) have been utilized by various researchers to evaluate the suitability of these coatings for various applications. The discussion reveals that the electroless coatings are mainly applied for wear resistance and corrosion resistance applications. By observing the expansion of the possibilities in case of electroless nickel coatings, more advanced tribological application using the coatings may be expected.

REFERENCES

2404J, A. (2018). Plating, Electroless Nickel. AMS, SAE International.

Acuña, C. J. and F. E. Echeverría (2015). A review with respect to electroless NiP (ENP) coatings: Fundamentals and properties. Part i.

Agarwala, R. C. and V. Agarwala (2003). "Electroless alloy/composite coatings: A review." *Sadhana* **28** (3): 475–493.

Alirezaei, S., S. M. Monir Vaghefi, M. Salehi and A. Saatchi (2007). "Wear behavior of NiP and Ni–P–Al_2O_3 electroless coatings." *Wear* **262**: 8.

Anik, M. and E. Körpe (2007). "Effect of alloy microstructure on electroless NiP deposition behavior on Alloy AZ91." *Surface and Coatings Technology* **201**(8): 4702–4710.

Arabani, M. R. and S. M. M. Vaghefi (2006). Effect of heat-treatment on hardness and adhesion of NiP coated 7075 aluminum alloys by electroless deposition process. *15th IFHTSE–International Federation for Heat Treatment and Surface Engineering Congress.*

Ashassi-Sorkhabi, H. and S. H. Rafizadeh (2004). "Effect of coating time and heat treatment on structures and corrosion characteristics of electroless NiP alloy deposits." *Surface and Coatings Technology* **176**(3): 318–326.

ASTM (2015). Standard Specification for Autocatalytic (Electroless) Nickel-Phosphorus Coatings on Metal. ASTM B733-15. West Conshohocken, ASTM International.

Balaraju, J. N., S. M. Jahan and K. S. Rajam (2006a). "Studies on autocatalytic deposition of ternary Ni–W–P alloys using nickel sulphamate bath." *Surface and Coatings Technology* **201**(3): 507–512.

Balaraju, J. N., T. S. N. S. Narayanan and S. K. Seshadri (2006b). "Structure and phase transformation behaviour of electroless NiP composite coatings." *Materials Research Bulletin* **41**(4): 847–860.

Balaraju, J. N., V. E. Selvi, V. K. W. Grips and K. S. Rajam (2006c). "Electrochemical studies on electroless ternary and quaternary NiP based alloys." *Electrochimica Acta* **52**(3): 1064–1074.

Balaraju, J. N. and S. K. Seshadri (1999). "Preparation and characterization of electroless NiP and NiP-Si3N4 composite coatings." *Transactions of the Institute of Metal Finishing* **77**(2): 3.

Ban, C. L., X. Shao and L. P. Wang (2014). "Ultrasonic irradiation assisted electroless NiP coating on magnesium alloy." *Surface Engineering* **30**(12): 880–885.

Baskaran, I., T. S. N. S. Narayanan and A. Stephen (2006). "Effect of accelerators and stabilizers on the formation and characteristics of electroless NiP deposits." *Materials Chemistry and Physics* **99**(1): 117–126.

Baudrand, D. W. (1978). *Metals Handbook*. Matreials Park, OH: American Society for Metals.

Berríos, J. A., M. H. Staia, E. C. Hernández, H. Hintermann and E. S. Puchi (1998). "Effect of the thickness of an electroless NiP deposit on the mechanical properties of an AISI 1045 plain carbon steel." *Surface and Coatings Technology* **108–109**: 466–472.

Biswas, A., S. K. Das and P. Sahoo (2017). "Correlating tribological performance with phase transformation behavior for electroless Ni-(high)P coating." *Surface and Coatings Technology* **328**(Supplement C): 102–114.

Cheng, Y., Z. Zhu and Z. Han (2011). "The effect of heat treatment on the microstructure of electroless NiP coatings." *Key Engineering Materials* **464:** 474–477.

Chitty, J. A., A. Pertuz, H. HInterman and E. S. Puchi (1999). "Influence of electroless nickel-phosphorus deposits on the corrosion-fatigue life of notched and unnotched samples of an AISI 1045 Steel." *Journal of Materials Engineering and Performance* **8**: 4.

Contreras, G., C. Fajardo, J. A. Berríos, A. Pertuz, J. Chitty, H. Hintermann and E. S. Puchi (1999). "Fatigue properties of an AISI 1045 steel coated with an electroless NiP deposit." *Thin Solid Films* **355–356**: 480–486.

Das, C. M., P. K. Limaye, A. K. Grover and A. K. Suri (2007). "Preparation and characterization of silicon nitride codeposited electroless nickel composite coatings." *Journal of Alloys and Compounds* **436**(1–2): 328–334.

Deng, H. and P. Moller (1993). "Effect of the substrate surface morphology." *Transactions of Institute of Metal Finishing* **71**(4): 142–148.

Deshmukh, P., M. Lovell and A. J. Mobley (2005). Friction and wear performance of surface coatings in brake applications. *2005 World tribology congress III.*

Don, B. (2018). "Electroless Nickel Plating—Where is it Going." Retrieved October 20, 2018, http://www.plateworld.com/editorial9.htm.

Doong, J. C., J. G. Duh and S. Y. Tsai (1986). "Corrosion behaviour of electroless nickel plating modified TiN coating." *Surface and Coatings Technology* **27**(1).

Duncan, R. N. (1981). "Properties and application of electroless nickel." *Finisher's Management* **26**(3): 5–26.

Duncan, R. N. (1983). "Effect of solution age on corrosion resistance of electroless nickel plating." *Surface Finishing* **10**: 5.

Duncan, R. N. (1996a). "Metallurgical structure of electroless nickel deposit, plating and surface finishing." *Plating and Surface Finishing* **83**(11): 65–69.

Duncan, R. N. (1996b). "Effect of solution age on corrosion resistance of electroless nickel deposits." *Plating and Surface Finishing* **83**(10): 64–68.

Elansezhian, R., B. Ramamoorthy and P. K. Nair (2008). "Effect of surfactants on the mechanical properties of electroless (Ni–P) coating." *Surface and Coatings Technology* **203**(5): 709–712.

Elansezhian, R., B. Ramamoorthy and P. K. Nair (2009). "The influence of SDS and CTAB surfactants on the surface morphology and surface topography of electroless NiP deposits." *Journal of Materials Processing Technology* **209**(1): 233–240.

Eraslan, S. and M. Ürgen (2015). "Oxidation behavior of electroless Ni–P, Ni–B and Ni–W–B coatings deposited on steel substrates." *Surface and Coatings Technology* **265**(Supplement C): 46–52.

Fields, W. and J. R. Zickearff (1984). "Electroless." Publication of ASM committee on electroless nickel plating.

Gao, R., M. Du, X. Sun and Y. Pu (2007). "Study of the corrosion resistance of electroless NiP deposits in a sodium chloride medium." *Journal of Ocean University of China* **6**(4): 349–354.

Gordani, G. R., R. ShojaRazavi, S. H. Hashemi and A. R. N. Isfahani (2008). "Laser surface alloying of an electroless NiP coating with Al-356 substrate." *Optics and Lasers in Engineering* **46**(7): 550–557.

Hari Krishnan, K., S. John, K. N. Srinivasan, J. Praveen, M. Ganesan and P. M. Kavimani (2006). "An overall aspect of electroless NiP depositions—A review article." *Metallurgical and Materials Transactions A: Physical Metallurgy and Materials Science* **37**(6): 1917–1926.

Hino, M., K. Murakami, Y. Mitooka, K. Muraoka, R. Furukawa and T. Kanadani (2009). "Effect of zincate treatment on adhesion of electroless NiP coating onto various aluminum alloys." *Materials Transactions* **50**(9): 2235–2241.

Hui, Z. and L. Rongli (2008). "Properties of (NiP)-SiO2 (nanometer) electroless composite coating on PET fabrics." *Sen'i Gakkaishi* **64**(12): 372–377.

Kanta, A. F., V. Vitry and F. Delaunois (2009). "Wear and corrosion resistance behaviours of autocatalytic electroless plating." *Journal of Alloys and Compounds* **486**(1–2): L21–L23.

Keong, K. G., W. Sha and S. Malinov (2003). "Hardness evolution of electroless nickel–phosphorus deposits with thermal processing." *Surface and Coatings Technology* **168**(2): 263–274.

Kerr, C., D. Barker and F. C. Walsh (1997). Physical and electrochemical characteristics of electroless nickel on carbon steel. *International Conference on Computer Methods and Experimental Measurements for Surface Treatment Effects.*

Klingenberg, M. L., E. W. Brooman and T. A. Naguy (2005). "Nano-particle composite plating as an alternative to hard chromium and nickel coatings." *Plating and Surface Finishing* **92**(4): 42–48.

Krishnan, K. H., S. John, K. N. Srinivasan, J. Praveen, M. Ganesan and P. M. Kavimani (2006). "An overall aspect of electroless NiP depositions—A review article." *Metallurgical and Materials Transactions A* **37**(6): 1917–1926.

Krishnaveni, K., T. S. N. Sankara Narayanan and S. K. Seshadri (2005). "Electroless Ni–B coatings: Preparation and evaluation of hardness and wear resistance." *Surface and Coatings Technology* **190**(1): 115–121.

Kundu, S., S. K. Das and P. Sahoo (2018). "Tribological behaviour of electroless NiP deposits under elevated temperature." *Silicon* **10**(2): 329–342.

Lee, C. K. (2008). "Corrosion and wear-corrosion resistance properties of electroless NiP coatings on GFRP composite in wind turbine blades." *Surface and Coatings Technology* **202**(19): 4868–4874.

Lee, C. K. (2009). "Structure, electrochemical and wear-corrosion properties of electroless nickel-phosphorus deposition on CFRP composites." *Materials Chemistry and Physics* **114**(1): 125–133.

Lee, M. G., T. J. Yang, C. A. Hsieh and C. C. Huang (2009). Corrosion performance of electroless nickel-plated steel. *The Nineteenth International Offshore and Polar Engineering Conference.*

Leisner, P. and M. E. Benzon (1997). "Porosity measurement of coatings." *Transactions of Institute of Metal Finishing* **75**(2): 5.

León, O. A., M. H. Staia and H. E. Hintermann (2003). "High temperature wear of an electroless Ni–P–BN (h) composite coating." *Surface and Coatings Technology* **163–164**: 578–584.

Li, Z.-H., Z.-Y. Chen, S.-S. Liu, F. Zheng and A. G. Dai (2008). "Corrosion and wear properties of electroless NiP plating layer on AZ91D magnesium alloy." *Transactions of Nonferrous Metals Society of China* **18**(4): 819–824.

Linka, G. and W. Riedel (1986). "Corrosion resistance of nickel-phosphorus alloy coatings deposited by chemical-reduction." *Galvanotechnik*: 6.

Liu, P., C. Li, X. Liang, G. Lu, J. Xu, X. Dong, W. Zhang and F. Ji (2014). "Recovery of high purity ferric phosphate from a spent electroless nickel plating bath." *Green Chemistry* **16**(3): 1217–1224.

Liu, W., Q. Liu, L. Xu, M. Qin and J. Deng (2018). "A zero-emission electroless nickel plating bath." *Surface Review and Letters* **26**(1): 1850130.

Liu, Z. and W. Gao (2006a). "The effect of substrate on the electroless nickel plating of Mg and Mg alloys." *Surface and Coatings Technology* **200**(11): 3553–3560.

Liu, Z. and W. Gao (2006b). "Scratch adhesion evaluation of electroless nickel plating on mg and mg alloys." *International Journal of Modern Physics B* **20**(25n27): 4637–4642.

Loto, C. A. (2016). "Electroless nickel plating—A review." *Silicon* **8**(2): 177–186.

Mallory, G. O. and J. B. Hadju (1991). *Electroless Plating: Fundamentals and Applications.* Orlando, FL: AESF.

Masoumi, F., H. R. Ghasemi, A. A. Ziaei and D. Shahriari (2012). "Tribological characterization of electroless Ni–10% P coatings at elevated test temperature under dry conditions." *The International Journal of Advanced Manufacturing Technology* **62**(9): 1063–1070.

MIL-C-26074 (1998). Military Specification: Coatings, Electroless Nickel, Requirements for. MIL-C-26074, Department of Navy.

Novák, M., D. Vojtěch and T. Vítů (2010). "Influence of heat treatment on tribological properties of electroless NiP and NiP-Al_2O_3 coatings on Al-Si casting alloy." *Applied Surface Science* **256**(9): 2956–2960.

Osifuye, C. O., A. P. I. Popoola, C. A. Loto and D. T. Oloruntoba (2014). "Effect of bath parameters on electroless NiP and Zn-P deposition on 1045 steel substrate." *International Journal of Electrochemical Science* **9**: 14.

Palaniappa, M. and S. K. Seshadri (2008). "Friction and wear behavior of electroless NiP and Ni–W–P alloy coatings." *Wear* **265**(5): 735–740.

Parkinson, R. (1997). Properties and applications of electroless nickel, Nickel Development Institute. NiDL Technical Series No. 10081.

Paunovic, M. and M. Schlesinger (2006). *Fundamentals of Electrochemical Deposition.* Hoboken, NJ: Wiley Interscience.

Prasad, P. B. S. N. V., S. Ahila, R. Vasudevan and S. K. Seshadri (1994). "Fatigue strength of nickel electrodeposits prepared in ultrasonically agitated bath." *Journal of Materials Science Letters* **13**(1): 2.

Puchi, E. S., M. H. Staia, H. Hintermann, A. Pertuz and J. Chitty (1996). "Influence of Ni-P electroless coating on the fatigue behavior of plain carbon steels." *Thin Solid Films* **290–291**(Supplement C): 370–375.

Rabinowicz, F. (1995). *Friction and Wear of Materials.* New York: John Wiley & Sons.

Rabizadeh, T., S. R. Allahkaram and A. Zarebidaki (2010). "An investigation on effects of heat treatment on corrosion properties of NiP electroless nano-coatings." *Materials & Design* **31**(7): 3174–3179.

Ramalho, A. and J. C. Miranda (2007). "Tribological characterization of electroless NiP coatings lubricated with biolubricants." *Wear* **263**(1): 592–597.

Riedel, W. (1991). *Electroless Nickel Plating.* Metal Parks, OH: ASM International.

Riedel, W. (1997). *Electroless Nickel Plating.* ASN International.

Sahoo, P. (2008). "Optimization of electroless NiP coatings based on multiple roughness characteristics." *Surface and Interface Analysis* **40**(12): 1552–1561.

Sahoo, P. (2009). "Wear behaviour of electroless NiP coatings and optimization of process parameters using Taguchi method." *Materials & Design* **30**(4): 1341–1349.

Sahoo, P. and S. K. Das (2011). "Tribology of electroless nickel coatings—A review." *Materials & Design* **32**(4): 1760–1775.

Sampath Kumar, P. and P. Kesavan Nair (1996). "Studies on crystallization of electroless Ni-P deposits." *Journal of Materials Processing Technology* **56**(1): 511–520.

Sankara Narayanan, T. S. N., I. Baskaran, K. Krishnaveni and S. Parthiban (2006). "Deposition of electroless NiP graded coatings and evaluation of their corrosion resistance." *Surface and Coatings Technology* **200**(11): 3438–3445.

Schlesinger, M. (2011). Electroless deposition of nickel. In *Modern Electroplating.* M. Schlesinger and M. Paunovic (Eds.). Hoboken, NJ: John Wiley & Sons.

Shaffer, S.J. (2008). "Wear and friction testing of hard and soft lubricious coatings in dry sliding for use in small arms action components." *Journal of ASTM International* **5**(2): 1–10.

Shipley Jr., C. R. (2018). "Historical highlights of electroless plating." *NASF Surface Technology White Papers* **82**(9): 12.

Singh, D., R. Balasubramaniam and R. K. Dube (1995). "Effect of coating time on corrosion behaviour of electroless nickel-phosphorus coated powder metallurgy iron specimens." *Corrosion* **51**(8): 581–585.

Srinivasan, K. N., R. Meenakshi, A. Santhi, P. R. Thangavelu and S. John (2010). "Studies on development of electroless NiB bath for corrosion resistance and wear resistance applications." *Surface Engineering* **26**(3): 153–158.

Staia, M. H., E. J. Castillo, E. S. Puchi, B. Lewis and H. E. Hintermann (1996). "Wear performance and mechanism of electroless Ni-P coating." *Surface and Coatings Technology* **86–87**(Part 2): 598–602.

Sullivan, F. (2017). "Frost perspectives—The future of the electroless plating market. Retrieved October 22, 2018. from https://ww2.frost.com/frost-perspectives/future-electroless-plating-market/.

Taheri, R. (2003). Evaluation of electroless nickel-phosphorus (electroless nickel) coatings. PhD thesis, University of Saskatchewan.

Taheri, R., I. N. A. Oguocha and S. Yannacopoulos (2001). "The tribological characteristics of electroless NiP coatings." *Wear* **249**(5): 389–396.

Tomlinson, W. J. and G. R. Wilson (1986). "The oxidation of electroless NiB and NiP coatings in air at 800°C to 1000°C." *Journal of Materials Science* **21**(1): 97–102.

Tsujikawa, M., D. Azuma, M. Hino, H. Kimura and A. Inoue (2005). "Friction and wear behavior of laser irradiated amorphous metal." *Journal of Metastable and Nanocrystalline Materials* **24–25**: 375–378.

Veronesi, P., R. Sola and G. Poli (2008). "Electroless Ni coatings for the improvement of wear resistance of bearings for lightweight rotary gear pumps." *International Journal of Surface Science and Engineering* **2**(3–4): 190–201.

Vitry, V., E. Francq and L. Bonin (2018). "Mechanical properties of heat-treated duplex electroless nickel coatings." *Surface Engineering* **35**(2): 1–9.

Wang, J., F. Yan and Q. Xue (2009). "Friction and wear behavior of ultra-High molecular weight polyethylene sliding against GCr15 steel and electroless NiP alloy coating under the lubrication of seawater." *Tribology Letters* **35**(2): 85–95.

Wang, W., W. Zhang, Y. Wang, N. Mitsuzak and Z. Chen (2016). "Ductile electroless NiP coating onto flexible printed circuit board." *Applied Surface Science* **367**: 528–532.

Wu, Y., B. Shen, L. Liu and W. Hu (2006a). "The tribological behaviour of electroless NiP-Gr-SiC composite." *Wear* **261**(2): 201–207.

Wu, Y. T., L. Lei, B. Shen and W. B. Hu (2006b). "Investigation in electroless NiP-Cg(graphite)-SiC composite coating." *Surface and Coatings Technology* **201**(1–2): 441–445.

Yamasaki, T., H. Izumi and H. Sunada (1981). "The microstructure and fatigue properties of electroless deposited Ni-P alloys." *Scripta Metallurgica* **15**(2): 177–180.

Ye, C. Y., Z. B. Wang, M. L. Yan and P. Q. Li (2005). "Resistance to CO_2 corrosion of electroless plating of Ni-P coating on steel N80." *Corrosion Science and Protection Technology* **17**(4): 265.

Zahid, R., H. H. Masjuki, M. Varman, R. A. Mufti, M. A. Kalam and M. Gulzar (2015). "Effect of lubricant formulations on the tribological performance of self-mated doped DLC contacts: A review." *Tribology Letters* **58**(2): 32.

Zhang, H., J. Zou, N. Lin and B. Tang (2014). "Review on electroless plating Ni–P coatings for improving surface performance of steel." *Surface Review and Letters* **21**(4): 1430002.

Zhang, Y. Z. and M. Yao (1999). "Studies of electroless nickel deposits with low phosphorus content." *Transactions of the IMF* **77**(2): 78–83.

Zhu, X.-H., F. Xie, J. Li and G.-P. Jin (2015). "Simultaneously recover Ni, P and S from spent electroless nickel plating bath through forming graphene/NiAl layered double-hydroxide composite." *Journal of Environmental Chemical Engineering* **3**(2): 1055–1060.

4A Electroless Nickel-Boron
Part A: As-Plated Electroless Nickel-Boron

Fabienne Delaunois

CONTENTS

4A.1 CHEMISTRY OF ELECTROLESS NICKEL-BORON (ELECTROLESS NICKEL-B) COATINGS

To remember, various reducing agents are suitable for electroless nickel plating, but only two lead to the formation of nickel-boron alloys (Delaunois et al. 2000): (i) amine-borane compounds, the most popular of which is dimethylamine borane (DMAB) and (ii) borohydride salts (either sodium or potassium, the most used being $NaBH_4$). The deposition rate from bath reduced with sodium borohydride is higher than that reduced with DMAB (Sudagar et al. 2013).

Due to numerous deposition bath compositions, principally depending on the reducing agent and stabilizer used, each particular coating presents well-defined properties, and it is thus difficult to describe the chemical composition, structure and morphology of electroless nickel-boron coatings univocally. In short, the boron content and the reduced stabilizer strongly influence these properties. Nevertheless, it is possible to group coatings with similar contents and properties and to determine trends in their evolution. (Vitry and Delaunois 2015b; Delaunois and Vitry 2015).

Metallic boron (B), formed in the plating bath as a result of the oxidation reaction of the reducing agent, is incorporated in the coating and its amount is depending on the bath chemistry and operating conditions.

Some stabilizers, generally heavy metal salts (ions such as Pb^{2+}, Cd^{2+}, Hg^{2+}, Tl^{3+} and Cu^{2+}) are also co-deposited together with boron into the coating, and their amount may become important. This co-deposition is the result of the fact that these agents stabilize the electroless nickel-boron bath by depositing on the active metal surface through a displacement reaction, thus inhibiting the occurrence of nickel reduction.

Consequently, variation in the chemistry of electroless nickel-boron plating baths, according to the reducing agent and the stabilizer used, is responsible for the variation of the chemistry of deposited electroless nickel-boron coatings.

The nature of the reducing agent and its concentration in the plating bath lead to variations in boron content in electroless nickel-boron coatings. The boron content of electroless nickel-boron coatings synthesized with DMAB is generally lesser (0.1–4.0 wt.%) than that of coatings reduced with $NaBH_4$ (4–9 wt.%) (Delaunois and Vitry 2015; Sudagar et al. 2013; Riedel 1989; Ohno et al. 1985). However, from recent studies, boron level up to 8 wt.% in both DMAB and sodium borohydride-based electroless nickel-boron baths was measured, as shown in Table 4A.1 (Riedel 1989; Delaunois 2002; Vitry and Bonin 2017; Anik et al. 2008; Contreras et al. 2006).

TABLE 4A.1
Composition of Various Electroless Nickel-Boron Coatings Obtained with either Borohydride or Amineborane Reducing Agent

References	Reducing Agent	Stabilizer Type (if known)	Boron Content (wt.%)	Alloying Element Content (if known) (wt.%)
Wang et al. (2012)	$NaBH_4$	Organic	6.25	
Gorbunova et al. (1973)	$NaBH_4$	Heavy metal (Tl)	4.3	
	$NaBH_4$	Heavy metal (Pb)	6.4	
	$NaBH_4$	—	5.7	
Selvaraj et al. (1989)	$NaBH_4$	Heavy metal (Tl)	3	
Srinivasan et al. (2010)	$NaBH_4$	Organic	4.94	
Saito et al. (1998)	DMAB	Organic	3–5	
Delaunois and Lienard	$NaBH_4$	Heavy metal (Tl)	4	6.9 (Tl)
(2002)	$NaBH_4$	Heavy metal (Pb)	5.75	1.25 (Pb)
Delaunois et al. (2000)	$NaBH_4$	Heavy metal (Tl)	8	5 (Tl)
Vitry et al. (2012c)	$NaBH_4$	Heavy metal (Pb)	6	1 (Pb)
Kaya et al. (2008)	$NaBH_4$	Heavy metal (Pb)	5	
Dadvand et al. (2003)	$NaBH_4$	Heavy metal (Tl)	>6	
Baskaran et al. (2006)	$NaBH_4$ (low T°C)	Heavy metal (Tl)	0.6–3.2	0.1–0.2 (Tl)
Narayanan et al. (2003)	$NaBH_4$	Heavy metal (Tl)	6.5	0.3 (Tl)
Contreras et al. (2006)	DMAB		8,34	
Anik et al. (2008)	$NaBH_4$	Heavy metal (Tl)	6.5	
Bonin (2018)	$NaBH_4$	Heavy metal (Pb)	6	1 (Pb)
Clerc (1986)	$NaBH_4$	Heavy metal (Pb)	6.4	1.8 (Pb)
Das and Sahoo (2011a)	$NaBH_4$	Heavy metal (Pb)	5.72–7.46	
Hamid et al. (2010)	DMAB	Heavy metal (Pb)	5–6	

An increase in the reducing agent content in the bath leads to an increase in the coating boron content (Anik et al. 2008; Saito et al. 1998; Dadvand et al. 2003; Baskaran et al. 2006; Kumar and Nair 1994; Zhang n.d.). The obtained amount of boron also depends on the reducing agent used and on the overall bath chemical composition. Baskaran et al. (2006) have shown that an increase in the concentration of $NaBH_4$ from 0.2, 0.4, 0.6, 0.8 to 1.0 g/L in the plating bath leads to an increase in boron content from 0.6, 1.2, 2.0, 2.6 to 3.2 wt.% in the deposit, respectively. Figure 4A.1 summarizes various studies on the evolution of boron content in the coating with the reducing agent concentration of the plating bath.

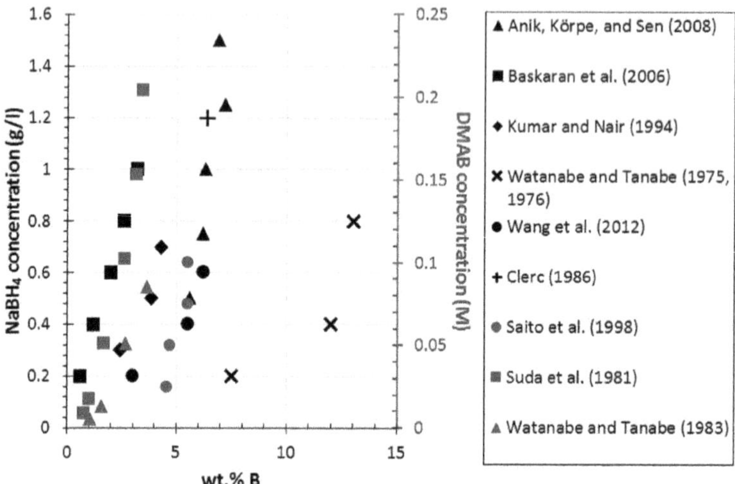

FIGURE 4A.1 Effect of the reducing agent concentration ($NaBH_4$ [black] or DMAB [red]) in the plating bath on the boron content of the coating.

Based on the following reactions describing borohydride oxidizing during the electroless process:

$$BH_4^- + 4OH^- = BO_2^- + 2H_2O + 2H_{2(g)} + 4e \tag{4A.1}$$

$$BH_4^- = B + 2H_{2(g)} + e \tag{4A.2}$$

Anik et al. (2008) have shown that the kinetics of reaction (4A.1) is not significantly affected by the increase in $NaBH_4$ content. In contrast, reaction (4A.2) is more sensitive to this increase in $NaBH_4$ content.

The increase in the boron content in the film is linked to the change in coating morphology with the increase in $NaBH_4$ concentration in the plating bath (Baskaran et al. 2006) (see paragraph 4A.3.2).

The stabilizer also influences the amount of boron in the deposit (Zhang n.d.). Delaunois and Lienard (2002) have investigated the influence of thallium or lead stabilizers. When 0.11 g/L $TlNO_3$ or 0.02 g/L $PbWO_4$ is added in the bath (chemical is different for both baths), they obtain a ternary alloy Ni-B-Tl with 3.99 wt.% B and 6.90 wt.% Tl and a NiB coating with 5.75 wt.% B and 1.25 wt.% Pb, respectively. In fact, $TlNO_3$ suppresses the reduction of B element, inducing a lower B content in the deposit (Zhang n.d.).

The boron content of the coating increases with the increase in complexing agent content (Zhang n.d.). Saito et al. (1998) have reported that the increase in sodium citrate, sodium tartrate or sodium malonate concentration increases the boron content. Anik et al. (2008) have shown that an increase in ethylenediamine concentration above 90 g/L, that is the minimum critical concentration below which the efficiency of nickel deposition decreases, causes a sharp decrease in the boron content of the coating. This is probably due to the changing surface charge of the metallic substrate with the increase in complexing agent concentration.

The decrease in sodium hydroxide concentration below 50 g/L causes a decrease in the boron content of the deposit due to the partial hydrolyzation of borohydride ions at lower sodium hydroxide concentration. Above 50 g/L, the boron content doesn't sensitively change (Anik et al. 2008).

The plating bath temperature has a strong influence on the boron content of the coating.

Vitry et al. (2017) and Anik et al. (2008) have studied the influence of bath temperature on the boron content of the deposit. The boron content of the coating slightly increases with increasing bath temperature: from approximately 6.7 to 7.7 wt.% B with an increase from 80°C to 95°C, respectively (Anik et al. 2008). For a bath reduced by $NaBH_4$ and stabilized with lead tungstate, the increase in temperature from 95°C, temperature usually used for this kind of bath (Delaunois et al. 2000; Vitry et al. 2008, 2011, Vitry et al. 2008), to 96.5°C leads to an increase in the boron content in the deposit, from approximately 6 to 8 wt.%, respectively (Vitry and Bonin 2017).

The amount of stabilizer co-deposited in the electroless nickel-boron coating varies also with the type of stabilizer used (nearly no incorporation for organic stabilizers) (Table 4A.1) and its concentration in the bath: for baths stabilized with thallium salts, thallium content varies from 0.1 to 6 wt.% (Delaunois et al. 2000; Krishnaveni et al. 2005; Narayanan et al. 2003; Baskaran et al. 2006); but with the use of lead salts, the concentrations of lead are more stable (approximately 1 wt.%) (Vitry et al. 2012c; Pal et al. 2011). The increase in bath temperature doesn't influence this lead content (Vitry and Bonin 2017). Anik et al. (2008) have shown that an increase in thallium acetate concentration in the plating bath corresponds to a sharp decrease in boron content in the coating and to an increase of its Tl content due to the following competitive deposition reaction:

$$Tl^{+} + e = Tl \tag{4A.3}$$

In fact, an increase in thallium acetate concentration from 12 to 35 mg/L leads to an increase in the Tl content in the deposit from 0.35 to 0.95 wt.%, respectively.

Finally, the most frequent chemistry encountered in electroless nickel-boron deposits prepared using baths reduced by $NaBH_4$ is 5–6 wt.% B, with or without co-deposited stabilizer.

The chemistry of electroless nickel-boron deposits is also not homogeneous due to their growth process. Pal et al. (2011) have shown the chemical heterogeneity of electroless nickel-boron coatings with the use of local analysis performed on the subsurface (Figure 4A.2). Variations of plating with position can be observed due to the catalytic nature of the reaction and the use of stabilizers that act as a catalytic poison and adsorb on the substrate to block parts of the active sites. The boron content is thus higher in the middle of the grains (A position—Figure 4A.2) than at their edges (B position—Figure 4A.2), with a very low concentration in the inter-grain spaces (C position—Figure 4A.2).

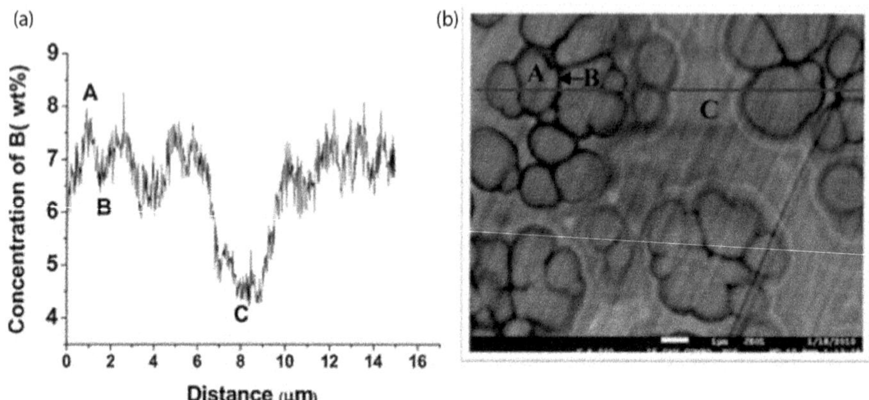

FIGURE 4A.2 Local chemical variations inside electroless nickel-boron as-deposited coating (From Pal S et al., Mat. Sci. Eng. A. 528, 8269-8276, 2011): (a) Variation of boron concentration on the polished top surface of the as-deposited coating. (b) The image of the top surface, profile taken along the marked line on the image. Regions A, B and C on the image show regions corresponding to the composition A, B and C marked in profile (a).

Moreover, this deposition process causes variations of plating with time (Vitry et al. 2012e). In fact, the electroless nickel-boron coatings are formed by a catalyzed chemical reaction in bath that can be considered (except for the injection of heat) as a closed system. The parts formed earlier contain more boron than that formed later (Pal et al. 2011). From X-ray photoelectron spectroscopy measurements, Vitry (2010, 2012) have also observed variations of boron content in samples with various plating time.

However, the local variations in the boron content from interface to surface are not observed from Glow Discharge Optical Emission Spectroscopy (GDOES) depth profile (Vitry et al. 2012; Vitry and Bonin 2017; Clerc 1986; Correa et al. 2013) (Figure 4A.3). This is because the coating does not form in a layered but in a columnar

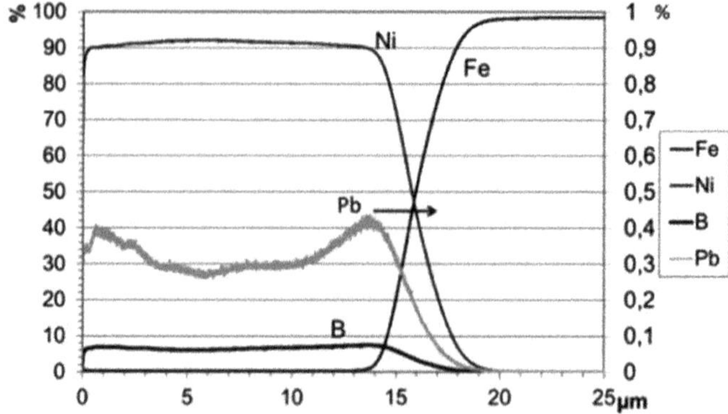

FIGURE 4A.3 GDOES depth profile of as-plated electroless nickel-boron coating. Left axis: Ni, B and Fe. Right axis: Pb. Lead content is multiplied 100 times to increase its visibility. (From Vitry, V. and Bonin, L., *Surf. Coat. Technol.*, 311, 164–171, 2017.)

manner, with deposited materials positioned not on a line parallel to the substrate but on the surface of a column. However, a higher concentration of boron at the interface between the coating and the substrate was found on a substrate presenting a honeycomb-like surface structure of the substrate (Wang et al. 2012).

The amount of stabilizer incorporated in the electroless nickel-boron deposits can also be heterogeneous through the coating. The thallium repartition in the deposit is not reported. However, GDOES analysis has detected variations in lead concentration in electroless nickel-boron coatings (Vitry et al. 2012c; Clerc 1986), with a higher lead content close to the interface with the substrate and also near to the free surface of the coating (Clerc 1986) (Figure 4A.3).

Moreover, the plating bath temperature also has an influence on the lead repartition in electroless nickel-boron coatings (Vitry and Bonin 2017). An increase in bath temperature from 95°C to 96.5°C leads to a continuous increase from the interface to the free surface (Figure 4A.4).

The non-replenishment or replenishment in a discontinuous manner of electroless nickel-boron baths increases these local chemical heterogeneities due to the decrease in reactives content in the plating bath. Vitry et al. (2012a) have studied the evolution of reactive concentration in the bath during plating process without replenishment to obtain information about reaction kinetics and to propose a replenishment process. The $NaBH_4$ consummation occurs in a smooth regular manner throughout the plating. After 30 min of plating, half the initial borohydride content is consumed and the consummation rate decreases after this time. The batch replenishment in reducing agent should thus be made no more than 30 min after the beginning of plating. For lead, this consummation evolves in a less smooth manner than borohydride. They have also shown that the $NaBH_4$ reaction is immediate when reactive arrives at the surface of the substrate and that rate-limiting step was the diffusion of reactive toward the reaction site. On the contrary, the consumption of lead is not limited by diffusion; the reaction itself is the rate-limiting step. Moreover, the composition of the coating is not intrinsically linked to the batch chemistry. The efficiency of the plating and the coating chemical composition are influenced by the bath load (that is

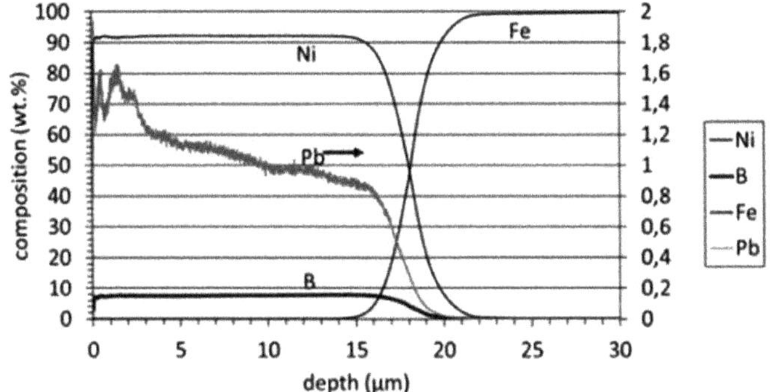

FIGURE 4A.4 GDOES depth profile of as-plated electroless nickel-boron coating at a plating temperature of 96.5°C. Left axis: Ni, B and Fe. Right axis: Pb. Lead content is increased 100 times for better visibility. (From Vitry, V. and Bonin, L., *Surf. Coat. Technol.*, 311, 164–171, 2017.)

the ratio of the bath volume to the sample surface). Finally, they proposed a way to predetermine the amount of each reactive to add to replenish a bath and to allow the synthesis of thick coatings (up to 36 μm).

4A.2 INITIATION MECHANISM OF ELECTROLESS NICKEL-BORON COATINGS

The initiation mechanism of electroless nickel-boron coating depends on the nature of the substrate and the reducing agent used in the plating bath. In fact, DMAB is usually used in acid bath, but some alkaline baths are also reduced by DMAB. Sodium borohydride is only used in alkaline bath because borohydride ions readily hydrolyze in acid or neutral solutions (Duncan and Arney 1984; Watanabe and Tanabe 1983; Sudagar et al. 2013). Thus, direct plating can be performed on steel (Vitry et al. 2012c), copper (Hamid et al. 2010) and magnesium (Wang et al. 2012) substrates from alkaline and/or acid baths. This is not the case for aluminum substrate. Unless aluminum is catalytically active for plating, direct plating is not possible due to the spontaneous formation of a passivation layer on its surface that impedes plating (Vitry and Delaunois 2015b). It is thus necessary to subject the aluminum substrate to a double zincate process (Delaunois et al. 2000). The porous zinc layer obtained on the substrate is later dissolved in the plating bath allowing electroless deposition. Moreover, with alkaline baths, it is not possible to directly plate the aluminum substrate due to the high pH of the plating solution. It is thus necessary to protect the substrate by a nickel strike or a thin electroless nickel-phosphorus layer from acid bath (Delaunois et al. 2000). The initiation mechanism of nickel deposition is thus influenced by those previous steps of electroless nickel-boron plating.

The initiation mechanism of electroless nickel-boron coating from bath reduced by sodium borohydride and stabilized with lead salt was extensively studied by Vitry et al. (2010, 2012c) on mild steel. As mild steel is less noble than nickel (−0.47 V vs. Normal Hydrogen Electrode (NHE) for Fe^{2+}/Fe against −0.27 V vs. NHE for Ni^{2+}/Ni), the initiation of deposition processes by a displacement reaction. The experiment was performed with a bath exempt of reducing agent. The reasoning was as followed: if the displacement is the actual initiation mechanism, it will occur whether or not there is sodium borohydride in the bath; if the process initiates by catalytic oxidation of sodium borohydride, the reaction will not occur without reactive agent. The results showed that no nickel deposition occurs after 4 min of immersion. However, lead was identified on the substrate surface. It reduced on steel substrate rather than nickel because its redox potential is higher than that of nickel (−0.13 V vs. NHE against −0.27 V vs. NHE, respectively). In conclusion, the initiation mechanism of electroless nickel-boron coating from bath reduced by sodium borohydride and stabilized with lead salt is a displacement reaction with the reduction of lead at the steel substrate.

Wang et al. (2012) have studied the initiation mechanism of electroless nickel-boron coating from bath reduced by sodium borohydride on magnesium alloy. After pickling, the substrate is partially covered by a porous magnesium film. The nucleation of Ni occurs around the edges of magnesia film pores through reaction (4A.4).

$$Mg + Ni^{2+} \rightarrow Mg^{2+} + Ni \tag{4A.4}$$

4A.3 STRUCTURE, MORPHOLOGY AND SURFACE ASPECT OF ELECTROLESS NICKEL-BORON COATINGS

4A.3.1 Structure

The structure of electroless nickel-boron deposits is linked to their chemistry and to further applied heat treatments.

In the as-deposited state, electroless nickel-boron is known to be a metastable supersaturated alloy (supersaturated solid solution of boron in nickel) (Delaunois and Lienard 2002; Vitry et al. 2010). The equilibrium nickel-boron phase diagram (Figure 4A.5) shows no solid solubility of boron in nickel at ambient temperatures (Vitry et al. 2012b).

The crystallographic data of NiB binary system including intermetallics are given in Table 4A.2 (Mut 2015).

Moiseev et al. n.d. have calculated the values of ΔH^o_{298}, S^o_{298}, $H^o_{298} - H^o_0$, T_m, ΔH of melting, $C_p(T)$ of crystals and C_p at $T>T_m$ for NiB, Ni_2B, Ni_3B and Ni_4B_3 phases (Table 4A.3).

The structure of electroless nickel-boron coatings depends on the boron content (Wang et al. 2012; Gorbunova et al. 1973; Baskaran et al. 2006; Zhang n.d.). For coatings with a boron content close to 6 wt.% B, nickel-boron alloys consist basically of pure nickel and Ni_3B phases; for more than 6 wt.% B, Ni_2B will also be found. However, in electroless plating, intermetallic compounds cannot be formed. Actually, the system does not achieve equilibrium conditions and boron atoms are trapped between nickel atoms, resulting in supersaturation (Weil and Parker 1990). These trapped boron atoms are perturbing the nickel face-centered-cubic (FCC) crystal structure and the arrangement is impossible for large surfaces. The structure

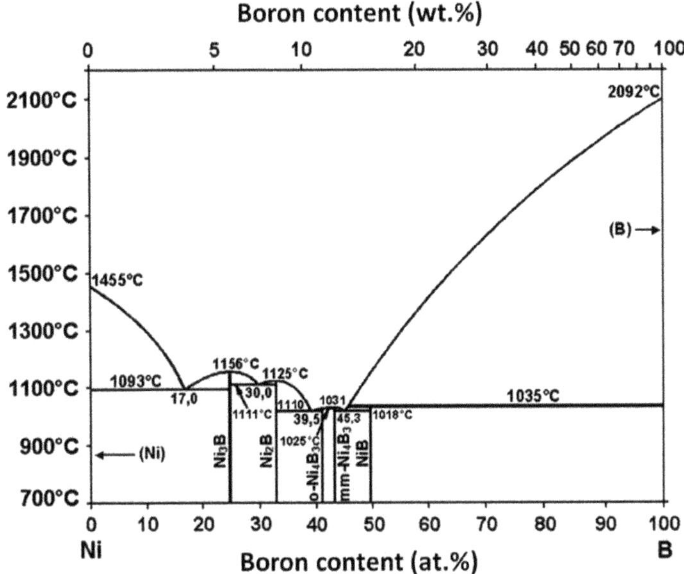

FIGURE 4A.5 Nickel-boron phase diagram.

TABLE 4A.2
Crystallographic Data of NiB Binary System Including Intermetallics

Phase	Structure	Type
(Ni)	Face-centered cubic (FCC)	Cu
Ni_3B	Orthorhombic	Fe_3C
Ni_2B	Tetragonal	Al_2Cu
Ni_4B_3 (o)	Orthorhombic	B_3Ni_4 (o)
Ni_4B_3 (m)	Monoclinic	B_3Ni_4 (m)
NiB	Orthorhombic	CrB
(B)	Rhombohedral	B

TABLE 4A.3
Values of ΔH^o_{298}, S^o_{298}, $H^o_{298} - H^o_0$, T_m, ΔH of Melting, $C_p(T)$ of Crystals and C_p at $T>T_m$ for NiB, Ni_2B, Ni_3B and Ni_4B_3 Phases

Phase	$-\Delta H^o_{298}$ (J/mol)	S^o_{298} (J/mol. K)	$H^o_{298} - H^o_0$ (J/mol)	T_m (K)	$-\Delta H$ (J/mol)	$C_p(T)$ (J/mol K)	C_p at $T>T_m$ (J/mol K)
NiB	82420	30.12	5270	2136	41290	$42.93+14.60\times10^{-3}\times T$ $-11.25\times10^{3}\times T^{-2}$	78.7
Ni_2B	110190	47.15	8200	1786	52610	$65.62+21.48\times10^{-3}\times T$ $-16.53\times10^{3}\times T^{-2}$	110.8
Ni_3B	98920	62.86	10860	1751	69310	$87.50+28.64\times10^{-3}\times T$ $-22.04\times10^{3}\times T^{-2}$	146.8
Ni_4B_3	293800	114.6	83100	1853	126220	$155.98+49.12\times10^{-3}\times T$ $-37.78\times10^{3}\times T^{-2}$	263.1

is thus considered amorphous (Delaunois and Lienard 2002; Baskaran et al. 2006; Hamid et al. 2010; Krishnaveni et al. 2005; Shakoor et al. 2016; Lu 1996), as shown in Figure 4A.6.

Theoretically, the observed broad peak in Figure 4A.6 corresponds to a disorder in the arrangement of atoms. Moreover, for electroless coating, the extent of segregation of metalloid alloy determines its crystallinity. For boron content of approximately 6 wt.%, the required boron segregation is relatively large, preventing Ni phase nucleation and resulting in an amorphous structure (Krishnaveni et al. 2005).

Consequently, low-alloy electroless nickel-boron deposits are considered microcrystalline or nanocrystalline due to the small number of boron atoms trapped in the nickel lattice, and electroless nickel-boron coatings containing a boron content

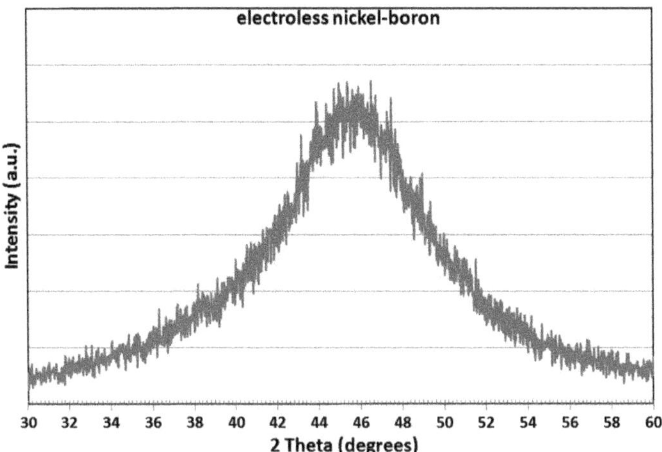

FIGURE 4A.6 XRD patterns of as-deposited electroless nickel-boron coating. (From Vitry, V. et al., *Surf. Coat. Technol.*, 206, 3444–3449, 2012d.)

of more than 5 wt.% are amorphous due to the decrease in grain size of electroless nickel-boron coatings with increasing boron content (Riedel 1989; Gorbunova et al. 1973; Baskaran et al. 2006; Watanabe and Tanabe 1983; Krishnaveni et al. 2005; Vitry et al. 2012d).

However, the notions of "amorphous" and "nanocrystalline" for electroless nickel-boron coatings are not obvious. A nanocrystalline material is a polycrystal with grain size in the nanometer range (less than 100 nm in at least one direction) (Vitry 2010). It is characterized by a large volume fraction of grain boundaries (or interface). For spherical grains, this volume fraction is approximately 50% for 5 nm grains and 3% for 100 nm grains (Lu 1996).

Many authors have discussed the amorphous nature of the electroless nickel-boron coating (Sudagar et al. 2013; Wang et al. 2012; Saito et al. 1998; Watanabe and Tanabe 1983; Krishnaveni et al. 2005; Narayanan and Seshadri 2004). And to summarize, many previous works that studied the structure of electroless nickel-boron coatings have wrongly described them as amorphous, while they are in fact nanocrystalline (Vitry 2010).

A recent transmission electron microscopy (TEM) studies have shown the presence of microcracks through the coating, and the analysis performed at the cracks/boundary regions indicated the presence of nanometer-sized crystals concentrated in these regions (Zhang n.d.; Narayanan and Seshadri 2004). The existence of a structural order in as-deposited electroless nickel-boron coatings with high boron content was shown by TEM (Figures 4A.7 and 4A.8) (Vitry et al. 2012d; Vitry 2010; Kumar and Nair 1994), with a grain size close to 1 nm for 5 wt.% boron content (Vitry et al. 2012d). The electroless nickel-boron coating is thus not fully amorphous (Vitry et al. 2012d). In fact, in the as-plated state, the electroless nickel-boron coating consists of two phases: the major part is amorphous, whereas a small portion is nanocrystalline (Zhang n.d.).

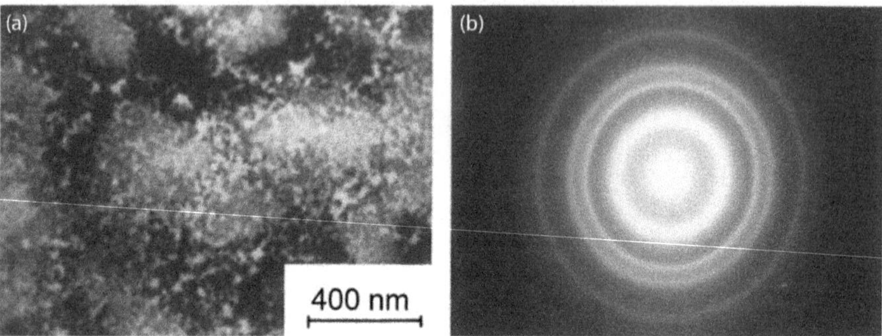

FIGURE 4A.7 (a) TEM micrograph of electroless nickel-boron deposit with 4.32 wt.% B in as-deposited state, parallel to the surface; (b) Selected Area Diffraction (SAD) pattern corresponding to micrograph presented in (a). (From Kumar, P.S. and Nair, P.S., *Nanostruct. Mater.*, 4, 183–198, 1994.)

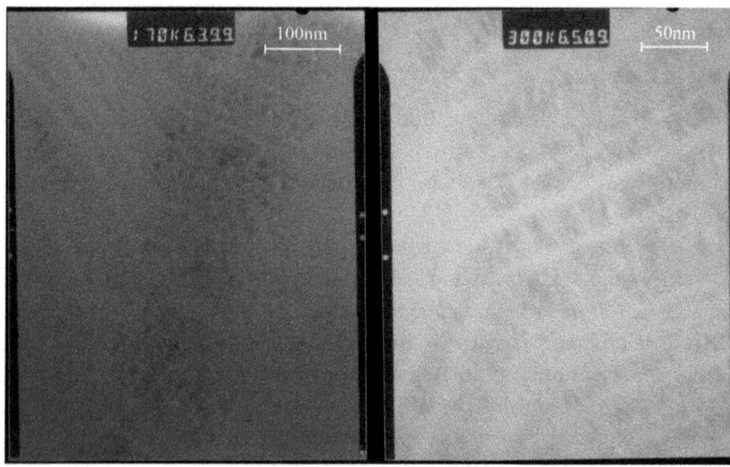

FIGURE 4A.8 TEM micrograph of as-plated electroless nickel-boron deposit with 6 wt.% B (along growth direction). (From Vitry, V. et al., *Surf. Coat. Technol.*, 206, 3444–3449, 2012d.)

Vitry (2010) has made a grain-size modeling in as-deposited electroless nickel-boron coatings using the following equation (Hentschel et al. 2000; Färber et al. 2000):

$$d = \frac{\delta}{1 - \sqrt[3]{\frac{C_0 - C_{gb}}{C_g - C_{gb}}}} \tag{4A.5}$$

The grain boundaries thickness (δ) was fixed at 0.8 nm and the boron concentration in the crystallites (C_b) at 1 at.%. The boron concentration in the grain boundaries (C_{gb}) was calculated using the law that links content and density of grain boundaries

TABLE 4A.4
Boron Content of the Grain Boundaries of Electroless Nickel-Boron Coatings with Various Boron Content

wt.% B in Coating	at.% B in Coating	Volume Fraction of Boundaries	at.% in Boundaries
3.85	17.86	87.5	20.26
4.32	19.69	96.15	20.43

Source: Vitry, V., Electroless Nickel-Boron Deposits: Synthesis, Formation and Characterization; Effect of Heat Treatments; Analytical Modeling of the Structural State, University of Mons, Belgium, 2010.

(results are given in Table 4A.4) and thus was fixed at 20 at.%. The term C_0, which is the average boron content of the material, is the variable of the model.

The results of the modeling of the evolution of the grain size with the boron content are given in Figure 4A.9. Vitry (2010) and Vitry et al. (2017) and other published results (Mut 2015; Gaevskaya et al. 1996; Bekish et al. 2010; Kostyanovskii et al. 1991) were used to validate the model.

From Figure 4A.9, it can be said that electroless nickel-boron deposits containing more than 4.5 wt.% B are amorphous. Deposits with 6 wt.% B or more are thus generally considered amorphous (Delaunois and Vitry 2015d; Wang et al. 2012).

In conclusion, the grain size of electroless nickel-boron deposits decreases with the increase in boron content; for high boron content, the grain size is so small that the coating, even pseudo-amorphous, keeps low distance order.

To summarize, deposits containing medium-to-high levels of boron content are amorphous, while deposits with lower levels are nanocrystalline.

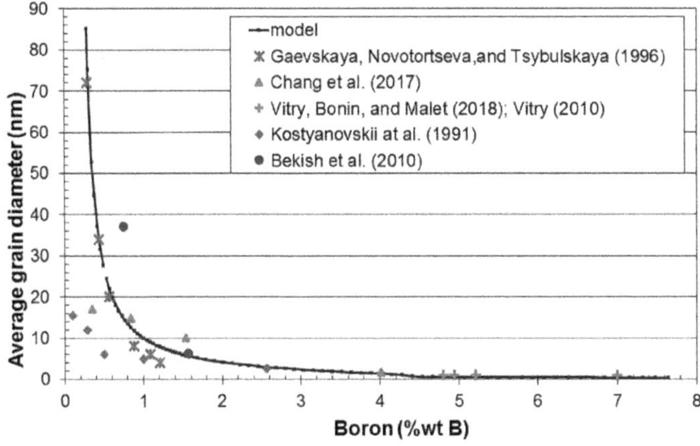

FIGURE 4A.9 Evolution of grain size with boron content in electroless nickel-boron coating, and comparison of model and experimental results.

The performing of heat treatments to electroless nickel-boron deposits bring enough energy to revert them to equilibrium state, leading to the crystallization of the coating structure with the appearance of a mixture of Ni and Ni_3B or Ni_3B and Ni_2B phases, depending on the coating chemistry (Delaunois and Lienard 2002; Hamid et al. 2010; Vitry et al. 2010).

Hamid et al. (2010) have studied the influence of plating temperature on the structure of electroless nickel-boron coating performed with bath reduced by DMAB. They showed that as-deposited electroless nickel-boron coatings are amorphous whatever the bath temperature (60°C or 80°C). After heat treatment at 250°C, electroless nickel-boron coatings performed at 60°C are fully crystallized, while electroless nickel-boron coatings performed at 80°C are a mixture of amorphous and microcrystalline. With heat treatment at 400°C, both are crystalline.

4A.3.2 Morphology and Surface Aspect

Electroless nickel process has a significant advantage to produce uniform thickness whatever the substrate geometry (Weil and Parker 1990). Moreover, the columnar morphology of electroless nickel-boron coatings allows retaining lubricants under conditions of adhesive wear (Gavrilov 1979; Baudrand 2002). Due to its granular structure, electroless nickel-boron coatings are naturally lubricious and can achieve a higher wear resistance by reducing the surface contact (Sudagar et al. 2013; Delaunois and Lienard 2002).

Actually, the morphology of electroless nickel-boron deposits is influenced by their synthesis process: reagents diffusion in the liquid phase and their adsorption onto the substrate surface are two important parameters. Due to the coating initiation into the island Volmer-Weber regime, observed for borohydride-reduced baths, the morphology is columnar (Vitry et al. 2012c). Thus, the surface aspect of the electroless nickel-boron coatings presents a typical cauliflower-like texture (also described as blackberry-like) (Figures 4A.10 and 4A.11) (Delaunois 2002; Contreras et al. 2006; Srinivasan et al. 2010; Krishnaveni et al. 2005; Vitry et al. 2010; Bülbül et al. 2012), which can also be observed, but at a smaller scale, on electroless nickel-phosphorus coatings (Vitry and Delaunois 2015c; Zhang n.d.).

This surface texture is linked to the coating growth mechanism: formation of nodules in the beginning of deposition and further growth in a columnar morphology (Figure 4A.11). In fact, in the first stage of deposition, a thick diffusion layer forms near to the deposit when the bath depletes in reactive species, slowing down the growth of the columns and inducing a new germination phase. Due to the presence of other growing columns, the diffusion layer is laterally confined. As it is protected from disruption by convection at the edge of columns, its thickness is higher than on their tops (Figure 4A.12) (Rao et al. 2005; Vitry 2010). As the coating growth is quicker at the top of the columns, this leads to the formation of the cauliflower-like texture.

Figure 4A.13 shows that, depending on the direction of observation, as-plated grains of X-ray amorphous electroless nickel-boron coating present two different morphologies: (i) a columnar texture with the grains length perpendicular to the

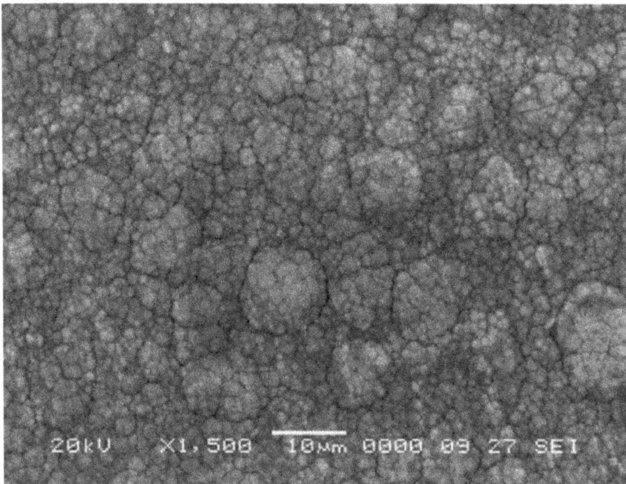

FIGURE 4A.10 Typical cauliflower-like surface texture of an electroless nickel-boron coating. (From Delaunois, F., Les Dépôts Chimiques de Nickel-Bore Sur Alliages d'aluminium, Faculté Polytechnique de Mons, 2002.)

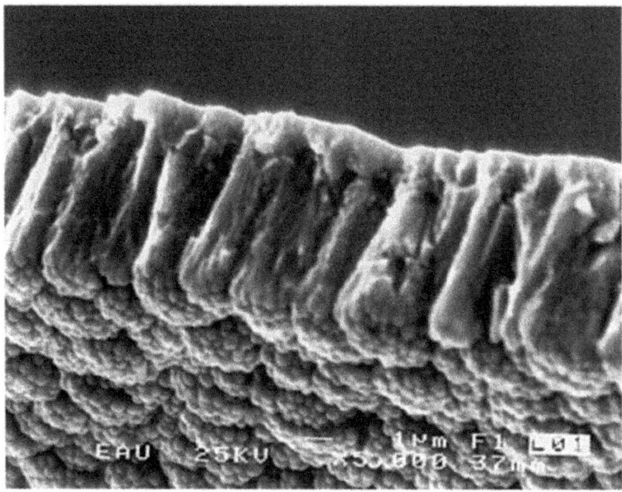

FIGURE 4A.11 Surface and cross-sectional view of electroless nickel-boron deposit. (From Bülbül, F. et al., *Met. Mater. Int.*, 18, 631–637, 2012.)

substrate surface due to the high spatial density of deposit nucleation on the substrate (Figure 4A.13a); (ii) semi-hemispherical (nodular) caps at the exposed surface leading to the typical cauliflower-like aspect (Figure 4A.13b). Moreover, the adsorption of some stabilizers onto the substrate surface can limit the lateral growth of electroless nickel-boron grains, resulting in a columnar deposit (Baskaran et al. 2006).

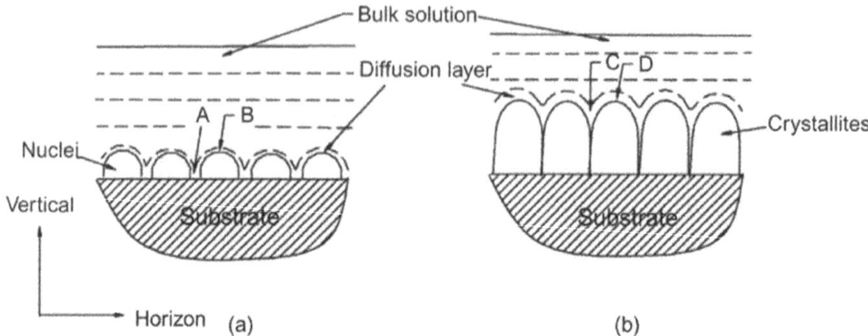

FIGURE 4A.12 Diffusion layer on top of electroless nickel-boron deposit during deposition and effect on the deposit formation: (a) nucleation period; (b) crystallite growing period. (From Rao, Q. et al., *Appl. Surf. Sci.*, 240, 28–33, 2005.)

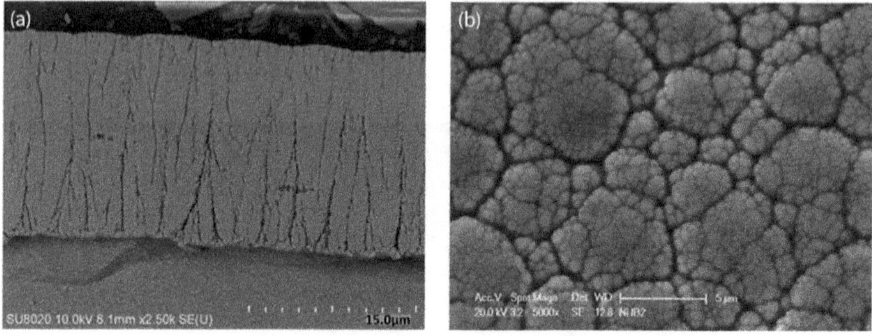

FIGURE 4A.13 SEM observation: (a) Cross-cut observations (columnar structure characteristic); (b) topographical observations (structure type: "cauliflower").

Bulbul (2011) has shown that, depending on the deposition parameters for electroless nickel-boron baths reduced with borohydride, six types of morphological structures can be observed: pea-like, maize-like, primary nodular, blackberry-like or grape-like, broccoli-like and cauliflower-like, which is the most amorphous one.

However, GDOES analysis has detected variations in the lead concentration in electroless nickel-boron coatings (Vitry et al. 2012c; Clerc 1986), with a higher lead content close to the interface with the substrate and also near to the free surface of the coating (Clerc 1986). The size of columns is thus slightly wider in zones with lower lead content (Figure 4A.14) (Vitry et al. 2012c).

4A.3.2.1 Influence of Bath Temperature

Bulbul (2011) has studied the effect of bath temperature with electroless nickel-boron reduced by borohydride. At 80°C, the structure is granular. The increase in bath

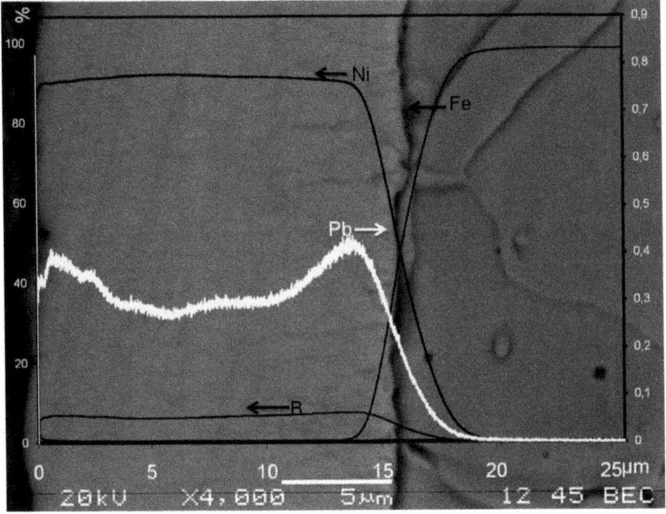

FIGURE 4A.14 Comparison of the composition (from GDOES) and morphology (from SEM) across electroless nickel-boron deposit. (From Vitry et al., Appl. Surf. Sci., 263, 640–647, 2012)

temperature to 95°C leading to an increase in diffusion, the grain size increases and the granular structure transforms into a primary nodular structure.

4A.3.2.2 Influence of Deposition Time

Due to the consumption of bath components, the nucleation and growth of crystal will vary, which will influence the morphology of the electroless nickel-boron coating (Zhang n.d.).

The results of the early coating deposition were obtained on various substrates such as steels (carbon and stainless steels) and magnesium (Wang et al. 2012; Vitry et al. 2010, 2012c; Clerc 1986; Correa et al. 2013; Vitry and Delaunois 2015d).

The formation process of electroless nickel-boron coatings obtained on mild steel (grounded up to 4000 Mesh SiC paper) from baths reduced by $NaBH_4$ as reducing agent was extensively studied by Vitry et al. (2008, 2010, 2011) (Figures 4A.15 and 4A.16).

The initiation of the process is the catalytic oxidation of $NaBH_4$, followed by the progressive formation of small nodules of nickel (10–20 nm in diameter) on the surface defects and scratches present on the substrate surface (after only 15 s of plating), that quickly colonize the while substrate. This step takes approximately 1 min. Actually, the initiation process occurs sooner on rougher substrates due to higher number of scratches and defects on the substrate surface, but it does not go faster (Vitry et al. 2010). The density of the nodules increases with the increase in roughness substrate and their size decreases, with a nodule size of approximately 50 nm for unpolished surface. Consequently, the nucleation of nodules is facilitated by a high roughness (Vitry et al. 2010).

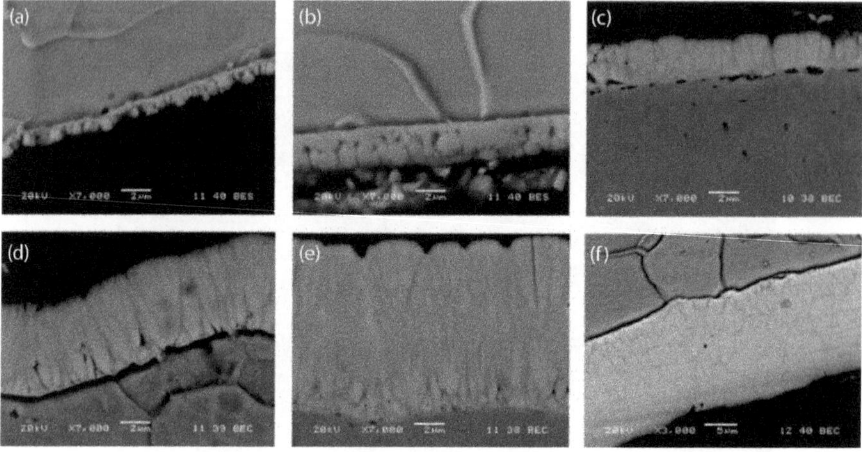

FIGURE 4A.15 Cross-sectional observation of mild steel (St-37) samples after an immersion for 90 s (a), 4 min (240 s) (b), 7 min (c), 10 min (d), 30 min (e) and 1 h (f) in the electroless nickel-boron plating bath. (From Vitry et al., Appl. Surf. Sci., 263, 640–647, 2012.)

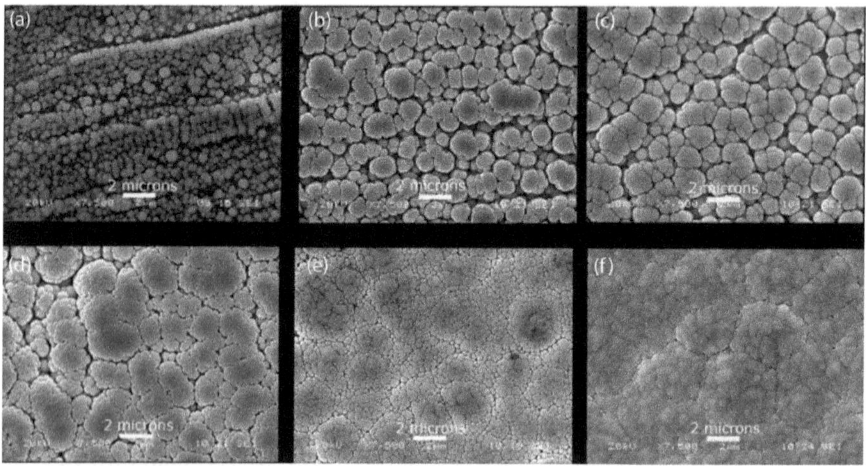

FIGURE 4A.16 Surface observation of mild steel (St-37) samples after an immersion for 90 s (a), 4 min (240 s) (b), 7 min (c), 10 min (d), 30 min (e) and 1 h (f) in the electroless nickel-boron plating bath. (From Vitry et al., Appl. Surf. Sci., 263, 640–647, 2012.)

Vitry et al. (2015d) have studied the formation of electroless nickel-boron coatings from bath reduced with $NABH_4$ and stabilized with lead tungstate on various steel substrates (mild steel, low-alloyed carbon steel, cryogenic steel, austenitic stainless steel and duplex stainless steel). The initial roughness R_a of all the substrates is approximately 0.15 μm. They have confirmed that the initiation of the deposition process is linked to spontaneous oxidation of the reducing agent on the substrate rather than a displacement process. They have also shown that the presence of nickel in the substrate chemical composition accelerated the initiation and the deposition process. The presence of multiple phases is not a favorable factor but the presence of numerous grain boundaries is an accelerator factor. On all the substrates, the deposition process always initiated on scratches and/or grain boundaries. The columnar morphology of the electroless nickel-boron coatings was due to the vertical growth of numerous islands that were refined during the plating process, providing a sheaf-like morphology. The refining episodes are accompanied by a decrease in the growth rate for all substrates (excepted for the low-alloyed carbon steel) between 4 and 10 min and can be linked to the lead content evolution in the deposit. The deposition process continued with continuous changes in the width of the columns and the growth rate appeared to be more regular during the later stages of the plating process. Finally, due to bath depletion, a decrease in plating rate is observed at the end of the process.

The deposition process was also studied by Clerc (1986) on Armco steel (polished with 1 µm diamond). On the surface, after 5 min of immersion, the deposit thickness is approximately 0.5 µm with a very fine columnar structure with an average column diameter of 0.1–0.3 µm. After 15 min of immersion, with a deposit thickness of approximately 2 µm, the structure evolves to a walled structure with large smooth plates with a size of 6 µm lined with cracks. This structure appears as a transitional structure between two columnar structures whose columns' average size increases from 0.1–0.3 µm to 0.5 µm. Thus, after 25 min of immersion and a thickness of 4 µm, the structure evolves as a columnar one. After 35 min and a thickness of 5 µm, the structure is a mixture of columnar (0.2–0.6 µm of diameter) and walled one (with plates size to 3 µm) with cracks. From 40 to 60 min of immersion, the structure is regularly columnar with a diameter of 0.5 to 1.0 µm. This observation on cross section shows that the colonization of the surface is due to the growth of numerous nuclei. Some columns are growing preferentially to others blocking their growth. A second stop phase of columns growth leads to the formation of a small number of increasingly larger columns. The more the deposit grows, the more the number of columns decreases but more their size increases. This mechanism of columns growth could explain the roughness increase of the deposit with its thickness. Clerc has also proved that the electroless nickel-boron coatings morphology is linked to the operating conditions and to the substrate roughness.

On magnesium substrate, the process begins slightly later (after 30 s) with less numerous and larger nodules, leading to a slower spreading and a complete coverage reached after 10 min (Correa et al. 2013; Wang et al. 2012). The germination rate is thus slower on magnesium and its alloys than on steel (Vitry and Delaunois 2015a).

After the initial coverage of the substrate surface, the growth of the coating leads to the formation of three layered zones: (i) a first layer of 2.5 µm thickness with columns of 0.1 µm thickness; (ii) a second layer nearly twice as thick containing wider columns; (iii) a top layer going up to the surface with columns of 1 µm thickness (Clerc 1986). The columnar structure is thus gradually formed, beginning with the presence of "balloon-like" features, then with the filling of the gaps between them (Figure 4A.15) (Vitry et al. 2012c). Figure 4A.16 shows variations in the size of the surface texture with deposition time, which lead to morphological modifications in cross sections (Vitry et al. 2012c).

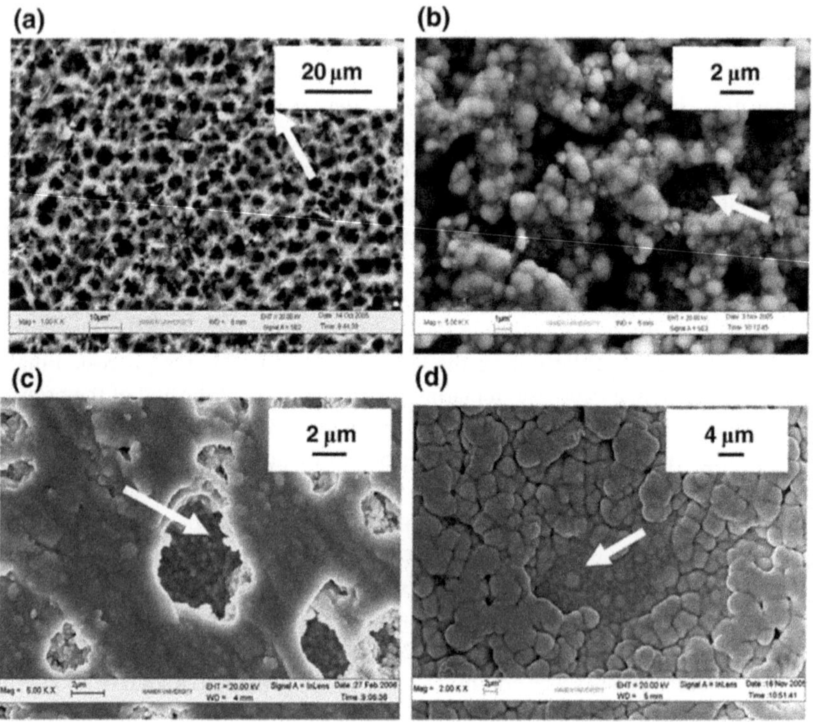

FIGURE 4A.17 SEM images of the electroless nickel-boron coating on AZ91D as a function of plating time: (a) 0 min, (b) 5 min, (c) 10 min and (d) 30 min. (From, Wang, Z.C. et al., *Surf. Coat. Technol.*, 206, 3676–3685, 2012.)

Wang et al. (2012) have studied the formation process of electroless nickel-boron coatings obtained on magnesium alloy from baths reduced by $NaBH_4$ as reducing agent (Figure 4A.17). After pickling (Figure 4A.17a), a honeycombed porous film is observed, as a result of Mg corrosion product film or magnesia film. After 5 min of immersion (Figure 4A.17b), clusters of a mixture of Ni plating and Mg corrosion products are formed on the substrate surface. After 10 min, the surface is overlaid by a film. Figure 4A.17c shows that crowdedly distributed small particles are still visible. After 30 min of plating, a complete coverage of the substrate by deposited Ni film is observed (Figure 4A.17d).

4A.3.2.3 Influence of the Reducing Agent Content

The morphology of electroless nickel-boron deposit depends on the borohydride concentration of the plating bath (Baskaran et al. 2006; Zhang n.d.). Actually, as previously said, the increase in $NaBH_4$ concentration in the bath leads to an increase in the boron content in the coating. The size of the nodules increases, combining to form a granular-type structure (Baskaran et al. 2006). Consequently, the amount of $NaBH_4$ is the most important parameter influencing the growth of electroless nickel-boron coating, and thus its morphology, among other process parameters such as bath temperature, deposition time and $NiCl_2$ concentration (Baskaran et al. 2006).

4A.3.2.4 Influence of the Stabilizer

The columnar-lamellar morphology is strongly related to the presence of a stabilizer that limits lateral growth of the coating, resulting in a finer column structure in areas where the stabilizer content of the coating is higher (Vitry and Bonin 2017).

Moreover, Bonin (2018) has demonstrated that the kind of stabilizer and its concentration strongly influence the coating morphology. In fact, as the simplest form of chemical plating is the so-called metal displacement reaction or galvanic displacement, this type of reaction is the first step toward the autocatalytic deposition process. Just after the stabilizer addition in the plating bath, the galvanic displacement occurs between the substrate and the stabilizer. The rate of this reaction will depend on two factors: the stabilizer concentration in relation to the substrate total surface and the redox potential difference (ΔE) between substrate and stabilizer. For example, the displacement reaction rate between iron and tin will be slower than that between iron and bismuth (element with a higher redox potential). The presence of high concentration of stabilizers on the catalytic surface of the substrate will block the lateral growth of nickel and leads to the formation of columns. Consequently, the columnar growth is correlated with the stabilizer concentration on the substrate surface: an increase in concentration is responsible for thinner columns. For example, she has compared the influence of Pb^{2+}, Tl^{+} and Bi^{3+} stabilizers on electroless nickel-boron typical columns size (Table 4A.5). As Bi^{3+} has a much higher redox potential, a more intense displacement reaction is observed. The columns size of the coating generated by this stabilizer is lower.

The size of the stabilizer atoms can also influence the morphology. During the first step of deposition, just after the galvanic displacement, the first layer of nickel is autocatalytically deposited between the stabilizer atoms. If the stabilizer atoms are much larger than the nickel atoms, when the second layer of nickel is plated, its lateral growth is blocked by the stabilizer atoms, giving rise to the formation of a columnar structure. If the size of stabilizer atoms is close to that of nickel, the second layer is just partially blocked and a partial lateral growth is possible.

TABLE 4A.5
Comparison of Redox Potential and Columnar Size of the Electroless Nickel-Boron Coating Stabilized with Thallium, Lead or Bismuth Salts

Stabilizer	Tl^{+}	Pb^{2+}	Bi^{3+}
Redox potential (V)	−0.336	−0.126	0.293
Columns size (μm)	0.5–3.5	0.1–2.0	0.1–1.5

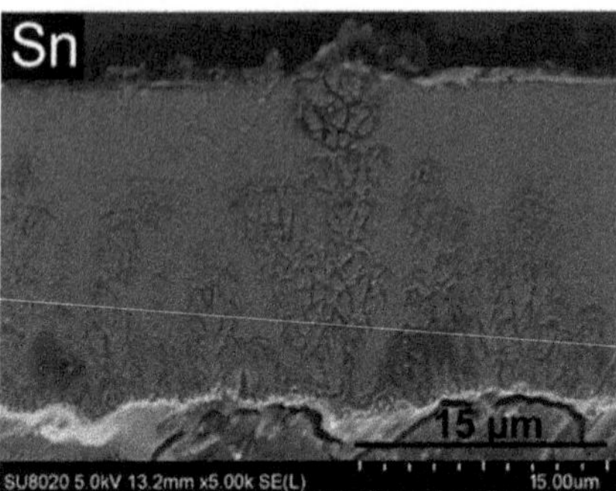

FIGURE 4A.18 Influence of the stabilizer atomic size on the coating morphology: Left—Sn salt (tree-like structure); right—Pb salt (columnar structure).

For example, Pb and Sn have the same redox potential (−0.126 V) but Pb atoms are larger than nickel and Sn atoms are smaller than nickel. Baths stabilized with Pb salts generate a columnar deposit morphology and Sn generates a tree-like structure (Figure 4A.18).

In conclusion, by controlling the type of stabilizer and its concentration in the plating bath and by controlling the ratio between the atomic sizes of the nickel and the stabilizer, different morphologies of electroless nickel-boron coatings can be obtained.

Bonin et al. (2018a) have also studied the influence of the anionic part of the lead stabilizer on the electroless nickel-boron coatings structure. They concluded that the use of various lead stabilizers had no effect on the structure (from X-ray diffraction [XRD] analyses) and little effect on the morphology and roughness of the coatings.

4A.3.2.5 Deposition Rate

The deposition rate is influenced by many plating bath parameters such as types and concentrations of reducing agent, stabilizer and complexing agent, the concentration

TABLE 4A.6
Action of Various Plating Bath Parameters on the Electroless Nickel-Boron Deposition Rate

Parameters of the plating bath. Action: INCREASE	Influence on the electroless nickel-boron deposition rate
Bath temperature	Quasi-linearly increase
Type of reducing agent	Higher deposition rate with $NaBH_4$ than with DMAB
Concentration of reducing agent	Increase
Concentration of stabilizer	Increase
Concentration of complexing agent	Maximum for a critical concentration
Concentration of NaOH	Increase
Immersion time	Decrease if no replenishment

of NaOH, plating bath temperature and immersion time. Table 4A.6 summarizes the action of these parameters on the deposition rate.

Moreover, the type of substrate is an important factor influencing the deposition rate. Actually, the deposition rate is higher on steel substrate than that on copper one (Hamid et al. 2010).

It is generally admitted that the deposition rate increases quasi-linearly with the bath temperature (Delaunois et al. 2000; Delaunois 2002; Srinivasan et al. 2010; Srivastava et al. 1992) and increases with the amount of reducing agent (Anik et al. 2008; Srinivasan et al. 2010; Baskaran et al. 2006; Oraon et al. 2007; Gorbunova et al. 1973) and with the amount of stabilizer (Anik et al. 2008). The increase in the deposition rate with temperature is due to the catalytic chemical reactions that are temperature-dependent (Srivastava et al. 1992). Moreover, the rate of deposition time as a function of plating time is not linear throughout and decreases with time (Hamid et al. 2010). This may be due to the decrease in reactants content in the plating bath but also due to the increase in the amount of oxidation products in the bath (Riedel 1989; Hamid et al. 2010; Gavrilov 1979).

The influence of the reducing agent content is the following: the deposition rate increases with an increase in sodium borohydride (Srivastava et al. 1992). At 95±1°C (temperature usually used for this kind of bath (Vitry and Bonin 2017), the electroless nickel-boron deposition rate is approximately 18–20 μm/h at 0.8 g/L of $NaBH_4$ and 25–30 μm/h at 1.05 g/L (Baskaran et al. 2006). The deposition rate for bath reduced with DMAB is approximately 6 mg/cm^2.h for 0.025 M of DMAB and 8.3 mg/cm^2.h for 0.10 M of DMAB (Saito et al. 1998). It was observed that in case of the use of DMAB, the deposition rate increases with DMAB content and remains constant after a certain DMAB concentration (Srivastava et al. 1992).

The threefold increase of thallium acetate concentration causes an almost two-fold increase of the deposition rate: the electroless nickel-boron deposition rate is approximately 13 μm/h at 12 mg/L of thallium acetate and 25 μm/h at 35 mg/L. This is due to the better stability of the plating bath with the increase in stabilizer concentration (Anik et al. 2008). Bonin et al. (2018a) have studied the influence of the anionic part of the lead stabilizer on the deposition rate. They concluded that the thickest electroless nickel-boron coating was obtained for bath stabilized with lead tungstate. The use of either lead nitrate or sulfate led to slightly thinner coatings containing more lead and boron than the tungstate-stabilized one. Lead chloride led to significantly thinner coatings with a significant content of lead.

The influence of the complexing agent concentration on the deposition rate was studied by (Anik et al. 2008). They found that below 90 g/L of ethylenediamine, which is the critical concentration to obtain a good efficiency of nickel deposition, the deposition rate reduces sharply; above this value, there is a slight decrease. Srinivasan et al. (2010) have also observed the same phenomenon. For DMAB-reduced bath, the increase in complexing agent content decreases the deposition rate due to the decrease in free nickel ions (Saito et al. 1998). However, an acceleration of deposition rate can be found at low concentration due to the buffering action inhibiting the pH change in the reaction layer. Moreover, the complexing agent having smaller stability constant to nickel ions leads to a higher deposition rate (bath complexed with sodium citrate has a lower deposition rate than with succinic acid) (Saito et al. 1998).

An increase in the sodium hydroxide concentration increases the deposition rate (Anik et al. 2008). This is predictable from Equation (4A.1) that is driven by hydroxide ions.

The deposition rate is also influenced by the immersion time due to the consumption of reducing agent and stabilizer (Delaunois 2002; Delaunois et al. 2000; Saito et al. 1998; Vitry and Delaunois 2015a) and the accumulation of oxidation product of borohydride in the plating bath (Krishnaveni et al. 2005).

The immersion time influences the coating thickness (Delaunois et al. 2000; Vitry et al. 2010, 2012c). In the first time of immersion, up to 4 min, the growth rate is close to 40 μm/min. After this first quick growth, between 4 and 7 min of immersion, the deposition rate decreases; the coating thickness doesn't increase due to the densification process of nickel nodules (which was proved by the evolution of weight gain with immersion time). For longer immersion time, a second quick deposition rate is observed; the thickness grows quasi-linearly for up to 1 h of immersion with an average growth rate of approximately 18 μm/h (Vitry et al. 2012d). This decrease in plating rate is due to the depletion of the bath (Delaunois et al. 2000; Bonin 2018).

There is a link between the evolution of the deposition rate and the lead content of electroless nickel-boron coatings. At a distance of 1 to 3 μm from the coating–substrate interface, the highest content of lead was found, linked to the low deposition rate observed between 4 and 7 min of immersion. The lowest lead content was found at a distance of 5 μm from the coating–substrate interface, where the columns are slightly wider, corresponding to the second phase of quick growth (after 7 min of immersion) (Vitry et al. 2012c).

Bonin (2018) has found an average thickness of electroless nickel-boron coatings obtained from bath reduced with sodium borohydride and stabilized with lead nitrate of approximately 16.25 μm.

4A.3.2.6 Bath Replenishment

The bath replenishment influences the thickness of the coating (Delaunois et al. 2000). Actually, the immersion time is linked to the coating thickness because of the decrease in reducing agent content in the bath. Bath replenishment with reducing agent and stabilizer after 30 min helps to maintain the deposition rate constant. This leads to the obtention of thicker and uniform coatings.

Bath replenishments also modify the chemistry of the plating bath, and thus similar modifications should be expected in the coating. As shown in Figure 4A.19a and b, corresponding to electroless nickel-boron coatings obtained from baths with the same chemistry, respectively, without and with replenishment, the replenishment process leads to the appearance of a sharp demarcation line in the coating, with the appearance of two successive layers, similar in aspect to what can be observed in coatings plated with two successive immersions in new electroless nickel-boron baths (Delaunois and Vitry 2015d; Vitry et al. 2012c). This was confirmed by GDOES analysis (Vitry et al. 2012c): a constant boron content was found inside each layer, with a slightly different boron content between the two layers.

The electroless nickel-boron bath replenishment corresponds to a readjustment of reactive level to the initial value that induces a new phase of germination of the coating.

This particular morphology limits the presence of transverse porosities, which would be interesting in some applications.

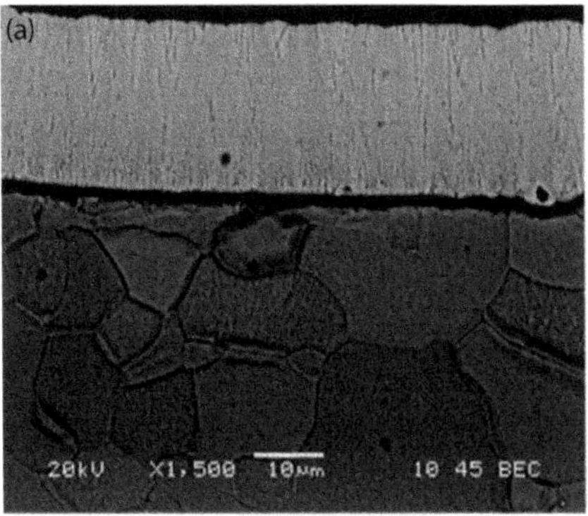

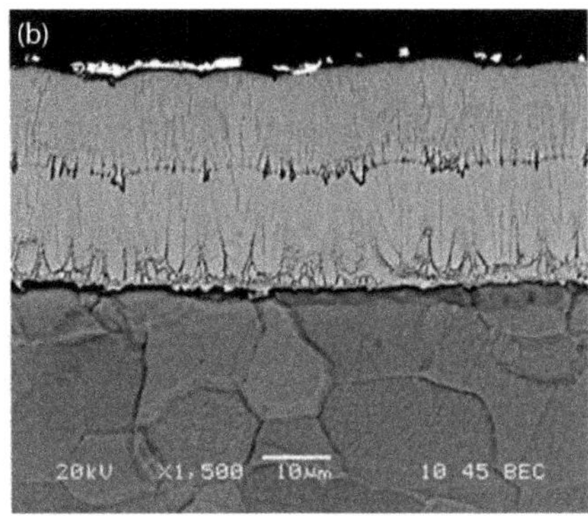

FIGURE 4A.19 Scanning electron microscope (SEM) cross-sectional images of electroless nickel-boron coatings (4 vol.% Nital etching): (a) coating obtained without replenishment; (b) coating obtained with bath replenishment after 30 min of plating time. (From Vitry, V. and Delaunois, F., *Appl. Surf. Sci.*, 359, 692–703, 2015d.)

4A.4 MECHANICAL PROPERTIES OF ELECTROLESS NICKEL-BORON COATINGS

Electroless nickel-boron deposits are the best candidate for mechanical and tribological properties. It is important to note that the properties of the electroless nickel-boron coatings are greatly influenced by their chemistry.

4A.4.1 Hardness

It is difficult to compare the hardness values given in literature because the electroless nickel-boron coatings hardness is strongly influenced by their chemistry (mainly boron content), the nature of substrate, the coatings grains size (for heat-treated coatings mainly) and even by the measurement method (influence of the indenter and the applied load) (Vitry 2010). Figure 4A.20 summarizes the evolution of microhardness measurements with boron content from various studies (Vitry and Bonin 2017; Srinivasan et al. 2010; Delaunois and Lienard 2002; Bonin 2018; Vitry et al. 2008; Krishnaveni et al. 2005; Vitry 2010; Mukhopadhyay et al. 2017; Chang et al. n.d.): for surface (HV) (Figure 4A.20a) and cross section (HK) (Figure 4A.20b) measurements.

As expected, Figure 4A.20 shows a high dispersion in the results from various studies. It is generally admitted that the increase in boron content of the deposit increases its hardness (Weil and Parker 1990) but this conclusion is not obvious from Figure 4A.20a. Moreover, the load used to perform microhardness measurements on cross section (Figure 4A.20b) influences the results: the higher the load, the lower is the hardness. However, it can be seen that electroless nickel-boron coatings show a high hardness in the as-plated state; values from approximately 500 to 950 HV_{100} have been reported by authors. This is in the range and often much harder than electroless nickel-phosphorus coatings that are close to 500–700 HV_{100} depending on the phosphorous content (Weil and Parker 1990) and very higher than as-deposited electrolytically deposited nickel

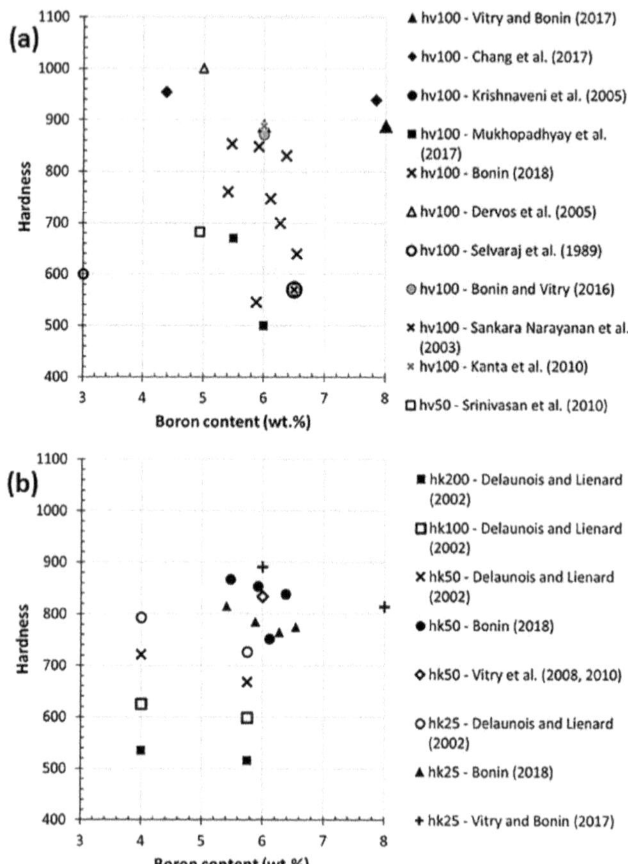

FIGURE 4A.20 As-plated electroless nickel-boron coating hardness with coating boron content: (a) on the surface; (b) on the cross section.

with a value of 150–400 HV_{100} (Keong et al. 2004). Actually, the values of 700 HV_{100} and above are generally reported in literature in the as-deposited state (Sudagar et al. 2013; Anik et al. 2008; Delaunois and Lienard 2002; Vitry et al. 2011; Krishnaveni et al. 2005; Pal et al. 2011; Sahoo and Das 2011; Dervos et al. 2004).

Moreover, the electroless nickel-boron coatings hardness can be improved after adequate heat treatment, to reach the same values than hard chromium (800–1000 HV) (Meyers and Lynn 1994). Dervos et al. (2004) have reported locally values of approximately 2000 HV after 5 min of thermal treatment in a high vacuum environment.

For Lekka et al. (2018) who have studied the evolution of microhardness with boron content until 0.16 wt.% in the electroless nickel-boron coatings, the noticeable increase in microhardness with boron content is due to the incorporation of B in the Ni matrix and due to the co-deposition of lead. Anik et al. (2008) have reported that electroless nickel-boron coating hardness is increasing with boron content up to at least 8 wt.% (from 500 HV_{100}, for 5 wt.% B, to nearly 800 HV_{100}, for 8 wt.% B) (Sahoo and Das 2011; Vitry and Bonin 2017). This evolution is not obvious regarding Figure 4A.20.

Some authors have also found low values for more than 6 wt.% B, which suggests that perhaps higher boron content could lead to a decrease in hardness as it is observed for electroless nickel-phosphorus coatings (Vitry and Delaunois 2015c).

Deposits from baths stabilized by lead salts are those that reach the highest hardness, with values higher than 800 HV_{100} for boron contents of 6–8 wt.% (Vitry and Bonin 2017; Anik et al. 2008; Srinivasan et al. 2010; Delaunois and Lienard 2002; Kaya et al. 2008; Narayanan et al. 2003; Vitry et al. 2008; Rao et al. 2005). Surface hardness and cross section hardness of 848 hv_{100} and 854 $hk_{50,}$ respectively, have been reported (Bonin 2018).

Berkovitch nanoindentations on cross section were performed by Vitry et al. (2008) on deposits from baths stabilized by lead salts. The hardness value at a load of 4000 μN for as-plated electroless nickel-coating with 6 wt.% B content is approximately 825 Berkovitch hardness values (which are equivalent to Vickers values). They also studied the nanoindentation hardness evolution through the coating thickness. They found that the hardness is relatively stable across the deposit.

Montagne et al. (2018) also performed nanoindentation tests on the cross section of an as-deposited electroless nickel-boron coating with a Berkovitch indenter and a 300 nm maximum penetration depth. Using the Oliver and Pharr method, they found no variation of hardness in the thickness of the coating, with a value of 10 GPa.

Moreover, electroless nickel-boron deposits possess a higher hot hardness than electroless nickel-phosphorus deposits (Weil and Parker 1990).

4A.4.2 Ductility, Tensile Strength and Young's Modulus

Electroless nickel-boron coatings in the as-deposited state have low ductility and medium tensile strength (approximately one-fifth that of high-phosphorous deposits for electroless nickel-5% B) (Baudrand 2002; Krishnan et al. 2006): elongation of approximately 0.2%–0.25% and tensile strength of approximately 110 MPa (Sudagar et al. 2013; Riedel 1989; Delaunois 2002; Duncan and Arney 1984; Gavrilov 1979; Baudrand 2002; Krishnan et al. 2006). Some authors reported values of tensile strength of approximately 550 or 700 MPa (Riedel 1989; Duncan and Arney 1984). The Young's modulus in the as-deposited state is close to 120 GPa for borohydrides-reduced coatings (Sudagar et al. 2013; Delaunois and Lienard 2002; Duncan and Arney 1984; Gavrilov 1979; Baudrand 2002; Krishnan et al. 2006). Using the Oliver and Parr method (Oliver and Pharr 1992) from nanoindentation tests on the cross section of an as-deposited electroless nickel-boron coating with a Berkovitch indenter and a 300 nm maximum penetration depth, Montagne et al. (2018) and Vitry (2010) found no variation of Young's modulus in the thickness of the coating, with a value of 190 GPa for electroless nickel-boron coating containing 6 wt.% B and 1 wt.% Pb. In 2008, Vitry et al. (2008) have found a value of 170 GPa.

4A.4.3 Internal Stresses

Various techniques can be used to calculate the residual stresses in coatings (Lekka et al. 2018): XRD (Tan et al. 2017), Raman (Yang et al. n.d.), nanoindentation (Tan et al. 2017; Fang et al. 2016), hole drilling (Jalali Azizpour 2017) and more recently the combination of focused ion beam micro-machining with scanning electron microscopy (SEM) imaging and digital image correlation (Song et al. 2012).

For NiB coatings, the presence of B as interstitial atom in the Ni elementary cell and external factors related to the plating conditions, bath composition and substrate create residual stresses that influence negatively the mechanical properties of coatings (Lekka et al. 2018).

High levels of internal tensile stress reduce the ductility of the coating and cause cracking or peeling of the film (Deckert and Andrus 1978), and a too high internal stress level in the coatings can lead to premature breakdown by fatigue of the substrate, loss of deposit adherence and cracks formation in the coating (Deckert and Andrus 1978; Weil and Parker 1990; Shakoor et al. 2016).

Deckert and Andrus (1978) have shown that for a plating bath reduced with DMAB, the film stresses are compressive during the early stages of deposition and become gradually highly tensile. The compressive stresses are due to the growth of hemispheric nuclei. The step from compressive to tensile stresses appears when the film starts to be continuous. The degree of compressive stress observed initially is a function of catalytic particle size and surface density. Higher densities of catalytic sites lead to early compressive stresses followed quickly by a shift to tensile stresses.

As for the phosphorous content of electroless nickel-phosphorus deposits, the internal stresses of electroless nickel-boron deposits are decreasing with the increase of boron concentration (Deckert and Andrus 1978; Shakoor et al. 2016). For example, the internal stress for electroless nickel-phosphorus coating with 2 wt.% P is approximately 10 MPa and becomes compressive with approximately −60 MPa for 12 wt.% P (Weil and Parker 1990). For electroless nickel-boron coatings, the internal stress values are always tensile and close to 110 MPa for 5 wt.% B, 310 MPa for 1.3 wt.% B and 480 MPa for 0.4 wt.% B for DMAB reducing agent (Sudagar et al. 2013; Riedel 1989; Delaunois 2002; Duncan and Arney 1984; Baudrand 2002). For borohydride-reduced coating, values of 110–200 MPa have been reported (Baudrand 2002). Montagne et al. (2018) have found a residual stress of 130 MPa for a 6 wt.% B. This value was explained by the presence of interfaces between "bunches" that seem to relax or accommodate the local residual stress.

To conclude, the internal stresses for electroless nickel-boron coatings are usually very high (Baudrand 2002) and higher than the stress level of electroless nickel-phosphorus coatings.

Moreover, the internal stress level increases with the increase in coating thickness, deposition bath temperature and reactants content (Deckert and Andrus 1978). Actually, as the rate of deposition increases with increasing bath temperature, a higher deposition rate leads to higher tensile stress.

The hydrogen entrapment also plays an important role because it can increase the internal stress level (Deckert and Andrus 1978). However, recent technological developments have allowed to substantially reduce the presence of entrapped hydrogen in electroless nickel-boron coatings.

4A.4.4 Fatigue Resistance

The surface tensile stresses are detrimental to the fatigue life of the base metal.

The fatigue resistance and endurance limit of the steel substrate can be reduced by 10–50% when covered with an electroless nickel-boron coating (Krishnan et al.

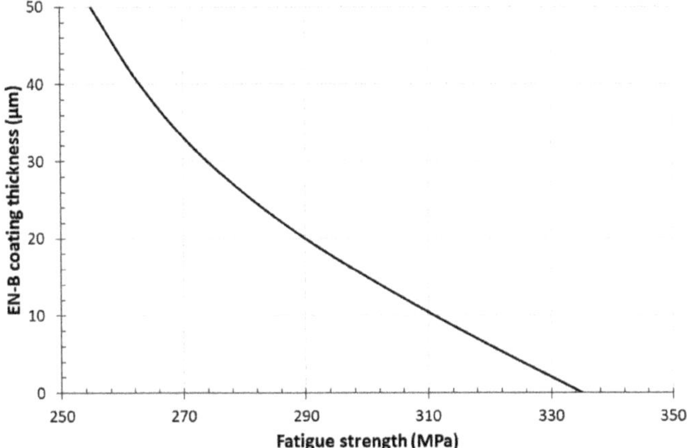

FIGURE 4A.21 Influence of electroless nickel-5wt.% B coating thickness on the fatigue resistance of a C-Mn steel.

2006). Due to their high levels of internal tensile stresses, these coatings tend to crack and initiate fatigue failures under cyclic load (Baudrand 2002).

However, the importance of this effect depends on various factors such as electroless nickel-boron deposit composition, thermal history of the plated system and initial fatigue resistance of the substrate (Baudrand 2002). The fatigue strength of a 0.42% C steel (C45, Werkstoff 1.0503) coated with 30 µm electroless nickel-5 wt.% B coating from borohydride-reduced bath is 270 MPa in the as-deposited state (Baudrand 2002).

Moreover, the thickness of the coating influences the fatigue resistance of the substrate (Figure 4A.21): the thicker the deposit, the higher the decrease of the fatigue limit (Baudrand 2002). It is thus necessary to adapt the coating thickness to the application.

4A.4.5 Creep

The normalized viscosity of an electroless nickel-boron coating deposited on Armco steel is approximately 0.4. From creeping tests with a load of 5 mN, results show that the tribological behavior of the coating is linked to elastoplastic flow mechanisms (Clerc 1986).

4A.5 TRIBOLOGICAL AND WEAR PROPERTIES OF ELECTROLESS NICKEL-BORON COATINGS

4A.5.1 Roughness

Electroless nickel-boron deposits present a very smooth surface (Sahoo and Das 2011) with a small roughness (R_a lower than 1 µm; Vitry 2010; Shaffer and Rogers 2007; Das and Sahoo 2009), with a valley depth lower than the peak height (Vitry 2010). This was confirmed by results from various studies (Table 4A.4): all R_a values are not more than 0.65, and values of approximately 0.15 are still reported. R_q values are in the same order than R_a in all cases.

However, a tribological study of the deposit roughness cannot be limited to simple geometrical parameters such as R_a and R_q (Vitry 2010; Delaunois 2002). Other parameters such as R_z, R_p, R_t and R_v give also interesting information about the surface state. For example, it can be seen from Table 4A.4 that coatings with a low R_a of 0.16 and 0.22 µm can have a high R_t that is a parameter very sensitive to high picks and deep valleys (2.3 and 3.33 µm, respectively).

Moreover, the roughness of electroless nickel-boron deposits is less uniform than electroless nickel-phosphorus coatings due to their typical cauliflower-like surface morphology (Vitry 2010; Delaunois et al. 2000).

The initial surface roughness of the substrate influences the electroless nickel-boron roughness deposit with the conservation of the magnitude order of the roughness before and after plating (Vitry 2010) accompanied by a slight decrease of the value (Dervos et al. 2004). Clerc (1986) has shown that the roughness of electroless nickel-boron deposits is depending on the initial substrate roughness because of the formation of columns locally perpendicular to the surface during the growth mechanism of the coating: a smooth substrate will produce a coating smoother than the substrate due to the formation of parallel columns (substrate R_a of 0.1 µm and 10 µm coating R_a deposited on the substrate of 0.07 µm); a rough substrate will produce a coating rougher than the substrate due to the "fan"-like disposition of the columns (substrate R_a of 0.71 µm and 10 µm coating R_a deposited on the substrate of 0.87 µm). However, Table 4A.7 shows that depending on the publication, the coating R_a can be lower (Vitry et al. 2017; Dervos et al. 2004) or higher (Shaffer and Rogers 2007; Vitry 2010; Vitry et al. 2017; Clerc 1986) than the initial R_a of the substrate.

The final coating thickness seems to have no influence on final roughness (Table 4A.7).

Asymmetry and flattening of the roughness profile are also influenced by substrate roughness (Vitry 2010).

The very smooth surface of electroless nickel-boron coatings will be favorable to wear resistance (Vitry et al. 2008).

4A.5.2 Scratch Test Resistance

The performing of scratch test allows testing the in-use coating performances under scratch solicitations by providing information about the behavior of the coating/substrate system under load. This test also permits to evaluate the coating adhesion onto the substrate, a parameter particularly difficult to study, with the demonstration of a lack of adhesion, if it exists, by generating detachments of layers in the scratch or at the edge of the scratch (Bull and Berasetegui 2006). The scratch test is standardized in EN 1071-3 and the adhesion is quantified by the critical load (L_c), which is the load where damage is detected (Vitry and Delaunois 2015c).

It is very difficult to compare the results obtained by various authors because the nature of the substrate and the coating and the surface preparation of the substrate and the thickness of the coating influence the scratch test response (Vitry and Delaunois 2015c). Moreover, the scratch width increases with the increase in applied load but the coating behavior during test seems not to be affected by the applied load (Clerc 1986).

Vitry et al. (2011) and Vitry (2010) have studied the resulting scratches obtained after scratch testing up to 150 N, on an as-deposited electroless nickel-boron coating (6 wt.%

TABLE 4A.7
Comparison of Roughness Values for Electroless Nickel-Boron Coatings Obtained from Various Authors

				Substrate Roughness				Electroless Nickel-Boron Coating Roughness					
References	Reducing Agent and Stabilizer (if known)	B wt.% in the Coating	Coating Thickness (μm)	R_a (μm)	R_t (μm)	R_z (μm)	R_p (μm)	R_a (μm)	R_q (μm)	R_t (μm)	R_z (μm)	R_v (μm)	R_p (μm)
Shaffer and Rogers (2007)	NanoCem™		14–16	0.05				0.62					
Bonin (2018)	$NaBH_4$ Pb	5.9	16					0.32				0.63	0.58
Bonin and Vitry (2016)	$NaBH_4$ Pb	6	20					0.16					0.37
Dervos et al. (2004)	$NaBH_4$ Pb	5	25	0.58	4.70	3.80		0.21		1.90	1.20		
Vitry (2010)	$NaBH_4$ Pb	6	16–20	0.18			0.51	0.43	0.52	3.51	2.22	1.25	0.97
Vitry et al. (2008)	$NaBH_4$	6	15					0.13	0.17	1.69			
	Pb		6					0.16	0.22	2.29			
Vitry and Bonin (2017)	$NaBH_4$	6		0.18				0.13	0.17	1.69	1.19	0.48	0.71
	Pb	8		0.18				0.21	0.29	2.44	1.62	0.90	0.72
Delaunois et al. (2000)	$NaBH_4$ Tl	4	30					0.22	0.31	3.33	1.69		0.76
	$NaBH_4$ Pb	5.75	25					0.14	0.18	1.27	0.88		0.45
Clerc (1986)	$NaBH_4$ Pb	1.4	11	0.02				0.11					
Kanta et al. (2010)	$NaBH_4$ Pb	6	30					0.90	1.14		6.32		

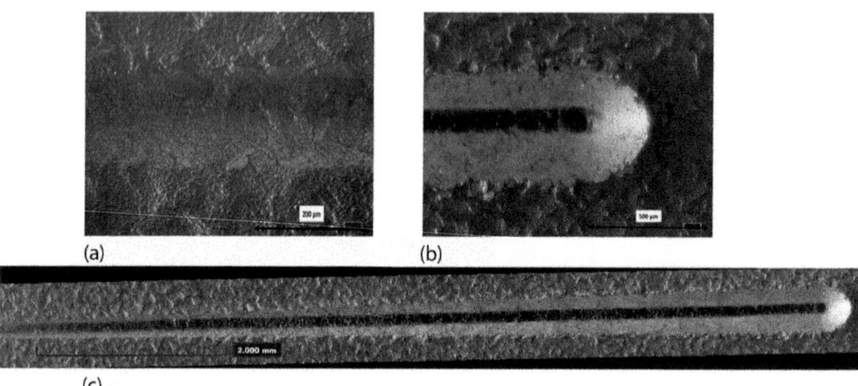

FIGURE 4A.22 Resulting scratch of scratch test up to 150 N on as-deposited electroless nickel-B: (a) middle of the scratch (load close to 60 N), (b) end of the scratch and (c) whole scratch. (From Delaunois, F., and Vitry, V., Electroless Nickel–Boron Coatings: Process, Composition, Microstructure, and Properties, In *Encyclopedia of Iron, Steel, and Their Alloys*, edited by George E Totten and Rafael Colas, CRC Press, Taylor & Francis Group, Boca Raton, FL, 2015).

B, 20 μm of thickness) on mild steel substrate with a electroless nickel-boron bath stabilized with lead salt. The critical load (L_c) obtained from scratch test was 25 N (Bonin 2018; Bonin and Vitry 2016). Only fine cohesive cracks are observed for loads close to 60 N; they disappear for higher loads for which the plastic deformation of the coating is then predominant (Figure 4A.22). No spalling, delamination or ductile failure event is observed at the highest load. The adhesion of this coating can thus be qualified to very good (Vitry et al. 2014). The damage observed on industrially usable thickness coatings (more than 15 μm) is generally limited. For thinner coating (6 μm thick), failure was observed and a ductile perforation occurred at a load of 21 N (Vitry et al. 2008, 2011).

In conclusion, electroless nickel-boron coatings have excellent scratch resistance, which attests to their outstanding adhesion.

Bond strength was studied for electroless nickel-boron coatings by Riddle and Bailer (2005). From epoxy-bonded test, they have found bond strength to the substrate that exceeds 68.9 MPa on steel alloys and is approximately 48.2 MPa on aluminum alloys. Delaunois et al. (2000) have found a lower limit of 15–20 MPa on aluminum alloy.

4A.5.3 Friction Coefficient (COF)

The COF is calculated from pin-on-disc test results.

Due to their hardness, together with a columnar morphology that allows retention of lubricants, electroless nickel-boron deposits are good candidates for wear resistance (Vitry 2010; Sahoo and Das 2011). This typical surface texture reduces the contact area up to 70%, retaining lubricants under conditions of adhesive wear. Moreover, their boron content can act as a solid lubricant and decrease adhesive, fretting and abrasive wear. The COF thus varies with boron content (Sudagar et al. 2013).

Electroless nickel-boron deposits present a low friction coefficient, in the same order than that to hard chromium or electroless nickel-phosphorus coatings (Table 4A.8).

TABLE 4A.8
Comparison of Friction Coefficient vs. Steel for Electroless Nickel-Boron, Electroless Nickel-Phosphorus and Hard Chromium Coatings Obtained from Various Studies

	Friction Coefficient vs. Steel					
	Electroless Nickel-Phosphorus		Electroless Nickel-Boron		Hard Chromium	
References	For Lubricated Conditions	For Unlubricated Conditions	For Lubricated Conditions	For Unlubricated Conditions	For Lubricated Conditions	For Unlubricated Conditions
Baudrand (2002)	0.13	0.4	0.12–0.13	0.43–0.44		
Das and Sahoo (2009)			n.a.	0.19		
Mukhopadhyay et al. (2017)			n.a.	0.643		
Bonin (2018)			n.a.	0.51		
Meyers and Lynn (1994)					0.16	n.a.

For electroless nickel-boron coatings, this coefficient is nearly constant during the abrasion testing (Krishnaveni et al. 2005; Mukhopadhyay et al. 2017). The COF found for an as-plated electroless nickel-boron coating and mild steel substrate using in an electroless nickel-boron bath stabilized with lead salt after 100 m of sliding wear test increases with the applied load is as follows: 0.45 with a 5 N load and 0.51 with a 10 N load (Bonin 2018); the same evolution of COF is found after a sliding distance of 2700 m: 0.742 with a 20 N load, 0.770 with a 30 N load and 0.784 with a 40 N load (Krishnaveni et al. 2005).

Das and Sahoo (2009) have optimized the electroless nickel-boron plating bath parameters to decrease the COF in non-lubricated conditions to reach values of 0.19 and have shown that the most plating parameter influencing the COF is the reducing agent concentration.

The COF is also affected by the test temperature (Mukhopadhyay et al. 2017). After a sliding duration of 300 s for electroless nickel-boron coatings obtained from baths reduced by $NaBH_4$ and stabilized with lead, the COF at room temperature is approximately 0.643 under a load of 10 N. At 100°C test temperature, a significant increase in the COF is observed with a value of approximately 0.864. At 300°C and 500°C, the COF decreases under 0.6.

The COF can also be obtained from scratch test (Vitry et al. 2008). For electroless nickel-boron coatings of 6 μm thickness, under the critical load of 21 N, the COF is close to 0.3, and above the critical load, it increases up to 0.45.

4A.5.4 Specific Wear Rate or Sliding Wear Resistance

Electroless nickel-boron coatings are more suitable for wear-resistant application than electroless nickel-phosphorus coatings due to their high hardness in the as-deposited state, which can be improved by adequate heat treatments (Sahoo and Das 2011).

The wear resistance of electroless nickel-boron coatings can be evaluated using several different testing geometries (Narayanan et al. 2003, 2005; Correa et al. 2013; Sahoo and Das 2011; Das and Sahoo 2009; Riddle and Bailer 2005; Ahmadkhaniha and Mahboubi 2012), but generally pin-on-disc tests is preferred (Krishnaveni et al. 2005; Lee and Chen n.d.). Under a load of 40 N, using a pin-on-disc test with a hardened steel as counter disk, Narayanan et al. (2003) have found a specific wear rate (W_s) of 2.45×10^{-10} kg N^{-1} m^{-1} (compared with 4.60×10^{-10} kg N^{-1} m^{-1} for electroless nickel-phosphorus coating). As the substantial attractive force between steel and electroless nickel-boron coating leads to a high mutual solubility of nickel and iron, an adhesive wear is most likely found to occur.

Krishnaveni et al. (2005) have found that the W_s increases with the increase in applied load: values of W_s of 0.52, 1.36 and 2.46×10^{-10} kg N^{-1} m^{-1} are found under a load of 20, 30 and 40 N, respectively, after a sliding distance of 2700 m. This was confirmed by Shaffer and Rogers (2007), who have found values of 7.3×10^{-6} mm^3 N^{-1} m^{-1} under a load of 4.5 N and 1.1×10^{-5} mm^3 N^{-1} m^{-1} under a load of 27.2 N.

However, Bonin (2018) have found the opposite from calculation following the European standard electroless nickel 1017-13:2008: under a load of 5 N, an impressive W_s of 0.63 $\mu m^2/N$ was obtained for electroless nickel-boron coating performed in bath reduced by $NaBH_4$ and stabilized by lead, and a lower value of 0.41 $\mu m^2/N$ was obtained under a load of 10 N.

Actually, specific sliding wear values of 1530 $mm^3 N^{-1} m^{-1}$ are usually measured (Krishnaveni et al. 2003, 2005; Bülbül et al. 2012; Vitry and Delaunois 2015c).

The wear rate is also affected by test temperature (Mukhopadhyay et al. 2017). Under a load of 10 N, it increases by 3–4 times from room temperature to 100°C. However, at 300°C, the wear rate is improved. It increases again when tested at 500°C.

4A.5.5 Abrasive Wear Resistance

The wear resistance of electroless nickel coatings is linked not only to their hardness, but also to their surface morphology (Vitry and Delaunois 2015c).

Many authors (Srinivasan et al. 2010; Sudagar et al. 2013; Vitry et al. 2011, 2012d; Krishnaveni et al. 2005; Sahoo and Das 2011; Krishnan et al. 2006; Riddle and Bailer 2005) have studied abrasive wear resistance of electroless nickel-boron coatings under various test conditions, and it is thus difficult to compare results. However, in all cases, authors claim that electroless nickel-boron deposits have a very good abrasive wear resistance (Riddle and Bailer 2005) with typical Taber Wear Index (TWI) values of 7–9 in the as-deposited state (Vitry et al. 2014; Krishnan et al. 2006; Sudagar et al. 2013). TWI of approximately 8–11 are obtained with wheels of reasonable abrasiveness (CS-10) and a load of 1 kg for coatings from electroless nickel-boron baths stabilized with lead salts (Sudagar et al. 2013; Vitry et al. 2014), while higher values are obtained under the same conditions for coatings containing thallium (TWI of approximately 24) (Vitry et al. 2014). Table 4A.9 compares abrasive wear resistance (TWI) of electroless nickel-boron coatings under comparable test conditions. In as-deposited state, NiB coatings obtained from bath reduced by sodium borohydride show a wear rate that increases with increasing applied load under dry sliding environment using a pin-on-disc apparatus during tests conducted in air at 25°C with a relative humidity(RH) of approximately 35%–45% RH (Krishnaveni et al. 2005). SEM

TABLE 4A.9
Abrasive Wear Resistance (TWI) of Electroless Nickel-Boron Coatings Under Comparable Test Conditions

References	Reducing Agent and Stabilizer (if known)	B wt.% in the Coating	Wheels	TWI
Vitry et al. (2011); Vitry et al. (2012f)	$NaBH_4$ Pb	6	CS-17	28.5
Bonin and Vitry (2016)	$NaBH_4$ Pb	6	CS-17	31.1
Kanta et al. (2010)	$NaBH_4$ Pb	6	CS-17	28.8
Vitry et al. (2014)	$NaBH_4$ Tl		CS-10	23.7
Vitry et al. (2014)	$NaBH_4$ Pb		CS-10	11.4
Baudrand (2002)	$NaBH_4$	5	CS-10	9

analysis of tested samples suggests an adhesive wear as prominent mechanism due to the presence of torn patches and some detachment of the coatings (Krishnaveni et al. 2005; Mukhopadhyay et al. 2017; Sahoo and Das 2011; Das and Sahoo 2011b). This is due to the attractive force that operates between the atoms of nickel from the coating and iron from the counter disc (Krishnaveni et al. 2005).

Moreover, the wear properties of these electroless nickel-boron coatings after adequate heat treatment (TWI of 3) are in the same order than hard chromium (TWI of 2) (Sudagar et al. 2013; Krishnan et al. 2006; Baudrand 2002).

4A.6 CORROSION RESISTANCE OF ELECTROLESS NICKEL-BORON COATINGS

There are many parameters that influence the corrosion resistance of electroless nickel coatings. It is thus important to be cautious when describing the corrosion evolution (Vitry and Delaunois 2015c). In general, electroless nickel coatings act as barrier coatings, protecting the substrate by sealing it off from corrosive environment. Moreover, the corrosion resistance of electroless nickel alloys depends on their ability to form a surface protective film (Sahoo and Das 2011). From Hamid et al.'s (2010) results, it was concluded that electroless nickel-boron layers decreased the samples' susceptibility to be attacked by the chloride ions as they formed a protective layer on the substrate.

The corrosion resistance of electroless nickel-boron coatings is an important parameter that conditions their applications. It is generally accepted that electroless nickel-boron coatings are less resistant to corrosion than electroless nickel-phosphorus alloys (Sudagar et al. 2013; Krishnan et al. 2006; Narayanan et al. 2003). They show corrosion resistance to non-oxidizing mineral acids, hydrofluoric acid and heavy metals (Riedel 1989). Moreover, they provide a protection against corrosion for 380 h, and even for 1000 h, using the ASTM B117-02 standard, with an undamaged coating after this test (Riddle and Mccomas 2005).

The corrosion potential is relatively high, generally in the range of −0.57 to −0.35 V/SHE in a neutral medium depending on the nature of the stabilizer and the boron content of the coatings (Srinivasan et al. 2010; Das and Sahoo 2011a; Narayanan and Seshadri 2004; Kanta et al. 2010; Dervos et al. 2004; Bonin et al. 2018b), and can even reach values higher than −0.3 V/SHE in an acid medium (Bonin et al. 2018b). Table 4A.10 compares corrosion potentials and corrosion currents measured by potentiodynamic polarization on various electroless nickel-boron coatings in the as-plated state.

Studies have found, from potentiodynamic polarization curves obtained in 3.5 vol.% NaCl, values of corrosion potential E_{corr} for as-deposited coatings of at least −150 mV/SHE, and even −50 mV/SHE for an optimized coating. It was also found that their corrosion resistance was moderate in 3.5% NaCl solution (Narayanan and Seshadri 2004; Das and Sahoo 2011a) with a preferential dissolution at the grains - and columns - boundaries (Kanta et al. 2010). Moreover, as boron and thallium are not homogeneously distributed throughout the electroless nickel-boron coating, the presence of area with various corrosion potentials on the surface leads to galvanic corrosion that accelerated the chloride ion attacks (Narayanan et al. 2003).

TABLE 4A.10
Corrosion Potential and Corrosion Current Measured by Potentiodynamic Polarization on Various Electroless Nickel-Boron Coatings in As-Plated State

References	Reducing Agent and Stabilizer (if known)	B and Stabilizer (if known) in the Coating (wt.%)	Substrate	Reference Electrode	Medium	E_{corr} (mV)	I_{corr} (μA/cm²)
Narayanan et al. (2004)	$NaBH_4$	6.4	Steel	SCE	3.5% NaCl	–516	9.15
	Tl	0.24					
Dervos et al. (2005); Dervos et al. (2004)	$NaBH_4$	5	Steel	SCE	3.5% NaCl	–536	4.86
	Pb						
Vitry et al. (2012f)	$NaBH_4$		Aluminum	Ag/AgCl (KCl sat'd)	3.5% NaCl	–245	3.53
	Pb						
Kanta et al. (2010)	$NaBH_4$		Steel	Ag/AgCl (KCl sat'd)	3.5% NaCl	–335	0.09
	Pb						
Srinivasan et al. (2010)	$NaBH_4$	4.94	Steel/copper	SCE	3.5% NaCl	–568.4	13.2
	Organic						
Das and Sahoo (2011a)	$NaBH_4$	5.72–7.46	Steel	SCE	3.5% NaCl	–381	5.04
	Pb						
Bonin et al. (2018b)	$NaBH_4$	8	Steel	Ag/AgCl (KCl sat'd)	3.5% NaCl	–396	7.81
	Pb	0.5					
Narayanan et al. (2003)	$NaBH_4$	6.5	Steel	SCE	3.5% NaCl	–508	9.15
	Tl	0.3					
Hamid et al. (2010)	DMAB	5 or 6	Steel	NHE	5% NaCl		7.1 or 19
	Pb						
Hamid et al. (2010)	DMAB	5 or 6	Copper	NHE	5% NaCl		3.1 or 5.6
	Pb						

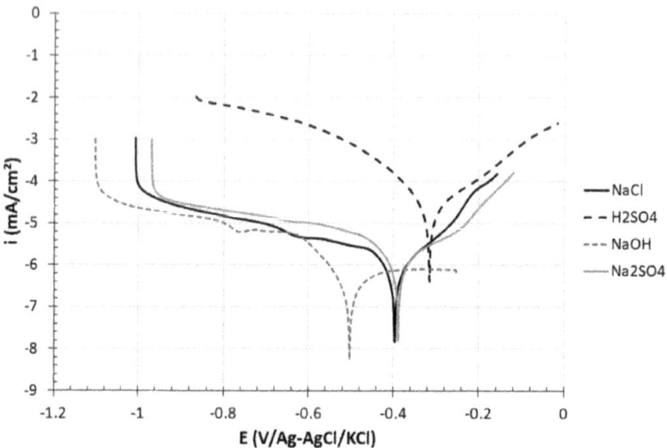

FIGURE 4A.23 Potentiodynamic polarization curves of electroless nickel-8 wt.% B coatings in various media.

Bonin et al. (2018b) have shown that the corrosion resistance of electroless nickel-boron coatings is excellent in alkaline environments in which these deposits passivate effectively. However, their resistance is less good in neutral and acidic media, even nonaggressive such as Na_2SO_4, in which their passivation is not clearly established (Figure 4A.23).

Many researchers have also studied the electrochemical properties of electroless nickel-boron coatings with electrochemical impedance spectroscopy (EIS) (Anik et al. 2008; Contreras et al. 2006; Srinivasan et al. 2010; Vitry 2010; Kanta et al. 2010; Narayanan and Seshadri 2004). Narayanan and Seshadri (2004) obtained, in 3.5% NaCl solution, Nyquist plots with a single semicircle in the high-frequency range indicating that the corrosion process involved a single time constant. This was confirmed by the presence of a single inflection point in the Bode graph. They concluded that the electroless nickel-boron coating–solution interface exhibited charge transfer behavior. Contreras et al. (2006) have shown that, in NaCl neutral environment (3% NaCl, pH 7), Nyquist diagrams show two loops at low and high frequencies. Actually, for metallic coatings, the high-frequency loop is related to the electrical capacitance of the film and the low-frequency loop to ion diffusion phenomena in the film. Bode diagrams indicate a diffusion phenomenon in the low-frequency range.

The electroless nickel-boron coating behavior during the first hours of immersion in sodium chloride solution can be represented by an equivalent circuit as shown in Figure 4A.24 (Vitry 2010).

Table 4A.11 presents a comparison of double layer capacitance (C_{dl}) and charge transfer resistance (R_{ct}) from various studies.

Contreras et al. (2006) have also studied the corrosion resistance of electroless nickel-boron coatings in acid environments (3% NaCl, pH 2) by EIS. They have shown that, in the acid media, the coating was more susceptible to corrosion effects due to its complex electrochemical behavior with three different time constants related to corrosion phenomena.

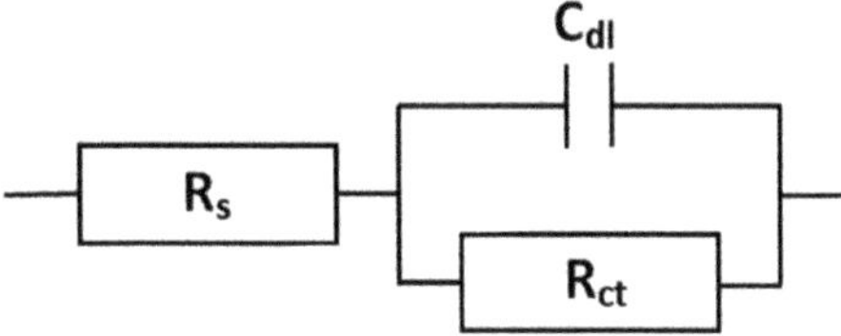

FIGURE 4A.24 Equivalent circuit for fresh electroless nickel-boron coating behavior in EIS.

TABLE 4A.11
Double Layer Capacitance and Charge Transfer Resistance from EIS on Various Electroless Nickel-Boron Coatings in As-Plated State

References	Reducing Agent and Stabilizer (if known)	B and Stabilizer (if known) in the Coating (wt.%)	Reference Electrode	Medium	C_{dl} ($\times 10^{-4}$ F/cm²)	R_{ct} ($\Omega\cdot cm^2$)
Srinivasan et al. (2010)	$NaBH_4$ Organic	4.94	SCE	3.5% NaCl	6.81	866.7
Narayanan et al. (2004)	$NaBH_4$ Tl	6.4 0.24	SCE	3.5% NaCl	3.41	3835
Narayanan et al. (2003)	$NaBH_4$ Tl	6.5 0.3	SCE	3.5% NaCl	3.53	3844
Vitry (2010)	$NaBH_4$ Pb	6 1	Ag/AgCl (KCl sat'd)	3.5% NaCl	14.9	17000
Dervos et al. (2004, 2005)	$NaBH_4$ Pb	5	SCE	3.5% NaCl		10690

There is an important number of parameters influencing the electroless nickel-boron coatings corrosion resistance (Cheong et al. 2004): the coating chemical composition (including the influence of the used stabilizer, the presence of impurities, the coating homogeneity, etc.), the surface morphology and microstructure of the coating, its thickness, its porosity, the nature and surface state of the substrate, the pretreatment applied to the substrate before plating, the use of replenishment during plating and so on.

In fact, the more uniform thickness and lower porosity of electroless nickel coating in comparison with electroplated nickel coating can explain its better corrosion protection. In addition, the amorphous nature of electroless nickel-boron deposits allows higher corrosion resistance in the as-plated state. Nevertheless, due to their particular morphology with a columnar structure, electroless nickel-boron deposits are less corrosive-resistant than electroless nickel-phosphorus deposits (Contreras et al. 2006; Krishnaveni et al. 2005; Narayanan and Seshadri 2004; Dervos et al. 2004). Dervos et al. (2004, 2005) have found a corrosion rate of 0.058 mm/year for electroless nickel-boron deposits in as-deposited state, which is good but greater than for electroless nickel-phosphorus deposits (0.019 mm/year). This was explained by Vitry (2010) from EIS measurements: the presence of boundaries between the columns allows the corrosive solution to reach the substrate.

To conclude, electroless nickel-boron coatings can increase the corrosion behavior of mild steels and magnesium alloys, completely isolating the substrate from the corrosive medium (Wang et al. 2013; Sahoo and Das 2011; Das and Sahoo 2011a), and also used as cathodic protection.

4A.7 CATALYTIC PROPERTIES

It is well known that nickel possesses catalytic properties. In fact, it can alter the velocity of a certain chemical reaction without any chemical change. It can be used as an electrocatalyst to interact with some species during a Faradaic reaction and still remaining unaltered. The comparison of current densities at a constant overpotential or overpotential at a constant current density allows to compare the catalytic activity of various electrode materials. A good electrocatalyst should show high current density at low overpotential.

A qualitative way to predict the activity of heterogeneous catalysts is given by the Sabatier Principle: to have high catalytic activity, the interaction between reactants and catalysts should neither be too strong nor too weak. Balandin volcano plots for metal-hydrogen bonding energy present the logarithm of exchange current densities for cathodic hydrogen evolution vs. the bonding adsorption strength of intermediate metal-hydrogen bonds formed during the reaction itself (Jaksic et al. 1998). The optimal metal catalyst for the hydrogen evolution reaction is located near the peak of the volcano curve. As expected, with a bonding adsorption strength of approximately 47 kcal/mol for exchange current densities for cathodic hydrogen evolution of approximately -0.7 A/cm^2, nickel can be used for the evolution of hydrogen and consequently for the hydrogenation-dehydrogenation reactions.

Electroless nickel-boron deposits can thus be used as a catalyst for the hydrogenation of various chemicals, such as soybean oil, sulfolene and p-chloronitrobenzene (Ma et al. 2003; Vitry 2010; Li et al. 1999, 2003, 2009; Shen and Chen 2007; Lee and Chen n.d.; Chen et al. 2005, 2008; Xu et al. 2009; Okamoto et al. n.d.; Amer 2008). They can be deposited on many substrates after adequate sensitization to plating, with a thickness depending on the in-use.

4A.8 PHYSICAL PROPERTIES OF ELECTROLESS NICKEL-BORON COATINGS

Physical properties such as density, melting temperature, electrical, magnetic and thermal properties influence the material choice for all applications.

As for other properties, the physical properties of electroless nickel-boron coatings vary with their boron content.

4A.8.1 DENSITY

When weight is an important factor that conditions the choice of materials, density is a relevant property.

The density of pure metallic nickel at room temperature is 8.91 g/cm^3 (Riedel 1989). The density (at 25°C) of electroless nickel-5 wt.% B coatings is 8.25 g/cm^3 (Sudagar et al. 2013; Baudrand 2002). It is lower than the density of

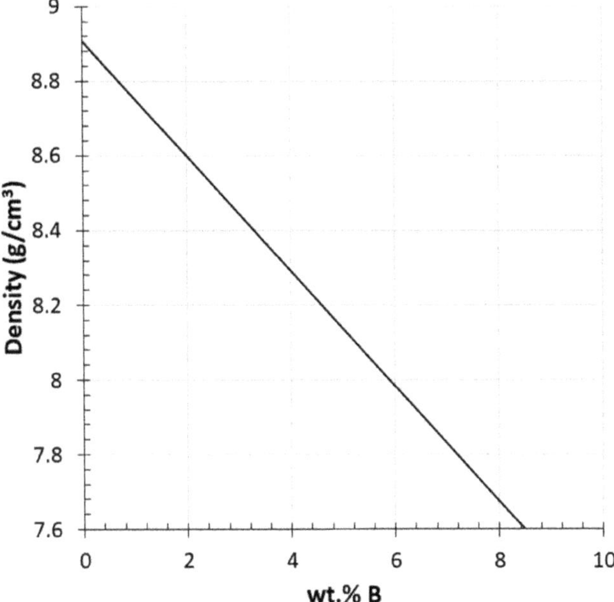

FIGURE 4A.25 Density of electroless nickel-boron alloys as a function of their boron content.

pure bulk nickel, and it decreases with increasing boron content (Sudagar et al. 2013; Riedel 1989; Delaunois 2002; Krishnaveni et al. 2005; Weil and Parker 1990) (Figure 4A.25).

4A.8.2 Porosity

The porosity of electroless nickel-boron coatings is an important parameter that conditions their corrosion resistance or chemical attack when deposited on a substrate due to its more noble character than the steel, aluminum or magnesium substrate and its action as corrosion barrier. It also influences density, solderability and ductility. Many tests exist but only a few are relevant for nickel on ferrous basis metal, such as ferroxyl test, hot water test and salt spray test (Riedel 1989).

Furthermore, development in deposition processes led to electroless nickel-boron coatings as compact as electroless nickel-phosphorus coatings, with a very low pore density (Riddle and Bailer 2005).

Bonin (2018) and Bonin et al. (2019a) have studied the corrosion resistance after salt spray tests of various electroless nickel-boron coatings depending on the stabilizer used. After 10 days of neutral salt spray test, for electroless nickel-boron coating of approximately 15 μm of thickness, the volume fraction of corroded surface is approximately 21% for lead tungstate and only approximately 6.5% for high tin chloride (Figure 4A.26).

In both cases, corrosion results showed impressive corrosion resistance and thus low deposit porosity.

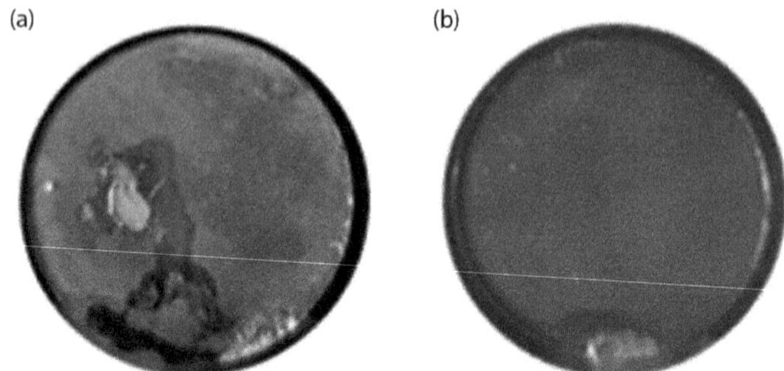

FIGURE 4A.26 Surface aspect of electroless nickel-boron coatings after 240 h of neutral salt spray test as a function of the type of stabilizer: (a) lead tungstate; (b) high tin chloride.

4A.8.3 Melting Point

Above 250°C, nearly all the electroless nickel coatings undergo a modification of their atomic structure, their microstructure and properties related to these coatings (Riedel 1989).

The melting point of electroless nickel-boron coatings is significantly higher than those of electroless nickel-phosphorus coatings (Riedel 1989) and varies between the melting temperature of pure nickel (1455°C) and the temperature of the eutectic solidification of Ni/Ni_3B (1093°C) (Riedel 1989; Vitry 2010), as predicted by the NiB alloy phase diagram, and thus decreases, just like the density, with the increase in boron content (Riedel 1989). It is, for example, 1350°C for electroless nickel-4.3 wt.% B coatings and 1080°C for electroless nickel-5 wt.% B coatings reduced by borohydride (Riedel 1989) and approximately 1350°C–1360°C for a DMAB-reduced bath (Sudagar et al. 2013; Baudrand 2002), to compare with 890°C for electroless nickel-phosphorus coating (Dervos et al. 2005).

4A.8.4 Electrical and Magnetic Properties

The electrical conductivity of electroless nickel-coatings in low and heavy current applications, especially where high or very high frequencies are used (Riedel 1989). The contact resistance is another important parameter because it can provide information about corrosion or tarnishing or abrasion resistance, when electroless nickel coating is substituted for precious metals in electrical contacts (Riedel 1989).

The presence of alloying elements in the electroless nickel coatings such as boron leads to an increase in its electrical resistivity: comparing with electroplated nickel (8 μΩ.cm), an electroless nickel-5 wt.% B deposit presents a resistivity of 90 μΩ·cm (Riedel 1989; Sudagar et al. 2013; Baudrand 2002), which is similar to that of electroless nickel-phosphorus coatings. This explains why major applications of electroless nickel-boron are in electronic industries in low resistivity field (Sudagar et al. 2013). However, much lower (13 to 15 μΩ·cm) or higher values (190 μΩ·cm) have been also reported (Riedel 1989).

As for density, an increase in the boron content increases the coating resistivity: for example, deposits with 1 wt.% B have a resistivity of approximately 10–20 μΩ·cm, and deposits with 7 wt.% B can reach values as high as 190 μΩ·cm (Baskaran et al. 2006). Hamid et al. (2010) have reported values of approximately 2.02–2.05 Ω·cm for electroless nickel-boron coatings on copper substrate from bath reduced by DMAB. As these coatings have high oxidation resistance, and thus high electrical conductivity, they can be used for applications as diffusion barriers and metal capping layers in the copper interconnect technology.

For contact resistance, values of approximately 15 μΩ·cm were reported for electroless nickel-5 wt.% B (Riedel 1989). The electroless nickel-boron coatings can be used to cover a carbon steel substrate and to drop the contact resistance by a factor of ten (compared with bare substrate) due to the lower resistivity of the coating and/or more effective heat dissipation properties (Dervos et al. 2005).

Nickel is a ferromagnetic material and the co-deposition of boron strongly affects the magnetic properties of deposits (Riedel 1989). Electroless nickel-5 wt.% B deposits are very weakly ferromagnetic (Riedel 1989). Baskaran et al. (2006) have observed that the saturation field increases with the decrease in their boron content and thus with the increase in crystallinity. This is substantiated by the nonmagnetic nature of a coating containing 6.4 wt.% B in as-plated condition (Narayanan and Seshadri 2004). Consequently, the magnetic properties of electroless nickel-boron deposits are influenced by their boron content (the higher the boron content, the lesser is the saturation magnetization). Saito et al. (1999) have also observed this dependence of magnetic property on the boron content of the coating.

4A.8.5 Thermal Properties

The thermal conductivity must be high to allow the evacuation of the calories generated by the contact friction and to increase the resistance to thermal shocks. The higher it is, the fewer is the temperature gradient and therefore thermal stress (Baudrand 2002). The thermal conductivity of NiB deposits is similar to that of Ni-10.5 wt.% P deposits (0.016 cal/cm.s.°C) (Riedel 1989). The thermal conductivity of electroless nickel-boron coatings is influenced by their boron content: an increase in the boron content leads to an increase in the thermal conductivity (Gavrilov 1979). The rare studies about this property are often limited to qualitative assessments (Riedel 1989).

The thermal expansion coefficient in the range of 20°C–100°C of electroless nickel-5 wt.% B coatings has been measured as 12.6 μm/m·°C, compared with 13.3 μm/m·°C for pure nickel (Riedel 1989; Baudrand 2002).

4A.8.6 Suitable for Soldering, Welding and Bonding

For electronics industries, solderability is an important property. However, solderability is not an intrinsic property but a function of the overall system (Riedel 1989).

Electroless nickel-boron deposits are easily solderable (Riedel 1989; Bielinski et al. 1990; Chow et al. 2001) and sometimes used as barrier against diffusion in flip chips if their boron content is low (Saito et al. 1999; Lee and Lin 1993). They can also be welded if attention is paid to risks of boron or nickel diffusion into the substrate or into the weld bead (Riedel 1989). A thickness of approximately 3 μm is recommended in soft-soldering (Riedel 1989). Electroless nickel-boron deposits from bath reduced with DMAB show good solderability immediately after deposition, irrespective of boron content, but this solderability degrades with time (Riedel 1989).

Their bondability is excellent (Riedel 1989).

4A.8.7 Adhesion

It is well known that the adhesion of a deposit on a metallic substrate increases with the substrate surface roughness. Moreover, an adhesion lost can be due to a large difference in coefficients of thermal expansion between the deposit and the substrate (Weil and Parker 1990). In fact, the adhesion of electroless nickel-boron coating on metallic substrates is good and greater than the cohesive resistance of the glue (68.9 MPa) (Delaunois 2002).

REFERENCES

Ahmadkhaniha, D., and F. Mahboubi. 2012. "Effects of Plasma Nitriding on Properties of Electroless Ni-B Coating." *Surface Engineering* 28 (3): 195–198. doi:10.1179/1743294411Y.0000000058.

Amer, J. 2008. "Développement de Membranes Métalliques de Nickel Déposées Sur Support Céramiques Par Electroless Plating: Etude Des Propriétés Particulières de Permsélectivité."

Anik, M., E. Körpe, and E. Şen. 2008. "Effect of Coating Bath Composition on the Properties of Electroless Nickel–boron Films." *Surface and Coatings Technology* 202 (9): 1718–1727. doi:10.1016/j.surfcoat.2007.07.031.

Baskaran, I., R. Sakthi Kumar, T. S. N. S. Narayanan, and A. Stephen. 2006. "Formation of Electroless Ni-B Coatings Using Low Temperature Bath and Evaluation of Their Characteristic Properties." *Surface and Coatings Technology* 200 (24): 6888–6894. doi:10.1016/j.surfcoat.2005.10.013.

Baudrand, D. W. 1994. "Electroless Nickel Plating". In *Surface Engineering*, Vol 5, ASM Handbook, ed. Cotell, C.M. et al, 5:290–310. ASM International Publication.

Bekish, Y. N., S. K. Poznyak, L. S. Tsybulskaya, and T. V. Gaevskaya. 2010. "Electrodeposited Ni-B Alloy Coatings: Structure, Corrosion Resistance and Mechanical Properties." *Electrochimica Acta* 55 (7): 2223–2231. doi:10.1016/j.electacta.2009.11.069.

Bielinski, J., J. Krol, and A. Bielinska. 1990. "Inorganic Additives in the Solution for Electroless Nickel-Boron Alloy Deposition." *Bulletin of Electrochemistry* 6: 828–831.

Bonin, L., and V. Vitry. 2016. "Mechanical and Wear Characterization of Electroless Nickel Mono and Bilayers and High Boron-Mid Phosphorous Electroless Nickel Duplex Coatings." *Surface and Coatings Technology* 307 (December): 957–962. doi:10.1016/j.surfcoat.2016.10.021.

Bonin, L., V. Vitry, and F. Delaunois. 2019. "The Tin Stabilization Effect on the Microstructure, Corrosion and Wear Resistance of Electroless NiB Coatings." *Surface and Coatings Technology* 357 (January): 353–363. doi:10.1016/j.surfcoat.2018.10.011.

Bonin, L. 2018. "Replacement of Lead Stabilizer in Electroless NiB Baths: New Composition of Green Baths with Properties Characterization." UMONS.

Bonin, L., V. Vitry, and F. Delaunois. 2018a. "Influence of the Anionic Part of the Stabilizer on Electroless Nickel-Boron Plating." *Materials and Manufacturing Processes* 33 (2). doi:10.1080/10426914.2017.1291949.

Bonin, L., V. Vitry, and F. Delaunois. 2018b. "Corrosion Behaviour of Electroless High Boron-Mid Phosphorous Nickel Duplex Coatings in the as-Plated and Heat-Treated States in NaCl, H_2SO_4, NaOH and Na_2SO_4 Media." *Materials Chemistry and Physics* 208 (C): 77–84. doi:10.1016/j.matchemphys.2017.12.030.

Bulbul, F. 2011. "The Effects of Deposition Parameters on Surface Morphology and Crystallographic Orientation of Electroless Ni-B Coatings." *Metals and Materials International* 17 (1): 67–75. doi:10.1007/s12540-011-0210-4.

Bülbül, F., H. Altun, Ö. Küçük, and V. Ezirmik. 2012. "Tribological and Corrosion Behaviour of Electroless Ni-B Coating Possessing a Blackberry like Structure." *Metals and Materials International* 18 (4): 631–637. doi:10.1007/s12540-012-4011-1.

Bull, S.J., and E. G. Berasetegui. 2006. "An Overview of the Potential of Quantitative Coating Adhesion Measurement by Scratch Testing." *Tribology International* 39 (2): 99–114. doi:10.1016/j.triboint.2005.04.013.

Chang, C.R., K.H. Hou, M.D. Ger, J.R. Wang, and undefined 2017. n.d. "Characteristics of Nickel Boron Coatings Prepared by Direct Current Electrodeposition Technique". *International Journal of Electrochemical Science* 12:2055–2069.

Chen, J., D. Ci, R. Wang, and J. Zhang. 2008. "Hydrodechlorination of Chlorobenzene over NiB/SiO_2 and NiP/SiO_2 Amorphous Catalysts after Being Partially Crystallized: A Consideration of Electronic and Geometrical Factors." *Applied Surface Science* 255 (5): 3300–3309. doi:10.1016/j.apsusc.2008.09.035.

Chen, Y.-Z., B.-J. Liaw, and S.-J. Chiang. 2005. "Selective Hydrogenation of Citral over Amorphous NiB and CoB Nano-Catalysts." *Applied Catalysis A: General* 284 (1–2): 97–104. doi:10.1016/j.apcata.2005.01.023.

Cheong, W. J., B. L. Luan, and D. W. Shoesmith. 2004. "The Effects of Stabilizers on the Bath Stability of Electroless Ni Deposition and the Deposit." *Applied Surface Science* 229: 282–300. doi:10.1016/j.apsusc.2004.02.003.

Chow, Y. M., W. M. Lau, and Z. S. Karim. 2001. "Surface Properties and Solderability Behaviour of Nickel-Phosphorus and Nickel-Boron Deposited by Electroless Plating." *Surface and Interface Analysis* 31 (4): 321–327. doi:10.1002/sia.980.

Clerc, M-A. 1986. "Morphologie, Propriétéss Tribologiques et Rhéologiques Des Alliages Nickel-Bore Amorphes Obtenus Par Voie Chimique." Université de Besanéon.

Contreras, A., C. León, O. Jimenez, E. Sosa, and R. Pérez. 2006. "Electrochemical Behavior and Microstructural Characterization of 1026 Ni–B Coated Steel." *Applied Surface Science* 253 (2): 592–599. doi:10.1016/j.apsusc.2005.12.161.

Correa, E., A.A. Zuleta, L. Guerra, M.A. Gómez, J.G. Castaño, F. Echeverría, H. Liu et al. 2013. "Coating Development during Electroless Ni–B Plating on Magnesium and AZ91D Alloy." *Surface and Coatings Technology* 232 (October): 784–794. doi:10.1016/j.surfcoat.2013.06.100.

Dadvand, N., G. J. Kipouros, and W. F. Caley. 2003. "Electroless Nickel Boron Plating on AA6061." *Canadian Metallurgical Quarterly* 42 (3): 349–364. doi:10.1179/cmq.2003.42.3.349.

Das, S. K., and P. Sahoo. 2009. "ICME09-RT-03 optimization of electroless Ni-B coatings based on multiple roughness characteristics" In *Proceedings of the International Conference on Mechanical Engineering 2009 (ICME2009)*. Dhaka, Bangladesh, pp. 26–28.

Das, S. K., and P. Sahoo. 2011a. "Study of Potentiodynamic Polarization Behaviour of Electroless Ni-B Coatings and Optimization Using Taguchi Method and Grey Relational Analysis." *Journal of Minerals and Materials Characterization and Engineering* 10 (14): 1307–1327.

Das, S. K., and P. Sahoo. 2011b. "Tribological Characteristics of Electroless Ni-B Coating and Optimization of Coating Parameters Using Taguchi Based Grey Relational Analysis." *Materials & Design* 32 (4): 2228–2238. doi:10.1016/j.matdes.2010.11.028.

Deckert, C. A., and J. Andrus. 1978. "Stress of Electroless Nickel Films on Stainless Steel by Proximity Measurement." *Plating and Surface Finishing* 65: 43–48.

Delaunois, F., and P. Lienard. 2002. "Heat Treatments for Electroless Nickel–Boron Plating on Aluminium Alloys." *Surface and Coatings Technology* 160 (2–3): 239–48. doi:10.1016/S0257-8972(02)00415-2.

Delaunois, F., J. P. Petitjean, M. Jacob-Dulière, and P. Liénard. 2000. "Autocatalytic Electroless Nickel-Boron Plating on Light Alloys." *Surface and Coatings Technology* 124: 201–209.

Delaunois, F. 2002. "Les Dépôts Chimiques de Nickel-Bore Sur Alliages d'aluminium." Faculté Polytechnique de Mons.

Delaunois, F., and V. Vitry. 2015. "Electroless Nickel–Boron Coatings: Process, Composition, Microstructure, and Properties." In *Encyclopedia of Iron, Steel, and Their Alloys*, edited by George E Totten and Rafael Colas. CRC Press, Taylor & Francis Group.

Dervos, C. T., P. Vassiliou, and J. Novakovic. 2005. "Electroless Ni-B Plating for Electrical Contact Applications." *Revista de Metalurgia* 41 (Extra): 232–238. doi:10.3989/revmetalm.2005.v41.iExtra.1031.

Dervos, C. T., J. Novakovic, and P. Vassiliou. 2004. "Vacuum Heat Treatment of Electroless Nickel-Boron Coatings." *Materials Letters* 58: 619–623.

Duncan, R. N., and T. L. Arney. 1984. "Operation and Use of Sodium-Borohydride-Reduced Electroless Nickel." *Plating and Surface Finishing* 71: 49:54.

Fang, X., G. Jin, X. F. Cui, and J. N. Liu. 2016. "Evolution Characteristics of Residual Stress in Metastable Ni-B Alloy Coatings Identified by Nanoindentation." *Surface and Coatings Technology* 305 (November): 208–214. doi:10.1016/j.surfcoat.2016.08.042.

Färber, B., E. Cadel, A. Menand, G. Schmitz, and R. Kirchheim. 2000. "Phosphorus Segregation in Nanocrystalline Ni-3.6 at.% P Alloy Investigated with the Tomographic Atom Probe (TAP)." *Acta Materialia* 48: 789–796. doi:10.1016/S1359-6454(99)00397-3.

Gaevskaya, T. V., I. G. Novotortseva, and L. S. Tsybulskaya. 1996. "The Effect of Boron on the Microstructure and Properties of Electrodeposited Nickel Films." *Metal Finishing* 94 (6): 100–103.

Gawrilov, G.G. 1979. *Chemical (Electroless) Nickel Plating*. Ed. Portcullis Press, LTD Redhimm, pp. 57–97.

Gorbunova, K. M., M. V. Ivanov, and V. P. Moiseev. 1973. "Electroless Deposition of Nickel-Boron Alloys: Mechanism of Process, Structure, and Some Properties of Deposits." *Journal of the Electrochemistry Society* 120: 613–618.

Hamid, Z. Abdel, H. B. Hassan, and A. M. Attyia. 2010. "Influence of Deposition Temperature and Heat Treatment on the Performance of Electroless Ni–B Films." *Surface and Coatings Technology* 205 (7): 2348–2354. doi:10.1016/j.surfcoat.2010.09.025.

Hentschel, T. H., D. Isheim, R. Kirchheim, F. Muller, and H. Kreye. 2000. "Nanocrystalline Ni-3.6 At% P and Its Transformation Sequence Studied by Atom-Probe Field-Ion Microscopy." *Acta Materialia* 48: 933–941.

Jaksic, J. M., N. M. Ristic, N. V. Krstajic, and M. M. Jaksic. 1998. "Electrocatalysis for Hydrogen Electrode Reactions in the Light of Fermi Dynamics and Structural Bonding FACTORS—I. Individual Electrocatalytic Properties of Transition Metals." *International Journal of Hydrogen Energy* 23 (12): 1121–1156. doi:10.1016/S0360-3199(98)00014-7.

Jalali Azizpour, M. 2017. "Evaluation of through Thickness Residual Stresses in Thermal Sprayed WC–Co Coatings." *Journal of the Brazilian Society of Mechanical Sciences and Engineering* 39 (2): 613–620. doi:10.1007/s40430-016-0523-9.

Kanta, A.-F., M. Poelman, V. Vitry, and F. Delaunois. 2010. "Nickel–boron Electrochemical Properties Investigations." *Journal of Alloys and Compounds* 505 (1): 151–156. doi:10.1016/j.jallcom.2010.05.168.

Kaya, B., T. Gulmez, and M. Demirkol. 2008. "Preparation and Properties of Electroless Ni-B and Ni-B Nanocomposite Coatings." In *Proceedings of the World Congress on Engineering and Computer Science (WCECS 2008)*, San Francisco, CA, pp. 44–48.

Keong, K. G., W. Sha, and S. Malinov. 2004. "Hardness Evolution of Electroless Nickel-Phosphorus Deposits with Thermal Processing." *Surface and Coatings Technology* 168: 263–274. doi:10.1016/S0257-8972(03)00209-3.

Kostyanovskii, M. A. Y. V. Prusov, N. V. Suvorov, N. M. Kulin, and V. N. Flerov. 1991. "Effect of Heat Treatment on the Structure of Chemically Deposited Nickel with a Low Boron Content." *Zashchita Metallov* 27: 317–320.

Krishnan, K. H., S. John, K. N. Srinivasan, J. Praveen, M. Ganesan, and P. M. Kavimani. 2006. "An Overall Aspect of Electroless Ni-P Depositions—A Review Article." *Metallurgical and Materials Transactions* 37A: 1917–1926.

Krishnaveni, K., T. S. N. S. Narayanan, and S. K. Seshadri. 2005. "Electroless Nickel-Boron Coatings: Preparation and Evaluation of Hardness and Wear Resistance." *Surface and Coatings Technology* 190: 115–121.

Kumar, P. S., and P. K. Nair. 1994. "X-Ray Diffraction Studies On The Relative Proportion And Decomposition of Amorphous Phase in Electroless Ni-B Deposits (Accepted March 1994) Produced on Mild Steel Substrates. The Deposits Had a Structure Consisting of a Mixture of Mic." *Nanostructured Materials* 4 (2): 183–198.

Lee, C., and K. Lin. n.d. "Preparation of Solder Bumps Incorporating Electroless Nickel-Boron Deposit and Investigation on the Interfacial Interaction Behaviour and Wetting Kinetics." *Journal of Materials Science: Materials in Electronics* 8 (6): 377–383.

Lee, C.-Y., and K.-L. Lin. 1993. "Materials Interaction in Pb-Sn/Ni-P/Al and Pb-Sn/Ni-B/Al Solder Bumps on Chips." *Thin Solid Films* 229 (1): 63–75. doi:10.1016/0040-6090(93)90411-H.

Lee, S. P., and Y. W. Chen. n.d. "Selective Hydrogenation of Furfural on Ni-PB Nanometals." *Studies in Surface Science and Catalysis* 130: 3483–3488.

Lekka, M., R. Offoiach, A. Lanzutti, M.Z. Mughal, M. Sebastiani, E. Bemporad, and L. Fedrizzi. 2018. "Ni-B Electrodeposits with Low B Content: Effect of DMAB Concentration on the Internal Stresses and the Electrochemical Behaviour." *Surface and Coatings Technology* 344 (June): 190–196. doi:10.1016/j.surfcoat.2018.03.018.

Li, H., H. Li, W. Dai, and M. Qiao. 2003. "Preparation of the Ni-B Amorphous Alloys with Variable Boron Content and Its Correlation to the Hydrogenation Activity." *Applied Catalysis A: General* 238 (1): 119–30. doi:10.1016/S0926-860X(02)00342-3.

Li, H., H. Li, W.-L. Dai, W. Wang, Z. Fang, and J.-F. Deng. 1999. "XPS Studies on Surface Electronic Characteristics of Ni-B and Ni-P Amorphous Alloy and Its Correlation to Their Catalytic Properties." *Applied Surface Science* 152 (1–2): 25–34. doi:10.1016/S0169-4332(99)00294-9.

Li, T., W. Zhang, R. Z. Lee, and Q. Zhong. 2009. "Nickel-Boron Alloy Catalysts Reduce the Formation of Trans Fatty Acids in Hydrogenated Soybean Oil." *Food Chemistry* 114 (2): 447–452. doi:10.1016/j.foodchem.2008.09.068.

Lu, K. 1996. "Nanocrystalline Metals Crystallized from Amorphous Solids: Nanocrystallization, Structure, and Properties." *Materials Science and Engineering: R: Reports* 16 (4): 161–221. doi:10.1016/0927-796X(95)00187-5.

Ma, Y., W. Li, M. Zhang, Y. Zhou, and K. Tao. 2003. "Preparation and Catalytic Properties of Amorphous Alloys in Hydrogenation of Sulfolene." *Applied Catalysis A: General* 243 (2): 215–223. doi:10.1016/S0926-860X(02)00470-2.

Meyers, B., and S. Lynn. 1994. "Chromium Elimination." In *Surface Engineering, Vol 5*, ASM Handbook, ed. Cotell, C.M. et al, 5:925-929. ASM International Publication.

Moiseev, G. K., N. I. Il'inykh, T. V. Kulikova. 2005. Calculation of the thermodynamic properties of the condensed phases NiB, Ni2B, Ni3B, and Ni4B3. *Russian Metallurgy (Metally)* 1:20–25.

Montagne, A., M. Z. Mughal, L. Bonin, M. Sebastiani, E. Bemporad, V. Vitry, A. Iost, and M. H. Staia. 2018. "Residual Stresses In Electroless Nickel Coatings." In *Proceedings of SMT32*.

Mukhopadhyay, A., T. K. Barman, and P. Sahoo. 2017. "Tribological Behavior of Sodium Borohydride Reduced Electroless Nickel Alloy Coatings at Room and Elevated Temperatures." *Surface and Coatings Technology* 321 (July): 464–476. doi:10.1016/j.surfcoat.2017.05.015.

Mut, L. S. 2015. "Synthesis and Structural Characterization of Nickel-Boron Nanoalloys." Graduate School of Applied and Natural Sciences of Middle East Technical University, Turkey.

Narayanan, T. S. N. S., K. Krishnaveni, and S. K. Seshadri. 2003. "Electroless NiP/NiB Duplex Coatings: Preparation and Evaluation of Microhardness, Wear and Corrosion Resistance." Materials Chemistry and Physics 82: 771–779.

Narayanan, T. S. N. S., and S. K. Seshadri. 2004. "Formation and Characterization of Borohydride Reduced Electroless Nickel Deposits." *Journal of Alloys and Compounds* 365: 197–205. doi:10.1016/S0925-8388(03)00680-7.

Ohno, I., O. Wakabayashi, and S. Haruyama. 1985. "Anodic Oxidation of Reductants in Electroless Plating." *Journal of the Electrochemical Society* 132: 2323–2330. doi:10.1149/1.2113572.

Okamoto, Y., Y. Nitta, T. Imanaka, S. Teranishi. 1980. "Surface State and Catalytic Activity and Selectivity of Nickel Catalysts in Hydrogenation Reactions: III. Electronic and Catalytic Properties of Nickel Catalysts." *Journal of Catalysis* 64:397–404.

Oliver, W. C., and G. M. Pharr. 1992. "An Improved Technique for Determining Hardness and Elastic Modulus Using Load and Displacement Sensing Indentation Experiments." *Journal of Materials Research* 7 (6): 1564–1583. doi:10.1557/JMR.1992.1564.

Oraon, B., G. Majumdar, and B. Ghosh. 2007. "Parametric Optimization and Prediction of Electroless Ni-B Deposition." *Materials and Design* 28: 2138–2147.

Pal, S., N. Verma, V. Jarayam, S. K. Biswas, and Y. Riddle. 2011. "Characterization of Phase Transformation Behaviour and Microstructural Development of Electroless Ni-B Coating." *Materials Science and Engineering A* 528: 8269–8276.

Rao, Q., G. Bi, Q. Lu, H. Wang, and X. Fan. 2005. "Microstructure Evolution of Electroless Ni-B Film during Its Depositing Process." *Applied Surface Science* 240: 28–33.

Riddle, Y. W., and T. O. Bailer. 2005. "Friction and Wear Reduction via an Ni-B Electroless Bath Coating for Metal Alloys." *Jom* 40–45. doi:10.1007/s11837-005-0080-7.

Riddle, Y. W., and C. E. Mccomas. 2005. "Advances in Electroless Nickel Boron Coatings: Improvements to Lubricity and Wear Resistance on Surfaces of Automotive Components." In *Proceeding of the 2005 SAE world Congress.* SAE Technical Paper 2005-01-0615. doi:10.4271/2005-01-0615.

Riedel, W. 1989. *Electroless Nickel Plating.* London, OH: Finishing Publication.

Sahoo, P., and S. K. Das. 2011. "Tribology of Electroless Nickel Coatings–A Review." *Materials and Design* 32 (4): 1760–1775. doi:10.1016/j.matdes.2010.11.013.

Saito, T., E. Sato, M. Matsuoka, and C. Iwakura. 1999. "Effect of Heat-Treatment on Magnetic Properties of Electroless Ni-B Films." *Plating and Surface Finishing* 86: 53–56.

Saito, T., E. Sato, M. Matsuoka, and C. Iwakura. 1998. "Electroless Deposition of Ni–B, Co–B and Ni–Co–B Alloys Using Dimethylamineborane as a Reducing Agent." *Journal of Applied Electrochemistry* 28 (5): 559–563. doi:10.1023/A:1003233715362.

Selvaraj, M., S. Azim, K. Chandran, and S. Guruviah. 1989. "Sodium Borohydride Reduced Electroless Nickel." *Transactions of the SAEST.* 24: 41–44.

Shaffer, S.J., and M.J. Rogers. 2007. "Tribological Performance of Various Coatings in Unlubricated Sliding for Use in Small Arms Action Components—A Case Study." *Wear* 263 (7–12): 1281–1290. doi:10.1016/j.wear.2007.01.115.

Shakoor, R. A., R. Kahraman, W. Gao, and Y. Wang. 2016. "Synthesis, Characterization and Applications of Electroless Ni-B Coatings—A Review." *International Journal of Electrochemical Science* 11: 2486–2512.

Shen, J.-H., and Y.-W. Chen. 2007. "Catalytic Properties of Bimetallic NiCoB Nanoalloy Catalysts for Hydrogenation of P-Chloronitrobenzene." *Journal of Molecular Catalysis A: Chemical* 273 (1–2): 265–276. doi:10.1016/j.molcata.2007.04.015.

Song, X., K. B. Yeap, J. Zhu, J. Belnoue, M. Sebastiani, E. Bemporad, K. Zeng, and A. M. Korsunsky. 2012. "Residual Stress Measurement in Thin Films at Sub-Micron Scale Using Focused Ion Beam Milling and Imaging." *Thin Solid Films* 520 (6): 2073–2076. doi:10.1016/j.tsf.2011.10.211.

Srinivasan, K. N., R. Meenakshi, A. Santhi, P. R. Thangavelu, and S. John. 2010. "Studies on Development of Electroless Ni–B Bath for Corrosion Resistance and Wear Resistance Applications." *Surface Engineering* 26 (3): 153–158. doi:10.1179/174329409×409468.

Srivastava, A., S. Mohan, V. Agarwala, and R. C. Agarwala. 1992. "Factors Influencing the Deposition Rate of Ni-B Electroless Films." *Zeitschrift fur Metallkunde* 83: 251–253.

Sudagar, J., J. L., and W. Sha. 2013. "Electroless Nickel, Alloy, Composite and Nano Coatings–A Critical Review." *Journal of Alloys and Compounds* 571 (September): 183–204. doi:10.1016/j.jallcom.2013.03.107.

Tan, N., Z. Xing, X. Wang, H. Wang, G. Jin, and B. Xu. 2017. "Investigation of Sprayed Particle Filling Qualities within the Texture on the Bonding Behavior of Ni-Based Coating." *Surface and Coatings Technology* 330 (December): 131–139. doi:10.1016/j.surfcoat.2017.09.079.

Vitry, V., and F. Delaunois. 2015d. "Formation of Borohydride-Reduced Nickel–Boron Coatings on Various Steel Substrates." *Applied Surface Science* 359 (December): 692–703. doi:10.1016/j.apsusc.2015.10.205.

Vitry, V., F. Delaunois, and C. Dumortier. 2008. "Mechanical Properties and Scratch Test Resistance of Nickel-Boron Coated Aluminium Alloy after Heat Treatments." *Surface and Coatings Technology* 202 (14): 3316–3324. doi:10.1016/j.surfcoat.2007.12.001.

Vitry, V., A.-F. F. Kanta, and F. Delaunois. 2010. "Initiation and Formation of Electroless Nickel-Boron Coatings on Mild Steel: Effect of Substrate Roughness." *Materials Science and Engineering B: Solid-State Materials for Advanced Technology* 175 (3): 266–273. doi:10.1016/j.mseb.2010.08.003.

Vitry, V., A.-F. F. Kanta, J. Dille, and F. Delaunois. 2012d. "Structural State of Electroless Nickel-Boron Deposits (5wt.% B): Characterization by XRD and TEM." *Surface and Coatings Technology* 206 (16): 3444–3449. doi:10.1016/J.SURFCOAT.2012.02.003.

Vitry, V., A. Sens, and F. Delaunois. 2012e. "Abrasion Resistance of Steel Coated with Various Chemical Nickel Deposits." In *Junior Euromat 2012, Lausanne.*

Vitry, V., A. Sens, A.-F. Kanta, and F. Delaunois. 2012f. "Wear and Corrosion Resistance of Heat Treated and As-Plated Duplex NiP/NiB Coatings on 2024 Aluminum Alloys." *Surface and Coatings Technology* 206 (16): 3421–3427. doi:10.1016/j.surfcoat.2012.01.049.

Vitry, V. 2010. "Electroless Nickel-Boron Deposits: Synthesis, Formation and Characterization; Effect of Heat Treatments; Analytical Modeling of the Structural State." University of Mons, Belgium.

Vitry, V., and L. Bonin. 2017. "Increase of Boron Content in Electroless Nickel-Boron Coating by Modification of Plating Conditions." *Surface and Coatings Technology* 311 (February): 164–171. doi:10.1016/j.surfcoat.2017.01.009.

Vitry, V., L. Bonin, and L. Malet. 2017. "Chemical, Morphological and Structural Characterisation of Electroless Duplex NiP/NiB Coatings on Steel." *Surface Engineering* 34 (6): 1–10. doi:10.1080/02670844.2017.1320032.

Vitry, V., and F. Delaunois. 2015a. "Electroless Nickel-Phosphorous Vs Electroless Nickel-Boron: Comparison of Hardness, Abrasion Resistance, Scratch Test Response and Corrosion Behavior." In *Comprehensive Guide for Nanocoatings Technology, Volume 1: Deposition and Mechanism*, edited by M Aliofkhazraei, pp. 145–173. Nova Science Publishers.

Vitry, V., and F. Delaunois. 2015b. "Nanostructured Electroless Nickel-Boron Coatings for Wear Resistance." In *Anti-Abrasive Nanocoatings*, pp. 157–199. Cambridge, UK: Elsevier. doi:10.1016/B978-0-85709-211-3.00007-8.

Vitry, V., and F. Delaunois. 2015c. "Nanostructured Electroless Nickel-Boron Deposition: Initiation Mechanism and Formation, Effect of Bath Chemistry Modificationhardness, Abrasion Resistance, Scratch Test Response and Corrosion Behavior." In *Comprehensive Guide for Nanocoatings Technology, Volume 3: Properties and Development*, edited by Mahmood Aliofkhazraei. New York: Nova Science Publishers.

Vitry, V., A.-F. F. Kanta, and F. Delaunois. 2011. "Mechanical and Wear Characterization of Electroless Nickel-Boron Coatings." *Surface and Coatings Technology* 206 (7): 1879–1885. doi:10.1016/j.surfcoat.2011.08.008.

Vitry, V., A.-F. F. Kanta, and F. Delaunois. 2012a. "Evolution of Reactive Concentration during Borohydride-Reduced Electroless Nickel-Boron Plating and Design of a Replenishment Procedure." *Industrial and Engineering Chemistry Research* 51 (1): 9227–9234. doi:10.1021/ie202687y.

Vitry, V., A.-F. F. Kanta, and F. Delaunois. 2012b. "Application of Nitriding to Electroless Nickel-Boron Coatings: Chemical and Structural Effects; Mechanical Characterization; Corrosion Resistance." *Materials and Design* 39 (August): 269–278. doi:10.1016/j.matdes.2012.02.037.

Vitry, V., A. Sens, and F. Delaunois. 2014. "Comparison of Various Electroless Nickel Coatings on Steel: Structure, Hardness and Abrasion Resistance." *Materials Science Forum* 783–786 (May): 1405–1413. doi:10.4028/www.scientific.net/MSF.783-786.1405.

Vitry, V., A. Sens, A.-F. Kanta, and F. Delaunois. 2012c. "Experimental Study on the Formation and Growth of Electroless Nickel-Boron Coatings from Borohydride-Reduced Bath on Mild Steel." *Applied Surface Science* 263 (December): 640–647. doi:10.1016/j.apsusc.2012.09.126.

Wang, Z.-C., L. Yu, Z.-B. Qi, and G.-L. Song. 2013. "Electroless Nickel-Boron Plating to Improve the Corrosion Resistance of Magnesium (Mg) Alloys." *Corrosion Prevention of Magnesium Alloys*, 370–392. doi:10.1533/9780857098962.3.370.

Wang, Z. C., F. Jia, L. Yu, Z. B. Qi, Y. Tang, and G. L. Song. 2012. "Direct Electroless Nickel-Boron Plating on AZ91D Magnesium Alloy." *Surface and Coatings Technology* 206: 3676–3685. doi:10.1016/j.surfcoat.2012.03.020.

Watanabe, T., and Y. Tanabe. 1975. Ni-B Amorphous Alloy Film Deposited by Electroless Deposition Method. *Journal of the Japan Institute of Metals and Materials* 39:831–836.

Watanabe, T., and Y. Tanabe. 1976. Formation and morphology of NiB amorphous alloy deposited by electroless plating. *Materials Science and Engineering* 23:97–100.

Watanabe, T., and Y. Tanabe. 1983. "The Lattice Images of Amorphous-like Ni-B Alloy Films Prepared by Electroless Plating Method." *Transactions of the Japan Institute of Metals* 24: 396–404.

Weil, R., and K. Parker. 1990. "The Properties of Electroless Nickel." In Electroless Plating: Fundamentals and Applications, ed. O. Mallory, and J.B. Hajdu, 111–137. American Electroplaters and Surface Finishers Society, INC., Orlando, Florida.

Xu, J., L. Chen, K. Tan, A. Borgna, and M. Saeys. 2009. "Effect of Boron on the Stability of Ni Catalysts during Steam Methane Reforming." *Journal of Catalysis* 261 (2): 158–165. doi:10.1016/j.jcat.2008.11.007.

Yang, J., L. Wang, D. Li, X. Zhong. n.d. "Stress Analysis and Failure Mechanisms of Plasma-Sprayed Thermal Barrier Coatings." *Journal of Thermal Spray Technology* 26:890–901.

Zhang, B. 2016. *Amorphous and Nano Alloys Electroless Depositions: Technology, Composition, Structure and Theory.* Chemical Industry Press. Elsevier Inc.

4B Electroless Nickel-Boron

Part B: Posttreatments for Electroless Nickel-Boron

Véronique Vitry and Luiza Bonin

CONTENTS

4B.1 INTRODUCTION

Electroless nickel-boron presents, as described in Chapter 4, excellent mechanical and tribological properties. However, like for nickel-phosphorus coatings, further treatment can be used to improve their performance.

The most well-known posttreatment for this kind of coatings is heat treatment but thermochemical treatments, mostly in the form of nitriding treatments, have also been investigated.

This chapter aims to describe the usual conditions for heat treating of electroless nickel-boron-coated samples and the effect of this posttreatment on the coating.

4B.2 HEAT TREATMENTS

As mentioned earlier, heat treatments are the most popular type of posttreatment that can be applied to nickel-boron coatings. These treatments are usually carried out in relatively low-temperature range (less than 500°C), under air or under a protective (neutral or slightly reducing) atmosphere. They induce significant modifications to the properties of the coating, mostly at the microscopic scale.

For this reason, this section will first examine the rationale for heat treating of electroless nickel-boron and the necessary conditions for heat treatment, and then the effect these treatments have on all the properties of the coating, from morphology and structure to functional properties such as hardness, wear and corrosion resistance.

4B.2.1 Rationale behind Heat Treatment of Electroless Nickel-Boron Coatings

The structure of as-plated electroless nickel-boron coatings is very far from equilibrium in most cases: when one looks at the NiB phase diagram (Sheng et al. 2015; Teppo and Taskinen 1993) (Figure 4B.1), the solubility of boron in solid nickel is close to zero (the maximal solubility, obtained at 1360 K, is 0.29 at.%) (Figure 4B.1b), which is negligible in terms of weight fraction. Under 700 K, the solubility is zero,

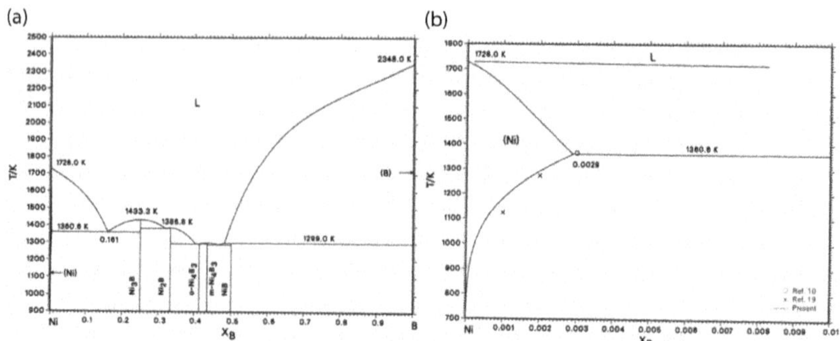

FIGURE 4B.1 (a,b) Nickel-boron phase diagram and details of the Ni-B phase diagram for low boron content. (From Teppo, O. and Taskinen, P., *Mater. Sci. Technol.*, 9, 205–212, 1993.)

and a mixture of nickel and nickel boride is expected even for very low boron content. However, electroless nickel-boron coatings do not usually contain nickel boride in the as-plated form and are usually considered as nanocrystalline supersaturated solutions of boron in nickel or amorphous nickel/boron alloys.

Given this huge difference between the structure that is predicted by the phase diagram and the one that is obtained, it is easily predicted that the material will tend to reach the equilibrium structure if some energy is provided to it in the form of heat and to crystallize as nickel and nickel boride (mostly Ni_3B and Ni_2B) phases, depending on the boron content (Teppo and Taskinen 1993). Some of the boride phases that are formed during the heat treatment process possess very high hardness due to their specific structure, similar in some cases to the hard orthorhombic crystal structure of Fe_3C cementite phase encountered in steel (Finch et al. 1984). Some nickel-boron coatings obtained when sodium borohydride is used as a reducing agent present boron content very close to that of the hard Ni_3B phase (Vitry et al. 2012a). It is thus specifically interesting to heat-treat those coatings and try to form hard nickel boride coatings from amorphous or nanocrystalline nickel-boron.

4B.2.2 Determination of the Minimal Heat-Treatment Temperature for Electroless Nickel-Boron

It is possible to assess the minimal temperature required to crystallize electroless nickel-boron coatings into their equilibrium form with the use of differential scanning calorimetry (DSC). Several authors have observed exothermic peaks in the 300°C–450°C range that are attributed to the formation of nickel boride crystals. The presence of one or several peaks and the temperature for which they are observed depend strongly on the boron content of the coating: in the 1–4.5 wt.% B range, only the formation of Ni_3B is observed by DSC—it appears between 300°C and 350°C (Li et al. 2001). For higher boron concentrations (up to 7 wt.% B), another peak is observed at higher temperature—slightly over 400°C—that corresponds to Ni_2B (Narayanan et al. 2004) while for over 7 wt.% boron, only the formation of Ni_2B is observed.

Some authors have observed peaks that are related to phases not predicted by the diagram for the average composition of the alloy they studied. They are related to local chemical heterogeneities in the as-plated coating (Pal et al. 2011).

It is also possible to assess the temperature at which various phases form by high-temperature X-ray diffraction (XRD) analysis. For a coating with 6 wt.% B, for example (see Figure 4B.2), crystallization begins between 250°C and 300°C and only the Ni_3B phase is formed (Vitry 2010).

From the DSC and high-temperature XRD results, it is possible to assess that heat treatment of electroless nickel-boron should be carried out at temperatures higher than 250°C (preferably 300°C) to ensure full crystallization. To decrease the treatment time as much as possible and stay under 4 h, slightly higher temperatures are chosen preferentially (350°C–450°C) (Watanabe and Tanabe 1983; Delaunois and Lienard 2002; Ziyuan 2006; Narayanan et al. 2004; Anik et al. 2008). Heat treatments are usually designed to reach an optimal hardness of the coating.

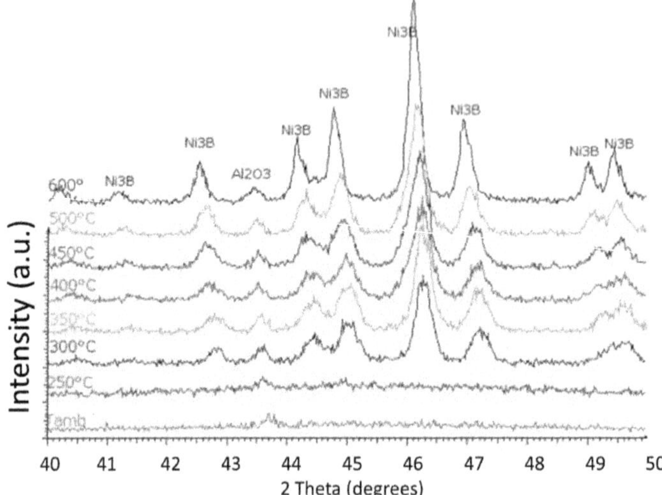

FIGURE 4B.2 High-temperature X-ray diffraction data from electroless nickel-boron coating with 6 wt.% B. (From Vitry, V., Electroless nickel-boron deposits: Synthesis, formation and characterization; effect of heat treatments; Analytical modeling of the structural state, University of Mons, Belgium, 2010.)

Alternately, low-temperature treatments that don't aim at full crystallization are also possible, mainly to release internal stress or to increase the hardness of coatings deposited on aluminum substrates that cannot be subjected to temperatures higher than 200°C (Delaunois and Lienard 2002; Yildiz 2017).

4B.2.3 Effect of Heat Treatment on Morphology, Microstructure and Structure of Electroless Nickel-Boron Coatings

The composition of electroless nickel-boron deposits is not modified macroscopically by heat treatment. However, local modifications can be observed, as shown in Figure 4B.3. These modifications are due to diffusion but it is not possible to extract a trend from them except in specific cases, such as aluminum alloys, in which significant interdiffusion has been observed (Fabienne Delaunois 2002; Delaunois and Lienard 2002) or for treatments carried out at temperatures over 450°C (Figure 4B.3f) (Arias et al. 2018).

Likewise, the morphology of electroless nickel-boron coatings is macroscopically not modified after conventional heat treatments (Narayanan et al. 2003; Delaunois and Lienard 2002; Dervos et al. 2004; Riedel 1989; Delaunois 2002; Vitry and Bonin 2017). The coatings retain thus their columnar aspect and "cauliflower-like" surface texture, as described in Section 4B.3.3 of Chapter 4. Macroscopic morphological modifications require high temperature (for example, Brunelli et al. (2009) treated samples at a temperature above 800°C).

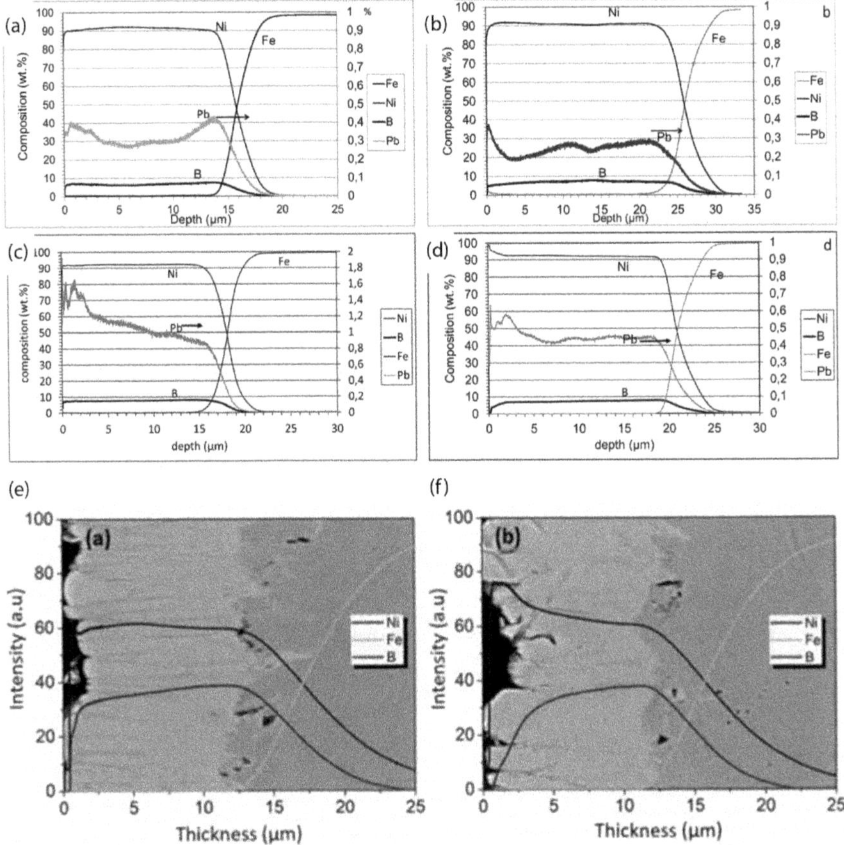

FIGURE 4B.3 GDOES depth profile composition analysis of (a) as-plated 6 wt.% B electroless nickel-boron coating; (b) heat-treated (400°C, 1 h) 6 wt.% B electroless nickel-boron coating; (c) as-plated 8 wt.% B electroless nickel-boron coating; (d) heat-treated (400°C, 1h) 8 wt.% B electroless nickel-boron coating. (From Vitry, V. and Bonin, L., *Surf. Coat. Technol.*, 311, 164–171, 2017.) Left axis: Ni, B, Fe. Right axis: Pb. Lead content is enlarged 100 times for better visibility. (e) As-plated electroless nickel-boron coating (unknown composition); (f) the same coating as in (e), after heat treatment at 450°C for 1 h. (From Arias, S. et al., *Rev. Fac. Ing.*, 27, 101, 2018.)

As mentioned in the previous sections of this chapter, heat treatment of electroless nickel-boron coating leads to the crystallization of nickel and/or nickel boride phases. These structural modifications can be described, depending on the nanocrystalline or amorphous description of the initial state, as either the growth of nanocrystalline grains or the precipitation of crystalline phases in an amorphous matrix.

X-ray diffraction studies carried out after heat treatment at temperatures higher than the crystallization onset temperature clearly show the formation of various crystalline phases in electroless nickel-boron depending on the boron content of the coating: for very low boron content (1.2 wt.% B, see Figure 4B.4a), nickel is the main crystalline phase after heat treatment and only small peaks of Ni_3B are observed

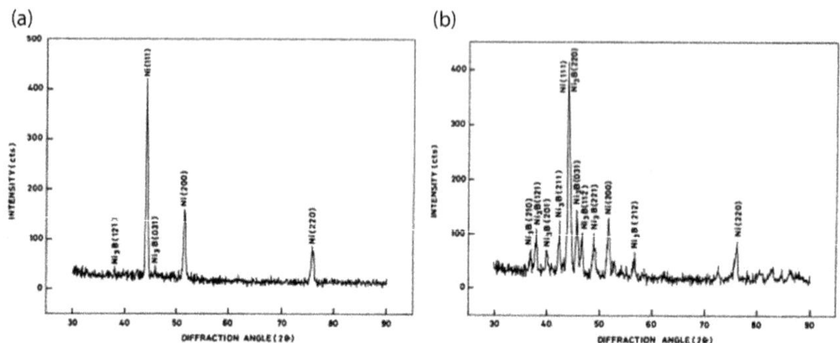

FIGURE 4B.4 XRD profile of electroless nickel-boron coatings with low boron content after heat treatment at 450°C for 1 h. (a) 1.2 wt.% B; (b) 3.2 wt.% B. (From Baskaran, I. et al., *Surf. Coat. Technol.*, 200, 6888–6894, 2006.)

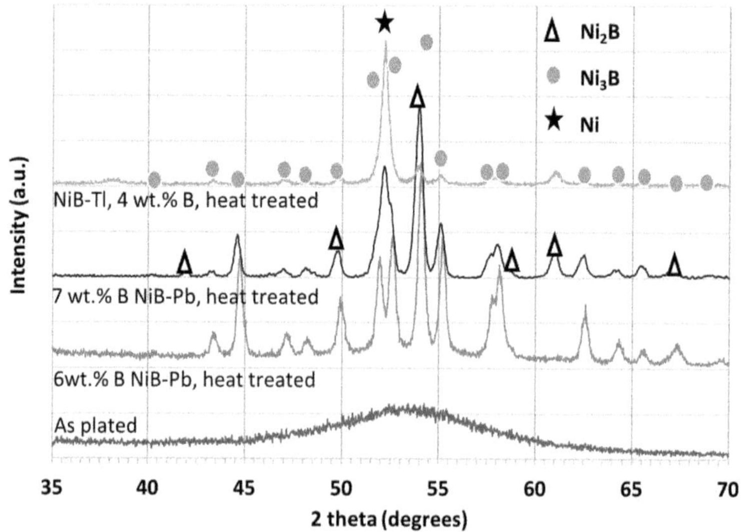

FIGURE 4B.5 XRD profile of electroless nickel-boron coatings with high boron content. (Adapted from Vitry, V. and Bonin, L., *Surf. Coat. Technol.*, 311, 164–171, 2017; Vitry, V. et al., *Mater. Sci. Forum*, 783–786, 1405–1413, 2014; Bonin, L. et al., *Techniques de l'ingénieur*, 2018a.)

(Baskaran et al. 2006). For slightly higher contents (>2 to 4.5 wt.% approximately, see Figures 4B.4b and 4B.5), both Ni and Ni_3B are observed in significant amounts (Baskaran et al. 2006; Kumar and Nair 1994). When the coating reaches 5–6 wt.% B, which corresponds to the composition of the Ni_3B phase, it can either crystallize fully in Ni_3B form (Vitry 2010, 2012a; Vitry and Bonin 2017) (see Figure 4B.5) or contain mostly Ni_3B with some Ni and Ni_2B (Delaunois and Lienard 2002; Das and Sahoo 2012; Anik et al. 2008; Mukhopadhyay et al. 2018b). The presence of Ni_2B is linked to both heat treatment conditions (temperature and time) and to the heterogeneities in the coating composition. Coatings with a boron content higher than 6 wt.% will crystallize as a mixture of Ni_3B and Ni_2B phase (Niksefat and Ghorbani 2015; Vitry and Bonin 2017) if heat-treated at sufficient temperature (see Figure 4B.5). However, as the onset temperature for the crystallization of the Ni_2B phase is higher than that for Ni_3B phase, it is possible to obtain only Ni and Ni_3B by using relatively low temperatures (usually between 325°C and 375°C) for the heat treatment (Narayanan et al. 2004; Pal and Jayaram 2018) (see Figure 4B.6). In some rare occasions, Ni_4B_3 has also been observed (Oraon et al. 2008).

It is usually accepted that the crystallinity of electroless nickel-boron coatings increases with the duration and temperature of the heat treatment, as can be easily attested from the observation of XRD profiles, where sharp peaks replace the low, wide and smooth domes usually observed in the case of as-plated electroless coatings (Thomas Evans et al. 1994). However, studies about the evolution of grain sizes are rare. From the few published results, the grain size observed in heat-treated coatings usually stays well within the nanocrystalline range: an average grain size of 38 nm was observed for coatings with 5 wt.% B (Vitry et al. 2012a) (Figure 4B.7a) and a value of 17–20 nm for coatings with 3 wt.% B (Krishnaveni et al. 2009) after a treatment for 1 h at 400°C. Likewise, 10–20 nm grains were observed after 30 min at 275°C for a boron content of 3.65 wt.% (Watanabe and Tanabe 1983). Even longer treatments (4 h at 385°C) did not lead to excessive growth of grain size, as evidenced by the value of 30–300 nm obtained after this treatment for a coating with 6.4 wt.% B (see Figure 4B.7b). Wang et al. (2012) measured grain sizes for various Ni_3B peaks after treatments at 300°C, 400°C and 500°C for 1 h and even the higher temperature did not lead to excess grain growth as all their sizes were below 100 nm.

As a conclusion, the effect of heat treatment on morphology and chemistry is negligible but the same heat treatment leads to extremely significant structural modifications, thus leading to the formation of new phases in the material. However, the grain size of the newly formed phases always stays in the nanometer range.

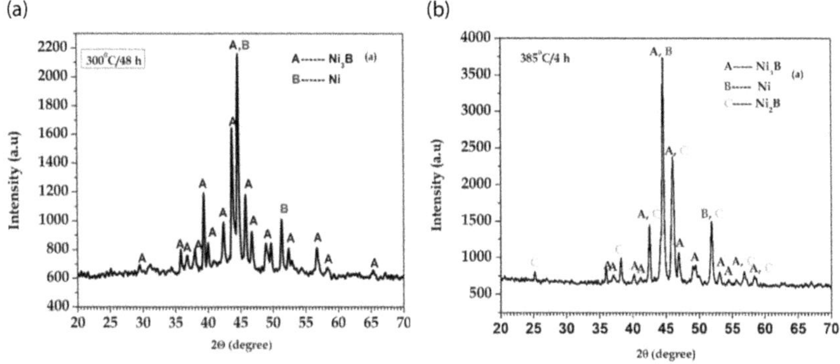

FIGURE 4B.6 XRD profile of electroless nickel-boron coating with 6.4 wt.% B after heat treatment at (a) 300°C for 48 h and (b) 385°C for 4 h. (From Pal, S. and Jayaram, V., *Materialia*, 2018.)

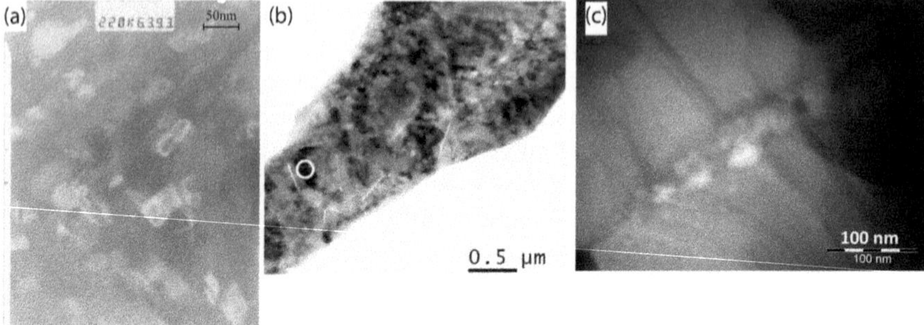

FIGURE 4B.7 TEM micrographs of heat-treated electroless nickel-boron coatings. (a) Coating with 5 wt.% B, heat-treated at 400°C for 1 h. (Adapted from Vitry et al. 2012a) (b) Coating with 6.4 wt.% B, heat-treated at 385°C for 4 h. (Adapted from Pal, S. and Jayaram, V., *Materialia*, 2018.) (c) Coating with 7 wt.% B, heat-treated at 400°C for 1 h. (Adapted from Vitry, V. et al., *Surf. Eng.*, 34, 1–10, 2017.)

4B.2.4 Effect of Heat Treatments on Hardness

The aim of heat-treating electroless nickel-boron coatings is usually to increase their mechanical properties so there's an abundant literature on the effect of heat treatment on hardness and optimization of treatment conditions.

The optimization of hardness has led several research groups to study the effect of heat-treatment temperature (for a set time) and heat-treatment time (for a set temperature) (Oraon et al. 2008; Pal et al. 2011; Delaunois and Lienard 2002; Ivanov 2001; Mukhopadhyay et al. 2018b; Vitry et al. 2008; Matik, n.d.; Krishnaveni et al. 2005; Hamid et al. 2010) on the coatings they obtained.

Several studies (Riedel 1991; Thomas Evans et al. 1994; Narayanan et al. 2004; Ziyuan et al. 2004; Anik et al. 2008; Oraon et al. 2008) attribute hardening of electroless nickel-boron to the creation of grain boundaries in the originally amorphous materials and (thus solely) to Hall-Petch effect. This may play a role, but this hypothesis cannot explain the huge increase in hardness observed in some cases. Moreover, some coatings are not fully amorphous in the as-deposited state (Watanabe and Tanabe 1983). These coatings may have grain sizes that are small enough to allow inverse Hall-Petch effect, but this effect could not explain the whole of the hardness increase. Moreover, this explanation does not account for variations of behavior with boron content.

As can be seen from Figure 4B.8, for a treatment duration of 1 h, hardness increases with heat-treatment temperature up to approximately 400°C for coatings with a boron content between 4 and 8 wt.%. However, the amplitude of this increase in hardness and the maximal hardness reached after 1 h of heat treatment vary with the boron content of the coating. As shown in Figure 4B.9a, coatings with a boron content close or equal to 6 wt.% present the highest hardness of all the electroless nickel-boron coatings after heat treatment with values exceeding 1300 $HV_{0.1.}$ This can be explained by the fact that these coatings present a boron content extremely close to the stoichiometric composition of the Ni_3B boride phase. This phase, whose structure

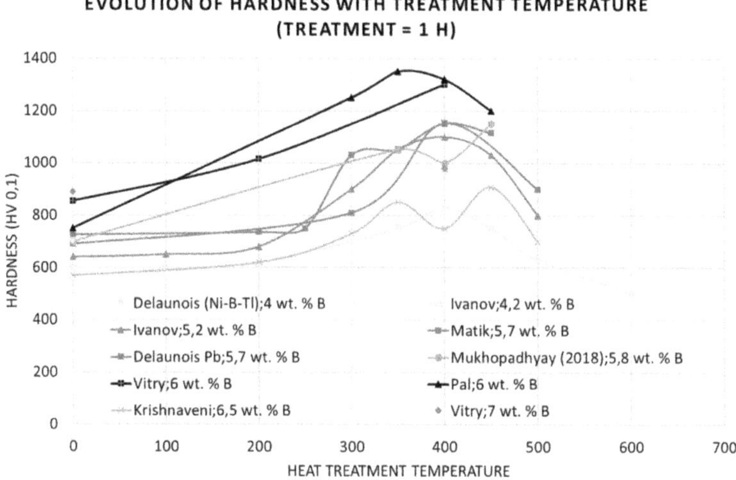

FIGURE 4B.8 Effect of heat-treatment temperature on hardness of electroless nickel-boron coatings with variable boron content after heat treatment for 1 h. (From Delaunois, F. and Lienard, P., *Surf. Coat. Technol.*, 160, 139–148, 2002; Ivanov, M.V., *Prot. Met.*, 37, 592–596, 2001; Matik, U., The effect of heat-treatment on the structure and hardness of electroless Ni-B coated ferrous PM parts, n.d.; Delaunois, F., Les Dépôts Chimiques de Nickel-Bore Sur Alliages d'aluminium, Faculté Polytechnique de Mons, 2002; Mukhopadhyay, A. et al., *Mater. Today*, 5, 3306–3315, 2018b; Vitry, V., Electroless nickel-boron deposits: Synthesis, formation and characterization; Effect of heat treatments; Analytical modeling of the structural state, University of Mons, Belgium, 2010; Pal, S. et al., *Mater. Sci. Eng. A*, 528, 8269–8276, 2011; Krishnaveni, K. et al., *Surf. Coat. Technol.*, 190, 115–121, 2005; Vitry, V. et al., *Surf. Eng.*, 2018.)

has been described by Kayser and Kayser (1996) presents a crystal structure similar to that of Fe_3C cementite. It is also known for its high hardness (values over 1000 HV have been measured (Finch et al. 1984) for single crystals). Adding the fact that electroless nickel-boron coatings stay nanocrystalline after heat treatment, the increased hardness compared to the single crystal hardness values is perfectly understandable. What is less easy to understand is the lowering of maximal hardness for coatings with a boron content higher than 6 wt.%. These coatings contain measurable quantities of crystalline Ni_2B. There is contradictory information about the hardness of this phase: some authors have calculated hardness values that are higher than that of Ni_3B (17.9 GPa (Xiao et al. 2010)) but values measured in boronized nickel (which may be closer to the actual situation observed in electroless nickel-boron coatings) propose a hardness close to 1000 HK for pure Ni_2B (Lou et al. 2009). The causes of the lower peak hardness of coatings with boron content higher than 6 wt.% may thus be linked to lower intrinsic hardness of the Ni_2B phase. They may also be linked to the higher crystallization temperature of that phase: at usual heat-treatment temperature for electroless nickel coatings, Ni_2B may not be able to crystallize fully. This is supported by data obtained by Krishnaveni et al. on 6.5 wt.% B coatings that showed peak hardness at 450°C for a treatment of 1 h.

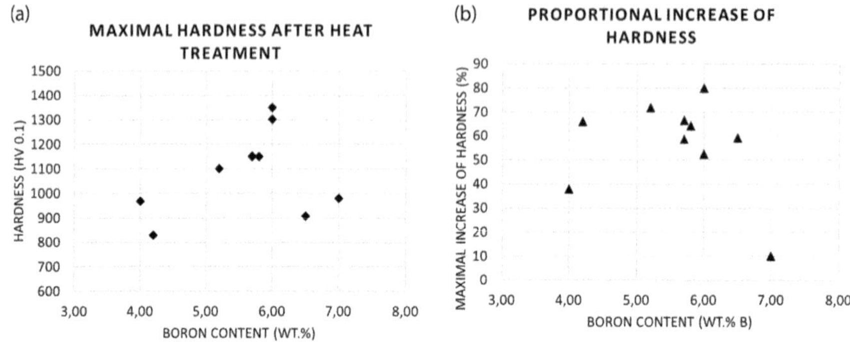

FIGURE 4B.9 Effect of boron content on (a) peak hardness—(b) proportional hardness increase—of electroless nickel-boron coatings after heat treatment for 1 h. (From Delaunois, F. and Lienard, P., *Surf. Coat. Technol.*, 160, 139–148, 2002; Ivanov, M.V., *Prot. Met.*, 37, 592–596, 2001; Matik, U., The effect of heat-treatment on the structure and hardness of electroless Ni-B coated ferrous PM parts, n.d.; Delaunois, F., Les Dépôts Chimiques de Nickel-Bore Sur Alliages d'aluminium, Faculté Polytechnique de Mons, 2002; Mukhopadhyay, A. et al., *Mater. Today*, 5, 3306–3315, 2018b; Vitry, V., Electroless nickel-boron deposits: Synthesis, formation and characterization; Effect of heat treatments; Analytical modeling of the structural state, University of Mons, Belgium, 2010; Pal, S. et al., *Mater. Sci. Eng. A*, 528, 8269–8276, 2011; Krishnaveni, K. et al., *Surf. Coat. Technol.*, 190, 115–121, 2005; Vitry, V. et al., *Surf. Eng.*, 2018.)

The proportional increase in hardness after a heat treatment of 1 h is in the range of 50–80% for most coatings (see Figure 4B.9b), exceptions being coatings with very low (4 wt.%) and very high (7 wt.%) boron content. This may be linked with the fact that selected heat-treatment temperatures (data are available only for treatment at 400°C for those coatings) are not optimal for those boron contents, which are quite outside the usual range.

Similar to the effect of treatment temperature, the effect of the duration of heat treatment on hardness has been investigated by several groups. As shown in Figure 4B.10, the hardness of electroless nickel-boron coatings increases at first, reaching a peak temperature after approximately 4 h at 300°C, and decreases later. This decrease can be explained by the increase in grain size observed for longer heat treatments, even if the grains remain nanometric. Once again, the coatings that have a boron content of 6 wt.% present the highest peak hardness.

It is thus possible to summarize the effect of heat treatment on hardness in the following manner: heat treatment carried out at temperatures higher than the crystallization temperature of Ni_3B phase will first bring an increase in the hardness due to the formation of very fine grains of Ni_3B (either alone or mixed with Ni and Ni_2B for coatings with a boron content different from that of the Ni_3B phase) that replace the softer boron-supersaturated amorphous nickel that is initially present. After peak hardness is reached, the hardness of the coating will thus decrease with increasing treatment time due to the increase in grain size (following the Hall-Petch effect). The time needed to reach peak hardness decreases when the heat-treatment temperature increases (from approximately 5 h at 300°C to 1 h at 400°C). The boron

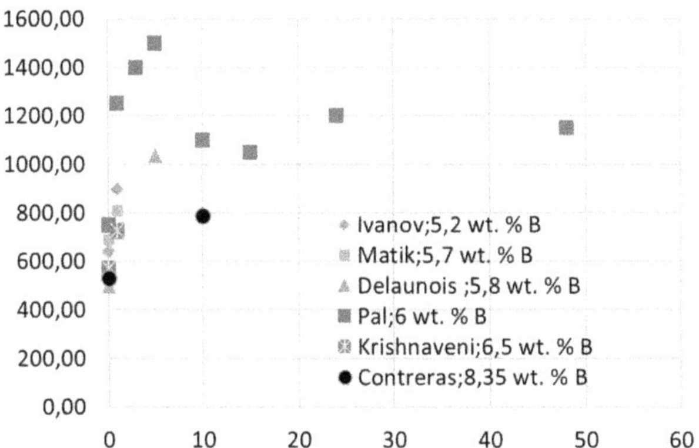

FIGURE 4B.10 Effect of treatment duration at 300°C on hardness of electroless nickel-boron coatings with variable boron content.

content of the coatings also influences peak hardness after heat treatment with the highest values obtained for coatings with a boron content close to 6 wt.% and lower values for coatings that are far from this composition.

In the search for optimal hardness, Oraon et al. (2008) and Das and Sahoo (2012) optimized the plating conditions and posttreatment of their coatings to reach the highest possible hardness after heat treatment. This allowed them to obtain high hardness levels: Das et al. report optimized hardness of 1252 HV_{100} for a coating with 7.5 wt.% B heat-treated at 350°C for 1 h and Oraon et al. report a value of 1402 HV_{100} for a coating of unknown composition after heat treatment at 300°C for 1 h. The optimized coating from Das et al. is not far, in terms of composition and heat treatment, from the best published results (obtained after heat treatment at 400°C for 1 h for a coating with 6 wt.% B) (Vitry and Delaunois 2015; Pal and Jayaram 2018). The hardness measured by Oraon et al. exceeds that but it's difficult to ascertain why as the coating's composition remains unknown.

Some authors (Delaunois and Lienard 2002; Yildiz et al. 2017) have investigated heat treatments that could be applied to electroless nickel-boron coatings deposited on light alloys (mainly aluminum alloys). Indeed, most engineering aluminum alloys present precipitation hardening that is obtained by heat treatment at low temperature (100°C–200°C usually). Submitting them, after electroless nickel plating, to heat treatment in the 300°C–400°C range would severely decrease the bulk properties of the alloy. Alternatively, low-temperature heat treatment has thus been investigated, aiming at improving the properties of electroless nickel-boron (in terms of adhesion and stress mainly), while keeping intact the mechanical properties of the aluminum alloy. Figure 4B.11 presents the effect of such treatments on electroless nickel-boron-plated 7075 aluminum alloy. The best properties can be obtained by a treatment at 150°C–175°C for 5 h or by a longer treatment 15–20 h at 125°C. Delaunois and Lienard (2002) obtained similar results, proposing a treatment at 180°C for 4 h for AS7G06 cast aluminum alloy and 1050 pure aluminum.

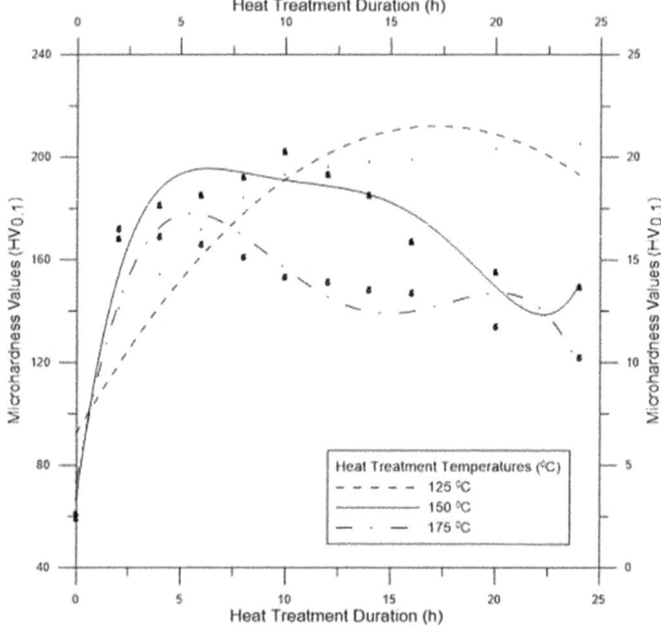

FIGURE 4B.11 Effect of low-temperature heat treatment on the properties of electroless nickel-boron-plated 7075 aluminum alloy. (From Yildiz, R.A. et al., *Int. J. Mater. Mech. Manuf.*, 5, 83–86, 2017.)

4B.2.5 Effect of Heat Treatment on Roughness

Roughness is an important parameter for wear applications. However, while several studies have investigated the roughness of as-deposited electroless nickel-boron coatings (Sahoo 2008; Das and Sahoo 2009; Vitry and Delaunois 2015; Vitry et al. 2008), the effect of heat treatment on this specific parameter has been rarely investigated.

It is expected from the significant structural reorganizations brought by heat treatment that roughness could be modified. However, as the morphology of coatings (which is strongly interlinked with roughness) is not significantly modified, the extent of those modifications is not expected to be huge.

Figure 4B.12 presents roughness results obtained before and after heat treatments for coatings with various boron contents and for different types of heat treatments (Vitry 2010; Vitry and Bonin 2017; Vitry et al. 2008; Celik et al. 2016; Dervos et al. 2004). It shows that the effect of heat treatment on roughness is not significant but seems to manifest in the form of a reduction of roughness that becomes more marked as heat-treatment temperature increases, except in the case of coatings with a high boron content (8wt.%) for which it's not possible to really observe a trend, the observed variations being smaller than the standard deviation of the data (Vitry and Bonin 2017).

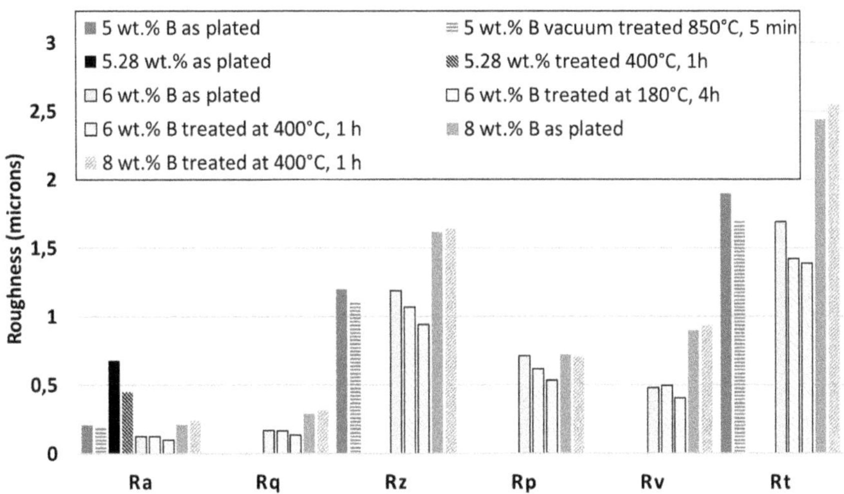

FIGURE 4B.12 Effect of heat treatment on roughness parameters of electroless nickel-boron coatings. (From Vitry, V., Electroless nickel-boron deposits: Synthesis, formation and characterization; effect of heat treatments; Analytical modeling of the structural state, University of Mons, Belgium, 2010; Vitry, V. and Bonin, L., *Surf. Coat. Technol.*, 311, 164–171, 2017; Vitry, V. et al., *Surf. Coat. Technol.*, 202, 3316–3324, 2008; Celik, I. et al., *J. Eng. Tribology*, 230, 57–63, 2016; Dervos, C.T. et al., *Mater. Lett.*, 58, 619–623, 2004.)

4B.2.6 Effect of Heat Treatment on Tribological Properties

Electroless nickel-boron coatings are often used in applications where wear resistance is required, together with high hardness. There is thus abundant literature on the effect of heat treatment on the wear performances of such coatings.

The literature usually investigates one (less often both) of these two parameters: abrasive wear resistance—usually measured through the Taber wear test—and sliding wear resistance—measured by pin-on-disc, ball-on-disc, reciprocal sliding, ball-on-plate, roller-on-plate and so on. The Section 3.4.4.1 of Chapter 3 of this book presents a thorough review of wear testing methods and the rationale for measuring wear resistance of electroless nickel coatings. In this section, the effect of heat treatment on abrasive wear, friction coefficient and sliding wear (including wear mechanisms) of NiB will be reviewed.

4B.2.6.1 Effect of Heat Treatment on Abrasive Wear of Electroless Nickel-Boron Coatings

Taber abrasion test is a popular method for testing the abrasive resistance of materials. However, the variety of test conditions (added to various boron contents and heat treatments) makes it difficult to make any quantitative comparison of results obtained by different groups (or even results from the same group over time). Nevertheless, it is

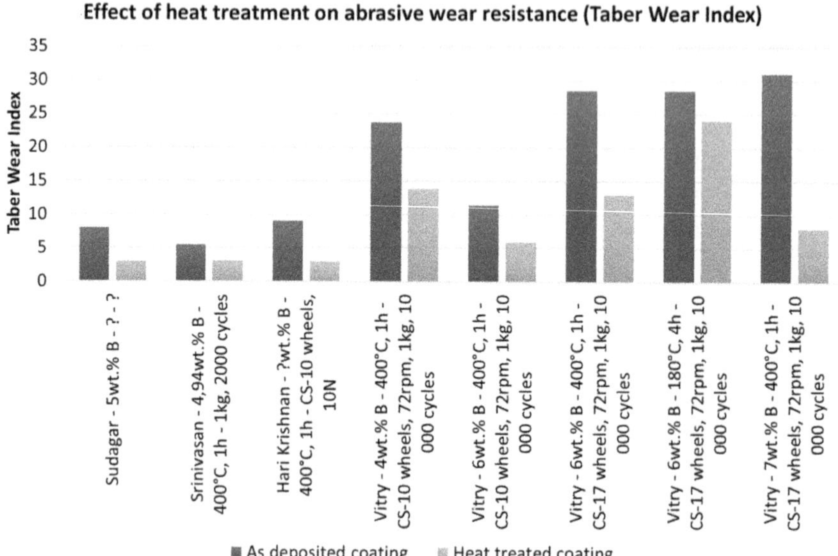

FIGURE 4B.13 Abrasive wear resistance of electroless nickel-boron coatings with and without heat treatment. (From Sudagar, J. et al., *J. Alloys Compd.*, 571, 183–204, 2013; Srinivasan, K.N. et al., *Surf. Eng.*, 26, 153–158, 2010; Krishnan, K.H. et al., *Metall. Mater. Trans.*, 37A, 1917–1926, 2006; Vitry, V., Electroless nickel-boron deposits: Synthesis, formation and characterization; effect of heat treatments; Analytical modeling of the structural state, University of Mons, Belgium, 2010; Vitry, V. and Bonin, L., *Surf. Coat. Technol.*, 311, 164–171, 2017; Vitry, V. et al., *Surf. Coat. Technol.*, 206, 3421–3427, 2012b.)

relatively easy to obtain qualitative information about the effect of heat treatment on abrasive wear of electroless nickel-boron, as shown in Figure 4B.13. Heat treatment improves the abrasive wear resistance of electroless nickel-boron coatings in all the available studies. This improvement of abrasive wear is linked to the increased hardness observed in all coatings after heat treatment (see Section 4B.2.4 of this chapter).

Some testing conditions are missing for all the best available Taber Wear Index (TWI) values in the published literature (Sudagar et al. 2013; Srinivasan et al. 2010; Krishnan et al. 2006); thus, let's focus the quantitative part of this studies on the results obtained by Vitry and Delaunois group (Vitry 2010; Vitry, et al. 2012b; Vitry and Bonin 2017). From this, it appears that an increase in the boron content (from 4 to 6 wt.% measured with CS-10 wheels and from 6 to 7 wt.% measured with CS-17 wheels) leads to a reduction in the TWI (and thus to an increase in abrasive wear resistance) after heat treatment at 400°C. The case of 7 wt.% boron coatings is specifically interesting because these coatings present higher TWI than those with 6 wt.% B in the as-deposited conditions, as well as a lower hardness after heat treatment. This behavior has been linked to the presence of a soft nickel layer on the surface of the heat-treated samples.

It also appears that the use of a low-temperature heat treatment (of the kind devised to be used on aluminum alloy substrates) already leads to a measurable decrease in TWI but that the full benefit of heat treating is only obtained for heat treatments leading to complete characterization of the coating (Vitry 2010).

4B.2.6.2 Effect of Heat Treatment Friction Coefficient of Electroless Nickel-Boron Coatings

The friction coefficient is an important parameter for applications in which sliding solicitations are involved. It is usually recorded during sliding wear tests, and thus, the recorded values are dependent on the load (see Figure 4B.14), speed and duration of the test. Moreover, even in similar conditions, the recorded values are prone to fluctuations, which makes their interpretation sometimes difficult.

There are not many studies that present complete information about the evolution of friction coefficient with heat treatment of electroless nickel-boron: most studies report only average values. Table 4B.1 presents the friction coefficient measured on coatings with varying boron content after heat treatment, in conditions of dry sliding. While it is not possible to describe the effect of boron content on this parameter (the testing conditions are too different for that), all the results present a similar trend: heat treatment decreases the friction coefficient. The extent of this decrease is, however, difficult to evaluate: according to Pal and Jayaram (2018), heat treatment can divide the value of friction coefficient by a factor of 2 while Krishnaveni et al. (2005), Arias et al. (2018) and Vitry and Bonin (2017) observe diminutions close to 20%.

Krishnaveni et al. (2005) attribute the decrease in friction coefficient after heat treatment to the ability of heat-treated coatings to present a surface that offers no compatibility with the material of the hard counterface piece.

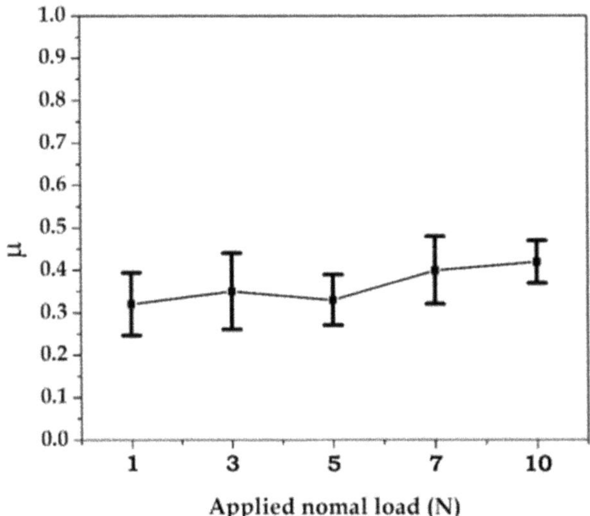

FIGURE 4B.14 Evolution of average friction coefficient on electroless-nickel-boron-coated titanium with applied load. (From Pal, S. and Jayaram, V., *Materialia*, 2018.)

TABLE 4B.1
Friction Coefficient Measured on Electroless Nickel-Boron Coatings with and without Heat Treatment

	Information			Friction Coefficient						
Author	**Boron Content**	**Load (N)**	**Counterpart**	**As-Plated**	**300°C, 10 h**	**300°C, 48 h**	**350°C, 1 h**	**385°C, 4 h**	**400°C, 1 h**	**450°C, 1 h**
Celik	5.28 wt.% B	2	Alumina ball	0.51					0.52	
Pal	6 wt.%	5	Steel	0.75	0.35	0.45		0.35		
Arias	5 (?) wt.% B	5	Alumina ball	0.33						0.24
Pal	6 wt.% B	10	Steel	0.9	0.45	0.75		0.42		
Vitry	6 wt.% B	10	Alumina ball	0.684					0.632	
Vitry	7 wt.% B	10	Alumina ball	0.593					0.579	
Krishnaveni	6.5 wt.% B	20	Steel	0.742			0.71			0.682
Krishnaveni	6.5 wt.% B	30	Steel	0.77			0.732			0.703
Krishnaveni	6.5 wt.% B	40	Steel	0.784			0.749			0.714

Source: Vitry, V. and Bonin, L., *Surf. Coat. Technol.*, 311, 164–171, 2017; Pal, S. and Jayaram, V., *Materialia*, 2018; Celik, I. et al., *J. Eng. Tribology*, 230, 57–63, 2016; Krishnaveni, K. et al., *Surf. Coat. Technol.*, 190, 115–121, 2005; Arias, S. et al., *Rev. Fac. Ing.*, 27, 101, 2018.

Some research group carried out optimization studies on pretreatment, coating and heat treatment conditions aiming at the production of coatings with low friction coefficient:

- Das and Sahoo (2009, 2011b) obtained for a coating with optimized plating conditions (including heat treatment at 350°C–450°C) a value of 0.129 for a heat-treated coating under dry sliding conditions against steel coated with titanium nitride. This may explain the low value compared to the other research groups as most available results were measured with either steel of alumina counterparts.
- Vijayanand and Elansezhian (2014) investigated the effect of pretreatment (for aluminum alloys) on the coefficient of friction of electroless nickel-boron against hardened steel (load 2.5 N) with and without heat treatment. Their main results are presented in Table 4B.2. They obtained a very low friction coefficient, probably due to the small load they used. Heat treatment once again seemed to have a beneficial effect on the friction coefficient of the coating. Their best results were obtained for a substrate pretreated by adsorbed hypophosphite layer.

The typical cauliflower-like surface texture of electroless nickel-boron coating is specifically favorable for the retention of lubricant. However, there are not many studies describing the friction behavior of the coatings in lubricated conditions.

Finally, Mukhopadhyay et al. (2018a) investigated high-temperature wear of as-plated electroless nickel-boron coatings. Some of their results can be used to assess the effect of heat treatment as the experiment can be considered as heat treatment carried out during the wear test (the coating will undergo the same transformation as during standard heat treatment). Figure 4B.15 shows the evolution of friction coefficient with time (and thus with degree of heat treatment) for various testing temperatures. The most interesting curve is the one obtained for 500°C: it shows

TABLE 4B.2
Friction Coefficient Measured on Electroless Nickel-Boron Coatings after Various Pretreatments on Aluminum Alloy, with and without Heat Treatment

Load = 2.5 N	Boron Content: Probably 6 wt.%	
	Friction Coefficient	
Pretreatment	**As-Plated**	**400°C, 1 h**
Nickel strike	0.167	0.143
Palladium strike	0.089	0.015
Absorbed hypophosphite layer	0.079	0.09

Source: Vijayanand, M. and Elansezhian, R., *Procedia Eng.*, 97, 1707–1717, 2014.

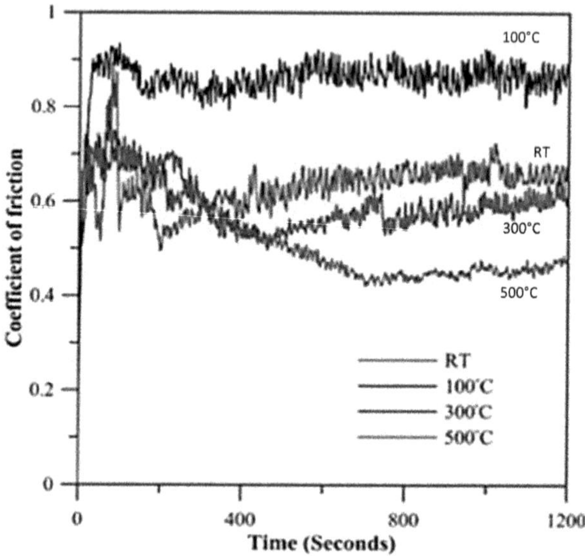

FIGURE 4B.15 Friction coefficient of electroless nickel-boron coatings with various heat treatments under oil-lubricated sliding condition (10 N, 5 Hz). (From Liang, Y. et al., *Surf. Coat. Technol.*, 264, 80–86, 2015.)

a decrease in the friction coefficient during the first 10 min of the experiment. This time corresponds approximately to the time required for crystallization of the coating at that temperature. These results clearly point to a decrease in friction coefficient due to heat treating. This decrease may be linked either to the decrease of roughness generally observed after heat treatment or to the modifications of crystalline structure.

4B.2.6.3 Effect of Heat Treatment on Sliding Wear of Electroless Nickel-Boron Coatings

Sliding wear behavior of electroless nickel-boron plating has been extensively studied. However, exactly like for abrasive wear, the variety of coating composition, heat treatment, testing method and testing condition makes comparison of results obtained by different teams extremely complicated. For example, Das and Sahoo (2009) used a plate-on-roller test, Riddle and Bailer (2005) used the Falex test, while Krishnaveni et al. (2005) and Vitry et al. (2017) preferred the pin-on-disc method. It is thus impossible to provide quantitative recapitulative data regarding the wear behavior of electroless nickel-boron coatings and its variations with boron content and heat treatment.

It is possible, however, to examine trends in the effect of heat treatment on this behavior by looking at the values provided by various research groups. Krishnaveni et al. (2005) observed a decrease in wear rate after heat treatment by 25–50% (depending on testing load) after heat treatment at 350°C for 1 h and 40–60% after heat treatment at 450°C for the same time. Ahmadkhaniha and Mahboubi (2012) and Vitry and Bonin (2017) obtained a reduction of 50%–70% of wear rate after a treatment of 1 h at 400°C. Pal and Jayaram (2018) investigated several heat-treating conditions and obtained the reduction of wear rate by close to 90% for treatments of 10 h at 300°C or 4 h at 385°C. However, a longer treatment at 300°C (48 h) did not lead to further improvement in wear resistance. On the contrary, the wear rate increased compared to previous treatments, as shown in Figure 4B.16.

It is not possible to assess how the effect of heat treatment on specific wear rate evolves for coatings with different boron contents because only one group published data obtained in similar conditions for coatings with different compositions (Vitry and Bonin 2017). Their results suggest that a coating with 6 wt.% is more favorably influenced by heat treatment than coating with 7 wt.%. It is noteworthy that reverse results were obtained for abrasive wear, with the coating with the highest boron content showing better performance than the coating with 6 wt.% B. This is due to the fact that abrasive wear is mostly predicted by hardness, while the H/E ratio (the ratio of hardness and elastic modulus) is a better predictor of sliding wear.

As it is difficult to compare the results of different groups, the evolution of wear rate is often predicted from hardness: Archard's law predicts that wear rate is inversely proportional to hardness. However, this is not the always the best predictor and other parameters, such as H/E ratio, are used by some authors (Vitry 2010; Liang et al. 2015). Coatings for which this ratio is higher will show better performance in terms of abrasion and fatigue fracture.

To conclude, the effect of heat treatment on wear of electroless nickel-boron coatings is complex, and there is a lack of reliable data in terms of the effect of boron content and heat treatment on wear behavior.

However, it appears that heat treatment can improve wear resistance of electroless nickel-boron coatings like it improves hardness, but this must be mitigated because

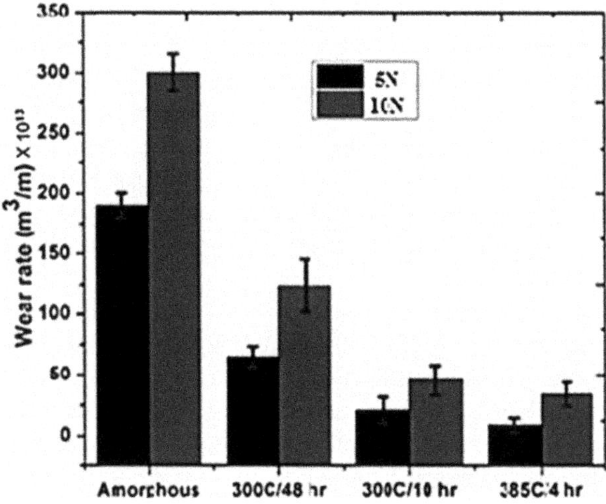

FIGURE 4B.16 Effect of heat treatment on wear rate of electroless nickel-boron coatings with 6wt.% Boron. (From Pal, S. and Jayaram, V., *Materialia*, 2018.)

wear behavior is also influenced by the value of the elastic modulus of the material, which is also modified by heat treatment, as shown by Liang for electrodeposited nickel-boron coatings (Liang et al. 2015). The H/E ratio is thus a good predictor of wear behavior of electroless nickel-boron coatings. It is however rather difficult to use because the value of the elastic modulus of electroless nickel-boron coatings is not always available.

The study of sliding wear of electroless nickel-boron coating has not been limited to quantitative evaluation: several groups examined the worn zones and the hard counterparts to assess the mechanisms that lead to wear of the coating.

For instance, Krishnaveni et al. (2005) observed that, while the main wear mechanism for as-deposited electroless nickel-boron coatings was adhesive wear, heat-treated coatings presented bright and smooth aspects in the wear track with grooves along the sliding direction, caused by the displacement of loose debris along the wear track. They attribute this behavior to increased plastic resistance of the coatings and the decrease in the true contact area (linked to the decrease in friction coefficient).

Pal and Jayaram (2018) also investigated wear mechanisms and observed a transition in wear mechanism with variations of microstructure. They observed three different mechanisms: abrasive, oxidative and adhesive, with a limited contribution from abrasive wear (which is caused by debris due to coating fracture). At low load, for short heat treatments, they observed oxidative wear, with low wear rate due to the presence of a thin oxide film preventing direct contact between the coating and counterpart. This has not been observed in the case of the coating heat-treated at 300°C for 48 h, for which oxidation is more pronounced, leading to lowered wear resistance.

For high loads, the soft oxide layer is not able to protect the coating from wear. This leads to adhesive wear of the coating.

4B.2.7 Effect of Heat Treatment on Adhesion and Scratch Test Resistance

Good adhesion of the coating is an essential need for it to provide protection against wear and corrosion (Das and Sahoo 2011c). However, it is difficult to quantitatively assess the adhesion of electroless nickel-boron coatings because most testing methods (pull-off test, ring shear test, peel test, three points bend) provide qualitative or semi-quantitative results (Sahoo and Das 2011). Moreover, they require significant preparation and are not easy to carry out, and it is not rare that the bonding strength of electroless coatings with the substrate exceeds the range of the testing methods.

Scratch testing is thus frequently used to assess adhesion of electroless nickel-boron coatings, even if it provides only indirect information because it is one of the rare methods that provide quantitative results on the behavior of the substrate/coating system. One of the preferred variations of the method involves continuous increase of the scratching load during the test. This allows to detect the critical load at which various damaging events happen.

There are not so many studies that investigate the effect of heat treatment on the scratch testing response of electroless nickel-boron coatings but Vitry and Delaunois' group used the method to test 20 μm thick coatings with various boron content

FIGURE 4B.17 Scratch test on electroless nickel-boron coatings with and without heat treatment at 400°C. (a) 4 wt.% B, as-deposited; (b) 4 wt.% B, heat-treated; (c) 6 wt.% B, as-deposited; (d) 6 wt.% B, heat-treated; (e) 7 wt.% B, as-deposited; (f) 7 wt.% B, heat-treated. (From Vitry, V. et al., *Mater. Sci. Forum*, 783–786, 1405–1413, 2014; Vitry, V. and Bonin, L., *Surf. Coat. Technol.*, 311, 164–171, 2017.)

(from 4 to 7 wt.%) with and without heat treatment (at 400°C for 1 h). The substrates used for these studies were mild steel. A summary of their results is presented in Figures 4B.17, 4B.18 and 4B.19. As shown in Figure 4B.17, only the coating with 4 wt.% boron in the as-deposited state presents extended chipping. The coatings with 7 wt. B present some matter accumulation at the edges of the scratch. The details of the end of the scratches shown in Figure 4B.18 show that the coating is usually still present and adherent to the substrate at the final load of testing (150 N in this case). Only the 7 wt.% B coating that has been heat-treated presents signs of serious damage at that load.

In terms of the damage types (see Figure 4B.19), the most common damage is transverse bottom cracks that appear in all coatings at high loads. Chevron cracks are also often observed at lower loads. Heat treatment effect seems to affect the scratch resistance of the various coatings in very different ways. The coatings with 4 and 6 wt.% B are affected positively, with an increase in the critical load at which the first type of damage is observed to values higher than 30 N (against 18 N for the coatings with 4 wt.% B and 25 N for the coatings with 6 wt.% B). The most positive effect appears for coatings with 4 wt.% B for which edge chipping is observed in the as-deposited state and disappears for heat-treated coatings. However, coatings with 7 wt.% B are severely detrimentally affected, with damage occurring even at very

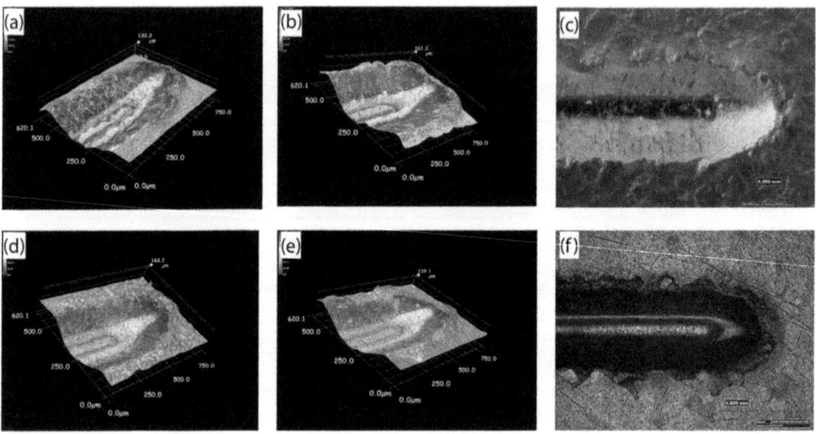

FIGURE 4B.18 Details of scratch tests on electroless nickel-boron coatings with and without heat treatment at 400°C. (a) 4 wt.% B, as-deposited; (b) 6 wt.% B, as-deposited; (c) 7 wt.% B, as-deposited; (d) 4 wt.% B, heat-treated; (e) 6 wt.% B, heat-treated; (f) 7 wt.% B, heat-treated. (From Vitry, V. et al., *Mater. Sci. Forum*, 783–786, 1405–1413, 2014; Vitry, V. and Bonin, L., *Surf. Coat. Technol.*, 311, 164–171, 2017.)

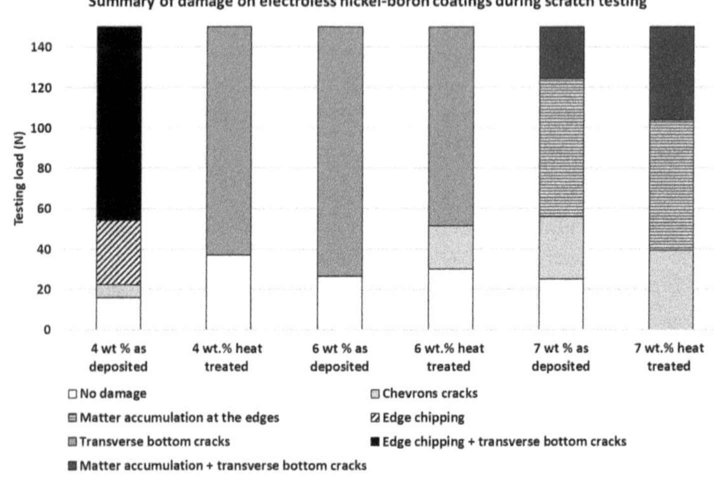

FIGURE 4B.19 Summary of damage types for electroless nickel-boron coatings with and without heat treatment at 400°C for 1 h. (From Vitry, V. et al., *Mater. Sci. Forum*, 783–786, 1405–1413, 2014; Vitry, V. and Bonin, L., *Surf. Coat. Technol.*, 311, 164–171, 2017.)

small loads for heat-treated coatings while it occurred at similar load levels than for other coatings in the as-deposited state.

The same group was also able to investigate the scratch test behavior of coatings heat-treated at 180°C. They presented similar behavior as the as-deposited coatings, with little damage observed and no adhesive failure (Vitry 2010). They managed, on such a coating with a reduced thickness (6 μm instead of 20 μm), to reach complete failure. It happened by plastic deformation followed by ductile perforation of the coatings (Vitry 2010), without any sign of adhesive failure. These findings follow Bull's model for the failure of a hard coating on a soft substrate (Bull 1997).

4B.2.8 Effect of Heat Treatment on Corrosion

The parameters that influence the corrosion resistance of electroless nickel coatings are many and is difficult to get information on for some of them. Some of the parameters are the chemical composition of the coating and its homogeneity, the presence of impurities, the surface morphology of the coating, its microstructure, the presence of pores and pits, the test medium (in terms of concentration and pH), the thickness of the coating, the nature of the substrate (Cheong et al. 2007), but also the surface state (roughness and cleanliness) of the substrate, the type of pre-treatment, the nature of the stabilizer, bath replenishment and so on.

Add to that the many available methods of investigating corrosion resistance of coatings—potentiodynamic polarization, electrochemical impedance spectroscopy (EIS), immersion tests, salt spray tests and so on—that all bring different types of information, and the comparative description of the corrosion behavior of electroless nickel-boron coatings becomes extremely difficult.

There are, however, some facts on which all authors agree: electroless coatings act as a barrier coating, unlike galvanizing that provides active protection. Electroless nickel-boron coatings are usually less resistant to corrosion than electroless nickel-phosphorus.

Several authors investigated the effect of heat treatment on various aspects of corrosion behavior of electroless nickel-boron. We'll describe those aspects separately.

First, some groups have determined the corrosion potential and the corrosion current on heat-treated coatings (Dervos et al. 2004; Srinivasan et al. 2010; Kanta et al. 2010a, 2010b; Vitry 2010; Bonin et al. 2018b; Das and Sahoo 2011a; Narayanan et al. 2004). Their findings are presented in Table 4B.3. As can be seen, the corrosion potential of electroless nickel-boron coatings appears to have been improved by heat treatment in most cases and when a decrease in corrosion potential was measured, it was very limited (Srinivasan et al. 2010). However, the same cannot be said of the corrosion current, which has been increased in most cases. The only observed decrease was obtained by Das et al. (Das and Sahoo 2011a) in a study aiming at optimizing the potentiodynamic polarization behavior of the coating and may not be representative of average behavior. Bonin et al. explain this increase in current densities during after heat treatment by the formation of grain boundaries in initially amorphous coatings. As for the improved corrosion potential, it is probably related to the crystallization phenomena occurring during heat treatment. This is confirmed by Abdel Hamid's work that did not provide values of corrosion potential but shows a shift of the potentiodynamic curves toward higher potentials after a treatment at 400°C and toward more negative ones for a treatment at 250°C (Hamid et al. 2010), as well as by Dadvand et al. (2004) who observed degradation of the potentiodynamic polarization behavior of electroless nickel-boron with increasing heat-treatment time at 220°C.

TABLE 4B.3
Corrosion Properties Obtained from Potentiodynamic Polarization on Various Electroless Nickel-Boron Coatings

References	wt.% B	Reference Electrode	Medium		E_{corr} (mV)	E_{corr} (mV against SHE)	I_{corr} (μA/cm²)	Heat Treatment
Dervos et al. (2004)	5	SCE	3.5% NaCl	AP	−536	−312	4.858	Vacuum, 850°C, 5 min
				HT	−483	−259	11.27	
Srinivasan (2010)	5	SCE	3.5% NaCl	AP	−568	−344	13.2	400°C, 1 h
				HT	−576,1	−352,1	14.7	
Vitry (2010)	6	Ag/AgCl (KCl Sat'd)	3.5% NaCl	AP	−335	−138	0.09	400°C, 1 h
				HT	−240	−43	0.178	
Narayanan (2004)	6.5	SCE	3.5% NaCl	AP	−516	−292	9.15	400°C, 1 h
				HT	−520	−296	14.60	
Bonin (2018a, 2018b)	7	Ag/AgCl (KCl Sat'd)	3.5% NaCl	AP	−400	−176	NA	400°C, 1 h
				HT	−290	−66	NA	
Das and Sahoo (2011a)	6	SCE	3.5% NaCl	AP	−381	−157	5.04	350°C, 1 h
				HT	−275	−51	0.11	

Source: Dervos, C.T. et al., *Mater. Lett.*, 58, 619–623, 2004; Srinivasan, K.N. et al., *Surf. Eng.*, 26, 153–158, 2010; Kanta, A.-F. et al., *J. Alloys Compd.*, 505, 151–156, 2010a; Kanta, A.-F. et al., *Mater. Sci. Forum*, 638–642, 846–851, 2010b; Bonin, L. et al., *Techniques de l'ingénieur*, 2018a; Vitry, V., Electroless nickel-boron deposits: Synthesis, formation and characterization; effect of heat treatments; Analytical modeling of the structural state, University of Mons, Belgium, 2010; Das, S.K., and Sahoo, P., *J. Miner. Mater. Char. Eng.*, 10, 1307–1327, 2011a; Narayanan, T.S.N.S. and Seshadri, S.K., *J. Alloys Compd.*, 365, 197–205, 2004.

Abbreviations: SCE—standard calomel electrode; SHE—standard hydrogen electrode; AP—as plated; HT—heat treated; NA—not available.

The influence of heat treatment on the corrosion behavior is thus influenced by two adverse effects: improvement of corrosion potential on one side and increase of corrosion current (and thus of corrosion rate) on the other.

Some researchers also investigated the effect of heat treatment on corrosion of electroless nickel-boron by EIS. The most extensive work in terms of influence of treatment temperature and time was carried out by Contreras et al. (2006) and Anik et al. (2008). As shown in Figure 4B.20, they observed a decrease in the size of the loop in the Nyquist diagram for all the heat-treated coatings, which means a decrease in the corrosion resistance. Anik et al. also determined the charge transfer resistance of their coatings, and it decreased from 1680 Ohm.cm² in the as-plated state to 1212 Ohm.cm² after heat treatment at 350°C for 1 h and to only 561 Ohm.cm² after heat treatment at 450°C for 1 h (Anik et al. 2008). Narayanan et al. (2004) observed

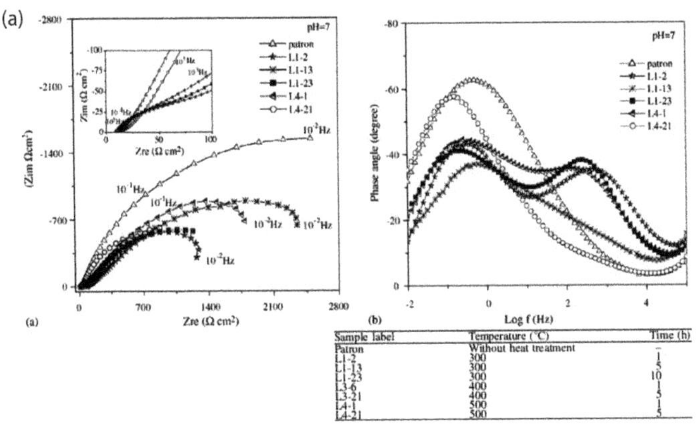

Sample label	Temperature (°C)	Time (h)
Patron	Without heat treatment	–
L1-2	300	1
L1-13	300	5
L1-23	300	10
L3-6	400	1
L3-21	400	5
L4-1	500	1
L4-21	500	5

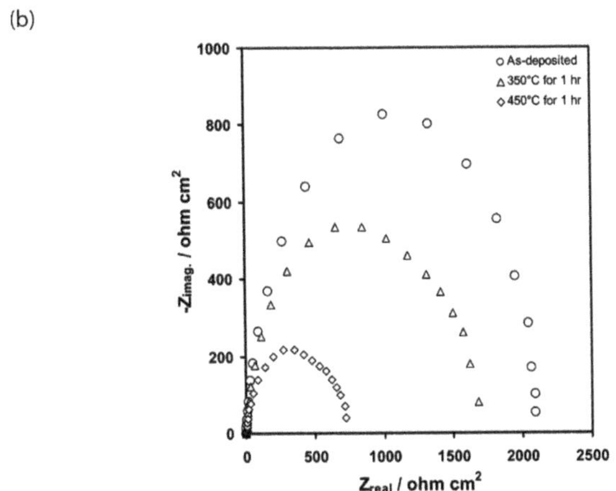

FIGURE 4B.20 Effect of heat treatment on electrochemical impedance spectroscopy behavior of electroless nickel-boron coatings: a) effect of treatment between 300 and 500°C for 1 to 10 hours (Contreras, A., Appl. Surf. Sci., 253, 592–599, 2006); b) Effect of treatment temperature for 1 hour of heat treatment (Anik, M. et al., Surf. Coat. Technol., 202, 1718–1727, 2008.)

a similar behavior. Only Das and Sahoo (2011c) obtained improved EIS behavior after heat treatment but it was for a coating specifically optimized for this parameter.

As can be seen from the results presented hereover, it's not easy to assess the corrosion behavior of electroless nickel-boron coatings from electrochemical testing methods. For this reason, some researchers have used other types of test.

Wang et al. (2012) have used immersion tests to assess the effect of heat treatment on electroless nickel-boron coatings with 6–7 wt.% B. Their results are shown in Figure 4B.21. They observed a limited effect of heat treatment up to 300°C, with slightly beneficial effects between 200°C and 300°C, which they attribute to stress release in the coating. However, for higher treatment temperatures, they observed a strong increase in the corrosion rate. Once again, these results confirm that heat treatment at temperatures higher than crystallization onset is detrimental to corrosion resistance of electroless nickel-boron coatings.

Finally, Bonin et al. (2018b) and Francq (2016) carried out salt spray tests on coatings with 7 wt.% B (see Figure 4B.22). After 7 days (268 h) and 14 days (336 h), the heat-treated samples always present a more important corroded surface than the as-plated samples. Those samples are identical to those on which potentiodynamic polarization was carried out, which confirms that while the corrosion potential was increased, the actual corrosion resistance of the samples was not improved by heat treatment.

The corrosion resistance of heat-treated electroless nickel-boron samples is not an easy topic to summarize due to the great variety of influencing parameters and testing methods.

However, all the published results tend to point that heat treating under the crystallization temperature may be beneficial to corrosion resistance due to stress relieve in the coating and that heat treatment above the crystallization onset point will lead to a decrease in the corrosion resistance. This decrease is observed even in cases where the corrosion potential of the sample has been increased after heat treatment.

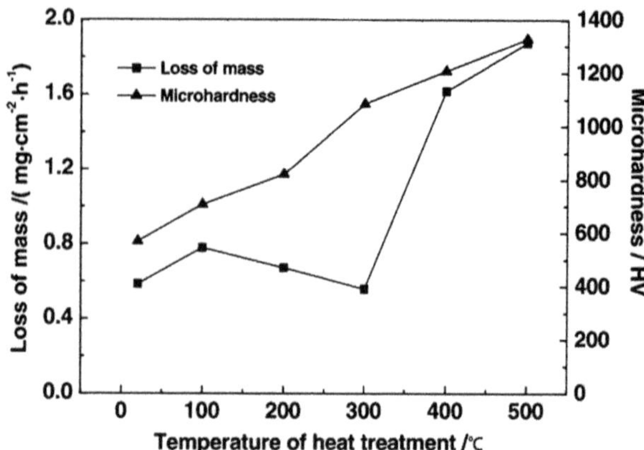

FIGURE 4B.21 Effect of heat treatment on mass loss of 6–7 wt.% boron electroless nickel-boron coating during immersion test in 3.5% NaCl solution (1 h). (From Wang, Z.-C. et al., *J. Electrochem. Soc.*, 159, D406–D412, 2012.)

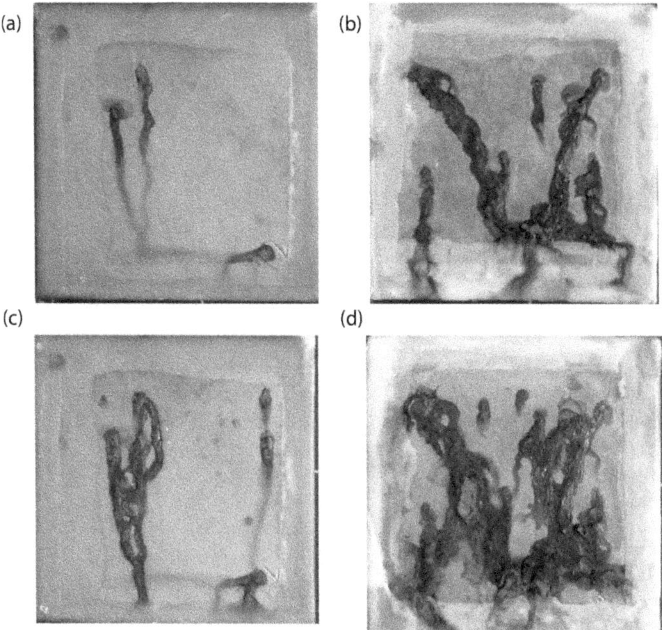

FIGURE 4B.22 Effect of heat treatment at 400°C for 1 h on the salt spray resistance of electroless nickel-boron coatings (20 μm) with 7 wt.% boron. (a) as-deposited coating, 168 h of exposure; (b) heat-treated coating, 168 h of exposure; (c) as-deposited coating, 336 h of exposure; (d) heat-treated coating, 336 h of exposure. (From Francq, E., Dépôts Chimiques Multicouches de NiP/NiB Synthèse et Caractérisation Après Traitement Thermique, Université de Mons, 2016.)

This behavior is usually attributed to structural modification in the coating (i.e., crystallization, which may be accompanied by the formation of defects and grain boundaries in otherwise amorphous or quasi-amorphous materials) during the heat treatment.

4B.2.9 Influence of Heat Treatment on Other Properties of Electroless Nickel-Boron

Hardness, wear and corrosion resistance are the most widely studied functional properties of electroless nickel-boron coatings. However, some authors have taken interest in other properties, such as electrical and magnetic properties, linked to potential applications of electroless nickel-boron in electrical or electronics applications or stress, elastic modulus and fatigue properties. In many cases, these were investigated only in the as-plated state but there are reports about the evolution of some of them with heat treatment.

Dervos et al. (2005) investigated the potential of electroless nickel-boron plating for electrical applications. They report that high-temperature vacuum heat treatment of electroless nickel-boron coatings, which leads to the crystallization of

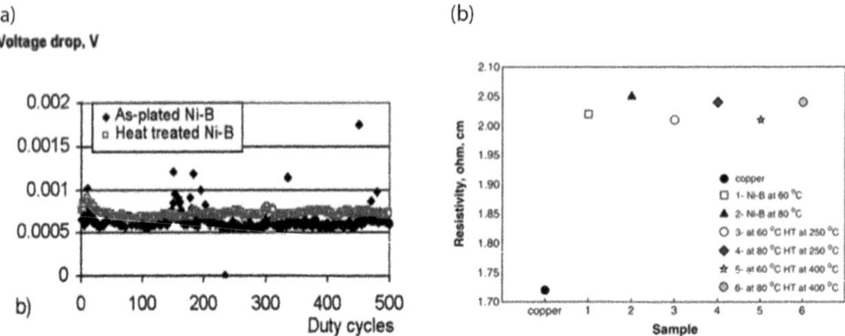

FIGURE 4B.23 (a) Effect of heat treatment (850°C, 5 min, under vacuum) on contact resistance of electroless nickel-boron. (From Dervos, C.T. et al., *Rev. Metal.*, 41, 232–238, 2005.) (b) Effect of heat treatment on resistivity of electroless nickel-boron with 5 wt.% B (60°C) and 6 wt.% B (80°C). (From Hamid, Z.A. et al., *Surf. Coat. Technol.*, 205, 2348–2354, 2010.)

nickel borides, regularized the behavior of electroless nickel-boron-plated contacts by inhibiting melting phenomena due to arcing, as can be seen in Figure 4B.23a. However, the contact resistance of as-plated electroless nickel-boron is slightly lower than that of heat-treated coatings, probably due to those melting effects.

Hamid et al. (2010) investigated the electrical resistivity of two different electroless nickel-boron coatings (one with 5 and one with 6 wt.% B). They observed a similar behavior with heat treatment: a slight decrease in resistivity that seemed to be amplified when heat-treatment temperature increased (see Figure 4B.23b).

The magnetic properties of electroless nickel-boron coatings were investigated by Baskaran et al. (2006). They observed that heat treatment increased the saturation magnetization, due to crystallization during the heat treatment process. Their results are presented in Table 4B.4.

The electrical properties of electroless nickel-boron appear thus to be differently influenced by heat treatment: there's little influence on resistivity but magnetic properties are strongly influenced by heat treatment.

TABLE 4B.4
Magnetic Properties of Electroless Nickel-Boron with Variable Boron Content with and without Heat Treatment at 400°C for 1 h

Boron Content	Thallium Content	Ms (emu/g) (As-Plated)	Ms (emu/g) (Heat-Treated)	Hc (Oe) (As-Plated)	Hc (Oe) (Heat-Treated)
0.6	0.1	16.8	96.12	26.15	258.4
1.2	0.1	15.01	98.16	31.98	257.8
2.0	0.1	12.56	105.2	35.93	257.75
2.6	0.2	11.81	108.0	37.13	254.48
3.2	0.2	7.53	109.1	41.49	253.33

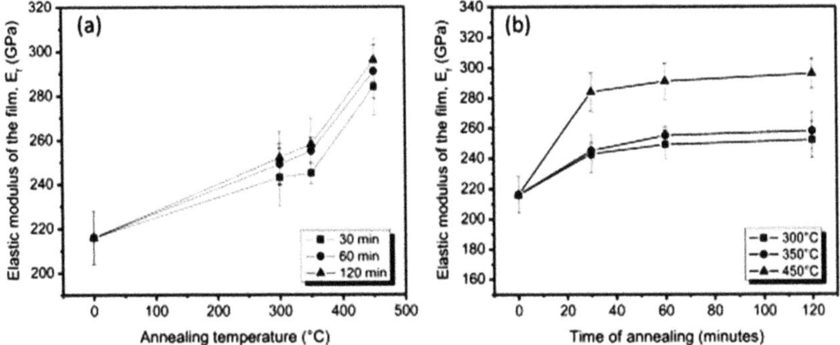

FIGURE 4B.24 Effect of (a) treatment temperature and (b) treatment time on the elastic modulus of electroless nickel-boron coatings. (From Domínguez-Ríos, C. et al., *Ind. Eng. Chem. Res.*, 51, 7762–7768, 2012.)

Finally, some authors have taken interest in the evolution of elastic modulus and fatigue resistance with heat treatment of electroless nickel-boron. Figure 4B.24 shows the evolution of elastic modulus of a coating with unmentioned boron content. The elastic modulus increases with heat-treatment temperature, and there seems to be a transition temperature in the 300°C–350°C range where the extent of the increase changes quite significantly. This temperature range matches the one for crystallization of the coatings; thus, the important variations observed for higher temperatures are most probably linked with the formation of hard boride phases. This is further evidenced by the evolution of modulus with treatment time, where most of the increase is observed in the first 30 min of treatment, whatever the temperature. The maximal value of elastic modulus was close to 300 GPa, against 220 GPa without heat treatment, which corresponds to an increase of close to 40%. The significant modification of elastic modulus probably plays a role in the sliding wear behavior of some of the coatings. However, as the boron content of the coatings described by Rios is not precisely given, it's difficult to assess how this could apply to an average coating.

There are no other systematic studies of the evolution of elastic modulus with heat treatment but some data are available: Arias et al. (2018) reported values of 277 GPa before treatment and 441 GPa after heat treatment at 450°C, Vitry et al. (2008) reported values of 170 GPa in the as-plated state and after treatment at 180°C for 4 h and 250 GPa after heat treatment at 400°C for 1 h for a coating with 6 wt.% B (which is a 50% increase) and Montagne reported similar values for coatings with 6 wt.% B with (260 GPa) and without (190 GPa) heat treatment at 400°C (Montagne et al. 2018).

Internal stress has also been investigated with somewhat contradictory results. Baudrand (2002) reports that heat treatment can increase internal stress on nickel-boron coatings, leading to a decrease in fatigue resistance of the substrate material. However, Montagne et al. (2018) reported only very low stress in electroless nickel-boron coatings with 6 wt.% B, which changed from slightly tensile (130 MPa) to slightly compressive (–120 MPa) after a heat treatment at 400°C for 1 h.

4B.3 THERMOCHEMICAL TREATMENTS

Thermochemical treatments that include carburizing, nitriding and other diffusion treatments are well-known methods for improving the surface properties of metals (Mittemeijer and Somers 1999; Steiner and Mittemeijer 2016). There are several methods for nitriding metals and alloys: gas nitriding (at atmospheric pressure), liquid phase nitriding (usually in solutions containing cyanide), vacuum nitriding, plasma nitriding (that can be direct or use an active screen), ion beam nitriding and so on (Steiner and Mittemeijer 2016; Mittemeijer and Somers 1999; Kogan 1974; Bell and Li 2007).

Nitriding has been used for nickel and nickel alloys in various applications (Makishi and Nakata 2004; He et al. 2003). It is thus possible to use it on the coatings. Moreover, there's a trend toward combined surface treatment that includes a coating and a heat treatment or thermochemical treatment (Kessler et al. 1998; Soares et al. 2017). It has been done for electroless nickel-phosphorus, but with a nitriding step before the plating (Soares et al. 2017).

The use of nitriding to improve the properties of electroless nickel-boron coatings derives from all this and from the fact that, as the nickel coating is oversaturated in interstitial boron atoms, it could be possible to form boron nitride upon nitriding. Some researchers even took things a step further by considering the catalytic properties of electroless nickel-boron and suggesting that they would allow to nitriding under a pure nitrogen atmosphere at low temperature (Vitry et al. 2012c).

In the next section, the effect of nitriding on electroless nickel-boron coatings, carried out by diverse methods, will be discussed.

4B.3.1 Types of Nitriding Treatments Applied to Electroless Nickel-Boron Coatings and Nitriding Conditions

Several nitriding treatments and conditions have been used by authors on electroless nickel-boron coatings.

- Classic ammonia-based nitriding has been used by Vitry et al. (2011) and Kanta et al. (2009). The treatment conditions were as follows: 540°C for 10 h at ambient pressure, with an atmosphere containing 60% of ammonia.
- Vacuum pure nitrogen nitriding has been used by the same group (Vitry et al. 2011, 2012a; Kanta et al. 2009). In this case, the coating was treated for 6 h at reduced pressure (100 mbar) in the 300°C–600°C range, in an atmosphere containing 95% N_2 and 5% H_2.
- Ahmadkhaniha and Mahboubi (2012) used a plasma nitriding treatment at a temperature of 450°C, for 5 h in an atmosphere containing 75% N_2 and 25% H_2 (Ahmadkhaniha and Mahboubi 2012). Yazdani et al. (2018) also used plasma nitriding, with the following treatment conditions: same gas composition as Ahmadkhaniha and Mahboubi (2012); 5 kW conventional direct current plasma-enhanced chemical vapor deposition chamber, pumped down to 10^{-2} torr; treatment at 400°C for 4 h with discharge of 500–700 V and 2–3 A current.

4B.3.2 Chemical and Microstructural Characterization of Nitrided Coatings

Yazdani et al. (2018) observed a "densification" of the surface of the nitride samples. Vitry et al. (2011) report a similar behavior for the vacuum-nitrided samples but not for the ammonia nitride samples on which a porous outer layer is formed over a dense inner layer due to nitriding. The cross-sectional observation allows to clearly see the porous outer layer, as shown in Figure 4B.25. That layer can be assimilated to the "combination" layer observed on gas nitrided steels. Figure 4B.25 also shows that plasma and vacuum nitriding do not significantly modify the cross-section appearance of electroless nickel-boron coatings.

In terms of crystalline structure, as nitriding includes exposure to high temperature for a relatively long time, it is of course expected that crystallization phenomena will be observed. However, the phases that are formed during the treatment significantly differ from those observed after standard heat treatment.

After plasma nitriding, Yazdani et al. (2018) observed mostly Ni_3B, but also observed some peaks for Ni_2B and Nickel, which can be explained by chemical heterogeneities in the coating. This corresponds fully to expectations for heat-treated electroless nickel-boron. However, they observed something much more significant, which is a peak at low angles that can be attributed to hexagonal boron nitride, that they claim is formed due to the reaction of sputtered boron atoms with nitrogen in the plasma and subsequent redeposition. They also observed a shift of nickel peaks toward lower angles that might be an indication of diffusion of nitrogen atoms in the coating. They did not provide chemical analysis that could confirm this hypothesis.

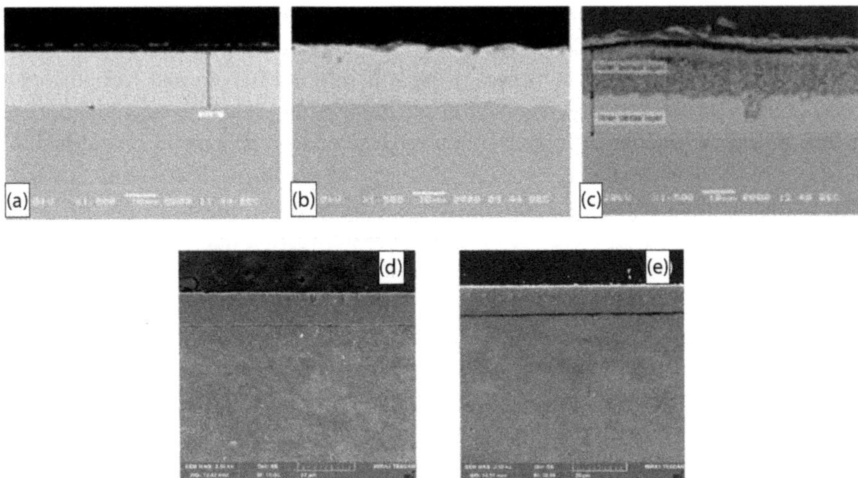

FIGURE 4B.25 Cross-section microstructure of electroless nickel-boron: (a,d) in the as-plated state; (b) after vacuum nitriding; (c) after ammonia nitriding; (e) after plasma nitriding. (From Vitry, V. et al., *Adv. Mater. Res.*, 409, 808–813, 2011; Yazdani, S. et al., *Appl. Surf. Sci.*, 45, 942–955, 2018.)

Ahmadkhaniha and Mahboubi (2012), however, did not observe boron nitride phases after plasma nitriding. They observed not only Ni_3B nickel boride phase and nickel, as expected from the heat treatment temperature, but also $(Fe, Ni)_xB$ phases and Fe_2N. They explained them by sputtering of the steel stage used to hold the sample during the nitriding process, which is confirmed by a high concentration of iron on the surface of the samples, compared to the bulk of the coating.

Vacuum nitriding leads to the formation of not only Ni_3B but also other boride phases such as Ni_2B and crystalline nickel (Vitry et al. 2011, 2012c). The nature and proportions of the formed phases vary with temperature (it has been shown earlier in this chapter that modification of the treatment temperature leads to the formation of different phases during heat treatment). No nitride phases were detected on the vacuum-nitrided coatings.

Ammonia nitriding of electroless nickel-boron led to the formation of mostly nickel and Ni_2B phases (Vitry et al. 2011, 2012c) with only a small amount of remaining Ni_3B. This can be explained by the high temperature and long treatment time used for this treatment. A dome was also observed at low angles that could be due to the presence of small amounts of amorphous or nanocrystalline boron nitride.

In terms of chemistry, only Vitry's works allowed to observe an increase in nitrogen content at the surface of coatings that were nitrided with ammonia (Vitry et al. 2012c). However, this does not mean that nitrogen was not incorporated in other nitride coatings as nitrogen concentration is notably difficult to measure.

4B.3.3 Mechanical and Tribological Properties of Nitrided Electroless Nickel-Boron Coatings

Nitriding causes an increase in the roughness of samples in all cases, contrary to heat treatment that is usually accompanied by a slight decrease of roughness. In the case of plasma nitriding, this phenomenon is explained by sputtering and redeposition of components of the coating (Yazdani et al. 2018; Ahmadkhaniha and Mahboubi 2012) and the porous coating formed on the surface of ammonia nitrided coatings is the origin of the increased roughness for this type of treatment (Vitry et al. 2012c). However, none of these explanations can be used for vacuum nitriding, and the only possible explanation in that case remains the crystallization due to heat treatment and diffusion of nitrogen in the coating.

Hardness of the nickel-boron coatings is also greatly influenced by nitriding treatments, as shown in Figure 4B.26. All nitriding treatments lead to a significant increase in hardness of nickel-boron coatings compared to not only the as-plated deposit but also heat-treated coatings. The magnitude of the hardness increase due to nitriding is in the 80%–100% range, compared to 35–50 for heat treatments. None of the groups were able to show that nitrogen was actually incorporated in the coating (except Yazdani et al); thus, it cannot be attributed to nitrogen diffusion with certainty. However, due to the temperature and duration of the treatments, the hardness increase cannot be only due to crystallization as the conditions were not the most favorable. Similar behaviors have been observed in NiP, but to a lesser extent (Zangeneh-Madar and Monir Vaghefi 2004).

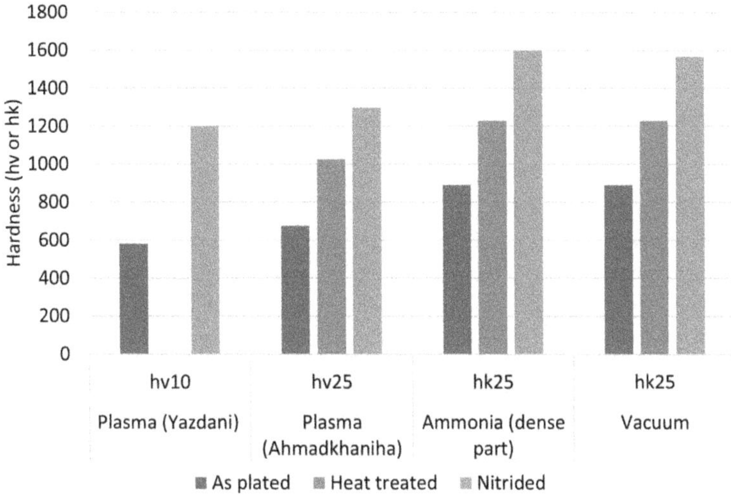

FIGURE 4B.26 Hardness of electroless nickel-boron coatings with and without nitriding treatment. (From Ahmadkhaniha, D. and Mahboubi, F., *Surf. Eng.*, 28, 195–198, 2012; Vitry, V. et al., *Surf. Coat. Technol.*, 206, 3421–3427, 2012b; Yazdani, S. et al.,. *Appl. Surf. Sci.*, 45, 942–955, 2018.)

Vitry's team was also able to study depth profile hardness by cross-section nanoindentation and observed that the hardness of nitrided coatings increased slightly from substrate to surface. The hardness of the porous part of the ammonia-nitrided coating could not be distinguished from the properties of the mounting resin (see Figure 4B.27).

Only Yazdani's group measured the friction behavior of nitrided electroless nickel-boron coatings. They observed a decrease in the friction coefficient that they

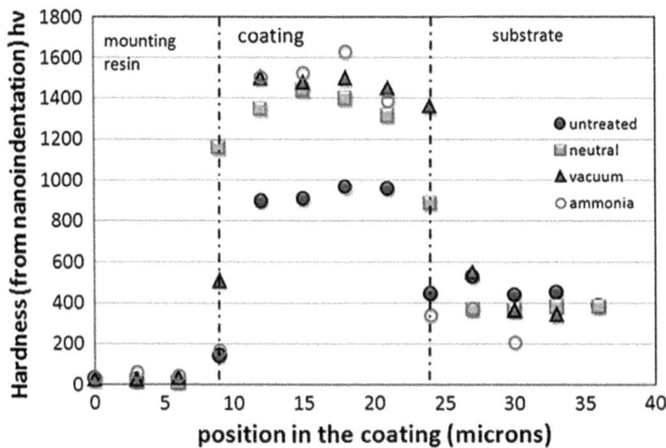

FIGURE 4B.27 Nanoindentation hardness profile of electroless nickel-boron coatings with and without nitriding treatment. (From Vitry, V. et al., *Surf. Coat. Technol.*, 206, 3421–3427, 2012b.)

attribute to an increase in hardness and a lower solubility of the crystalline phases formed during the treatment (Yazdani et al. 2018).

Yazdani and Ahmadkhaniha have measured wear behavior of plasma-nitrided electroless nickel-boron coatings. They both report improved behavior that they attribute to higher hardness and lower solubility of the crystalline phase in the wear counterpart (Yazdani et al. 2018; Ahmadkhaniha and Mahboubi 2012).

Vitry's team, on the other hand, investigated scratch test resistance of the vacuum and ammonia nitrided coatings and observed that vacuum nitrided coatings presented a fragile behavior with evidence of spalling at high loads while ammonia nitrided coatings, probably due to the presence of the porous outer layer, presented limited damage, similar to as-plated or heat-treated nickel-boron coatings (Vitry et al. 2012c).

4B.3.4 Corrosion Properties

None of the groups were able to determine the corrosion resistance of nitrided electroless nickel-boron coatings but Vitry et al. carried out potentiodynamic polarization on vacuum and ammonia nitrided samples (Vitry et al. 2011, 2012c; Kanta et al. 2009). They were able to show that the polarization curve was shifted to higher potential after nitriding, with a similar amplitude of shift for both methods. The shape of the curve remained similar to the that observed for as-plated and heat-treated samples. Due to the lack of information about nitrogen distribution in the coatings, they were not able to explain this phenomenon.

4B.4 CONCLUSION

Posttreatment of electroless nickel-boron coatings is a popular method for further improvement of their mechanical properties. The suitable heat treatment leads to a significant increase in hardness, and nitriding, using plasma, vacuum or ammonia, leads to further hardening of the coatings.

These phenomena are mostly due to the formation of hard crystalline phases in the coating that remain nanocrystalline even after several hours of heat treatment.

Wear resistance is also positively influenced by heat treatment and nitriding but roughness, while it is usually decreased by heat treatment, is increased by nitriding.

However, the effect of posttreatment on some properties, such as scratch and corrosion resistance, is not so easy to define: scratch resistance is beneficially affected by heat treatment for some coatings but not for all, and nitriding does damage scratch resistance.

Corrosion resistance on the other hand is negatively affected by heat treatment when studied by neutral salt spray test but the potentiodynamic polarization curves of heat-treated and nitrided coatings are often shifted to more positive potentials.

Nevertheless, even taking into account the detrimental effect on some parameters such as scratch resistance and corrosion, heat treatment and nitriding of electroless nickel-boron coatings remain very beneficial treatments for a number of applications.

REFERENCES

Ahmadkhaniha, D., and F Mahboubi. 2012. "Effects of Plasma Nitriding on Properties of Electroless Ni–B Coating." *Surface Engineering* 28 (3): 195–198. doi:10.1179/1743294411Y.0000000058.

Anik, M., E. Körpe, E. Şen, E. Korpe, and E. Sen. 2008. "Effect of Coating Bath Composition on the Properties of Electroless Nickel-Boron Films." *Surface and Coatings Technology* 202: 1718–1727. doi:10.1016/j.surfcoat.2007.07.031.

Arias, S., M. Gómez, E. Correa, F. Echeverría-Echeverría, J. G. Castaño. 2018. "Effect of Heat Treatment on the Tribological Properties of Nickel-Boron Electroless Coating." *Revista Facultad de Ingeniería* 27 (47): 101. doi:10.19053/01211129.v27.n47.2018.7927.

Baskaran, I., R. Sakthi Kumar, T. S. N. S. Narayanan, and A. Stephen. 2006. "Formation of Electroless Ni-B Coatings Using Low Temperature Bath and Evaluation of Their Characteristic Properties." *Surface and Coatings Technology* 200 (24): 6888–6894. doi:10.1016/j.surfcoat.2005.10.013.

Baudrand, D. W. 2002. "Electroless Nickel Plating." In *ASM Handbook*, ASM International. Vol. 5.

Bell, T., and C. X. Li. 2007. "Active Screen Plasma Nitriding of Materials." *International Heat Treatment and Surface Engineering* 1 (1): 34–38. doi:10.1179/174951407X169231.

Bonin, L., F. Delaunois, and V. Vitry. 2018a. "Revêtements Chimiques de Nickel-Bore." *Techniques de l'ingénieur.*

Bonin, L., V. Vitry, and F. Delaunois. 2018b. "Corrosion Behaviour of Electroless High Boron-Mid Phosphorus Nickel Duplex Coatings in the as-Plated and Heat-Treated States in NaCl, H_2SO_4, NaOH and Na_2SO_4 Media." *Materials Chemistry and Physics* 208 (C): 77–84. doi:10.1016/j.matchemphys.2017.12.030.

Brunelli, K., M. Dabalà, F. Dughiero, M. Magrini. 2009. "Diffusion Treatment of Ni-B Coatings by Induction Heating to Harden the Surface of Ti-6Al-4V Alloy." *Materials Chemistry and Physics* 115 (1): 467–472. doi:10.1016/j.matchemphys.2009.01.016.

Bull, S. J. 1997. "Failure Mode Maps in the Thin Film Scratch Adhesion Test." *Tribology International* 30 (7): 491–498. doi:10.1016/S0301-679X(97)00012-1.

Celik, I., M. Karakan, and F. Bulbul. 2016. "Investigation of Structural and Tribological Properties of Electroless Ni-B Coated Pure Titanium." *Proceedings of the Institution of Mechanical Engineers, Part J: Journal of Engineering Tribology* 230 (1): 57–63. doi:10.1177/1350650115588568.

Cheong, W.-J., B. L. Luan, and D. W. Shoesmith. 2007. "Protective Coating on Mg AZ91D Alloy–The Effect of Electroless Nickel (EN) Bath Stabilizers on Corrosion Behaviour of Ni-P Deposit." *Corrosion Science* 49 (4): 1777–1798. doi:10.1016/j.corsci.2006.08.025.

Contreras, A., C. León, O. Jimenez, E. Sosa, and R. Pérez. 2006. "Electrochemical Behavior and Microstructural Characterization of 1026 Ni–B Coated Steel." *Applied Surface Science* 253 (2): 592–599. doi:10.1016/j.apsusc.2005.12.161.

Dadvand, N., W. Caly, and G. Kiporous. 2004. "Investigation of the Corrosion Behavior of Electroless Nickel Coatings in Basic Solutions." 205th Meeting of the Electrochemical Society, 2004.

Das, S. K., and P. Sahoo. 2011c. "Electrochemical Impedance Spectroscopy of Ni-B Coatings and Optimization by Taguchi Method and Grey Relational Analysis." *Portugaliae Electrochimica Acta* 29 (4): 211–231. doi:10.4152/pea.201104211.

Das, S. K., and P. Sahoo. 2009. "ICME09-RT-03 Optimization of Electroless Ni-B Coatings Based on Multiple Roughness Characteristics." 2009 (December): 26–28.

Das, S. K., and P. Sahoo. 2011a. "Study of Potentiodynamic Polarization Behaviour of Electroless {Ni-B} Coatings and Optimization Using Taguchi Method and Grey Relational Analysis." *Journal of Minerals and Materials Characterization and Engineering* 10 (14): 1307–1327.

Das, S. K., and P. Sahoo. 2011b. "Tribological Characteristics of Electroless Ni-B Coating and Optimization of Coating Parameters Using Taguchi Based Grey Relational Analysis." *Materials & Design* 32 (4): 2228–2238. doi:10.1016/j.matdes.2010.11.028.

Das, S. K., and P. Sahoo. 2012. "Influence of Process Parameters on Microhardness of Electroless Ni-B Coatings." *Advances in Mechanical Engineering* 2012: 1–11. doi:10.1155/2012/703168.

Delaunois, F., and P. Lienard. 2002. "Heat Treatments for Electroless Nickel-Boron Plating on Aluminium Alloys." *Surface and Coatings Technology* 160: 139–148.

Delaunois, F. 2002. "Les Dépôts Chimiques de Nickel-Bore Sur Alliages d'aluminium." Faculté Polytechnique de Mons.

Dervos, C. T., P. Vassiliou, and J. Novakovic. 2005. "Electroless Ni-B Plating for Electrical Contact Applications." *Revista de Metalurgia* 41 (Extra): 232–38. doi:10.3989/revmetalm.2005.v41.iExtra.1031.

Dervos, C. T., J. Novakovic, and P. Vassiliou. 2004. "Vacuum Heat Treatment of Electroless Nickel-Boron Coatings." *Materials Letters* 58: 619–623.

Domínguez-Ríos, C., A. Hurtado-Macias, R. Torres-Sánchez, M. A. Ramos, and J. González-Hernández. 2012. "Measurement of Mechanical Properties of an Electroless Ni–B Coating Using Nanoindentation." *Industrial & Engineering Chemistry Research* 51 (22): 7762–7768. doi:10.1021/ie201760g.

Finch, C. B., O. B. Cavin, and P. F. Becher. 1984. "Crystal Growth and Properties of Trinickel Boride, Ni_3B." *Journal of Crystal Growth* 67: 556–558. doi:10.1016/0022-0248(84)90050-2.

Francq, E. 2016. "Dépôts Chimiques Multicouches de NiP/NiB Synthèse et Caractérisation Après Traitement Thermique." Université de Mons.

Hamid, Z. A., H. B. Hassan, and A. M. Attyia. 2010. "Influence of Deposition Temperature and Heat Treatment on the Performance of Electroless Ni–B Films." *Surface and Coatings Technology* 205 (7): 2348–2354. doi:10.1016/j.surfcoat.2010.09.025.

He, H., T. Czerwiec, C. Dong, and H. Michel. 2003. "Effect of Grain Orientation on the Nitriding Rate of a Nickel Base Alloy Studied by Electron Backscatter Diffraction." *Surface and Coatings Technology* 163–164 (January): 331–338. doi:10.1016/S0257-8972(02)00611-4.

Ivanov, M. V. 2001. "Electroless Nickel–Boron–Phosphorus Coatings: Protective and Functional Properties." *Protection of Metals* 37 (6): 592–596. doi:10.1023/A:1012827932615.

Kanta, A.-F., V. Vitry, and F. Delaunois. 2009. "Effect of Thermochemical and Heat Treatments on Electroless Nickel-Boron." *Materials Letters* 63: 2662–2665. doi:10.1016/j.matlet.2009.09.031.

Kanta, A.-F., M. Poelman, V. Vitry, and F. Delaunois. 2010a. "Nickel–Boron Electrochemical Properties Investigations." *Journal of Alloys and Compounds* 505 (1): 151–156. doi:10.1016/j.jallcom.2010.05.168.

Kanta, A.-F., V. Vitry, and F. Delaunois. 2010b. "Wear and Corrosion Resistance of Electroless Nickel-Boron Coated Mild Steel." *Materials Science Forum* 638–642 (January): 846–851. doi:10.4028/www.scientific.net/MSF.638-642.846.

Kayser, G. F., and F. X. Kayser. 1996. "Ni_3B: Powder Diffraction Pattern and Lattice Parameters." *Journal of Alloys and Compounds* 233: 74–79.

Kessler, O. H., F. T. Hoffmann, and P. Mayr. 1998. "Combinations of Coating and Heat Treating Processes: Establishing a System for Combined Processes and Examples." *Surface and Coatings Technology* 108–109 (October): 211–216. doi:10.1016/S0257-8972(98)00558-1.

Kogan, Y. D. 1974. "A Brief Historical Review." *Metal Science and Heat Treatment* 16 (3): 197–199. doi:10.1007/BF00663054.

Krishnan, K. H., S. John, K. N. Srinivasan, J. Praveen, M. Ganesan, and P. M. Kavimani. 2006. "An Overall Aspect of Electroless Ni-P Depositions–A Review Article." *Metallurgical and Materials Transactions* 37A: 1917–1926.

Krishnaveni, K., T. S. N. S. Narayanan, and S. K. Seshadri. 2009. "Corrosion Resistance of Electrodeposited Ni–B and Ni–B–Si_3N_4 Composite Coatings." *Journal of Alloys and Compounds* 480 (2): 765–770. doi:10.1016/J.JALLCOM.2009.02.053.

Krishnaveni, K., T. S. N. S. Narayanan, and S. K. Seshadri. 2005. "Electroless Nickel-Boron Coatings: Preparation and Evaluation of Hardness and Wear Resistance." *Surface and Coatings Technology* 190: 115–121.

Kumar, P. S., and P. K. Nair. 1994. "X-ray diffraction studies on the relative proportion and decomposition of amorphous phase in electroless Ni-B deposits (Accepted March 1994) Produced on Mild Steel Substrates. The Deposits Had a Structure Consisting of a Mixture of Mic." *Nanostructured Materials* 4 (2): 183–198.

Li, H., H. Li, and J.-F. Deng. 2001. "The Crystallization Process of Ultrafine Ni–B Amorphous Alloy," *Materials Letters* August: 41–46.

Liang, Y., Y.-S. Li, Q.-Y. Yu, Y.-X. Zhang, W.-J. Zhao, and Z.-X. Zeng. 2015. "Structure and Wear Resistance of High Hardness Ni-B Coatings as Alternative for Cr Coatings." *Surface and Coatings Technology* 264: 80–86. doi:10.1016/j.surfcoat.2015.01.016.

Lou, D. C., J. K. Solberg, O. M. Akselsen, and N. Dahl. 2009. "Microstructure and Property Investigation of Paste Boronized Pure Nickel and Nimonic 90 Superalloy." *Materials Chemistry and Physics* 115 (1): 239–244. doi:10.1016/j.matchemphys.2008.11.055.

Makishi, T., and K. Nakata. 2004. "Surface Hardening of Nickel Alloys by Means of Plasma Nitriding." *Metallurgical and Materials Transactions A* 35 (1): 227–238. doi:10.1007/s11661-004-0123-7.

Matik, U. "The Effect of Heat-Treatment on the Structure and Hardness of Electroless Ni-B Coated Ferrous PM Parts." , proceedings of World PM2016, 2016.

Mittemeijer, E. J., and M. A. J. Somers. 1999. "Thermodynamics, Kinetics and Process Control of Nitriding." IOM Communications Ltd.

Montagne, A., M. Z. Mughal, L. Bonin, M. Sebastiani, E. Bemporad, V. Vitry, A. Iost, and M. H. Staia. 2018. "Residual Stresses in Electroless Nickel Coatings." In *Proceedings of SMT32*.

Mukhopadhyay, A., T. K. Barman, and P. Sahoo. 2018a. "Effect of Operating Temperature on Tribological Behavior of As-Plated Ni-B Coating Deposited by Electroless Method." *Tribology Transactions* 61 (1): 41–52. doi:10.1080/10402004.2016.1271929.

Mukhopadhyay, A., T. K. Barman, and P. Sahoo. 2018b. "Effect of Heat Treatment on the Characteristics of Electroless Ni-B, Ni-B-W and Ni-B-Mo Coatings." *Materials Today: Proceedings* 5 (2): 3306–3315. doi:10.1016/j.matpr.2017.11.573.

Narayanan, T. S. N. S., K. Krishnaveni, and S. K. Seshadri. 2003. "Electroless {NiP/NiB} Duplex Coatings: Preparation and Evaluation of Microhardness, Wear and Corrosion Resistance." *Materials Chemistry and Physics* 82: 771–779.

Narayanan, T. S. N. S., S. K. Seshadri. 2004. "Formation and Characterization of Borohydride Reduced Electroless Nickel Deposits." *Journal of Alloys and Compounds* 365: 197–205. doi:10.1016/S0925-8388(03)00680-7.

Niksefat, V., and M. Ghorbani. 2015. "Mechanical and Electrochemical Properties of Ultrasonic-Assisted Electroless Deposition of Ni–B–TiO_2 Composite Coatings." *Journal of Alloys and Compounds* 633 (June): 127–136. doi:10.1016/j.jallcom.2015.01.250.

Oraon, B., G. Majumdar, and B. Ghosh. 2008a. "Improving Hardness of Electroless Ni-B Coatings Using Optimized Deposition Conditions and Annealing." *Materials & Design* 29 (7): 1412–1418. doi:10.1016/j.matdes.2007.09.005.

Pal, S., and V. Jayaram. 2018. "Effect of Microstructure on the Hardness and Dry Sliding Behavior of Electroless Ni-B Coating." *Materialia*. doi:10.1016/j.mtla.2018.09.004.

Pal, S., N. Verma, V. Jarayam, S. K. Biswas, and Y. Riddle. 2011. "Characterization of Phase Transformation Behaviour and Microstructural Development of Electroless Ni-B Coating." *Materials Science and Engineering A* 528: 8269–8276.

Riddle, Y. W., and T. O. Bailer. 2005. "Friction and Wear Reduction via an {Ni-B} Electroless Bath Coating for Metal Alloys." *JOM* 40–45. doi:10.1007/s11837-005-0080-7.

Riedel, W. 1991. *Electroless Nickel Plating Riedel*. Metals Park, OH: ASTM International.

Riedel, W. 1989. *Electroless Nickel Plating*. London, OH: Finishing Publication.

Sahoo, P. 2008. "Optimization of Electroless Ni-P Coatings Based on Surface Roughness" *Tribology Online* 1: 6–11. doi:10.2474/trol3.1.

Sahoo, P., and S. K. Das. 2009. "ICME09-RT-02 Friction Behaviour of Electroless Ni-B Coatings" Proceedings of the International Conference on Mechanical Engineering 2009 (ICME2009) 26–28 December 2009, Dhaka, Bangladesh, 2009.

Sahoo, P., and S. K. Das. 2011. "Tribology of Electroless Nickel Coatings–A Review." *Materials and Design* 32 (4): 1760–1775. doi:10.1016/j.matdes.2010.11.013.

Sheng, N., J. Liu, T. Jin, X. Sun, and Z. Hu. 2015. "Precipitation Behaviors in the Diffusion Affected Zone of TLP Bonded Single Crystal Superalloy Joint." *Journal of Materials Science & Technology* 31 (2): 129–134. doi:10.1016/J.JMST.2014.11.008.

Soares, M. E., P. Soares, P. R. Souza, R. M. Souza, and R. D. Torres. 2017. "The Effect of Nitriding on Adhesion and Mechanical Properties of Electroless Ni–P Coating on AISI 4140 Steel." *Surface Engineering* 33 (2): 116–121. doi:10.1080/02670844.2016.1148831.

Srinivasan, K. N., R. Meenakshi, A. Santhi, P. R. Thangavelu, and S. John. 2010. "Studies on Development of Electroless Ni–B Bath for Corrosion Resistance and Wear Resistance Applications." *Surface Engineering* 26 (3): 153–158. doi:10.1179/174329409X409468.

Steiner, T., and E. J. Mittemeijer. 2016. "Alloying Element Nitride Development in Ferritic Fe-Based Materials Upon Nitriding: A Review." *Journal of Materials Engineering and Performance* 25 (6): 2091–2102. doi:10.1007/s11665-016-2048-x.

Sudagar, J., J. Lian, and W. Sha. 2013. "Electroless Nickel, Alloy, Composite and Nano Coatings–A Critical Review." *Journal of Alloys and Compounds* 571 (September): 183–204. doi:10.1016/j.jallcom.2013.03.107.

Teppo, O., and P. Taskinen. 1993. "Thermodynamic Assessment of Ni-B Phase Diagram." *Materials Science and Technology* 9: 205–212.

Thomas Evans, W., M. Schlesinger. 1994. "The Effect of Solution PH and Heat-Treatment on the Properties of Electroless Nickel-boron Films." *Journal of the Electrochemical Society* 141 (1): 78–82. doi:10.1149/1.2054713.

Vijayanand, M., and R. Elansezhian. 2014. "Effect of Different Pretreatments and Heat Treatment on Wear Properties of Electroless Ni-B Coatings on 7075-T6 Aluminum Alloy." *Procedia Engineering* 97: 1707–1717. doi:10.1016/j.proeng.2014.12.322.

Vitry, V., and F. Delaunois. 2015. "Formation of Borohydride-Reduced Nickel–Boron Coatings on Various Steel Substrates." *Applied Surface Science* 359 (December): 692–703. doi:10.1016/j.apsusc.2015.10.205.

Vitry, V., F. Delaunois, and C. Dumortier. 2008. "Mechanical Properties and Scratch Test Resistance of Nickel-Boron Coated Aluminium Alloy after Heat Treatments." *Surface and Coatings Technology* 202 (14): 3316–3324. doi:10.1016/j.surfcoat.2007.12.001.

Vitry, V., E. Francq, and L. Bonin. 2018. "Mechanical Properties of Heat-Treated Duplex Electroless Nickel Coatings." *Surface Engineering*. doi:10.1080/02670844.2018.1463679.

Vitry, V., A.-F. F. Kanta, J. Dille, and F. Delaunois. 2012a. "Structural State of Electroless Nickel-Boron Deposits (5wt.% B): Characterization by XRD and TEM." *Surface and Coatings Technology* 206 (16): 3444–3449. doi:10.1016/J.SURFCOAT.2012.02.003.

Vitry, V., A. Sens, A.-F. F. Kanta, and F. Delaunois. 2012b. "Wear and Corrosion Resistance of Heat Treated and As-Plated Duplex NiP/NiB Coatings on 2024 Aluminum Alloys." *Surface and Coatings Technology* 206 (16): 3421–3427. doi:10.1016/j.surfcoat.2012.01.049.

Vitry, V., and L. Bonin. 2017. "Increase of Boron Content in Electroless Nickel-Boron Coating by ModiFiCation of Plating Conditions." *Surface & Coatings Technology* 311: 164–171. doi:10.1016/j.surfcoat.2017.01.009.

Vitry, V. 2010. "Electroless Nickel-Boron Deposits: Synthesis, Formation and Characterization; Effect of Heat Treatments; Analytical Modeling of the Structural State." University of Mons, Belgium.

Vitry, V., L. Bonin, and L. Malet. 2017. "Chemical, Morphological and Structural Characterisation of Electroless Duplex NiP/NiB Coatings on Steel." *Surface Engineering* 34 (6): 1–10. doi:10.1080/02670844.2017.1320032.

Vitry, V., A.-F. Kanta, and F. Delaunois. 2012c. "Application of Nitriding to Electroless Nickel-Boron Coatings: Chemical and Structural Effects; Mechanical Characterization; Corrosion Resistance." *Materials and Design* 39 (August): 269–278. doi:10.1016/j.matdes.2012.02.037.

Vitry, V., A. F. Kanta, A. Sens, and F. Delaunois. 2011. "Tribological Characterization of Electroless Nickel-Boron Coatings." *Advanced Materials Research* 409 (November): 808–813. doi:10.4028/www.scientific.net/AMR.409.808.

Vitry, V., A. Sens, and F. Delaunois. 2014. "Comparison of Various Electroless Nickel Coatings on Steel: Structure, Hardness and Abrasion Resistance." *Materials Science Forum* 783–786 (May): 1405–1413. doi:10.4028/www.scientific.net/MSF.783-786.1405.

Wang, Z.-C., L. Yu, F. Jia, and G.-L. Song. 2012. "Effect of Additives and Heat Treatment on the Formation and Performance of Electroless Nickel-Boron Plating on AZ91D Mg Alloy." *Journal of the Electrochemical Society* 159 (7): D406–D412. doi:10.1149/2.012207jes.

Watanabe, T., and Y. Tanabe. 1983. "The Lattice Images of Amorphous-like Ni–B Alloy Films Prepared by Electroless Plating Method." *Transactions of the Japan Institute of Metals* 24: 396–404.

Xiao, B., J. Feng, C. T. Zhou, J. D. Xing, X. J. Xie, Y. H. Cheng, and R. Zhou. 2010. "The Elasticity, Bond Hardness and Thermodynamic Properties of X_2B (X=Cr, Mn, Fe, Co, Ni, Mo, W) Investigated by DFT Theory." *Physica B: Condensed Matter* 405 (5): 1274–1278. doi:10.1016/J.PHYSB.2009.11.064.

Yazdani, S., R. Tima, and F. Mahboubi. 2018. "Investigation of Wear Behavior of As-Plated and Plasma-Nitrided Ni-B-CNT Electroless Having Different CNTs Concentration." *Applied Surface Science* 457 (May): 942–955. doi:10.1016/j.apsusc.2018.07.020.

Yildiz, R. A., K. Genel, and T. Gulmez. 2017. "Effect of Heat Treatments for Electroless Deposited Ni-B and Ni-WB Coatings on 7075 Al Alloy." *International Journal of Materials, Mechanics and Manufacturing* 5 (2): 83–86. doi:10.18178/ijmmm.2017.5.2.295.

Zangeneh-Madar, K., and S. M. M. Vaghefi. 2004. "The Effect of Thermochemical Treatment on the Structure and Hardness of Electroless Ni-P Coated Low Alloy Steel." *Surface and Coatings Technology* 182 (3): 65–71. doi:10.1016/S0257-8972(03)00810-7.

Ziyuan, S., W. Deqing, and D. Zhimin. 2004. "Surface Strengthening of Pure Copper by Ni-B Coating." *Applied Surface Science* 221: 62–68.

Ziyuan, S., W. Deqing, and D. Zhimin. 2006. "Nanocrystalline Ni–B Coating Surface Strengthening Pure Copper." *Applied Surface Science* 253 (3): 1051–1054. doi:10.1016/j.apsusc.2005.12.169.

5 The Effect of Ultrasound on Electroless Plating of Nickel

Shakiela Begum and Andrew J. Cobley

CONTENTS

5.1 INTRODUCTION TO ULTRASOUND/SONOCHEMISTRY

The potential for ultrasound to influence electrochemical processes has been realized for at least 60 years. In that time, the effects of ultrasonic irradiation on electroless systems in general and electroless nickel (Ni) in particular have been widely studied and a number of benefits have been identified, at least at the laboratory scale. Before the various studies are discussed, a short introduction is given to the effect of ultrasound on a liquid medium, i.e. sonochemistry.

When low-frequency ultrasound is applied to a liquid medium, transient or inertial gas cavities (or bubbles) are created inside the liquid that gradually grow in subsequent compression and rarefaction cycles (Wood, Lee & Bussemaker, 2017). The temperature of gases inside the cavity has been calculated to reach approximately 5000 K–7000 K (Crum & Suslick, 1995). The expansion of a gas cavity continues until it reaches a size where the vapor pressure inside the bubble becomes unstable whereupon it implodes (Figure 5.1a). When a cavity implodes, very high temperatures and pressures occur locally depending on the frequency of the ultrasound. The surrounding liquid in close proximity to the imploding cavity can reach 2343.15 K, and the pressure on cavity implosion has been estimated to be close to 500 atm (Suslick & Doktycz, 1990). After cavity collapse, there is a rapid drop in temperature at rates faster than 10^{10} degrees per second (Crum & Suslick, 1995). If a solid surface is present (for example, an electrode), then microjetting can occur due to asymmetric bubble collapse as shown in Figure 5.1b, producing microjets which, again depending on the frequency and power of the ultrasound, can hit the surface at speeds of up to >100 m/s.

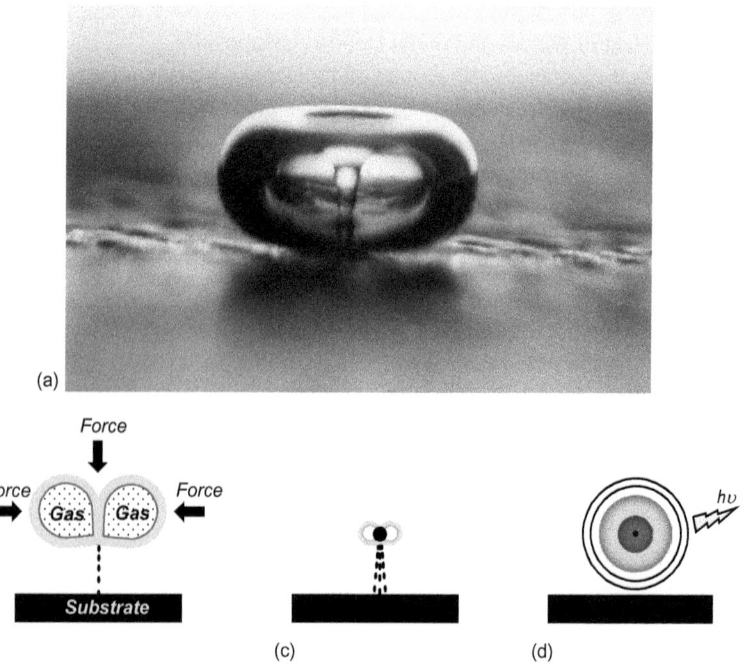

FIGURE 5.1 Cavity collapse: (a) high-speed flash photomicrograph of an unstable imploding cavity with a diameter of 150 μm in a liquid irradiated with ultrasound near a solid substrate surface, image provided by Lawrence Crum (copyright owner). Illustration of the (b) formation of a microjet from an imploding cavity, (c) a microjet striking the surface, and (d) pressure wave and light emitted after cavity collapse. (From Suslick, K.S. and Doktycz, S.J., Effects of ultrasound on surfaces and solids, In T.J. Mason (Ed.), *Advances in Sonochemistry*, vol. 1, pp. 197–230, JAI Press, Greenwich, CT, 1990.)

5.1.1 Influence of Ultrasonic Waves on Chemical Systems

The factors that influence sonoelectrochemical processes that result in a variety of reactions have been briefly summarized in Table 5.1. Primary factors that directly influence an ultrasonic system are (Wood et al., 2017):

1. *Power (pressure) amplitude*, determines the number of cavities produced, including motion, coalescence, and agglomeration of cavities by acoustic streaming.
2. *Frequency of sound field*, cavity size increases with decreasing applied frequency (<100 kHz) and collapse of a cavity accelerates. It may also affect cavity distribution and population and extent of chemical reaction.
3. *Reactor design*, encompasses transducer type (probe or plate), signal type (pulsed or continuous), liquid height, liquid temperature, reflective plate, diameter of vessel: diameter of transducer ratio, and flow of liquid.

The effect of ultrasound on a liquid can vary in different solutions because secondary factors have an influence on the effects of primary factors. Some examples of secondary factors include solution temperature and pressure, solution viscosity, concentration of gases dissolved in solution, reactor geometry, mechanical stirring speed, and so on.

The aim of this chapter is to review and present results from research studies mainly focused on the ultrasonic deposition of Ni on solid surfaces from an electroless Ni solution. By comparing different methods and results from electroless Ni deposition research studies, we intend to highlight trends and describe new recent developments in this field. Most sonochemical research for electroless Ni deposition has been conducted at a low-frequency range of 20 Hz to 100 kHz, although some examples are provided where frequency is extended up to 1 MHz.

5.2 EFFECT OF ULTRASOUND ON ELECTROLESS Ni PLATING RATE/WEIGHT GAIN

Acoustic streaming and turbulent flow of a plating solution agitated by ultrasound increase mass transport of ionic species to the electrode surface, thus increasing the rate of deposition. Many studies have looked at measuring electroless Ni plating rates; a detailed analysis of this topic led Bangwei Zhang to conclude that insufficient data are available in the literature to derive a mechanism (Bangwei, 2016). This was due to the multitude of electroless solution composition or physical parameters (e.g. plating time, temperature, and sonication power/frequency) used by various researchers and each time these are changed, they can influence the plating rate (Table 5.2). Another complication is that some studies have rotated the substrate in the electroless plating solution while applying ultrasound (to reduce turbulent flow and cavitation on the substrate), while others have kept the substrate stationary.

A good example of how the flow of electroless plating solution affects the deposit is the study (Sharifalhoseini & Entezari, 2015) in which it was found that

TABLE 5.1
Ultrasonic Factors That Affect a Liquid Solution and Solid Sample in an Electrochemical System

Factor	Description	Beneficial Effects	Adverse Effects
Acoustic streaming	Steady flow of solution. Controlled by the amplitude of power applied to solution. Asymmetry of cavities gives rise to this.	• Increase in mass transfer of active chemicals. • Reduces the diffusion boundary layer.	• Coalescence of cavitation bubbles at higher powers (W).
Turbulent flow	Disorderly movement of solution. Influenced by vessel designs.	• Similar benefits to acoustic streaming.	• Induces cavitation on solid surface. • Cavitation bubbles are moving randomly.
Microjets	Jet from collapsed cavity strikes the surface directly.	• Enhances mass transport of active species to surface. • Prevents fouling at surface by cleaning. • Remove or thin oxides/coatings on solid surfaces.	• Erosion of materials when cavity forms close to surface of solid. • Depassivation of metal surface increases corrosion risk. • Nonuniform work hardening of extended metal surface.
Emission of shock waves	Pressure wave follows violent collapse of cavity.	• Increased mass transport. • Causes collisions of particles.	• Wave can damage surface if cavity collapses very close to it. • Particles colliding at high speeds may fragment or fuse.

(Continued)

TABLE 5.1 (*Continued*)
Ultrasonic Factors That Affect a Liquid Solution and Solid Sample in an Electrochemical System

Factor	Description	Beneficial Effects	Adverse Effects
Chemical effects	Changes in solution or solute chemistry.	• Radicals are generated depending on solvent, e.g. HO•, HO_2•, O• from water, and recombined to form reactive molecules, e.g. H_2O_2, O_2, and H_2 that lead to redox reactions. • Cracking of alkanes at ambient temperature to form acetylene. • Controlled polymer degradation. • Amino acids alanine, ethylglycine, glycine form in acetic acid solutions or under CH_4 or CO_2 with N_2.	• Adverse reactions occur from radicals. • Can cause agglomeration of high melting point metal particles. • Mechanisms not studied in detail.
Heat	Cavity implosion releases heat that removes ligands on metal ion.	• Heat dissipates quickly. • Solution is warmed <373.15 K.	• Deformation/softening of metals very close to imploding cavity.
Sonoluminescence	Light is emitted when excited molecules return to ground state. This effect is more prominent in organic solvents.	• Increases by 29% when TiO_2 powder (20 mg) is added in water (100 cm^3). • Intensity increases with salt concentration in solution to a certain limit, due to gas solubility change.	• Can destabilize or reorder ligands on an organometallic molecule to form cluster compounds. • Sonoluminescence quenched by long-chain alcohols in solution at 515 kHz. • Absent in highly viscous liquids.

Source: Mason, T.J. and Bernal, V.S., An introduction to sonoelectrochemistry. In B.G. Pollet (Ed.), *Power Ultrasound in Electrochemistry: From Versatile Laboratory Tool to Engineering Solution*, pp. 21–40, John Wiley & Sons, Chichester, UK, 2012; Suslick, K.S. and Doktycz, S.J., Effects of ultrasound on surfaces and solids, In T.J. Mason (Ed.), *Advances in Sonochemistry*, vol. 1, pp. 197–230, JAI Press, Greenwich, CT, 1990; Wood, R.J. et al., *Ultrason. Sonochem.*, 38, 351–370, 2017.

TABLE 5.2
Ni Coatings Deposited by Ultrasound and Characterization Results Reported in Selected Journal Articles

Deposit Surface (Citation)	Label	Ultrasonic Parameters: kHz/Power	Results: Thickness/ mass/rate	Results: Δ (%)	Results: P % (B %)	Results: P (B) Δ (%)	Results: Hardness (HVN)	Results: E_{corr} (V)	Results: I_{corr} (μA/ cm²)	Conditions: Temp., K (Time, s)	Conditions: pH	Solution Concentration (mol/dm³): Ni²⁺/Reducing Agent	Solution Concentration (mol/dm³): Chelating Agent	Solution Concentration (mol/dm³): Other
NiB-Pb	1	0/0	16.22 μm		(5.92)		843	Not stated	Not stated	368.15 (3600)	12.0	0.1/0.016	Ethylenediamine (0.9)	Lead tungstate (4.61E-5)
Mild steel	2	20/0.058 Wcm⁻³ (p)	23.03 μm	42	(6.11)	(3.2)	751							
(Bonin et al., 2017)	3	35/0.065 Wcm⁻³	25.13 μm	55	(5.47)	(−7.6)	844							
NiB	4	0/0	7 μm		(6.8)		372	−0.859	31.16	358.15 (3600)	14.0	0.13/0.026	Ethylenediamine (1.5)	Lead acetate trihydrate (0.036)
Mild steel	5	60/150 W	8 μm	14	(6.9)	(1.5)	~380	−0.785	16.53					
NiB-Ce	6	0/0	10 μm		(5.9)		~730	−0.720	11.11					
Mild steel (Qian et al., 2017)	7	60/150 W	11 μm	10	(6.2)	(5.1)	956	−0.710	3.65					
NiB	8	0/0	~25.0 μm		(2.2)		Not stated	−0.502	0.0052	341.15 (1800)	4.0	0.02/0.034	Sodium acetate (0.054)	
Copper (Chiba et al., 2003)	9	45/100 W	~65.0 μm	160	(2.2)	(0)		−0.647	0.0086					
NiP	10	0/0	0.000 μm		0		Not stated	Not stated	Not stated	323.15 (Not stated)	—	Not stated	Not stated	Not stated
Copper	11	40/40 W	0.482 μm	48	14.51									
(Hu et al., 2011)	12	40/70 W	1.090 μm	109	11.80									
	13	40/100 W	1.430 μm	143	10.46									
NiP	14	0/0	0.1264 g		Not stated		395 Hk	Not stated	Not stated	363.15 (1800)	5.5	0.11/0.24	Citric acid (0.078)	Lead nitrate (1.21E-6)
Steel	15	28/200 W	0.0910 g	−28			262 Hk							
(Park et al., 2002)	16	40/200 W	0.1455 g	15			442 Hk							
	17	68/200 W	0.1239 g	−2			371 Hk							
NiP	18	0/0	0.8 μm/h		28		60	Not stated	Not stated	323.15 (3600)	5.2	0.13/0.45	Sodium citrate (0.12)	Ammonium chloride (0.75)
Cu clad epoxy	19	20/42 Wcm⁻² (p)	4.8 μm/h	*500*	21	−25	80							
(Cobley & Saez, 2012)	20	0/0	1.0 μm/h		27		80			343.15 (3600)				
	21	20/42 Wcm⁻² (p)	14.5 μm/h	1350	18.5	−31.5	102							
	22	0/0	9.2 μm/h		19.5		103			363.15 (3600)				
	23	20/42 Wcm⁻² (p)	15.8 μm/h	72	18	−7.7	139							

(Continued)

TABLE 5.2 (*Continued*)
Ni Coatings Deposited by Ultrasound and Characterization Results Reported in Selected Journal Articles

Deposit Surface (Citation)	Label	Ultrasonic Parameters kHz/Power	Results Thickness/ mass/rate	Δ (%)	P % (B %)	P (B) Δ (%)	Hardness (HVN)	E_{corr} (V)	I_{corr} (μA/cm^2)	Conditions Temp., K (Time, s)	pH	Solution Concentration (mol/dm^3) Ni^{2+}/Reducing Agent	Chelating Agent	Other
NiP	24	0/0	5 μm/h		14.2		Not stated	Not stated	Not stated	360.15 (Not stated)	4.8	0.10/0.28	Citric acid (0.2)	Thiourea (7.88E-6)
Not stated	25	49/30 W	5 μm/h	0	13.1	−7.8								
(Mallory, 1978)	26	0/0	26 μm/h		8.0								Lactic acid (0.3)	
	27	49/30 W	42 μm/h	61	6.8	−15.0								
	28	0/0	20 μm/h		9.2								Glycolic acid (0.3)	
	29	49/30 W	34 μm/h	70	7.8	−15.2								
	30	0/0	14 μm/h		4.9								Glycine (0.3)	
	31	49/30 W	20 μm/h	43	5.6	14.3	485							
	32	0/0	11 μm/h		10.0		622						Citric (0.07) + lactic acid (0.3)	
	33	49/30 W	23 μm/h	109	8.6	−14.0								
NiP	34	0/0	20 μm		9.6		474	−0.72	18	353.15 (7200)	6.4–6.5	0.06/0.14	Sodium acetate (0.16)	NH_4HF_2 (0.14)
Mg-Li alloy (Zou et al., 2014)	35	40/100 W	20 μm	0	8.9	−7.3	556	−0.34	8	341.15 (7200)				HF 40% v/v (12)
NiP	36	0/0	7.5 mg/cm^2·h		10.49		550	Not stated	Not stated	351.15–355.15 (Not stated)	4.8	0.08/0.19	Sodium acetate (0.24)	Thiourea (1.31E-5)
Mild steel (Yang, Hou & Wu, 1997)	37	0.033/1 Wcm^{-2}	10.3 mg/cm^2·h	37	8.39	−20.0	603							
NiP	38	0/0	1.6 μm/h				Not stated	Not stated	Not stated	323.15 (3600)	7.0	0.15/0.33	Sodium citrate dihydrate (0.17)	Not stated
Copper	39	28/3.15 Wcm^{-2}	3.8 μm/h	138										
(Kobayashi,	40	45/3.15 Wcm^{-2}	4.7 μm/h	194										
Chiba & Minami, 2000)	41	100/3.15 Wcm^{-2}	2.4 μm/h	50										

(*Continued*)

TABLE 5.2 (*Continued*)
Ni Coatings Deposited by Ultrasound and Characterization Results Reported in Selected Journal Articles

Deposit Surface (Citation)	Label	Ultrasonic Parameters	Results							Conditions		Solution Concentration (mol/dm³)		
		kHz/Power	Thickness/ mass/rate	Δ (%)	P % (B %)	P (B) Δ (%)	Hardness (HVN)	E_{corr} (V)	I_{corr} (μA/ cm²)	Temp., K (Time, s)	pH	Ni^{2+}/Reducing Agent	Chelating Agent	Other
Ni-Co-P	42	0/0	Not stated		6		Not stated	Not stated	Not stated	363.15 (3600)	6.4	Not stated	Not stated	Not stated
Mild steel	43	0/0	5.38 μm/h		4.89		459							
(Sun et al., 2011a)	44	28/not stated	14.87 μm/h	176	5.44	11.3	650			343.15 (3600)				
NiP	45	0/0			8.92		Not stated	−0.408	1.66	353.15 (7200)	4.8	9.5/29	Sodium citrate (5.1)	Sodium dodecyl sulfate (Not stated)
Mild steel	46	Not stated (p:v)	7.8 μm/h	24	8.50	−4.7		−0.391	1.60					
(Sharifalhoseini & Entezari, 2015)	47	Not stated (p:h)	9.4 μm/h	49	8.66	−2.9		−0.326	1.02					
NiP	48	0/0	17 μm		Not stated		Not stated	−1.2	13.1	353.15 (4500)	6.5	0.04/0.28	Citric acid monohydrate (0.04)	Thiourea (1.97E-5)
Mg AZ91D (Ban, Shao & Wang, 2014)	49	530/5 W	15 μm*	−12				−0.3	6.45					
Ni-Cu-P	50	0/0	Not stated		11.61		655	Not stated	Not stated	343.15 (3600)	6.4	Not stated	Not stated	Not stated
Mild steel (Sun et al., 2011b)	51	28/not stated			9.13	−21.4	765							
NiP	52	0/0	15 μ/h		Not stated		Not stated	Not stated	Not stated	366.15 (3000)	8.0–9.0	0.35/0.13	tri-Sodium citrate dihydrate (0.34)	Not stated
Mild steel (Vasudevan, Narayanan & Karthik, 1998)	53	25/400 W	45–50 μ/h	200–233										

Note: p = transducer is a probe/horn; v or h = sample was held vertical or horizontal with respect to probe (p), respectively.
E_{corr} and I_{corr} derived from electrochemical polarization curves where NaCl (aq) is an electrolyte.

*Thickness not including transition layer (17 μm).

the NiP deposition rate onto mild steel was higher when the sample was positioned horizontally with respect to the ultrasonic probe as opposed to vertical position (see Table 5.2). In addition to this, from the present literature survey, no two studies using similar power (W) and frequency (kHz), and different plating solution conditions (i.e. pH, composition, and concentration) have quoted a similar deposition rate/weight gain. The rate of deposit growth is thought to be influenced by the following factors when applying ultrasound:

1. Reduction of the diffusion layer, and the increase in mass transport to/from electrode surface (Abyaneh, Sterritt & Mason, 2007).
2. Increase in crystal nucleation rate by increasing active sites (Cobley, Mason & Saez, 2011).
3. Sonochemical activation of inactive hypophosphite (Mallory, 1978).

There is no existing evidence on which route dominates and influences the process the most. The representative research studies from the past 50 years that describe NiP or NiB deposition (rate/thickness/weight) with and without ultrasound are listed in Table 5.2. The results show the change Δ (%) in the amount of Ni deposited is evident when using ultrasound.

$$\text{Change by ultrasound, } \Delta(\%) = \frac{\text{Thickness} - \text{Thickness at 0 kHz}}{\text{Thickness at 0 kHz}} \times 100$$

Equation 5.1 Calculation for change in thickness when applying ultrasound. The calculation is also used for change in mass, rate or composition.

The results clearly indicate that in the majority of cases, sonication causes the amount of Ni deposited to increase. As has been noted elsewhere in this book, a standard still or stirred electroless Ni electrolyte typically operates at 353.15 K–368.15 K and reducing this temperature (approximately 323.15 K–343.15 K) would be beneficial from the perspective of overall process costs, reducing water consumption (due to evaporation) as well as bringing about environmental benefits. Sonication has been used as a means to enable the reduction in electroless Ni operating temperatures. For example, a copper substrate was electroless NiP plated using a constant solution temperature of 323.15 K with/without ultrasonication at 40 kHz frequency while the power was varied from 40 to 100 W (Hu et al., 2011). A still or stirred solution failed to plate a copper surface at 323.15 K, but when sonication was applied, NiP was deposited and the thickness increased with power. In a similar study (Cobley & Saez, 2012), NiP was deposited onto copper comparing different solution temperatures (323.15 K–363.15 K) at a constant ultrasonic frequency (20 kHz) and power (42 Wcm^{-2}). This work showed that the plating rate increased with temperature and that the application of ultrasound always caused a higher deposition rate. In addition, it was found that, at an electroless Ni solution temperature of 343.15 K, the plating rate observed when ultrasound was applied was at least 1.5× higher than the rate obtained from a "silent" electroless Ni solution at 363.15 K (see Figure 5.2a). Although these results suggested that sonication may enable low-temperature electroless Ni

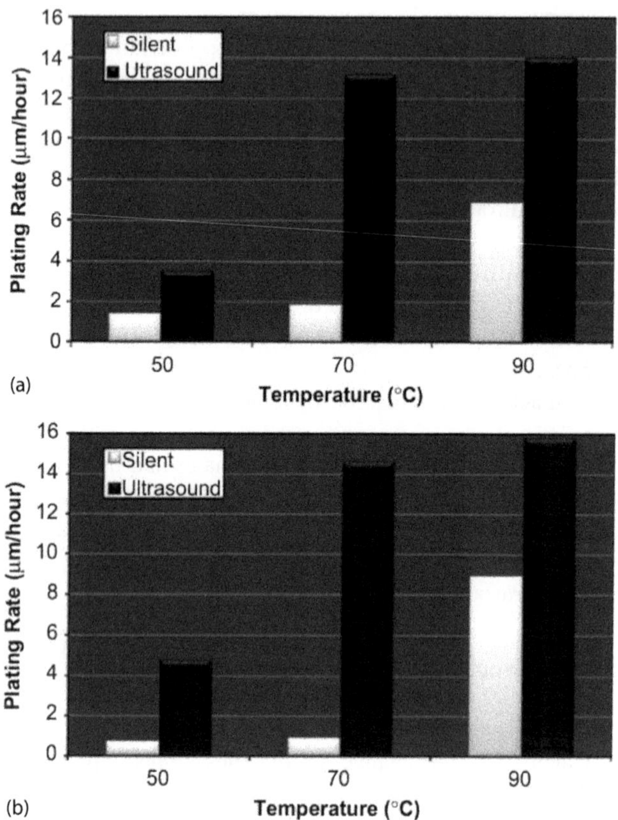

FIGURE 5.2 Effect of ultrasound (20 kHz) on the plating rate for an electroless Ni solution at different operating temperatures determined using (a) mixed potential theory and (b) "weight gain," on plated test coupons. (From Cobley, A.J. and Saez, V., *Circuit World*, 38, 12–15, 2012.)

deposition and therefore reduce energy consumption, it must be remembered that the application of ultrasound also requires energy. To our knowledge, no research has been conducted to calculate whether there is an overall energy benefit from the application of ultrasound in this situation.

As indicated in Table 5.2, there is no accepted value as the optimum ultrasonic frequency to get the maximum benefits from sonication. However, some researchers have suggested that the optimum electroless NiP plating frequency is 40 kHz using the weight gain method (Park et al., 2002) because they found that, at 28 or 68 kHz, the deposition rate becomes slower than that using no ultrasound (Table 5.2). An earlier study (Kobayashi et al., 2000) concluded that optimum electroless Ni plating frequency is 45 kHz in a solution containing a citrate complexing agent, but they found that the plating rate at 28 and 100 kHz is still higher than with no ultrasound. They also reported that an ultrasound-agitated electroless Ni is less stable at pH > 8 or temperature >333.15 K than under silent conditions. There is no accepted value for the optimum plating frequency; even after the studies by Park et al. and Kobayashi et al., researchers have continued to use lower/higher frequencies (40–45 kHz, Table 5.2). Inconsistencies are noticed when comparing studies that indicate electroless Ni plating is not primarily influenced by ultrasonic parameters and other factors (i.e. pH, solution temperature and composition) influence deposit thickness, compactness, and composition.

Experiments have been conducted (Mallory, 1978) where only the complexing agent for Ni^{2+} was varied; sonication power/frequency, time of exposure and solution composition were not changed. Results of this study demonstrated that plating rate can increase or decrease when the type of complexing agent is changed (Table 5.2, labels 24–33) but in all cases (except when citric acid is used as a complexing agent), the application of ultrasound increased the amount of Ni deposited. However, this and other studies also showed that increasing the amount of chelating agent in the Ni^{2+} electrolyte stabilizes the plating solution, and the deposition rate drops when applying ultrasonication (Abyaneh et al., 2007; Mallory, 1978). Figure 5.3 is an example of this effect, where the deposition rate dropped in the presence of ultrasound in a Ni^{2+} electroless electrolyte containing CH_3COONa (0.24 M) chelating agent, whereas at low chelate concentrations (0.04 M), sonication had a positive effect. This effect was explained by an increase in the solution stability when the concentration of chelating agent was high.

The molar ratio of the reducing agent to Ni^{2+} is also an important factor; Abyaneh and coworkers (Abyaneh et al., 2007) demonstrated that an increase of 89% in NiP deposition rate can be achieved (Figure 5.3a) using $NiCl_2$ (0.13 M) and $NaPO_2H_2$ (0.093 M). However, the solution was unstable because the complexing agent concentration was too low. This group investigated how the plating rate is affected by the amount of thiourea stabilizer at different pH values (Figure 5.4). No NiP coating was deposited below pH 3.6, which is in agreement with other studies listed in Table 5.2, and the most significant increase in plating rate with pH was using 2 ppm thiourea. An optimum pH of 5.25 (Figure 5.4) for plating rate is in close agreement with pH 5.5 selected using a statistical analysis modeling method in a different study (Park et al., 2002).

Another influence to consider is the use of surfactants in an ultrasonic solution. Surfactants will influence the surface tension of the electrolyte and will therefore have an impact on cavity growth and stability, which in turn affects the rate. A detailed review was compiled about studies on parameters that influence ultrasonic cavitation (Wood et al., 2017) and the use of surfactants (among other chemicals) is considered as a parameter that enhances the effects of a sonochemical system. Some findings of Wood et al. are detailed in Table 5.3. The effects of transient cavitation and resistance to mass transport are carefully balanced by surfactant and salt concentration in solution. There is evidence that higher concentrations of surfactants in solution form micelles and reduce mass transport.

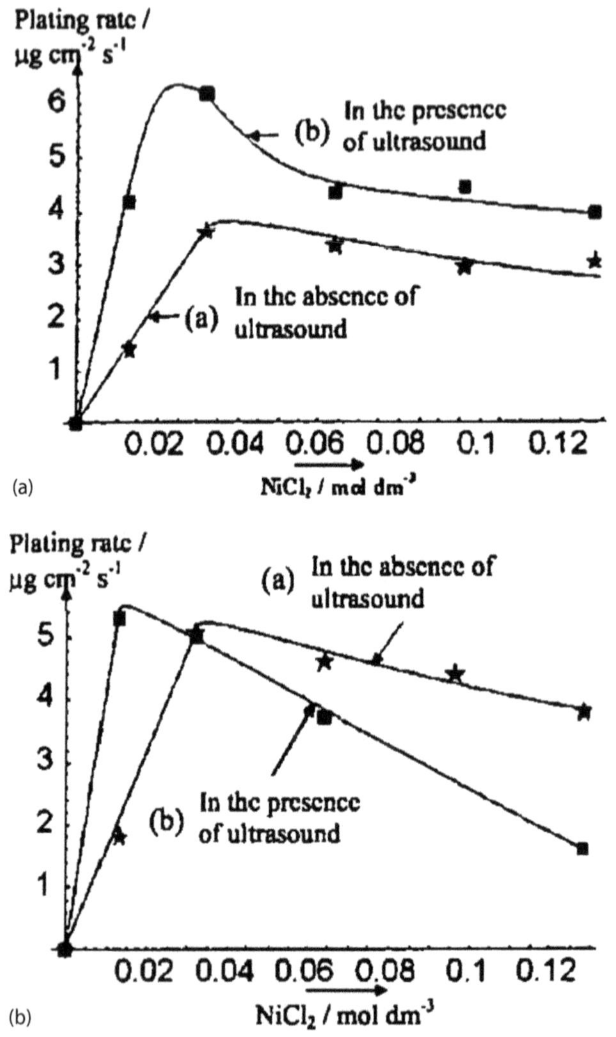

FIGURE 5.3 Rate of NiP deposition vs. $NiCl_2$ concentration for a silent and ultrasonic solution (35 kHz/0.80 W/cm^2) containing 0.19 M sodium hypophosphite, (a) 0.04 M sodium acetate, and (b) 0.24 M sodium acetate (plating at 363.15 K and 15 min). (From Abyaneh, M.Y. et al., *J. Electrochem. Soc.*, 154, D472, 2007.)

Chiba et al. prepared electroless NiB-plated copper substrate with/without ultrasonic waves and compared the plating rate to an alternative method in which the plating solution is still but the Cu sheet substrate is vibrated while plating. Their results show that the thickest coating and therefore the fastest deposition rate were achieved by ultrasonication (45 kHz, 100 W). Deposition rate by mechanical agitation (without ultrasound) was faster than the vibration of substrate (Chiba et al., 2003).

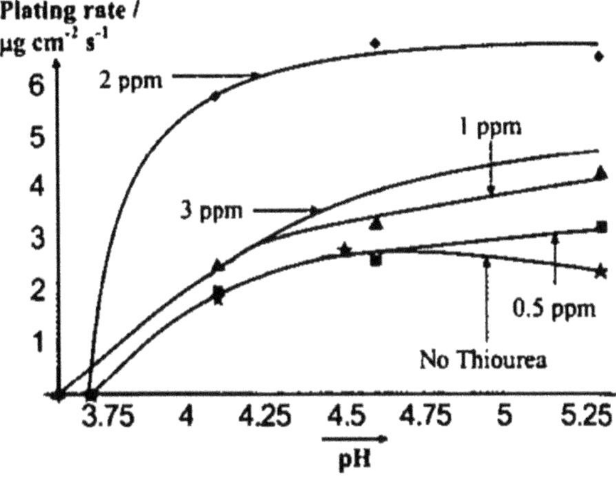

FIGURE 5.4 The effects of pH on the plating rate of NiP in the presence of ultrasound at a series of various thiourea concentrations (0–3 ppm). (From Abyaneh, M.Y. et al., *J. Electrochem. Soc.*, 154, D472, 2007.)

TABLE 5.3
Effect of Different Types of Surfactants on Cavitation

Characteristic	Surfactant	Effect on Cavitation
Bulky head group	Dodecyl dimethyl ammonium propane sulfonate	• Induce bubble surface instabilities. • Increase rectified diffusion.
Long-chain tail	Dodecyl trimethyl ammonium chloride (DTAC)	• Reduces resistance to mass transfer. • Chance surfactant will adsorb to cavity surface decreases with increasing chain length.
Ionic head group	Sodium dodecyl sulfate	• Reduces bubble coalescence by electrostatic repulsion. • Electrostatic interaction of surfactant is inhibited by salts.
	DTAC	• Increase in cavity growth rate and cavity surface instability.
	Perfluorocarboxylic acid	• Surface tension rises on cavity.

Source: Wood, R.J. et al., *Ultrason. Sonochem.*, 38, 351–370, 2017.

5.3 INFLUENCE OF ULTRASOUND ON STRUCTURE, MORPHOLOGY, AND PROPERTIES OF ELECTROLESS Ni COATINGS

5.3.1 Effect of Ultrasound on Ni Deposit Composition

The control of phosphorus or boron content in Ni coatings is very important because the composition influences the properties of the coating. ASTM B733–15 standard (ASTM Committee, 2015) definitions of changes in coating properties with changing P (%) are summarized in Table 5.4.

TABLE 5.4
Ni Deposits Containing Different Levels of P Content, and Applications That Match Their Properties

Type	P (%)	Microstructure	Applications	Properties of NiP
Lower	1–3	Crystalline	Electronics	Solderability, bondability, increased electrical conductivity, and resistance to strong alkali solutions.
Low	2–4	Crystalline	Abrasion and wear resistance	High as-plated hardness (620–750 HK 100).
Medium	5–9	Mixture of crystalline and amorphous	General purpose	Wear and corrosion resistance.
High	>10	Amorphous	Beryllium and titanium parts for low stress	Longer salt-spray and acid resistance. Considered non-ferromagnetic.

Source: ASTM Committee, ASTM B733-15 Standard Specification for Autocatalytic (Electroless) Nickel-Phosphorus Coatings on Metal, *ASTM International*, 2015.

It is generally accepted that the ultrasonication of the plating solution tends to cause a drop in P(%) in the deposited coating at higher temperatures compared to a similar silent plating solution (Cobley et al., 2011). For example (Cobley & Saez, 2012), a decrease in P(%) content was observed with increasing temperature (323.15 K–363.15 K) using a 20 kHz/42 Wcm^{-2} probe, and the difference between silent and ultrasonic deposition increases as the temperature is lowered. The content of P(%) in Figure 5.5 is much higher than the standard 4%–15% reported by most studies (Table 5.2).

A plot of P(%) vs. rate of deposition (μm/h) comparing four different research studies shows that generally, ultrasonically assisted electroless NiP coatings tend

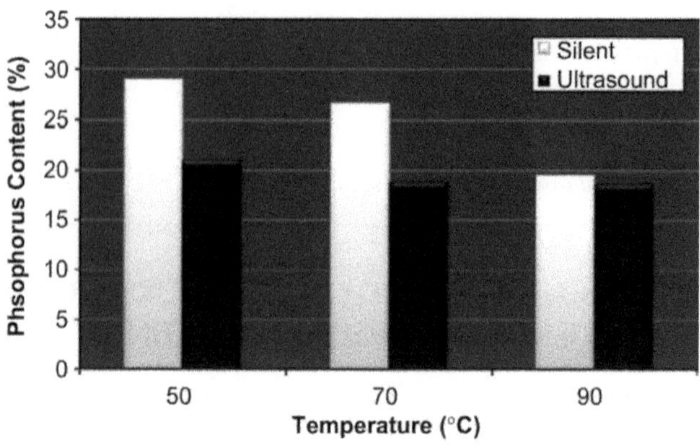

FIGURE 5.5 P content (%) in NiP deposited using 1 Ni^{2+}: 1.33 NaH_2PO_2 (323.15 K–363.15 K) at 20 kHz/42 W/cm^2 electroless plating solution. (From Cobley, A.J. and Saez, V., *Circuit World*, 38, 12–15, 2012.)

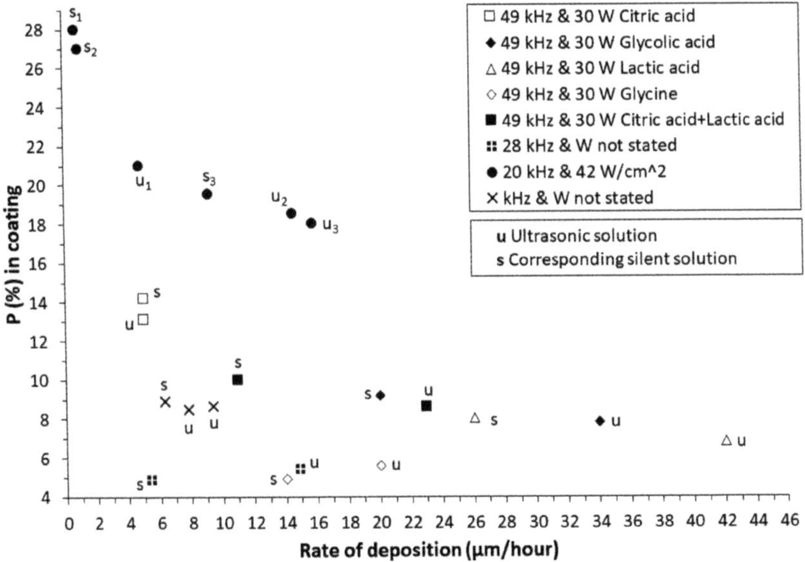

FIGURE 5.6 Plot of P (%) vs. rate of deposition of electroless NiP coating (µm/h). Ultrasonic and corresponding silent electroless plating solutions are indicated by u and s labels, respectively. Data are taken from four separate studies (see Table 5.2).

to have higher deposition rates and lower P content (Figure 5.6). However, there are some exceptions where P(%) is higher than the deposit P content from the silent solution due to secondary factors.

In the present literature survey, only two studies were found that showed increases in coating P content when comparing ultrasonic with silent agitation methods (Mallory, 1978; Sun et al., 2011a, 2011b). Mallory observed that P content in NiP deposit is higher (P 5.6%), than in a silent solution deposit (P 4.9%), only when using glycine chelating agent and applying ultrasound: 49 kHz/30 W (Mallory, 1978). From this evidence, Mallory concluded that the drop in P content when using ultrasound is caused not only by the increase in solution temperature but also the type of chelating agents in the solution. In parallel experiments, Mallory observed a decrease in P(%) in the coating deposited when using the same ultrasound settings and chelating agents containing only oxygen functional groups, i.e. citric acid, lactic acid or glycolic acid (Figure 5.7a–c). In addition

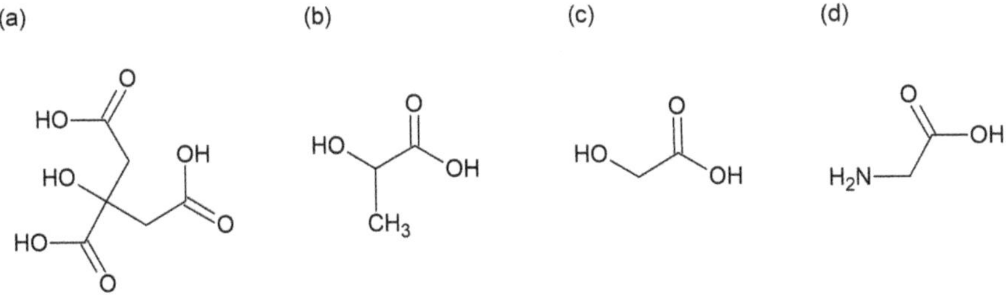

FIGURE 5.7 Chemical structures of (a) citric acid, (b) lactic acid, (c) glycolic acid, and (d) glycine.

to this, all NiP depositions listed in Table 5.2 from separate studies have used chelating agents with only oxygen functional groups at acidic pH. Although no theory/mechanism has been proposed to explain how glycine (Figure 5.7d) chelating agent increases P content in a binary NiP coating deposited by ultrasound at pH 4.8. Other studies have shown that glycine cannot form a stable complex with Ni^{2+} in an acidic pH range (Sotskaya & Dolgikh, 2008). However, when glycine is mixed with other complexing agents, i.e. alanine and citric acid at pH 5.4, to form a composite coating of NiP-SiC by ultrasound (Ashassi-Sorkhabi & Es′haghi, 2013), the P content decreases (see Table 5.5, page 279).

Sun et al. also reported higher P content when deposited from an electroless NiP electrolyte under ultrasonic irradiation (28 kHz/W not stated) compared to a silent solution (Sun et al., 2011a, 2011b), but did not indicate the type of chelating agent used, and their electroless solution temperature was lower (343.15 K) than Mallory's study (1978). However, a recent study (Zou et al., 2014) using a similar temperature (341.15 K) to Sun et al. has shown that using sodium acetate chelating agent, a –7.3% decrease in P(%) content is observed when sonication is applied. Therefore, in the study by Sun and coworkers, it is likely that P content was affected by cobalt sulfate concentration, as Co was co-deposited in Ni with phosphorus (Sun et al., 2011a, 2011b). They also showed that the Co content in the coating was higher when deposited under ultrasound compared to those coatings deposited from a silent electrolyte. A similar trend was shown in a study (Qian et al., 2017) that co-deposited Ce in NiB (Table 5.2, labels 4–7) where the NiB-Ce deposited by sonication contained more B than the binary NiB deposit prepared using the same ultrasonic parameters (60 kHz/150 W at 358.15 K). The Ce content in this study was also higher than the Ce deposited using a still solution.

On the other hand in a separate study (Sun et al., 2011a, 2011b), Cu was co-deposited with P from a NiP electroless solution using the same electroless plating parameters as in the study for NiP-Co; a trend in which P content decreases (compared to binary NiP) as Cu content increases in Ni-Cu-P when using ultrasound was observed. Matsuoka and Hayashi (1985) reported that increasing the concentration of thallous nitrate stabilizer in the solution (2–22 wt%) and pH (5–9) causes a decrease in the amount of P in a NiP coating deposited by ultrasound (25 kHz/150 W) at 353.15 K. This was due to the co-deposition of thallium in the Ni coating, and the thallium content increased in the deposit while the P content decreased.

5.3.2 The Influence of Ultrasound on Ni Deposit Morphology

Many authors have reported that the NiP deposit has a somewhat rounded nodular surface morphology (Figure 5.8a), while a NiB coating has cauliflower-type structures on the surface (e.g. Figure 5.9). It has been found (Zou et al., 2014) that the nodule size on the surface of an electroless NiP coating decreases when ultrasound is used during deposition (Figure 5.8b), and this can increase the hardness and wear resistance of the coating. In this example, it was also observed that, without sonication, there were pores at the NiP and substrate interface. It was proposed that the use of ultrasound during electroless deposition minimizes pin-hole gaps in the Ni deposit surface by efficient removal of hydrogen gas formed during the deposition reaction. The difference in coating porosity through the thickness of the coating is evident when applying ultrasonication. For example, the morphology of a sectioned crystalline NiB coating is

TABLE 5.5
Examples of Ni Composite Coatings Deposited Using Ultrasonic Electroless Solution

			Response						Plating Parameters					
Coating Surface (Citation)	Label	Sonication Parameters	Thickness (μm)	Hardness (H_V)	P % (B %)	i_{corr} (μA/cm²)	E_{corr} (V)	Particles through Thickness	Reducing Agent (M)/Ni^{2+} (M)	Chelating Agent (M)	Other	Particle g/dm³ (Size)	pH (Temp, K)	Time (s)
NiP-SiC	1	Silent	29.2	640	Not stated	Not stated	Not stated	Uniform	0.24/0.095	Sodium pyrophosphate decahydrate (0.11)	Not stated	5 (20 nm)	10 (298.15)	3600–5400
NiP-SiC Steel (Zhou et al., 2016)	2	40 Hz/100 W + Magnetic 7.5 T/m	29.6	680										
NiP	3	Silent	~15	Not stated	Not stated	3.55	–0.655	Uniform	0.19/0.095	Sodium citrate dehydrate (0.034) + glycine (0.33)	SDS (10 ppm)	0.1 (4 nm)	9 (358.15)	3600
NiP-diamond Steel (Ashassi-Sorkhabi & Es'haghi, 2013)	4	35 kHz/ <450 W	~15			0.85	–0.380							
NiP	5	Silent	15	Not stated	Not stated	1.412	–0.404	Uniform	0.22/0.036	Sodium citrate (0.046)	KF (0.14 mol/dm³)	5 (20 nm)	6 (353.15)	1800 –3600
NiP-ZrO_2 AZ91D Mg alloy (Song et al., 2008)	6	40 kHz (60% amplitude)	15			0.9674	–0.361							
NiP	7	Silent	14.3	Not stated	Not stated	Not stated	Not stated	Uniform	0.23/0.10	Lactic acid (0.28)	None	0.04 (5 μm)	4.7 (348.15)	7200
	8	40 kHz[a]	29.9	772.7										

(Continued)

TABLE 5.5 (*Continued*)
Examples of Ni Composite Coatings Deposited Using Ultrasonic Electroless Solution

			Response						Plating Parameters					
Coating Surface (Citation)	Label	Sonication Parameters	Thickness (μm)	Hardness (H_V)	P % (B %)	i_{corr} (μA/cm²)	E_{corr} (V)	Particles through Thickness	Reducing Agent (M)/Ni^{2+} (M)	Chelating Agent (M)	Other	Particle g/dm³ (Size)	pH (Temp, K)	Time (s)
NiP-GO Stainless steel (Yu et al., 2018)	9	40 kHz[a]	29.9	746.8							SDS (1200 ppm)			
NiP	10	kHz Not stated	35	560	8.75	1.75	−0.383	Gradient	0.19/0.076	Citric acid (0.013) Glycine (0.016) Alanine (0.011)	40% HF (12 mL/L) MBT (0.25 ppm) SDS (150 ppm) Thiourea (1 ppm)	4 (0.5–0.7 μm)	5.4 (353.15)	5400
NiP-SiC AZ91D Mg (Wang et al., 2013)	11	kHz Not stated	35	620	7.25	1.36	−0.382							
NiB	12	Silent	30	890	(6.6)	1.2	−0.64	Uniform	0.026/0.13	Ethylenediami ne (1.5)	CH_3COOTl (18 ppm)	2 (0.2–0.3 μm)	14 (358.15)	3600
NiB -TiO_2 Mild steel (Niksefat & Ghorbani, 2015)	13	40 kHz/150 W + rotation (300 rpm)	15	1263	(6.1)	0.2	−0.59							

Note: SDS = Sodium dodecyl sulphate; MBT= 2-Mercaptobenzothiazole; M = mol/dm³; GO = graphene oxide flakes.
60 wt% suspension.
[a]Intermittent repeats = 60 s sonication and 600 s silent.

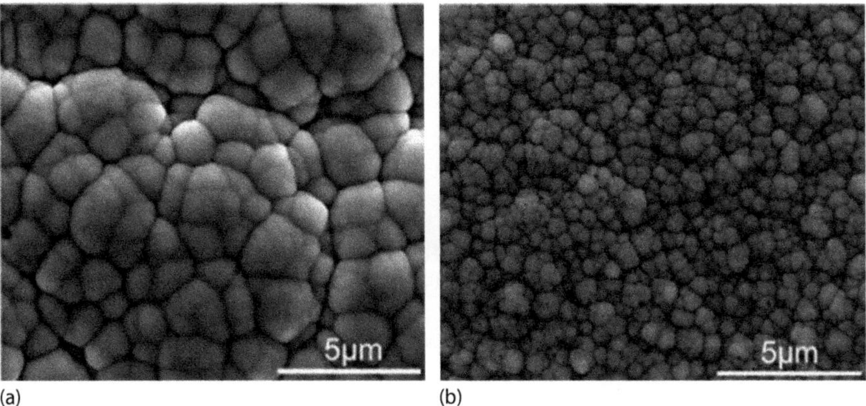

FIGURE 5.8 SEM image showing a change in surface morphology and nodule size of NiP deposit from an electroless Ni solution that is (a) silent and (b) agitated by ultrasound (40 kHz/100 W). (From Zou, Y. et al., *J. Electrochem. Soc.*, 162, C70, 2014.)

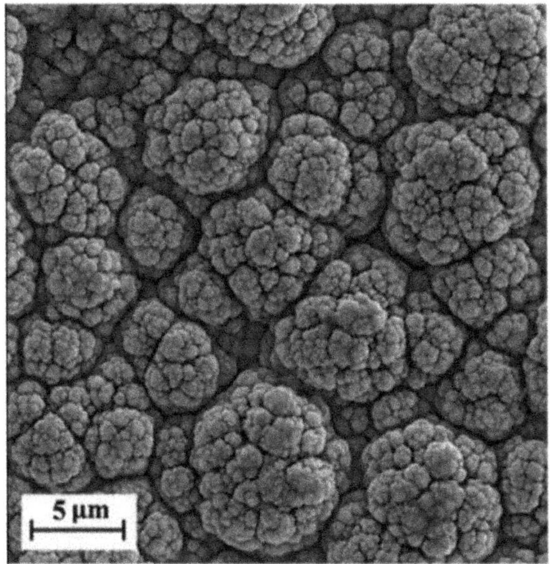

FIGURE 5.9 SEM image showing a cauliflower-type surface morphology and nodule size of NiB deposit from a silent electroless Ni solution with agitation by rotation of the sample at 300 rpm. (From Niksefat, V. and Ghorbani, M., *J. Alloy Compd.*, 633, 127–136, 2015.)

columnar and porous (e.g. Figure 5.10a), whereas a coating deposited with ultrasonication is columnar and compact, e.g. Figure 5.10b (Bonin et al., 2017).

X-ray diffraction (XRD) analysis of ultrasonically deposited Ni coatings shows a strong broad single Ni [111] peak for an amorphous structure, and ultrasound may result in some crystalline phases appearing.

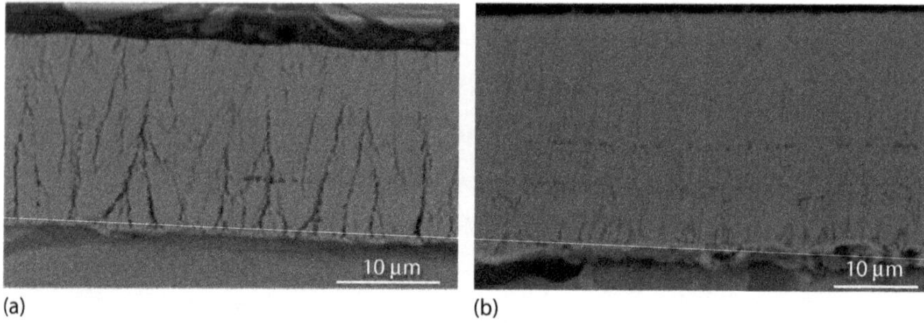

FIGURE 5.10 SEM images of cross-section morphology of electroless NiB coating deposited (a) without ultrasound and (b) with ultrasound (35 kHz/0.065 W/cm^3). (From Bonin, L. et al., *Ultrasonics*, 77, 61–68, 2017.)

Mallory reported crystal sizes from Ni [111] peaks from NiP plating deposited from silent and ultrasonic solutions were similar, 1.7 and 1.9 nm, respectively. Both deposits were amorphous and were distinguished by a minor Ni [200] shoulder on Ni [111], which was absent in as-plated NiP from a silent solution (Mallory, 1978).

Ban and coworkers deposited NiP onto Mg alloy from an ultrasonic (530 kHz/5 W) electroless Ni solution, and the coating had a dendritic transition layer (17 µm thick) where the Mg and NiP were intermixed (Figure 5.11, Ban et al., 2014). A transition layer formed when high-frequency ultrasonic irradiation deformed the Mg alloy surface while the electroless Ni solution penetrated and deposited NiP in the deformed Mg layer. The XRD pattern of NiP coating deposited from a silent solution was amorphous, and when ultrasound irradiation

FIGURE 5.11 SEM image of a sectioned NiP plating deposited on Mg alloy AZ31 at a constant frequency of 530 kHz/5 W (500 mL electroless solution, pH 6.5) for 75 min (353.15 K). (From Ban, C.L. et al., *Surf. Eng.*, 30, 880–885, 2014.)

used, the Ni phase was crystalline, indicating that ultrasound at 530 kHz frequency caused a phase transformation.

Sharifalhoseini and Entezari (2015) deposited NiP onto mild steel by positioning the sample vertically or horizontally with respect to the ultrasonic probe in electroless Ni, and they concluded that placing the sample horizontally below the probe tip gave an improved surface finish by comparing scanning electron microscopy (SEM) images (Figure 5.12).

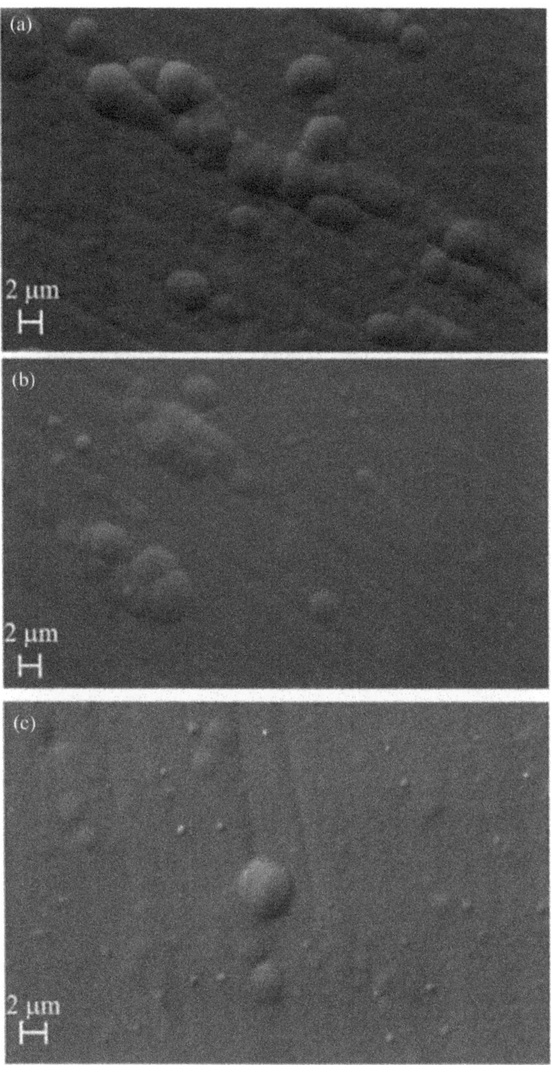

FIGURE 5.12 SEM images of electroless Ni deposits from: (a) silent solution, (b) sample in vertical position vs. ultrasonic probe, and (c) sample in horizontal position vs. ultrasonic probe. (From Sharifalhoseini, Z., and Entezari, M.H., *Appl. Surf. Sci.*, 351, 1060–1068, 2015.)

5.3.3 Effect of Ultrasound on Ni Deposit Mechanical/Physical Properties

Ni coatings are removed by adhesive or scratch wear; spall wear is usually observed when the Ni coating is a composite containing particles. The important factors in reducing wear are strong adhesion to substrate, high hardness, and low friction coefficient. Very few studies were found, in which the wear of NiP or NiB coating deposited by ultrasonication were reported because wear resistance tends to increases with hardness (Bonin et al., 2017; Sun et al., 2011a, 2011b). Most studies assume a coating with high hardness has excellent resistance to wear.

However, the mechanism of wear can vary depending on the type of ultrasonic agitation method/parameters. Bonin et al. deposited NiB using (i) silent mechanically agitated solution, (ii) ultrasonic cleaner (35 kHz/0.065 W mL^{-1}), and (iii) probe (20 kHz/0.058 W mL^{-1}). They established that NiB is removed via abrasive and adhesive wear for NiB deposited by silent and 35 kHz ultrasonic agitation (Bonin et al., 2017). Whereas, for NiB deposited using a 20 kHz probe, the adhesive wear mechanism dominated. In their study, they did not observe a significant improvement in hardness and wear when using ultrasound during electroless plating.

Electroless Ni plating deposited in the presence of ultrasound has been shown to increase the hardness of Ni coatings due to (1) an increase in the number of crystalline phases, i.e. Ni_2P_5, Ni_3P and $Ni_{2.55}P$ (Yang et al., 1997), (2) a decrease in the phosphorus or boron content, and (3) smaller Ni crystal grains.

Usually, the hardness is higher when a lower amount of B or P in the coating is present (Cobley et al., 2011). Figure 5.13 is a plot of hardness vs. composition, from research studies summarized in Table 5.2, and from this plot, it is evident that when NiP is deposited under ultrasound, the hardness is higher than that under equivalent silent conditions, and as has already been stated, the P % is lower (Figure 5.13). There is only

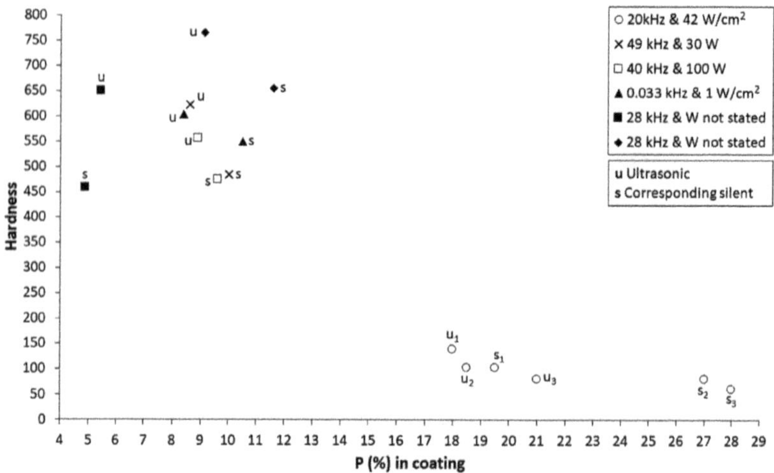

FIGURE 5.13 Hardness vs. P (%) content in electroless NiP coating. Ultrasonic and corresponding silent electroless plating solutions are indicated by u and s labels, respectively. Data are taken from six separate studies (see Table 5.2).

one study (Sun et al., 2011a, 2011b), see Figure 5.13 (■, 28 kHz/W not stated), that showed an increase in hardness with P content; in this study, the coating is NiP-Co. An increase in the hardness can be explained by the increase in Co content from 1.91% (silent) to 3.27% (28 kHz ultrasound), and this means more Co atoms are in the Ni crystal lattice (Srinivasan et al., 2011). A similar trend of increasing hardness with B % is observed for NiB-Ce coatings (Qian et al., 2017) when B 5.9% and Ce 6% (silent) are increased to B 6.2% and Ce 6.6% when applying ultrasound (60 kHz/150 W).

Interestingly, when copper is co-deposited in a NiP coating (Sun et al., 2011a, 2011b), the hardness is not affected in the same way as when using Ce or Co (Figure 5.13 [•], 28 kHz), and the same applies to the NiB coating (Bonin et al., 2017) containing trace amounts of Pb (see Table 5.2, labels 1–3).

The deposition method, and not just composition, affects hardness. One study described NiB (25 μm thickness, 5.47 wt% B) deposited at 368 K (pH 12) by ultrasound (35 kHz/0.065 Wcm^{-3} by probe) with no stirring (Bonin et al., 2017). In a separate study, NiB (30 μm thickness, 6.6 wt% B) was deposited from a silent solution (358.15 K, pH 14) while rotating (300 rpm) the substrate (Niksefat & Ghorbani, 2015). The hardness from Niksefat & Ghorbani method was higher (890 HV_{100}) than that from Bonin et al. (844 HV_{100}).

5.3.4 Effect of Ultrasound on Ni Deposit Adhesion to Substrate

There are few studies on the effect of ultrasonic irradiation on adhesion of electroless Ni coatings. The adhesion strength of NiP to a magnetic substrate (Nd-Fe-B) from an electroless alkaline solution (pH 8 and 353.15 K) improves from 12 MPa (silent solution) to a maximum of 56 MPa when ultrasound (40 kHz/150 W) is applied (Yan, Ying & Ma, 2009). Figure 5.14 shows a plot

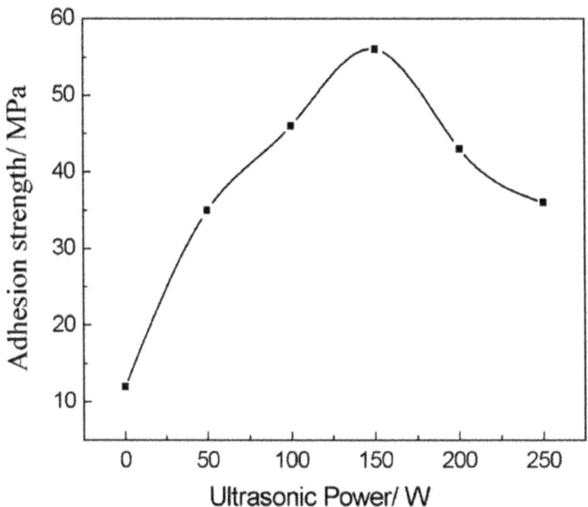

FIGURE 5.14 Adhesion strength of electroless NiP coating to Nd-Fe-B magnets deposited at 40 kHz and different ultrasonic powers (W). Solution pH = 8.5 and temperature = 353.15 K. (From Yan, M. et al., *Mater. Chem. Phys.*, 113, 764–767, 2009.)

of adhesion strength vs. ultrasonic power (W) at 40 kHz, and this shows that power affects adhesion; 150 W is the optimum power for an Nd-Fe-B magnet surface.

Zhu et al. conducted thermal shock tests on NiP-plated Mg-Li alloy to show that NiP deposited via ultrasonic assistance, 40 kHz/100 W (341.15 K, 7200 s), exhibits no evidence of poor adhesion, i.e. blistering, delamination, cracking, and peeling. Whereas, the Ni coating prepared without ultrasonic irradiation delaminated. The differences in the two coatings were noted by visual inspection only (Zou et al., 2014).

5.3.5 Corrosion Resistance of Ni Coatings Deposited in an Ultrasonic Field

Data from all studies on electroless Ni deposition by ultrasound indicate that the key to very good corrosion resistance in salts and acids is that the coating structure has the following features:

1. Amorphous phase (very little or no crystals);
2. Compact with little or no porosity/voids;
3. Lower surface area (or roughness);
4. Uniform thickness.

Polarization curves are often used to characterize corrosion resistance of Ni-plated samples by ultrasound because they can be used to derive the theoretical rate of corrosion. In all cases (Table 5.2), very low I_{corr} values in NaCl (aq) are reported because Ni forms a thin passive film that slows the corrosion process; E_{corr} increased and I_{corr} decreased when comparing Ni coatings deposited under ultrasound with those deposited under silent solutions (Ban et al., 2014; Bonin et al., 2017; Chiba et al., 2003; Qian et al., 2017; Sharifalhoseini & Entezari, 2015; Yan et al., 1997, 2009; Zou et al., 2014). For ultrasonically deposited NiP, E_{corr} is at −0.3 to −0.35 V, and for NiB, E_{corr} is at lower potentials approximately −0.65 to −0.75 V (Table 5.2).

Yang et al. prepared a NiP coating using mild ultrasonic agitation (0.033 kHz/1 Wcm^{-2}), which contained crystalline phases, and it was found that the deposit was less corrosion-resistant to 0.2 mol/dm^3 H_2SO_4 (aq) than a deposit plated under silent conditions (amorphous structure). In addition to this, micro-cracks from stress corrosion cracking were visible in the ultrasound-assisted NiP deposit (Yang et al., 1997). Other authors report Ni coatings deposited by ultrasound are likely to exhibit pitting (Bonin et al., 2017) and galvanic corrosion (Zou et al., 2014) when analyzed after corrosion tests in NaCl (aq).

There is evidence that NiP or NiB coatings co-deposited with other metals and plated using ultrasonic irradiation have greater corrosion resistance in NaCl than a binary NiP or NiB coating prepared using the same agitation method. Yan et al. deposited NiP coatings with varying [Cu^{2+}] in an electroless solution (pH 8.5, 353.15 K, and 40 kHz/150 W) and compared these coatings with a binary NiP deposited under the same conditions. Their results show a Cu/Ni ratio of 0.02 increases the time it takes for rust to appear on the coating from approximately 216 to 1843.2 kiloseconds in neutral salt spray compared to binary NiP. Polarization curve results also showed that 0.02 Cu/Ni in NiP coating has the highest corrosion resistance than other Cu/Ni ratios and binary NiP. The coating containing 0.02 Cu/Ni ratio was less porous than coatings prepared using higher ratios of Cu/Ni, and the 0.02 Cu/Ni ratio gave an amorphous coating. Coatings deposited using Cu/Ni ratios higher than 0.03 were polycrystalline, and phase boundaries weakened the coatings corrosion resistance (Yan et al., 2009).

Recently, a coating of NiB-Ce (Table 5.2, labels 4–7) on mild steel has been shown to enhance corrosion resistance compared to NiB (Qian et al., 2017); both exhibited an amorphous structure (by XRD) and were deposited by ultrasound (60 kHz, 150 W). However, a NiB-Ce deposited from a silent solution was still more corrosion-resistant than binary NiB deposited by ultrasonication.

5.4 ELECTROLESS Ni COMPOSITE PLATING ENABLED BY ULTRASONICALLY DISPERSED PARTICLES

Particle-reinforced Ni metal matrix composite (MMC) coatings have enhanced wear and corrosion resistance and can enable thinner coatings to be deposited in some cases. It has been proposed that particles distributed uniformly throughout the coating thickness (e.g. Figure 5.15) inhibit the growth of corrosion paths and dislocation movement in the NiP matrix. Generally, the morphology of Ni MMC is rougher than binary NiP or NiB. Particles in Ni coatings dispersed by ultrasonication are refractory ceramics, oxides (e.g. TiO_2, ZrO_2, and Al_2O_3) or graphene-based, and no studies were found in which polymer particles, e.g. polytetrafluoroethylene, were co-deposited in an electroless Ni coating using ultrasonic irradiation.

Particles in an electroless Ni MMC are deposited using similar ultrasonic parameters to standard electroless Ni solutions, the main difference is that, in some studies, surfactants are added to the plating solution to reduce particle agglomeration. The type of surfactant used depends on the particle surface chemistry, and this is

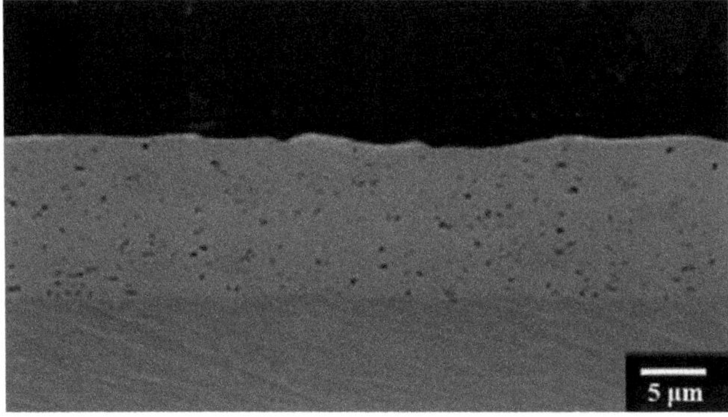

FIGURE 5.15 SEM image of sectioned sample showing 4 nm sized diamond nanoparticles co-deposited through thickness of NiP coating (35 kHz/450 W [358.15 K, pH 9]). (From Ashassi-Sorkhabi, H. and Es'haghi, M., *Corros. Sci.*, 77, 185–193, 2013.)

balanced with ultrasonication parameters. As stated earlier, surfactants can significantly influence the efficacy of sonication but no in-depth study has been conducted on the influence of surfactants on ultrasound when co-depositing particles in electroless Ni composites.

5.4.1 Distribution of Particles in a Ni Matrix

A study by Song et al. demonstrated that one benefit of ultrasound is that the weight ratio of ZrO_2 particles increases in a NiP deposit (see Figure 5.16a, Song, Shan & Han, 2008). This is due to pressure released from cavity collapse in an ultrasonic solution increasing the particle movement and collisions (Crum & Suslick, 1995). The maximum weight ratio of ZrO_2 (approximately 5.25%) in NiP was achieved at 60% amplitude of 40 kHz frequency (Figure 5.16b).

The effect of solution temperature (323.15 K–343.15 K) on co-deposition of Al_2O_3 in NiP at 40 kHz/60 W was researched by Fan et al; surfactants and stabilizers were used in the plating solution (Fan, Ma & Cao, 2011). The findings from this study concluded that Al_2O_3 particles do not agglomerate in the NiP alloy matrix and that the corrosion resistance (spot test) and hardness of NiP-Al_2O_3 coating are greatest at plating temperature of 333.15 K (3000 s at pH 7). Other authors have deposited composite coatings using different reinforcing phases, and there seems to be no fixed pH range across the different studies (Table 5.5). In a later study, Fan et al. prepared low-phosphorus (2.1 wt%) NiP-Al_2O_3 coating on a magnesium alloy from an electroless Ni solution by ultrasonic agitation and measured the rate of plating for different pH (5–11) values (Fan, Qiu & Ma, 2014). Their findings show that a NiP-Al_2O_3 coating prepared from a pH 7 solution has the fastest deposition rate, lowest roughness, and the thickest coating.

Wang et al. described a method of co-depositing SiC particles (0.5–0.7 μm size) in NiP with a gradient of SiC through the coating thickness (35 μm). A pre-made

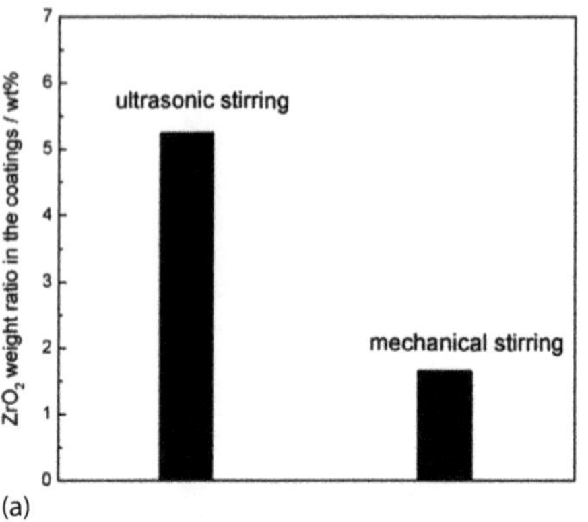

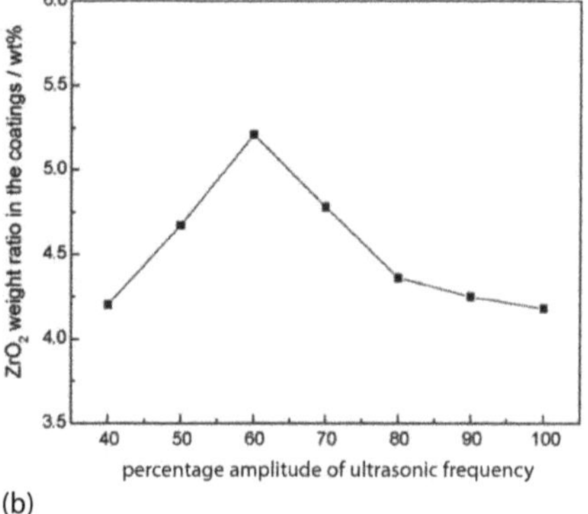

FIGURE 5.16 (a) A bar chart of ZrO_2 (wt%) in NiP coating with/without ultrasound. (b) Amount of ZrO_2 (wt%) in NiP coating when different percentages of ultrasonic frequency (40 kHz) are applied. (From Song, Y.W. et al., *Electrochim. Acta*, 53, 2135–2143, 2008.)

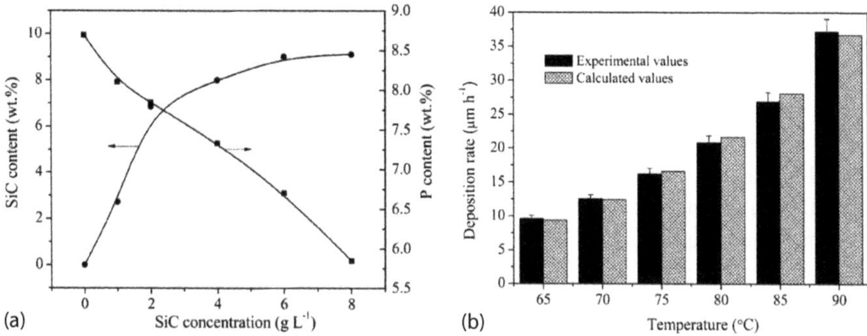

FIGURE 5.17 (a) A plot showing the effect of increasing SiC concentration in electroless plating solution on SiC and P content in the coating; (b) a plot of deposition rate (μm/h) vs. temperature showing experimental values from weight gain data and values calculated from a kinetic model. (From Wang, H. et al., *Appl. Surf. Sci.*, 286, 319–327, 2013.)

electroless Ni plating solution containing SiC (4 g/L) was added dropwise to the main electroless plating solution, at 353.15 K (pH 5.4) over 5400 s while applying ultrasound (Wang et al., 2013). SiC concentration added to the electroless plating solution affects SiC and P content in NiP-SiC deposit (Figure 5.17a), and the rate of deposition increases with temperature (Figure 5.17b). Lowering of P (wt%) in the deposit with increasing SiC (g/dm^3) in the plating solution was explained by reduction in available H^+ ions, which adsorb to SiC particles instead of reducing $H_2PO_2^-$ ions to form P. Adhesion test results indicated the coating does not blister or crack after aging the coating at 493.15 K (3600 s) and quenching in water (ambient temperature). Other tests comparing NiP with NiP-SiC gradient coating, i.e. corrosion in 3.5 wt% NaCl by polarization curves and microhardness, also show an improvement.

The preferred orientation of the NiP grain growth and an increased compactness of the deposit can be achieved using a combined electromagnetic-ultrasonic field (Zhou et al., 2016). SiC particles (20 nm size) were added to an electroless Ni electrolyte to prepare a NiP-SiC deposit by ultrasound, and the optimum deposit was formed using 20 kHz/300 W (magnetic field strength of 0.7 T). The coating was comprised of 200 nm NiP/SiC particles distributed evenly. Coatings prepared with a magnetic field and ultrasound had lower nanoindentation hardness (than just ultrasound).

5.4.2 Adhesion of Co-deposited Particles to the Ni Matrix

Very large diameter (150 μm) particles of cubic boron nitride (cBN) were co-deposited in NiP coating to form an abrasive grinding wheel (Okumiya et al., 2003). The optimal frequency and amplitude for the deposition of this composite were found to be 15.5 kHz and 11 μm, respectively. The transport of Ni ions to the cBN grains was improved by ultrasonication, which resulted in a stronger Ni-cBN bond. In the follow-up study, adhesion of the cBN to the NiP matrix in the abrasive wheel was characterized

(Okumiya et al., 2005) by measuring the Vickers indent load (N) required to push particles out of the Ni matrix. Their results showed that when NiP-cBN was deposited with 11 μm ultrasonic amplitude, there is a 20% increase in the load required for particles to drop out of the matrix compared to a deposit from a silent bath.

Yu and coworkers prepared NiP-graphene oxide (GO) coatings; GO flakes had a very high aspect ratio of length to thickness and deposited like a thin film (Yu et al., 2018). A composite coating of NiP-GO also had a very high surface roughness (2.84 μm) compared to a standard NiP deposit (0.26 μm) because GO flakes acted as barriers preventing an even growth of Ni grains. The rate of deposition with ultrasound was faster and the coating was more compact; without ultrasound, the coating had pores on the surface.

A monolayer of graphene has a Young's modulus of 1.0 TPa and a breaking strength of 42 N/m (Huang et al., 2012). One of the benefits of using GO in a composite is that the material becomes tougher. The Young's modulus of as-plated NiP derived from load-displacement curves (164.0 GPa) by Yu et al. is not significantly lower than NiP-GO MMC (167.1 GPa). Annealing at 673.15 K improved the mechanical properties of both types of coatings, but there was no significant change as observed in polymer-GO composites described by Huang et al. The maximum achievable strength of a composite can be realized when GO flakes are covalently bonded to the surrounding matrix (Huang et al., 2012). Therefore, when Yu et al. increased the compactness of the NiP-GO deposit by ultrasound (40 kHz, power not stated), the strength of the bond between the NiP matrix and GO flakes was not considerably improved.

5.4.3 Corrosion of Ni Composite Coatings

Song et al. focused on enhancing the corrosion resistance of NiP by ZrO_2 composite coatings. A NiP-ZrO_2 composite plating on an AZ91D Mg alloy deposited by ultrasound, withstood 720 kiloseconds in 3.5 wt% NaCl (aq) spray. This is two times the duration of a NiP plating (360 kiloseconds) from a silent electrolyte. However, they also demonstrated that the corrosion resistance is increased to 3600 kiloseconds by coating the ultrasonically deposited NiP-ZrO_2 with an interlayer of electroplated pure Ni (20 μm) followed by NiP (5 μm). In the multilayer coating of NiPZrO_2-Ni-NiP, only the NiP-ZrO_2 layer on the electrode was deposited by ultrasound (40 kHz at 60% amplitude/power not stated). The multilayer coating corroded more slowly because the NiP (the sacrificial layer) was the net anode, and the pure Ni interlayer acted as a net cathode because it had a higher E_{corr} (−0.3 V) than NiP (−0.4 V). The ultrasonically deposited compact composite layer (NiP-ZrO_2) under areas where the corrosive NaCl eventually pitted the Ni interlayer further delayed the corrosion of the Mg substrate (Song et al., 2008).

Sharifalhoseini and Entezari deposited a compact NiP coating by ultrasound, and a second layer of just ZnO was deposited by ultrasound to provide corrosion protection on the NiP surface. They found that ZnO particle-coated NiP surface has a lower corrosion rate than a bare NiP coating that is also deposited by ultrasound (Sharifalhoseini & Entezari, 2015). Their method can be improved by reducing/controlling the size of the particles as the ZnO particles they deposited were not capped with functionalizing agents.

A multilayer coating of NiP and 20 nm sized diamond nanoparticles (DNP) was deposited in the following order (Mazaheri & Allahkaram, 2012): (1) two layers of NiP on substrate, (2) NiP-DNP MMC inner layer, (3) NiP interlayer, and (4) final NiP-DNP MMC outer layer. Only the NiP-DNP layers were deposited by ultrasound (kHz/W not stated), whereas the NiP layers were deposited from a silent solution. Multilayer coatings were prepared from three DNP concentrations (0.5, 1, and 4 g/dm^3), and the optimum was 1 g/dm^3. Mazaheri and coworker concluded that the NiP-DNP coating had a higher corrosion resistance and hardness than a binary NiP deposit. The corrosion resistance improved because DNP in NiP deposit acted as pore fillers, thus reducing the density of pores on the coating surface.

Ashassi-Sorkhabi and Es'haghi optimized corrosion resistance of NiP-diamond nanoparticle (DNP) MMC (Figure 5.15) by adjusting the 4 nm sized DNP concentration in ultrasonically assisted electroless Ni solution. Their data show that the optimum level of DNP in the NiP electroless electrolyte is 100 mg/dm^3, but how much of this concentration is co-deposited with NiP has not been stated. They also compared binary NiP deposited using the same ultrasonic conditions as NiP-DNP MMC (35 kHz/450 W at 358.15 K, pH 9), and both types were amorphous. All coatings prepared with DNP had I_{corr} values from Tafel analysis of polarization curves that were lower than that of binary NiP (Ashassi-Sorkhabi & Es′haghi, 2013).

Niksefat et al. rotated (300 rpm) the mild steel substrate in an electroless plating solution for NiB-TiO_2 composite coating while applying ultrasound (40 kHz/150 W). The NiB-TiO_2 coating was compared to binary NiB deposited from a silent solution with agitation by rotation (300 rpm) of substrate only, and no comparison was performed with ultrasonically deposited NiB. Binary NiB has a cauliflower-type surface morphology, and the nodules on the cauliflower-type structures were finer in NiB-TiO_2. The structure of as-plated NiB-TiO_2 was crystalline, because the nanocrystalline size calculated from XRD patterns was 55 nm. The hardness of as-plated NiB-TiO_2 (1263 HV) was higher than binary NiB (890 HV), and the friction coefficients of the two coatings increased from 0.41 (NiB) to 0.43 (NiB-TiO_2). Corrosion current density, from polarization curves, of binary NiB (I_{corr} = 1.2 μA/cm^2) was improved in ultrasonically deposited NiB-TiO_2 (I_{corr} = 0.2 μA/cm^2) suggesting an increase in the corrosion resistance (Niksefat & Ghorbani, 2015).

5.5 ELECTROLESS Ni DEPOSITION ONTO MICRO AND NANOPARTICLES IN THE PRESENCE OF ULTRASOUND

Sonication can be used to disperse micron and nanometer-sized particles in an electroless Ni electrolyte, enabling them to be coated to form core-shell particles. Electroless Ni plating of particles is performed using a similar bath composition to plating used for standard substrates, except the temperature of plating solution has been varied between ambient temperature and 323.15 K, and this is lower than the temperatures used for plating engineering components (see Table 5.2). Consequently, the thickness of the deposit on micro or nanostructures is much thinner (<1 μm).

An example of the benefits of coating particles in Ni is the study by Wu et al., in which CaF_2 (5 μm) solid lubricant powders at 318.15 K–323.15 K (900–1200 s) were coated in a NiP electroless solution (pH 9.5–10.0) agitated using ultrasound (40 kHz/150 W). The Ni-CaF_2 particles (10 vol%) were hot pressed with alumina and carbide particles to form a self-lubricating high wear cutting tool (Wu et al., 2016). A comparison of cutting tools containing bare CaF_2 or Ni-CaF_2 particles shows that the cutting tool containing Ni-CaF_2 had higher flexural strength, Vickers hardness, fracture toughness, and friction coefficient at cutting speeds (80 and 170 m/min).

Luo et al. investigated the effect of Cr_3C_2 particle (150–200 μm) concentration, in an electroless NiP plating electrolyte, on the morphology of the deposited Ni coating by ultrasound (CCW-50W generator). Particles were ultrasonically roughened, and NiP (7.32 wt%) was deposited by ultrasonication at room temperature followed by drying in vacuum (453.15 K–473.15 K). The optimal weight of Cr_3C_2 was 40 g/L based on uniformity and compactness of the NiP deposit, and concentrations higher than 40 g/L in the NiP plating solution yielded incomplete and uneven coatings. A plot of the plating rate vs. Cr_3C_2 powder concentration demonstrated a trend in which NiP rate of deposition (g/min) increased linearly with Cr_3C_2 powder (g/dm^3) concentration in the electroless plating solution, thereby reducing the reaction time (Luo et al., 2010).

Xu et al. compared electroless NiP-coated graphite microparticles by ultrasound (25 kHz/250 W) and mechanical stirring (200 rpm) from an alkaline solution (pH 12) at 313.15 K. Their results show a significant improvement in uniformity, surface roughness and compactness of deposit when ultrasound is applied. The main difference was that the NiP coating prepared by mechanical stirring had spherical grains on the surface with areas of exposed graphite (Figure 5.18a); Ni deposited by ultrasound was an amorphous thin continuous film (Figure 5.18b). Samples of the powders were weighed at different plating stages to obtain a sigmoidal curve of weight gain vs. time; from this curve, Xu et al. deduced four stages of the plating process: (1) incubation, (2) initial deposition, (3) uniform deposition, and (4) retarded deposition. In the incubation stage, nuclei of Ni seeds are deposited on the graphite surface; in the mechanically stirred Ni deposition method, graphite has less seeding of Ni nuclei on the graphite surface than ultrasonically deposited Ni. Therefore, in the mechanically stirred Ni deposition method, when Ni seeds grow over time, the

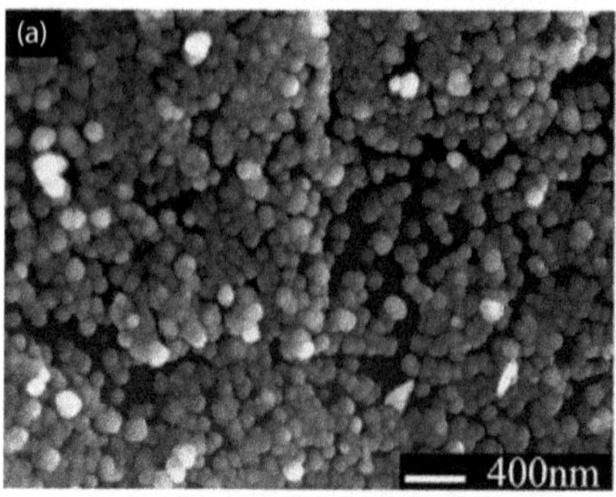

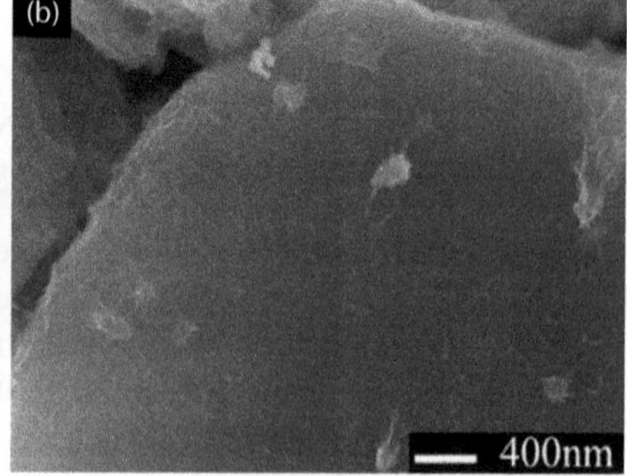

FIGURE 5.18 SEM images of NiP-coated graphite prepared at 313.15 K (pH 12) under (a) mechanical agitation (200 rpm for 900 s) and (b) ultrasonic agitation (25 kHz/250 W for 1200 s). (From Xu, X. et al., *Surf. Coat. Technol.*, 240, 425–431, 2014.)

grains do not merge leaving gaps in the coating (Xu et al., 2014). Other research studies have also described an incubation period before Ni plating proceeds, a similar mechanism of Ni film growth.

The mechanism by which Ni is deposited also depends on particle roughness and/or sub-micrometer defects on the surface. Luo et al. deposited NiP (4.59 wt%) at ambient temperature (pH 10–11) on WC particles (2–5 μm) that have a step-like surface morphology on some areas of the particle. In the early stages of plating, Ni particles seed preferentially on the step-like features (Figure 5.19a), and a continuous NiP film (Figure 5.19b) is eventually formed by further depositions on Ni seed particles (Luo et al., 2011).

Zheng et al. coated NiP on nano-Al_2O_3 particles (60 nm sized) that were ultrasonically coarsened in HF and activated by $SnCl_2/PdCl_2$ catalyst solution. The electroless Ni plating solution (pH 9) was agitated using intermittent mechanical stirring and ultrasonication (KQ-600DE generator) at 318.15 K. The particles were characterized by transmission electron microscopy (TEM) and high-resolution TEM. The nano-Al_2O_3 particles that were NiP-coated without ultrasound were agglomerated and the coating thickness varied (3–12 nm); some nano-Al_2O_3 particles had Ni nanoparticles on the NiP layer. NiP film deposited on nano-Al_2O_3 particles under ultrasonic irradiation and stirring was amorphous and had a uniform thickness of 5 nm (Zheng, Mo & Liu, 2012).

Gui et al. prepared NiP plating on soft Nylon 12 (50–70 μm size) polymer particles. Nylon 12 was functionalized with amine (NH_2) groups and activated by the deposition of $SnCl_2/PdCl_2$. The NiP was deposited by applying ultrasound (40 kHz/120 W) continuously at 333 K (1800 s) and pH of 9. Particles or shells of NiP formed on the surface rather than a continuous film when depositing NiP from a mechanically stirred or ultrasonic bath. The results of X-ray photoelectron spectroscopy analysis demonstrated that Pd adsorption to Nylon 12 was not hindered unlike the deposition of Ni^0. The loading of Pd catalyst on Nylon 12 was sufficient to enable NiP deposition on the surface. However, the method of plating was unsuccessful with/without ultrasound, and there was no explanation on how NiP adhesion to Nylon 12 was weakened (Gui et al., 2018).

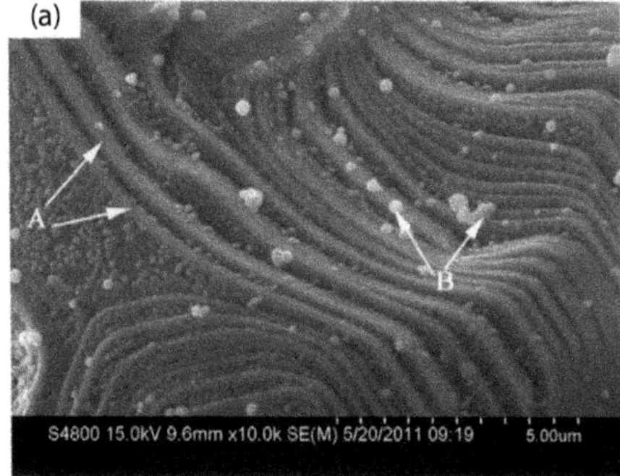

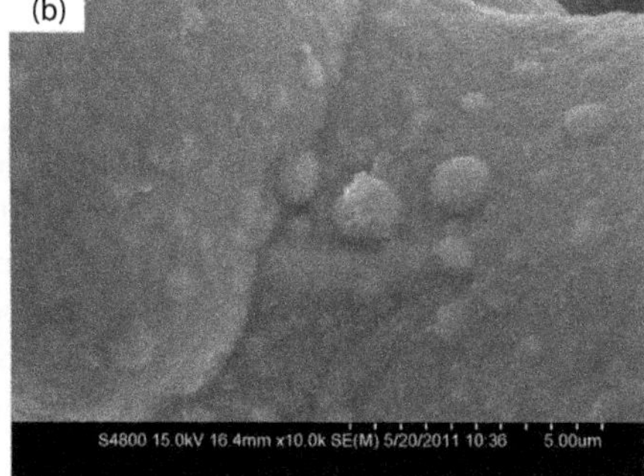

FIGURE 5.19 SEM images of (a) Ni particles deposited preferentially on step-like features and (b) NiP-coated WC powder with ultrasound (frequency/power not stated) at room temperature (pH 10–11). (From Luo, L. et al., *Surf. Coat. Technol.*, 206, 1091–1095, 2011.)

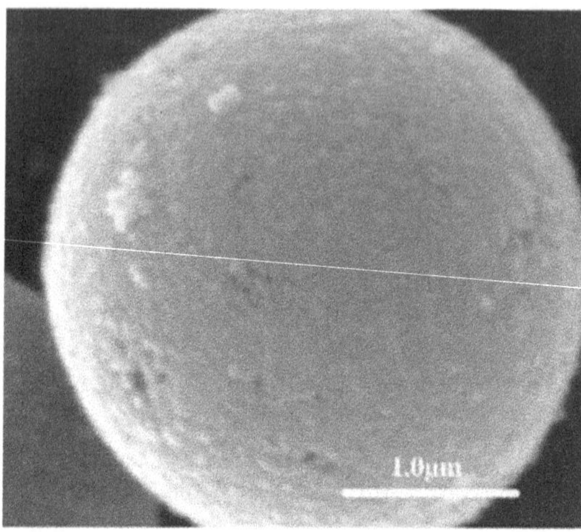

FIGURE 5.20 SEM image of an as-plated polystyrene bead in electroless NiP solution agitated by ultrasound (40 Hz/100 W). (From Jiang, J. et al., *Surf. Coat. Technol.*, 201, 7174–7179, 2007.)

Jiang et al. synthesized hollow Ni microspheres (170 nm or 3 μm) by the deposition of Ni on polystyrene (PS) beads with low-frequency ultrasonic agitation (40 Hz/100 W) followed by the removal of the PS template. The PS beads were activated using a $SnCl_2/PdCl_2$ catalyst, and the ultrasonic agitation was applied intermittently to the electroless plating solution (pH 8.5) for 10 s every 30 s at 313.15 K. PS particles NiP-coated by stirring only were agglomerated and particles of NiP deposited on the NiP coating. Ni on the PS beads coated by ultrasound consisted of a compact layer of Ni nanoparticles (Figure 5.20), and the XRD phase of as-plated NiP on PS was amorphous (Jiang et al., 2007).

5.6 THE USE OF ULTRASOUND TO IMPROVE COVERAGE OF ELECTROLESS Ni DEPOSITS ON COMPLEX STRUCTURES

Electroless deposition is advantageous when plating complex structures because generally a more uniform thickness can be achieved when compared to using an electrolytic approach. However, ultrasound can be used to enhance this uniformity still further and assist when coating complex shapes.

NiP coating (Saito et al., 2006) has been deposited on needle-shaped AFM/STM probes using 1 MHz frequency and power densities of 0, 1.6, and 2.2 Wcm^{-2} (Figure 5.21). It was established that a power density of 1.6 W/cm^2 enables the fabrication of a probe with a tip radius curvature of 25 nm (Figure 5.21c), and this radius size could not be achieved without ultrasonication. Starting from the base of the tip, the radial thickness of the Ni coating on the probe decreased gradually toward the tip. Figure 5.21d shows that Ni deposited at 1 MHz

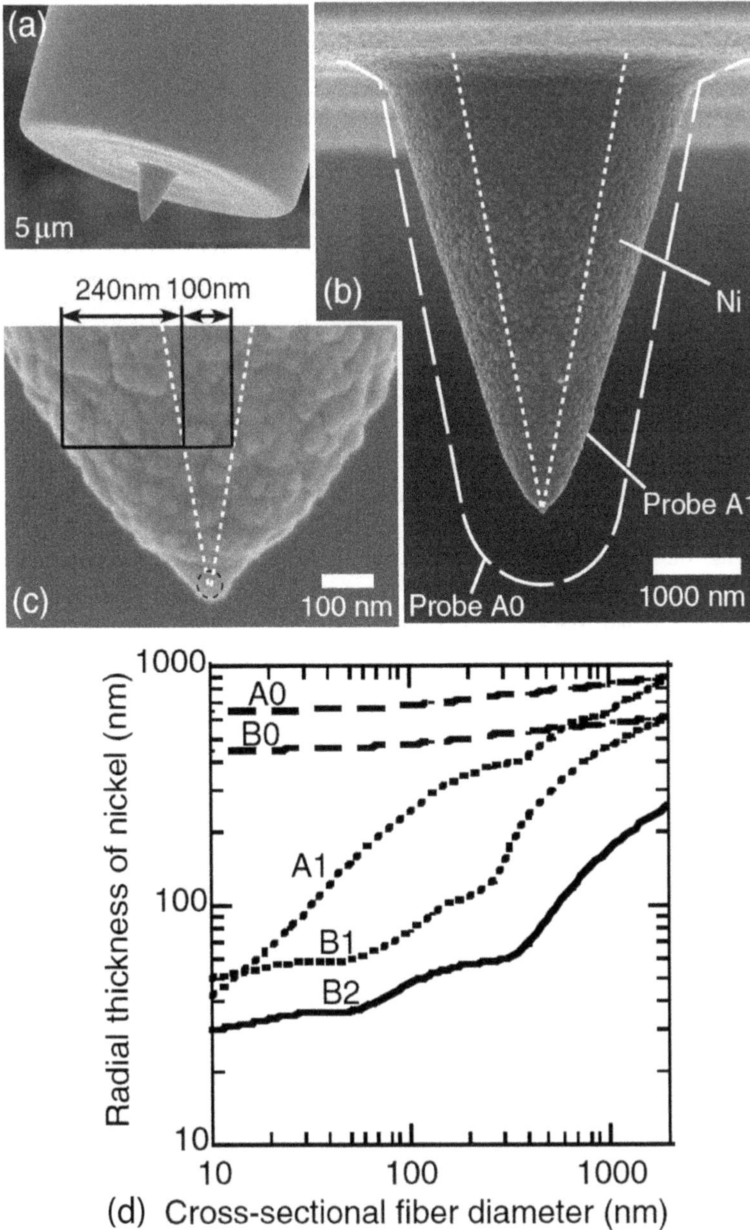

FIGURE 5.21 SEM images of (a) fabricated probe labeled as A1 and its (b) conical taper (dashed line A0 is the profile of a Ni coating deposited from a silent solution). (c) Magnified tip plated in electroless Ni solution at 1 MHz and 1.6 Wcm^{-2}. (d) Plot of radial thickness vs. fiber diameter; electroless Ni plating solution with a temperature of 333.15 K (900 s) was agitated by ultrasound using power densities of 0, 1.6, and 2.2 W/cm^2 (B0–B2, respectively). (From Saito, Y. et al., *Opt. Rev.*, 13, 225–227, 2006.)

frequency and 1.6 Wcm^{-2} power density (B1) has a profile closest to the bare probe (labeled as A1) at the tip.

Membrane structures for purifying gases can be difficult to coat because of the surface tension of the plating solution that, for this reason, cannot penetrate pores or flow out easily, and this slows the rate of deposition. Bulasara et al. conducted a detailed study of how the rate of deposition of electroless Ni is affected by ultrasonic deposition in a ceramic membrane (Bulasara et al., 2013). Four different (1:2) molar ratios of Ni^{2+} and $NaH_2PO_2{\cdot}H_2O$ were deposited on ceramic membranes (pore size 275 nm), at 243.15 K (3600 s) by ultrasound (37 kHz) or stirring (100–200 rpm). From this study, an optimal electroless Ni^{2+} solution concentration and loading ratio were derived: 0.08 M Ni^{2+} and 393 cm^2/L, which increased plating efficiency. Bulasara et al. concluded that sonication instead of stirring improves transport through membrane pores enabling improved plating efficiency, deposition rate, film thickness, and pore densification.

Chiba et al. ultrasonically deposited NiB (approximately 65 μm) on a phosphor bronze net (440 mesh) substrate from an acid solution (Chiba et al., 2003). It was found that the coating was harder than when deposited from a silent solution, but a higher density of micro-cracks was present (Figure 5.22). Plating deposited from a silent solution had $I_{corr} = 5.2 \times 10^{-6}$ mA/cm^2 and $E_{corr} = -0.502$ V; whereas ultrasonic plating had reduced corrosion resistance ($I_{corr} = 8.6 \times 10^{-6}$ mA/cm^2 and $E_{corr} = -0.647$ V). Consequently, the corrosion rate was accelerated in the ultrasonic deposit because the cracks allowed the penetration of 3% NaCl (aq). This was borne out by the Tafel analysis where NiB deposited from a silent solution had

FIGURE 5.22 SEM image showing micro-cracks in NiB coating deposited from: Ni^{2+} ion, 0.020 mol/dm^3; reducing agent, 0.034 mol/dm^3; pH = 4; deposition time, 1800 s; solution temperature, 341.15 K (45 kHz/100 W). (From Chiba, A. et al., *Surf. Coat. Technol.*, 169, 104–107, 2003.)

$I_{corr} = 5.2 \times 10^{-6}$ mA/cm^2 and $E_{corr} = -0.502$ V; whereas the ultrasonic plating had reduced corrosion resistance ($I_{corr} = 8.6 \times 10^{-6}$ mA/cm^2 and $E_{corr} = -0.647$ V).

Lu et al. coated bamboo fabric (BF) fibers with NiB to make the fabric conductive. The electroless plating process was catalyzed by reducing Ag^+, Cu^{2+}, or Ni^{2+} onto 3-aminopropyltrimethoxysilane-functionalized BF. Electroless NiB deposition on BF was accomplished at room temperature (3600 s, pH 7) by ultrasound, and all Ni deposits had excellent interfacial adhesion. Of the three catalysts, Ag^+ resulted in the smoothest Ni deposit on BF that was not cracked (Figure 5.23), but the rate of deposition was slowest. XRD patterns of the Ni-coated fabric show three crystalline peaks of face-centered cubic Ni (Lu, Liang & Xue, 2012).

Afshari and Montazer deposited pure Ni nanoparticle clusters with hedgehog-type morphology on polyester from a Ni sulfate and hydrazine electroless solution by ultrasound (50 kHz/50 W) at 348.15 K (7200 s). A central composite design method was used to identify the optimum electroless Ni solution concentrations of hydrazine hydrate: sodium hydroxide: nickel sulfate based on the weight gain of the fabric and the fastest response to a powerful magnet. XRD analysis of the fabrics shows that the Ni nanostructures are crystalline and SEM analysis shows that the average size of nanoparticles in the hedgehog cluster was 40 nm (Figure 5.24c). The Ni-hedgehog polyester fabric samples were washed in non-ionic detergent at 323.15 K for 2700 s with stirring (3000 rpm) and the change in weight and magnetic properties of the Ni-hedgehog fabric was not significant (Afshari & Montazer, 2018).

FIGURE 5.23 SEM image of Ni-plated bamboo fabric (BF) fiber coated from ultrasonic electroless Ni solution using Ag catalyst instead of $PdCl_2/SnCl_2$. (From Lu, Y. et al., *Surf. Coat. Technol.*, 206, 3639–3644, 2012.)

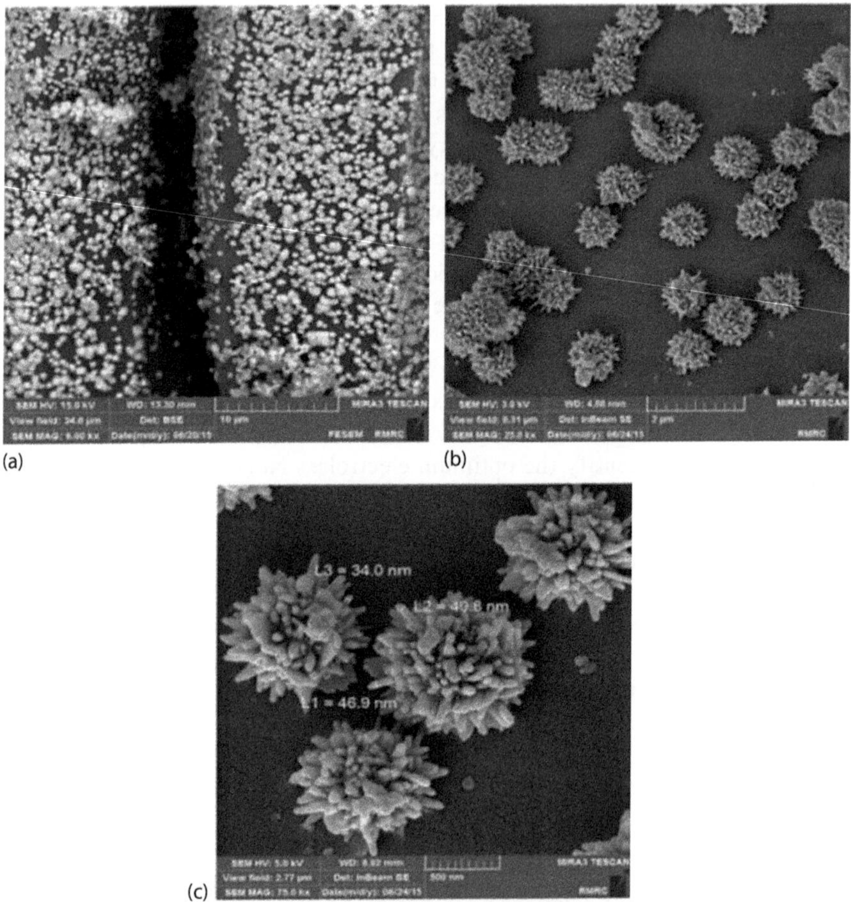

FIGURE 5.24 SEM images of Ni-hedgehog nanoparticles on fibers of polyester (polyethylene terephthalate) fabric deposited using optimum solution concentration at different magnifications scale bars: (a) 10 µm, (b) 2 µm, and (c) 500 nm. (From Afshari, S. & Montazer, M., *Ultrason. Sonochem.*, 42, 679–688, 2018.)

5.7 SUMMARY OF BENEFITS OF ULTRASOUND-ASSISTED ELECTROLESS Ni DEPOSITION

It can be seen from this chapter that many researchers have studied the effects of ultrasound on electroless Ni deposition. The sheer volume of work does throw up some contradictory results but there are some clear trends. Most research agrees that Ni can be deposited at a higher rate from an electroless bath agitated by ultrasound compared to a silent bath, because:

- Ultrasound reduces the diffusion boundary layer of ions on the electrode surface and increases mass transport of ionic species to the electrode via acoustic streaming.

- After cavities collapse, microjetting and this enhances mass transport of active species to the electrode surface.
- Heat released from cavity implosion helps to destabilize ligands on Ni^{2+} complex to free Ni^{2+} that are reduced on the electrode.

There is no set optimized method of ultrasonic deposition of Ni coatings from electroless solutions. Nevertheless, there is some agreement across all the studies that, apart from increased deposition rate, the effects of using ultrasound in electroless Ni deposition of a binary alloy are:

- Increase in thickness and compactness of deposit.
- Reduction of pores and voids in the deposit.
- Decreased P content in NiP coatings at higher plating temperatures.
- Decrease in hardness of NiP coatings with increase in P(%) content.
- Changes in phases, e.g. amorphous to crystalline.
- Enhanced corrosion resistance and hardness.

The corrosion resistance and hardness of coatings have been shown to further increase by depositing Ni coatings alloyed with other metals, e.g. Cu, Ce, Co, P, or B.

Areas where very little research has been conducted in this field include adhesion of coating, understanding wear mechanisms, and mechanisms of Ni growth.

In recent years, more studies have been published on Ni MMC coatings. The benefit of applying ultrasound is that it enables excellent dispersion of particles, resulting in higher co-deposition of reinforcing micro- or nanoparticles in a Ni MMC coating. Adhesion of the reinforcing phase to the Ni matrix and corrosion resistance of the coating have been shown to improve when using ultrasound. In addition, ultrasonic dispersion generally leads to a more uniform distribution of particles throughout the coating thickness. Sonication is also useful in the formation of core-shell particles, enabling particles to be efficiently dispersed in an electroless Ni electrolyte so that they can be uniformly coated. In addition, ultrasound can enable more uniform plating distribution on complex structures.

The use of ultrasound in electroless Ni deposition clearly has many benefits and has been and still is an active area of research. However, the vast majority of this work has been small-scale laboratory studies. The question of whether these benefits of using ultrasound can be employed at an industrial scale is still open.

REFERENCES

Abyaneh, M. Y., Sterritt, A., & Mason, T. J. (2007). Effects of ultrasonic irradiation on the kinetics of formation, structure, and hardness of electroless nickel deposits. *Journal of the Electrochemical Society, 154*(9), D472.

Afshari, S., & Montazer, M. (2018). In-Situ sonosynthesis of Hedgehog-like nickel nanoparticles on polyester fabric producing magnetic properties. *Ultrasonics Sonochemistry, 42*, 679–688.

Ashassi-Sorkhabi, H., & Es′haghi, M. (2013). Corrosion resistance enhancement of electroless Ni–P coating by incorporation of ultrasonically dispersed diamond nanoparticles. *Corrosion Science, 77*, 185–193.

ASTM Committee. (2015). ASTM B733-15 standard specification for autocatalytic (electroless) Nickel-Phosphorus coatings on metal. *ASTM International.* doi:10.1520/B0733-15.

Ban, C. L., Shao, X., & Wang, L. P. (2014). Ultrasonic irradiation assisted electroless Ni–P coating on magnesium alloy. *Surface Engineering, 30*(12), 880–885.

Bangwei, Z. (2016). Chapter 6 – Impact parameters and deposition rate. In Z. Bangwei (Ed.), *Amorphous and Nano Alloys Electroless Depositions: Technology, Composition, Structure and Theory* (pp. 323–381). St. Louis, MO: Elsevier.

Bonin, L., Bains, N., Vitry, V., & Cobley, A. J. (2017). Electroless deposition of nickel-boron coatings using low frequency ultrasonic agitation: Effect of ultrasonic frequency on the coatings. *Ultrasonics, 77*, 61–68.

Bulasara, V. K., Uppaluri, R., & Purkait, M. K. (2013). Surface engineering characteristics of ultrasound assisted hypophosphite electroless plating baths. *Surface Engineering, 29*(7), 489–494.

Chiba, A., Haijima, H., & Kobayashi, K. (2003). Effect of sonication and vibration on the electroless Ni–B deposited film from acid bath. *Surface & Coatings Technology, 169*, 104–107.

Cobley, A. J., Mason, T. J., & Saez, V. (2011). Review of effect of ultrasound on electroless plating processes. *Transactions of the IMF, 89*(6), 303–309.

Cobley, A. J., & Saez, V. (2012). The use of ultrasound to enable low temperature electroless plating. *Circuit World, 38*(1), 12–15.

Crum, L. A., & Suslick, K. S. (1995). Bubbles hotter than the Sun. *New Science Publications*, pp. 36–40. https://www.newscientist.com/article/mg14619754-300-bubbles-hotter-than-the-sun/

Fan, Y. Z., Ma, L., & Cao, X. M. (2011). Effect of ultrasonic wave on NiP-Al_2O_3 electroless composite coating on magnesium alloy. *Advanced Materials Research, 383–390*, 953–957.

Fan, Y. Z., Qiu, J., & Ma, R. N. (2014). Influence of pH on electroless NiP-γAl_2O_3 composite plating on AZ91D magnesium alloy by ultrasonic wave. *Applied Mechanics and Materials, 665*, 95–98.

Gui, C., Yao, C., Huang, J., & Yang, G. (2018). Effect of ultrasonic vibration on activation-electroless nickel plating on Nylon 12 powders. *Journal of Materials Science: Materials in Electronics, 29*(7), 5561–5565.

Hu, Y. J., Hu, G. H., Cheng, X. L., Xiao, X. T., Zhang, H. Y., Li, F., & Xu, Q. S. (2011). Study on the effect of ultrasonic waves on the low temperature electroless NiP plating on copper substrate. *Advanced Materials Research, 189–193*, 1142–1145.

Huang, X., Qi, X., Boey, F., & Zhang, H. (2012). Graphene-based composites. *Chemical Society Reviews, 41*(2), 666–686.

Jiang, J., Lu, H., Zhang, L., & Xu, N. (2007). Preparation of monodisperse Ni/PS spheres and hollow nickel spheres by ultrasonic electroless plating. *Surface and Coatings Technology, 201*(16), 7174–7179.

Kobayashi, K., Chiba, A., & Minami, N. (2000). Effects of ultrasound on both electrolytic and electroless nickel depositions. *Ultrasonics, 38*(1), 676–681.

Lu, Y., Liang, Q., & Xue, L. (2012). Electroless nickel deposition on silane modified bamboo fabric through silver, copper or nickel activation. *Surface and Coatings Technology, 206*(17), 3639–3644.

Luo, L., Wu, Y., Li, J., & Zheng, Y. (2011). Preparation of nickel-coated tungsten carbide powders by room temperature ultrasonic-assisted electroless plating. *Surface and Coatings Technology, 206*(6), 1091–1095.

Luo, L., Yu, J., Luo, J., & Li, J. (2010). Preparation and characterization of Ni-coated Cr_3C_2 powder by room temperature ultrasonic-assisted electroless plating. *Ceramics International, 36*(6), 1989–1992.

Mallory, G. O. (1978). The effects of ultrasonic irradiation on electroless nickel plating. *Transactions of the IMF, 56*(1), 81–86.

Mason, T. J., & Bernal, V. S. (2012). An introduction to sonoelectrochemistry. In B. G. Pollet (Ed.), *Power Ultrasound in Electrochemistry: From Versatile Laboratory Tool to Engineering Solution* (pp. 21–40). Chichester, UK: John Wiley & Sons.

Matsuoka, M., & Hayashi, T. (1985). Influence of ultrasonic radiation on chemical Ni-P plating. *Metal Finishing, 84*(3), 27–31.

Mazaheri, H., & Allahkaram, S. R. (2012). Deposition, characterization and electrochemical evaluation of Ni–P–nano diamond composite coatings. *Applied Surface Science, 258*(10), 4574–4580.

Niksefat, V., & Ghorbani, M. (2015). Mechanical and electrochemical properties of ultrasonic-assisted electroless deposition of Ni–B–TiO_2 composite coatings. *Journal of Alloys and Compounds, 633*, 127–136.

Okumiya, M., Tsunekawa, Y., Saida, T., & Ichino, R. (2003). Creation of high strength bonded abrasive wheel with ultrasonic aided composite plating. *Surface & Coatings Technology, 169*, 112–115.

Okumiya, M., Tsunekawa, Y., Ueno, A., Imada, Y., Ichino, R., Tamura, S., & Saida, T. (2005). Fabrication of high strength bonded abrasive wheel with ultrasonic composite plating. *Materials Transactions, 46*(9), 2047–2051.

Park, Y. S., Kim, T. H., Lee, M. H., & Kwon, S. C. (2002). Study on the effect of ultrasonic waves on the characteristics of electroless nickel deposits from an acid bath. *Surface and Coatings Technology, 153*(2), 245–251.

Qian, W., Chen, H., Feng, C., Zhu, L., Wei, H., Han, S., … Jiang, J. (2017). Microstructure and properties of the Ni–B and Ni–B–Ce ultrasonic-assisted electroless coatings. *Surface Review Letters, 25*(6), 1950006.

Saito, Y., Mononobe, S., Ohtsu, M., & Honma, H. (2006). Electroless nickel plating under continuous ultrasonic irradiation to fabricate a near-field probe whose metal coat decreases in thickness toward the tip. *Optical Review, 13*(4), 225–227.

Sharifalhoseini, Z., & Entezari, M. H. (2015). Enhancement of the corrosion protection of electroless Ni–P coating by deposition of sonosynthesized ZnO nanoparticles. *Applied Surface Science, 351*, 1060–1068.

Song, Y. W., Shan, D. Y., & Han, E. H. (2008). High corrosion resistance of electroless composite plating coatings on AZ91D magnesium alloys. *Electrochimica Acta, 53*(5), 2135–2143.

Sotskaya, N., & Dolgikh, O. (2008). Nickel electroplating from glycine containing baths with different pH. *Protection of Metals, 44*(5), 479–486.

Srinivasan, K. N., Selvaganapathy, T., Meenakshi, R., & John, S. (2011). Electroless deposition of nickel–cobalt–phosphorus nanoalloy. *Surface Engineering, 27*(1), 65–70.

Sun, H., Guo, X. F., Liu, K. G., Ma, H. F., & Feng, L. M. (2011a). Influence of ultrasonic on the microstructure and properties of electroless plating Ni-Co-P coating at low temperature. *Advanced Materials Research, 314–316*, 259–262.

Sun, H., Liu, K. G., Ma, H. F., & Feng, L. M. (2011b). Influence of ultrasonic on the property of electroless plating Ni-Cu-P alloy coating at low temperature. *Advanced Materials Research, 239–242*, 1292–1295.

Suslick, K. S., & Doktycz, S. J. (1990). Effects of ultrasound on surfaces and solids. In T. J. Mason (Ed.), *Advances in Sonochemistry* (vol. 1, pp. 197–230). Greenwich, CT: JAI Press.

Vasudevan, R., Narayanan, S., & Karthik, P. R. (1998). Some investigations on the effect of ultrasonic agitation on the properties of nickel-phosphorus and nickel-boron electroless deposits. *Transactions of the Indian Institute of Metals, 51*(5), 445–448.

Wang, H., Liu, L., Dou, Y., Zhang, W., & Jiang, W. (2013). Preparation and corrosion resistance of electroless NiP/SiC functionally gradient coatings on AZ91D magnesium alloy. *Applied Surface Science, 286*, 319–327.

Wood, R. J., Lee, J., & Bussemaker, M. J. (2017). A parametric review of sonochemistry: Control and augmentation of sonochemical activity in aqueous solutions. *Ultrasonics Sonochemistry, 38*, 351–370.

Wu, G., Xu, C., Xu, L., Xiao, G., Yi, M., & Chen, Z. (2016). Self-lubricating ceramic cutting tool material with the addition of nickel coated CaF_2 solid lubricant powders. *International Journal of Refractory Metals and Hard Materials, 56*, 51–58.

Xu, X., Cui, Z. D., Zhu, S. L., Liang, Y. Q., & Yang, X. J. (2014). Preparation of nickel-coated graphite by electroless plating under mechanical or ultrasonic agitation. *Surface and Coatings Technology, 240*, 425–431.

Yan, M., Ying, H. G., & Ma, T. Y. (2009). Preparation of coatings with high adhesion strength and high corrosion resistance on sintered Nd–Fe–B magnets through electroless plating. *Materials Chemistry and Physics, 113*(2), 764–767.

Yang, L. X., Hou, W. T., & Wu, Y. S. (1997). A study of the corrosion behavior of electroless ultrasonically deposited nickel. *WIT Transactions on Engineering Sciences, 17*, 103–111.

Yu, Q., Zhou, T., Jiang, Y., Yan, X., An, Z., Wang, X., …Ono, T. (2018). Preparation of graphene-enhanced nickel-phosphorus composite films by ultrasonic-assisted electroless plating. *Applied Surface Science, 435*, 617–625.

Zheng, H. Z., Mo, Y., & Liu, W. B. (2012). Fabrication of nano-Al_2O_3/Ni composite particle with core-shell structure by a modified electroless plating process. *Advanced Materials Research, 455–456*, 49–54.

Zhou, H. Z., Wang, W. H., Gu, Y. Q., Liu, R., & Zhao, M. L. (2016). Preparation research of Nano-SiC/NiP composite coating under a compound field. *IOP Conference Series: Materials Science and Engineering, 137*, 12066.

Zou, Y., Zhang, Z., Liu, S., Chen, D., Wang, G., Wang, Y., …Chen, Y. (2014). Ultrasonic-assisted electroless Ni-P plating on dual phase Mg-Li alloy. *Journal of the Electrochemical Society, 162*(1), C70.

6 Polyalloyed Electroless Nickel Plating

Véronique Vitry and Luiza Bonin

CONTENTS

6.1 INTRODUCTION

As mentioned in Chapter 1, several metals can be deposited by electroless plating. Table 6.1 provides a list of metals and metalloïds that can be deposited alone and co-deposited with nickel and cobalt. Of course, most of the metals that can be deposited alone can also be added to some extent to nickel. The idea of adding various alloying elements in electroless nickel comes naturally to anyone knowing the process: the presence of phosphorus or boron in the nickel coating causes significant

TABLE 6.1
List of Metals that Can Be Deposited by Electroless Plating or Co-Deposited with Nickel of Cobalt

Metals that Can Be Deposited by Electroless Plating	Metals that Can Be Co-Deposited with Co or Ni
Ni	P
Co	B
Cu	V
Cd	Mo
Pb	W
Sb	Mn
Bi	Re
Ag	Fe
Au	Zn
Pt	Tl
Pd	
Rh	
Ru	
Sn	
In	

Source: Popov, K.I. et al., *Fundamental Aspects of Electrometallurgy*, 249–270, 2005.

modification in its properties in a positive way. It is thus expected that adding other metals and metalloids to the plating bath will improve the behavior of coatings.

In this chapter, multi-alloys based on electroless nickel will be described. Ternary alloys will be the main topic and alloys based on nickel-phosphorus (NiP), nickel-boron (NiB), and nickel-phosphorus-boron (Ni-P-B) alloys will be reviewed. Some considerations on more complex alloys will also be presented.

6.2 TERNARY ELECTROLESS NICKEL ALLOYS

Adding a third alloy element to electroless nickel is considered as a very effective method to alter the chemical and physical properties of the binary (NiP and NiB) alloy deposit. It is interesting to note that even coatings that are reputedly binary may contain significant ternary alloy element due to the co-deposition of the stabilizing agent. For example, Delaunois et al. (2000) reported up to 5 wt.% Tl in their electroless NiB coating. However, as co-deposited metal from the stabilizing agent is not added voluntarily and it is not possible (or very difficult) to study a coating exempt of it and to modify its concentration, it is generally accepted that this does not constitute a ternary alloy.

Ternary electroless nickel alloys can thus be defined as alloys in which a third alloying element (other than phosphorus or boron) can be added voluntarily in (relatively) controllable amounts to an electroless nickel coating.

In this section, ternary electroless coatings will be examined according to the initial coating type they are based on: electroless NiP and electroless NiB, as these coatings usually maintain properties closer to the binary alloy with the same element. A specific section (Section 6.2.3) will be dedicated to electroless Ni-P-B, as it combines the two usually alloying elements.

6.2.1 Ni-Me-P Alloy

As electroless NiP is the most popular electroless nickel alloy, NiP-based ternary alloys are the most frequent and documented among electroless nickel ternary coatings. Table 6.2 lists NiP-based ternary alloys that were deposited on various ferrous alloy substrates to improve surface performance and some properties (pH) of the plating solution used to synthesize them (Zhang et al. 2014). It is clearly evident that there are a great variety of ternary electroless NiP alloys that have been investigated. This list is not exhaustive as some alloys are investigated for some very specific properties such as magnetic properties. Pearlstein (1990) also reports other ternary alloys such as Ni-Re-P and Ni-Pd-P but they were not fully investigated and their description is thus of relatively limited interest.

The incorporation of allying elements usually has beneficial effect on hardness, friction, wear, and/or corrosion properties. The following section will investigate these effects for various alloying elements.

6.2.1.1 Ni-Fe-P

Iron-nickel alloys are known to be used in various fields due to their excellent magnetic properties. Electroless plating of Ni-Fe-P films can be used to form thick, hard, uniform, and corrosion-resistant magnetic films on various substrates. Electroless

TABLE 6.2
Nickel-Phosphorous-Based Electroless Alloy Coatings, with the Substrates on which they were Deposited and Plating Conditions

Coatings	Substrate	Solution Properties (pH)
Ni-W-P	Mild steel sheets	9.5
Ni-Cu-P	Low carbon steel	4.7
Ni-Cu-P	Foam samples	9.5
Ni-Mo-P	Duplex stainless steel	4.6–8
Ni-W-P	Mild steel, tool steel, stainless steel	9.0
Ni-Cr-P	Steel	4.0–5.0
Ni-Fe-P, Ni-Fe-P-B	Carbon steel	11
Ni-P-Re (Ce, La)	Mild steel coupon	4.5
Ni-P-La	Q235 carbon steel	4.5–5.0
Ni-Ce-P	Mild steel	10
Ni-P-Re (La, Pr, Nd)	Q235 carbon steel	2.5–4.5

Source: Zhang, H. et al., *Surf. Rev. Lett.*, 21, 1430002, 2014.

Ni-Fe-P was first reported in 2000 (Zhao et al. 2000) and has been the subject of some studies (Huang et al. 2007a, 2008; Wang 2004; Zhao et al. 2000).

These coatings are obtained by replacing of the nickel salts (usually nickel sulfate) by iron salts (sulfate as well) in the electroless plating bath. The baths that were used were stabilized and complexed by sodium or ammonium citrate, potassium sodium tartrate, boric acid, and/or sucrose. Iron content from 2 to nearly 40 at.% has been reported for the ternary alloy coatings. Wang (2004), for a bath containing sucrose, boric acid, and potassium sodium tartrate, observed that the presence of ferrous sulfate in the plating bath had detrimental (inhibitory) effects on the deposition process that led to the iron content of the deposit being limited to 15.62 at.%. He also observed that alkalinization of the plating bath increased co-deposition of iron in the coating.

Electroless Ni-Fe-P deposits are usually amorphous (Huang et al. 2007a; Wang 2004; Zhao et al. 2000). However, this depends on the P content that is notably influenced by the complexing agent. Huang et al. (2008) observed that, with increasing [NHCit]/[NaCit + NHCit] ratio, the P content increased, while the Fe content decreased, resulting in the increased amorphous character of the coatings.

Wang (2004) investigated the hardness of electroless Ni-10.7Fe-20.3P deposits (at.%). He observed hardness values of as-plated deposits close to 800 HV that decreased slightly with heat treatment up to 350°C but increased significantly (up to nearly 2500 HV) for a treatment at 400°C–500°C, before decreasing again. He attributed this to crystallization of the coating: Ni_3P was formed at 377°C (beginning of the increased hardness zone) while cubic $FeNi_3$ phase appeared at 491°C (end of the increased hardness zone), see Figure 6.1.

The corrosion resistance of the Ni-Fe-P coatings was investigated by some researchers. They observed that, when the P content was kept relatively constant (in that case approximately 3 wt.%), the corrosion resistance of Ni-Fe-P increased with

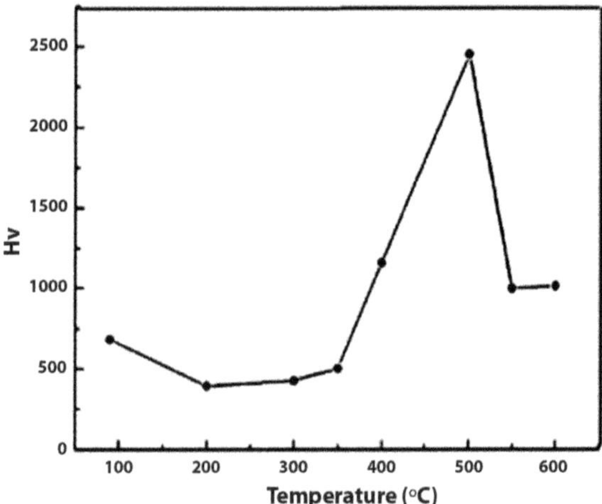

FIGURE 6.1 Evolution of hardness of a Ni-10.7Fe-20.3P deposit (at.%) with heat treatment temperature. (From Wang, S.-L., *Surf. Coat. Technol.*, 186, 372–376, 2004.)

Fe content (Zhao et al. 2000). Huang et al. (2007a), in a different bath containing both sodium and ammonium citrate, observed that the corrosion resistance of the alloy increased with the [NHCit]/[NaCit + NHCit] ratio, which leads to decrease in iron and increase in phosphorus content in the alloy. This is not contradictory with the other results, as it is known that phosphorus content has a significant effect on corrosion resistance. They also observed a better corrosion resistance in alkaline medium (NaOH) than in acidic medium (HCl), which is not unexpected as nickel and its alloys easily passivate in alkaline environment.

6.2.1.2 Ni-Co-P

The properties of cobalt are very close to those of nickel, and it is also a ferromagnetic element. Moreover, the element can be deposited by electroless plating in solutions very similar to nickel. For these reasons, cobalt is one of the most popular addition elements in electroless nickel coatings and has been the object of many studies (Armyanov and Sotirova 1989; Huang et al. 2007b; Kim et al. 1995; Kumar et al. 2012; Matsubara and Yamada 1994; Pang et al. 2011; Sankara Narayanan et al. 2003; Toda et al. 2013). Moreover, like Ni-Fe-P, the magnetic properties of electroless Ni-Co-P are of great interest, and they find applications mostly in thin magnetic recording media (Armyanov and Sotirova 1989; Kim et al. 1995; Matsubara and Yamada 1994) and as microwave absorber due to their high saturation magnetization, high permeability, and low coercive force (Pang et al. 2011).

As nickel and cobalt have similar electrochemical behaviors with standard oxidoreduction potential of −0.28 V vs. SHE for cobalt and −0.26 V for nickel, it is possible to modify the composition of the coating easily by modifying the cobalt and nickel salts content of the plating bath.

Like all electroless coatings, the properties of electroless Ni-Co-P alloys vary with thickness, chemical composition, and microstructure of the coating, the latter being directly influenced by the other two and operating conditions of the plating process (Toda et al. 2013).

The plating solutions used for the deposition of ternary Ni-Co-P coatings are based on sulfate metallic salts and contain usually sodium tricitrate. Some of them also contain sodium tartrate and/or ammonium sulfate or chloride. The amount of reducing agents in those plating baths is in the 10–60 g/L range and their pH is in the 6–10 range (Huang et al. 2007b; Kim et al. 1995; Kumar et al. 2012; Pang et al. 2011; Sankara Narayanan et al. 2003; Toda et al. 2013). The solutions operate in a wide temperature range (from 55°C to 90°C approximately), and some of them were used on polymers at even lower temperatures (30°C) (Matsuda et al. 1993).

The ratio of nickel to cobalt salts in the plating solution is one of the key parameters to tune the chemical composition of the coatings. This ratio and the associated concentrations vary widely between research groups. As a result, coatings with a large amount of nickel, cobalt and phosphorous content were synthesized. Logically, the plating rate and phosphorous content of the coatings are reported to increase with the amount of hypophosphite present in the plating bath (Pang et al. 2011). Likewise, the nickel and cobalt contents vary according to the ratio of their cations in the solution (which can be expressed as Ni^{2+}/Co^{2+} or $Co^{2+}/Ni^{2+} + Co^{2+}$) (Kim et al. 1995; Pang et al. 2011; Sankara Narayanan et al. 2003),

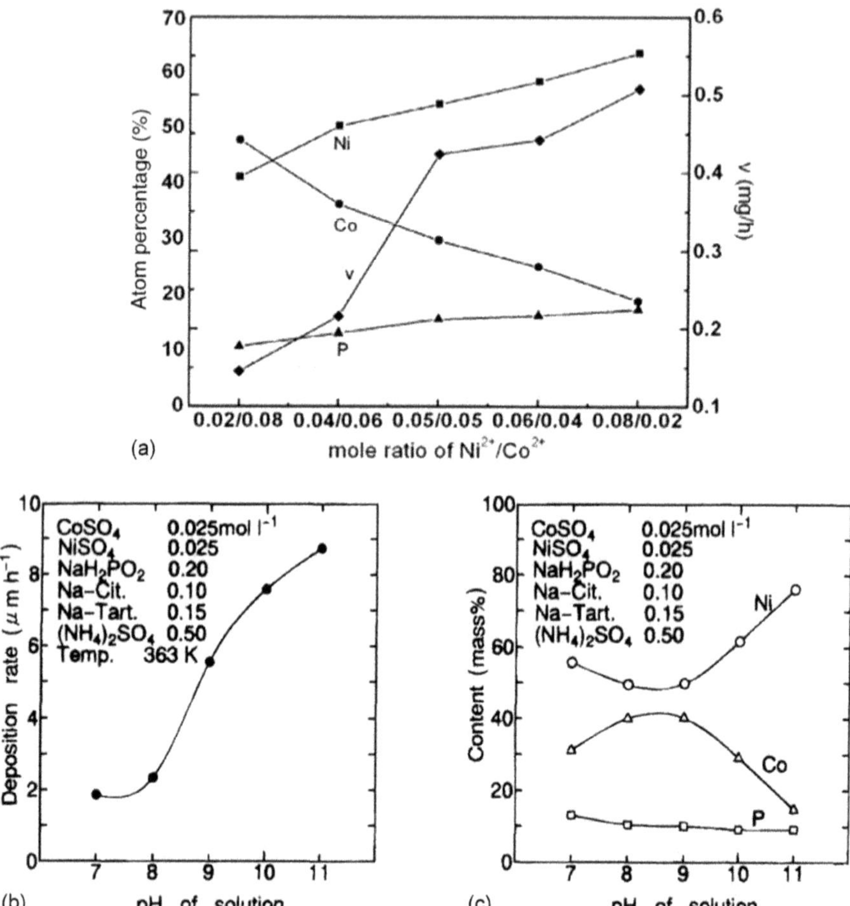

FIGURE 6.2 Effect of Ni^{2+}/Co^{2+} ratio (a) and pH (b) and (c) on plating rate and composition of electroless Ni-Co-P coating. (Reprinted from *Surf. Coat. Tech.*, 205, Pang, J. et al., Surface & coatings technology preparation and characterization of electroless Ni-Co-P ternary alloy on fly ash cenospheres, 4237–4242, Copyright 2011, with permission from Elsevier; Kim, D.-H. et al., *J. Electrochem. Soc.*, 142, 3763–3767, 1995.)

as shown in Figure 6.2a. The plating rate also increases when nickel is the main metallic salt in the plating bath and reaches nearly 10 μm/h for baths containing only nickel against 2–3 μm/h when only cobalt is present.

The pH of the solution also influences plating rate and, like for every type of electroless plating, increases with increasing pH. The increase in pH is also accompanied by a modification in the composition of coatings, with a decrease in phosphorous content and a modification in the nickel and cobalt concentration, with a maximized cobalt content in the 8–9 range, see Figure 6.2b and c. The effect of pH on the Co and Ni contents is explained by the difference in the stability of nickel and cobalt complexes present in the solution (Kim et al. 1995; Kumar et al. 2012; Pang et al. 2011).

A few authors discussed the morphology of electroless Ni-Co-P coatings. However, those were reported to be columnar morphologies with granular surface textures that became smoother as phosphorous content increased (Huang et al. 2007b; Itakura et al. 1999). Studies on the structure of the coatings are however more frequent, with several authors using transmission electron microscopy and electron diffraction to assess the crystallinity of coatings (Armyanov and Sotirova 1989; Kim et al. 1995; Matsuda et al. 1993; Toda et al. 2013). Most authors have observed that the coatings are amorphous in the as-plated state (Huang et al. 2007b; Pang et al. 2011; Sankara Narayanan et al. 2003) but they are crystallized when subjected to heat treatment (Pang et al. 2011; Sankara Narayanan et al. 2003). Sankara Narayana et al. (2003) investigated the crystallization temperature of the coatings by differential scanning calorimetry and observed three exothermic peaks that are attributed to strain relaxation during the phase separation, formation of nickel and nickel phosphides from the amorphous phase, and the decomposition of metastable phases in stable nickel phosphide. All these peaks are observed between 250°C and 400°C.

Because Ni-Co-P coatings are mostly used for their excellent magnetic properties, there is no report in the literature of mechanical testing or corrosion resistance of these materials. However, there are many studies that investigated their electrical and magnetic properties and the influence of Ni, Co, and P content and heat treatment on these properties. Electroless Ni-Co-P exhibit soft magnetic properties at room temperature, which are improved by increasing the cobalt content and heat treatment (Pang et al. 2011). Pang et al. (2011) reported values of saturation magnetization from 5.52 to 14.6 emu/g in the as-plated state depending on the Ni^{2+}/Co^{2+} ratio and an increase from 6 to 10.7 emu/g with heat treating at 450°C, with a similar behavior of remanence, and values of 416–296 kA/m for the coercivity with variation of nickel/cobalt ratio and from 336 to 426 kA/m for heat treatment in similar conditions. Sankara Narayanan reported similar values of magnetization (Sankara Narayanan et al. 2003) and measured the Curie temperature that was found to be in the 315°C–330°C range for the first Curie transition and between 525 and 630°C for the second transition. Matsuda reported lower values of coercivity (approximately 120 kA/m, Matsuda et al. 1993) for coatings that were deposited at room temperature and investigated the effect of pH and the $Co^{2+}/Ni^{2+} + Co^{2+}$ ratio on coercivity (see Figure 6.3). They obtained the highest coercivity for a pH of 9 and a $Co^{2+}/Ni^{2+}+Co^{2+}$ ratio of 0.8, i.e. for an optimal cobalt content of the coating.

Kumar et al. (2012) studied the diffusion barrier properties of Ni-Co-P alloys by heat treating Si/Ni-Co-P/Cu assemblies and observed that the film with the lowest Co content (around 11 wt.%) presented the best diffusion barrier properties at high temperature and that its sheet resistance only started to increase at a temperature of approximately 500°C.

6.2.1.3 Ni-Cu-P

Copper can be added to electroless nickel plating baths but only in limited amounts due to the large difference in their reduction potential (Copper being more noble than nickel). Tarozaitė and Selskis (2006) carried out an in-depth investigation of the effect of copper and carboxylic acid additions on electroless NiP plating and

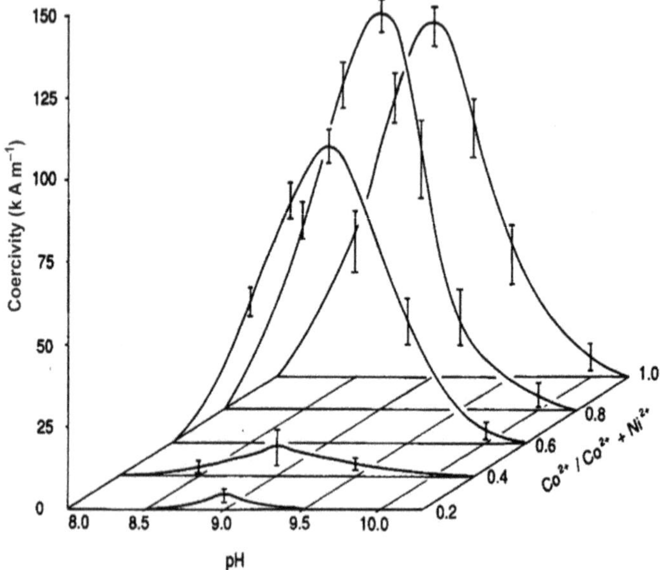

FIGURE 6.3 Effect of $Co^{2+}/Ni^{2+}+Co^{2+}$ ratio and pH on coercivity of Ni-Co-P alloys deposited at room temperature. (From Matsuda, H. et al., *J. Magn. Magn. Mater.*, 120, 338–341, 1993.)

observed that copper showed a stabilizer-like behavior: addition of small amounts led to an increase in plating rate of approximately 7%–15% and co-deposition of limited amounts of copper (approximately 1 wt.% in the coating).

For these reasons, additions of copper should always be in a limited range of composition. Copper is reported to improve smoothness (Balaraju and Rajam 2005a; Tarozaitė and Selskis 2006), brightness (Chen et al. 2006a; Tarozaitė and Selskis 2006) and corrosion resistance (Abdel Hameed and Fekry 2010; Liu and Zhao 2004; Zhao et al. 2004) of the coatings. It's also known to improve the thermal stability of the NiP alloy, i.e. not only to preserve their amorphous state and paramagnetic behavior after heat treatment (Krasteva et al. 1994) but also to decrease the magnetic behavior of the crystalline phases that form during heat treatment (Georgieva and Armyanov 2007). Finally, Ni-Cu-P electroless formed alloys have been used as catalysts for the electrocatalytic oxidation of methanol (Hameed and El-khatib 2010).

6.2.1.4 Ni-P-Mo

Molybdenum is a refractory metal. It has thus been added to electroless NiP with the aim to increase thermal stability. This addition is usually made by the use of sodium molybdate as a molybdenum source in alkaline plating baths (Bai et al. 2008; Mendoza et al. 2006; Song et al. 2017). The redox potential of molybdenum limits the amount of metal that can be incorporated: values of 3–6.5 wt.% have been reported (Bai et al. 2008; Liu et al. 2016) and Mo content in the coating increases with the pH of the bath. However, this increase is accompanied by a strong decrease in P content (Bai et al. 2008).

Electroless Ni-Mo-P is described as having a compact and pore-free morphology with cells larger than those observed on NiP (Song et al. 2017) but those coatings are shown to be quite rougher than coating exempt of molybdenum (Liu et al. 2016).

The structure of electroless Ni-Mo-P and the way it is influenced by heat treatment has been investigated by several groups. They generally observe that the coatings are amorphous (Bai et al. 2008; Liu et al. 2016; Mendoza et al. 2006) in the as-plated state and tend to crystallize in Ni_3P, $MoNi_4$, and Ni when heat-treated (Bai et al. 2008; Liu et al. 2016; Mendoza et al. 2006). However, some authors have observed crystalline structures in as-plated coatings but these coatings had very low P content (approximately 1 wt.%) that were not sufficient to distort the nickel crystal lattice to the point of amorphization (Song et al. 2017).

The hardness of Ni-Mo-P was not much investigated but a range of values from 600 (Mendoza et al. 2006) to 1100 HV (Song et al. 2017) has been reported for as-plated coatings. Heat treatment seems to be very beneficial to those coating because Mendoza et al. (2006) reported values as high as 1450 HV_{100} after heat treatment at 500°C for 2 h. Song et al. (2017) also measured the elastic modulus of the coating and obtained a value of 211 GPa, against 150 GPa for electroless NiP.

The corrosion resistance of electroless Ni-Mo-P coatings appears to be relatively similar to that of NiP, without any passive behavior (Bai et al. 2008; Liu et al. 2016).

6.2.1.5 Ni-Sn-P

Ternary Ni-Sn-P electroless coatings have been produced by additions of small amounts of sodium stannate in electroless nickel plating baths (Shimauchi et al. 1994; Zou et al. 2010). Due to the difference in redox potential between tin and nickel, tin, like copper, is easily co-deposited and the tin content in the coatings can increase up to 30 at.% (Shimauchi et al. 1994). Haowen et al. (1999) carried out a parametric study of the plating process that showed that the plating rate and tin content could be tuned by modification of the plating parameter. They also obtained a tin content up to 30 at.%.

In terms of morphology, electroless Ni-Sn-P coatings exhibit a cauliflower-like surface up to at least 20 at.% (Shimauchi et al. 1994; Zou et al. 2010) but coatings with Sn content over 30 at.% exhibit a wrinkle-like surface (Shimauchi et al. 1994).

Most authors report that electroless Ni-Sn-P coatings are amorphous (Haowen et al. 1999; Zou et al. 2010); however, Shimauchi reported that crystallinity of the coatings increased with tin content (Shimauchi et al. 1994). This is probably due to a strong decrease in the phosphorous content of coatings when tin content increased. However, crystallization temperature decreased for Ni-Sn-P coatings compared to NiP.

The hardness of electroless Ni-Sn-P was slightly lower than that of pure NiP (approximately 700 HV_{100}) but the main difference results from the effect of heat treatment that is much less significant for Ni-Sn-P coatings, with a value of 800 HV_{100} reached for heat treatments at 300°C (Georgieva and Armyanov 2007; Zou et al. 2010). Solderability was reported to be improved in electroless Ni-Sn-P coatings (Shimauchi et al. 1994).

The most interesting feature of electroless Ni-Sn-P coatings is their corrosion resistance that has been unanimously reported as being improved compared to NiP

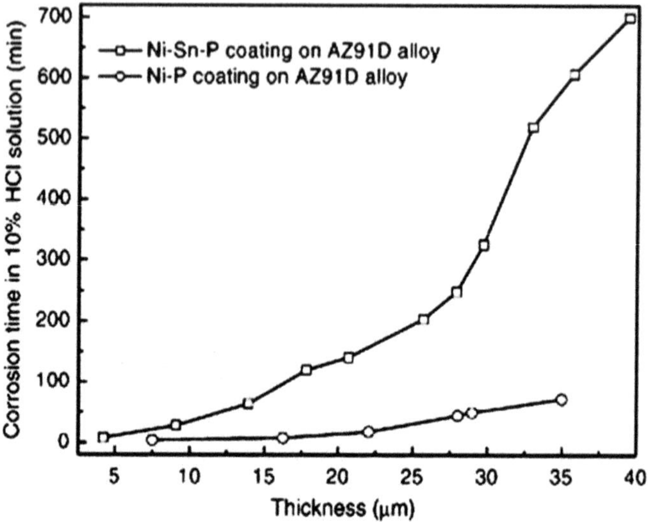

FIGURE 6.4 Time duration between immersion in 10 vol.% HCl and the formation of the first gas bubble as a function of coating thickness for NiP and Ni-Sn-P. (Reprinted from *Surf. Coat. Technol.*, 202, Zhang, W.X. et al., Electroless Ni–Sn–P coating on AZ91D magnesium alloy and its corrosion resistance, 2570–2576, Copyright 2008, with permission from Elsevier.)

coatings (Haowen et al. 1999; Mallory 1974; Mallory and Horn 1979; Shimauchi et al. 1994; Zhang et al. 2008; Zou et al. 2010) and is clearly shown in Figure 6.4. The corrosion potential and corrosion current recorded on Ni-Sn-P are improved compared to NiP, even for low Sn content (0.55 wt.%) (Georgieva and Armyanov 2007). Some authors attribute this to a decrease of porosity of the coating, even for very limited thickness (Zhang et al. 2008), others attribute this to the increase in the amorphous character of the Ni-Sn-P coatings (Haowen et al. 1999).

6.2.1.6 Ni-Zn-P

The incorporation of zinc in electroless NiP has been studied by several researchers (Bouanani et al. 1999a, 1999b; Oulladj et al. 1999; Ranganatha et al. 2010; Schlesinger and Meng 1990; Tai et al. 2009; Valova et al. 2001b; Veeraraghavan et al. 2004), mostly with the aim of increasing thermal stability. Zinc is usually introduced in the plating bath by the way of zinc sulfate (Bouanani et al. 1999a, 1999b; Ranganatha et al. 2010; Tai et al. 2009; Veeraraghavan et al. 2004) or zinc chloride (Schlesinger and Meng 1990). Bouanani et al. (1999a) carried out a parametric study of electroless Ni-P-Zn plating and observed that, while temperature had limited influence on the Zn and P content, modifications of pH lead to significant changes in coating composition, with a continuous decrease in P content for increasing pH and a maximal Zn concentration for a pH of 10. The increase in the amount of Zn salts in the bath led logically to an increase in Zn in the coating and decrease in P content. They observed that the presence of zinc in the plating bath had an inhibiting effect on plating, leading to lower deposition rates.

Veeraraghavan et al. (2004) observed a similar inhibiting effect that they attribute to modification of the surface coverage of the adsorbed electroactive species on the electrode surface.

The zinc content of electroless Ni-Zn-P stays usually lower than 12 wt.% (Bouanani et al. 1999b; Schlesinger and Meng 1990; Tai et al. 2009) and the obtained coatings present a cauliflower-like aspect. Their structure is a mixture of amorphous and nanocrystalline phases, with grain sizes ranging from 65 Å for coatings with the highest P content and lowest Zn content to 5000 Å for coatings with low P (Bouanani et al. 1999b). Zn and P are preferentially concentrated along grain boundaries in the as-plated film (Valova et al. 2001b).

Heat treatment of the coatings increases the crystallinity (Ranganatha et al. 2010), but it is more difficult to crystallize electroless Ni-Zn-P than NiP, which indicates its superior thermal stability and its suitability to be used in under-bump metallization processes (Tai et al. 2009).

The corrosion properties of electroless Ni-Zn-P are similar, if not better, to those of electroless NiP with a comparable P content (Schlesinger and Meng 1990).

6.2.1.7 Ni-W-P

Like Molybdenum, Tungsten is a refractory and difficult to reduce metal that is added to electroless NiP coatings to increase their thermal stability. It is added in the electroless plating bath as sodium tungstate to make its incorporation easier (Abdullah et al. 2009; Chen et al. 2006b, 2008; Du and Pritzker 2003; Liu et al. 2010) While additions of molybdenum present inhibiting properties, additions of tungsten to the plating bath increase the plating rate (Lu and Zangari 2003). The most significant plating parameter to adjust the W and P content of the coating is pH: higher pH leads to a decrease in P content and an increase in W content and plating rate (Chen et al. 2006b). A wide range of tungsten content has been reported, from 0.5 to 20 wt.% (Koiwa et al. 1988). The morphology of electroless Ni-W-P coatings is cauliflower-like with a grainy sub-texture (Liu et al. 2010) and the coatings are pore-free even for relatively limited thickness (Chen et al. 2008).

The presence of tungsten modifies the structure of the coating by making the formation of crystalline nickel during the plating process more difficult (Chen et al. 2008), and the addition of increasing amounts of tungsten maintain an amorphous state of the coating, even for lowering P content (Koiwa et al. 1988). High-resolution studies showed that P and W were concentrated at the grain boundaries of the material in the as-plated state (Valova et al. 2004). Like all amorphous electroless coatings, Ni-W-P crystallizes upon heat treatment. However, its thermal stability is greater than that of NiP and complete crystallization is not obtained before 500°C–600°C (against 350°C–400°C for NiP) and even at such temperatures, the Ni and Ni_3P grains never reach 100 nm (Liu et al. 2010). Koiwa et al. (1988) attribute the better thermal stability to the inhibition of Ni crystallization before the formation of Ni_3P. The optimized coatings (for composition and thickness) were able to sustain the contact with molten Sn-Bi solder for more than one month (Chen et al. 2006b).

The hardness of electroless NiP coatings is increased by additions of tungsten (Chen et al. 2008; Du and Pritzker 2003; Ranganatha et al. 2012). Ni-W-P is also reported by most authors as having a better corrosion resistance than NiP

(Chen et al. 2008; Ranganatha et al. 2012) that increases with W content; however, early work on the ternary alloy did not report improvement (Bangwei et al. 1996). Ni-W-P coatings with a thickness of 18 μm were able to sustain immersion in 10% HCl for nearly 3 h without showing any sign of corrosion (Chen et al. 2008).

6.2.1.8 Ni-Pt-P

Aoki et al. (1995) added platinum to an electroless plating bath in the form of $H_2PtCl_6{\cdot}2H_2O$ and were able to synthesize alloys with a Pt content from 7.8 to 11 wt.% that were fully crystalline. The addition of Pt to the plating bath was limited because the plating rate decreased strongly as soon as the concentration of $H_2PtCl_6{\cdot}2H_2O$ exceeded 0.25 g/L.

6.2.1.9 Ni-Cr-P

Very few authors have studied Ni-Cr-P electroless coatings (Shashikala et al. 2007; Wu and Lou 2011). The plating bath used for this contained chromium in the $CrCl_3$ form. Electroless Ni-Cr-P coatings presented a bright surface and were free of porosity. Hardness values of 707 HV_5 in the as-plated state and 810 HV_5 after heat treatment were reported, as well as increased corrosion resistance and thermal stability.

6.2.2 Ni-Me-B Alloy

Electroless NiB coatings are not studied as much as their phosphorus-based counterparts but several research groups have carried out an investigation of the effect of alloying on their properties. The elements that were added to NiB alloys are elements that were also added to electroless NiP alloys: Co, Cu, Fe, Mo, Sn, and W.

6.2.2.1 Ni-Co-B

While the addition of cobalt in electroless NiP is unanimously carried out to improve magnetic properties of the coating, the same cannot be said of its addition in NiB coating. Several authors have synthesized electroless Ni-Co-B materials, mostly as coatings (Campillo et al. 2002; Gamboa et al. 2006; Luo et al. 2001; Onoda et al. 1993; Saito et al. 1998; Sankara Narayanan et al. 2004; Ueda et al. 2004; Wang 2007; Wang et al. 1995; Wei et al. 2017; Yu et al. 1997; Xie et al. 2008), but some as powder in view of investigating the catalytic properties of the ternary alloy (Li et al. 2013; Yamauchi et al. 2004).

Most authors used borohydride-based reducing agents, but Saito et al. (1998) investigated the formation of electroless Ni-Co-B with dimethylamine borane, using cobalt sulfate in substitution of nickel sulfate in their plating bath. They observed that cobalt was preferentially deposited to nickel and that nickel concentration had to be nine times that of cobalt to form nickel-rich coatings. As cobalt presents a slightly more negative reduction potential than nickel, they concluded that anomalous deposition was occurring due to a low overvoltage for the reduction of cobalt. However, they also observed that nickel-based systems had lower deposition rates due to the high catalytic activity of NiB for the oxidation of dimethylamine borane.

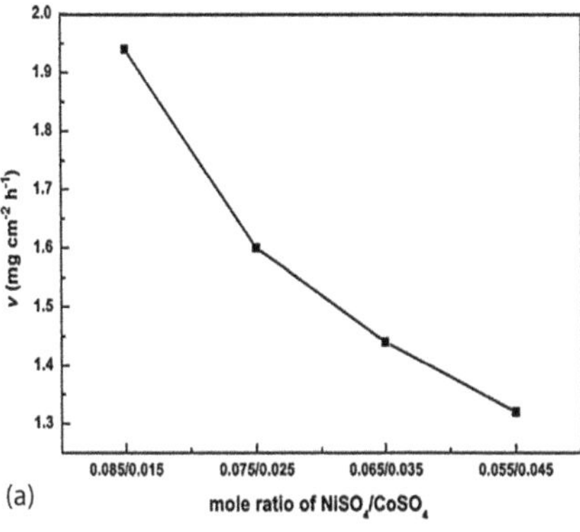

(a)

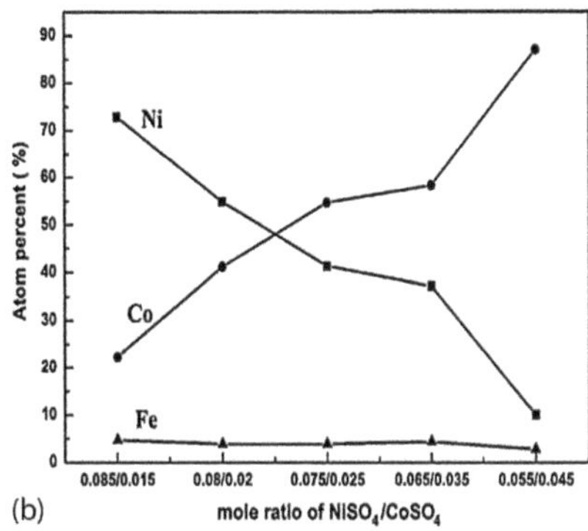

(b)

FIGURE 6.5 Effect of the mole ratio of nickel and cobalt sulfate on the plating rate (a) and the chemical composition (b) of electroless Ni-Co-B coatings. (Reprinted from *Thin Solid Films*, 515, Wang, S.-L., Electroless deposition of Ni–Co–B alloy films and influence of heat treatment on the structure and the magnetic performances of the film, 8419–8423, Copyright 2007, with permission from Elsevier.)

Wang (2007) and Sankara Narayanan et al. (2004) observed a similar effect for an electroless Ni-Co-B on borohydride, as shown in Figure 6.5. Sankara Narayana reported a Ni content of 24% for a bath with a $CoSO_4/NiSO_4 + CoSO_4$ ratio of 0.3, which shows the amplitude of the preferential deposition of cobalt.

In terms of properties, a few authors investigated hardness, and the reported values show that Ni-Co-B coatings are not harder than electroless NiB (Wang 2007). However, the efficiency of electroless Ni-Co-B coatings for corrosion protection has been reported (Gamboa et al. 2006; Saito et al. 1998), as well as their excellent catalytic activity (which is considered as even better than that of Nickel Mond) (Luo et al. 2001; Sankara Narayanan et al. 2004; Wang et al. 1995; Wei et al. 2017; Xie et al. 2008; Yu et al. 1997).

The thermal stability of electroless Ni-Co-B has also been investigated. They are reported as amorphous in the as-plated state (Wang 2004; Sankara Narayanan et al. 2004) and to have superior thermal stability compared to Ni-B but to crystallize at approximately 350°C, thus leading to an increase in hardness due to the formation of boride phases (Wang 2007).

The most significant feature about this type of material is their magnetic properties that are excellent in the as-plated condition and are improved by heat treating (Li et al. 2013; Onoda et al. 1993; Sankara Narayanan et al. 2004; Ueda et al. 2004; Wang 2007; Yamauchi et al. 2004).

6.2.2.2 Ni-Fe-B

There are not many reports on electroless Ni-Fe-B deposition, and they are mostly focused on magnetic (Richardson et al. 2015; Takai et al. 1995) and catalytic (Nie et al. 2012) applications.

Richardson et al. (2015) investigated the optimal conditions for the formation of Ni-Fe-B nanotubes by template electroless method from a bath using dimethylamine borane as reducing agent. They found that operating at high pH (9, just below the destabilization point) allowed to obtain a B content of 26 at.%, regardless of iron and nickel concentration in the plating bath. Using this method, they formed nanocrystalline nanotubes. These nanotubes presented soft magnetic properties and an increase in their Fe content led to a decrease in coercivity. Takai et al. (1995) also investigated magnetic properties and reported that the NiFeB film with 27 at.% Fe had the lowest coercivity (0.5 Oe) with a saturation magnetic flux density of 1.0 T.

Nie et al. (2012) investigated the catalytic properties of the films and obtained optimal results for a Fe/(Fe + Ni) ratio of 0.3.

6.2.2.3 Ni-Cr-B

There are few reports of Ni-Cr-B electroless nickel deposition, and they are not investigated for any of the usual applications of electroless NiB (hardness, wear, and catalysis).

Wang and coworkers (Wang et al. 2017) have synthesized electroless Ni-Cr-B coatings to be used as diffusion barrier and anti-corrosion layer on copper interconnections for the electronics industry. They used dimethylamine borane and added $CrCl_3$ to the plating bath to incorporate chromium in the coating. They obtained a material showing excellent thermal stability that could prevent copper diffusion up to 900°C.

On the other hand, Hamid et al. (2013) used Ni-Cr-B on diamonds to optimize the bonding between the diamond and the copper matrix. Coating the diamonds with electroless Ni-Cr-B produced a better coupling of the matrix and reinforcement and an increase in thermal conductivity to nearly 450 $Wm^{-1}K^{-1}$.

6.2.2.4 Ni-Mo-B

Electroless Ni-Mo-B was synthesized from bath using borohydride (Serin et al. 2015; Wang et al. 2009a, 2009b) as well as dimethylamine borane (Koiwa et al. 1988; Krutskikh et al. 2007) but the addition of Mo in the plating bath was always in the form of molybdates.

Krutskikh et al. (2007) observed that the additions of small amount of molybdates enhanced plating rate and that the plating rate decreased at higher concentrations due to the formation of intermediate products on the active surface.

The catalytic properties of Ni-Mo-B were investigated on amorphous powders (Wang et al. 2009a, 2009b).

When synthesized as coating, electroless Ni-Mo-B presents the usual cauliflower-like morphology of NiB and is usually amorphous (Osaka et al. 1987; Serin et al. 2015) and presents a better thermal resistance than NiB (Osaka et al. 1987; Yoshino et al. 2006), even if it can be crystallized at 400°C in some cases (Serin et al. 2015). However, Osaka reported thermal stability up to more than 600°C (Osaka et al. 1987).

The corrosion resistance of electroless Ni-Mo-B coatings has been reported as excellent and improves when the coatings are heat-treated, as much in terms of weight loss than in terms of corrosion potential (Serin et al. 2015). It may be linked to the presence of oxidized molybdenum species on the surface of the material (Zheng et al. 2004).

The thermal stability for copper diffusion of Ni-Mo-B coatings was also evaluated and the films proved to be stable up to 400°C (Yoshino et al. 2006).

6.2.2.5 Ni-W-B

The addition of tungsten to electroless NiB coatings aims at increasing hardness, wear and corrosion resistance, and thermal stability (Aydeniz et al. 2013; Drovosekov et al. 2005; Eraslan and Ürgen 2015; Mukhopadhyay et al. 2018; Yildiz et al. 2017). Drovosekov et al. (2005) investigated the plating process and obtained coatings with approximately 66 wt.% Ni and 33 wt.% W that were amorphous in the as-plated state and did not crystallize fully at 500°C (but were fully crystallized at 700°C). They also recorded hardness values that were sensibly higher than those of electroless NiB, from 950 HV in the as-plated state to more than 1500 HV after heat treatment at 500°C, which attests to the improvement of both hardness (and thus expectedly of wear resistance) and thermal stability caused by W additions in electroless NiB coatings.

Aydeniz et al. (2013) also reported improved hardness and increased corrosion resistance and better tribological properties. The improved properties are attributed to the presence of a tungsten oxide film of the surface of the material (Yildiz et al. 2017).

Mukhopadhyay et al. (2018) reported a denser morphology for electroless Ni-W-B, which they thought was a factor for improved corrosion resistance. They also reported relatively high hardness values before but mainly after heat treatment.

6.2.2.6 Ni-Cu-B

There are only a few reports of electroless Ni-Cu-B deposits. Hu's group (Hu et al. 1995; Wangyu et al. 1994) synthesized Ni-Cu-B coatings with a bath containing borohydride. They observed that the increase in Cu content led to an increase in the deposition rate up to a Cu^{2+}/Ni^{2+} ratio of 0.032, and then led to a decrease in the plating rate. They synthesized coatings with Cu content up to 8 at.%, with a boron content close to 20 at.%. They observed a modification in the surface appearance that became less bright when copper content increased. They did not report a significant improvement in hardness due to Cu incorporation in the coating, which was amorphous in the as-plated state. Copper increased slightly the crystallization temperature of the electroless Ni-Cu-B compared to NiB but the difference was of the order of 20°C. They recorded a longer time to wet Ni-Cu-B deposits than NiB deposits and much longer time than that required to wet gold wire.

Richardson and Rhen (2014) synthesized Ni-Cu-B nanotubes using template-assisted electroless plating, in view of obtaining tunable magnetic properties. They were able to form $(Ni_x\text{-}Cu_{100-x})_{60}\text{-}B_{40}$ alloys with a boron content nearly independent of the copper and nickel content. The nanotubes presented an amorphous character and their coercivity did not vary with the copper content (7.0 mT [5570 A/m]). However, the associated anisotropy increased with the copper content of the nanotubes.

6.2.2.7 Ni-Sn-B

The use of tin as the third element in electroless NiB has been the object of few studies (Aoki et al. 1993; Shimauchi et al. 1994) based on DMAB but it is becoming popular again with borohydride-based plating baths (Abdel-Gawad et al. 2019; Bonin et al. 2019).

The studies on DMAB reduced baths reported (Aoki et al. 1993; Shimauchi et al. 1994) tin content of approximately 30 at.% and an increase in the crystallinity of the coating with additions of tin, as well as improved corrosion resistance compared to electroless NiB. They also reported refining of the surface features with increasing tin content at first, followed by an increase in size for higher content, as shown in Figure 6.6. The hardness and electrical resistivity of the coatings decreased with increasing tin content.

A recent work has shown that tin was not only co-deposited with nickel but also had stabilizing action and enhanced the plating rate (Bonin et al. 2019). The addition of tin had a similar effect on the morphology to those reported in earlier works (i.e. decrease of the size of features). The electroless Ni-Sn-B coatings presented improved corrosion resistance due to the presence of tin at the surface of the material and mechanical behavior (hardness: 842 HK_{50}; specific wear rate (W_s): 0.11 μm 2/N; and first damage by scratch test (Lc): 30 N) close to electroless NiB coatings with a similar boron content (Bonin et al. 2019).

Abdel-Gawad et al. (2019) reported a similar morphological effect of the incorporation of tin in electroless NiB but observed that the coatings were more crystalline when they contained tin and had larger grains after crystallization at 400°C. They observed an improved hardness compared to the NiB coatings they synthesized

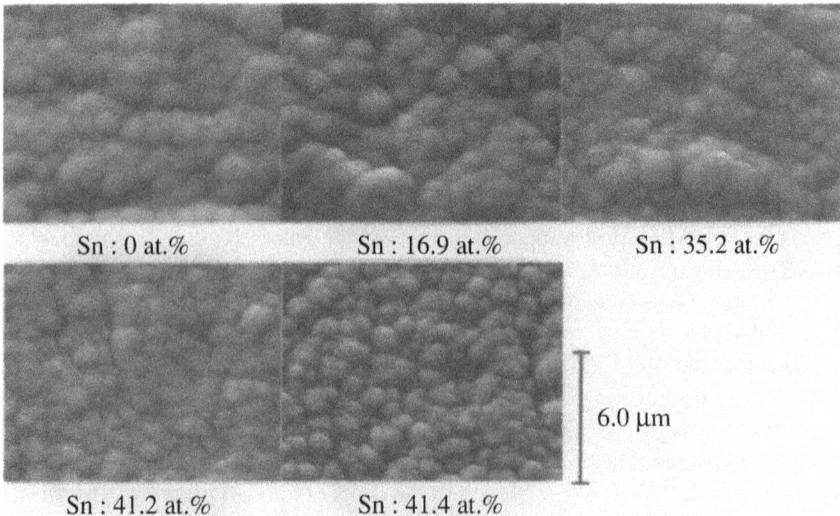

FIGURE 6.6 Morphology of electroless Ni-Sn-B coatings with varying Sn content obtained from bath using dimethylamine borane as reducing agent. (From Shimauchi, H. et al., *J. Electrochem. Soc.*, 141, 6–11, 1994.)

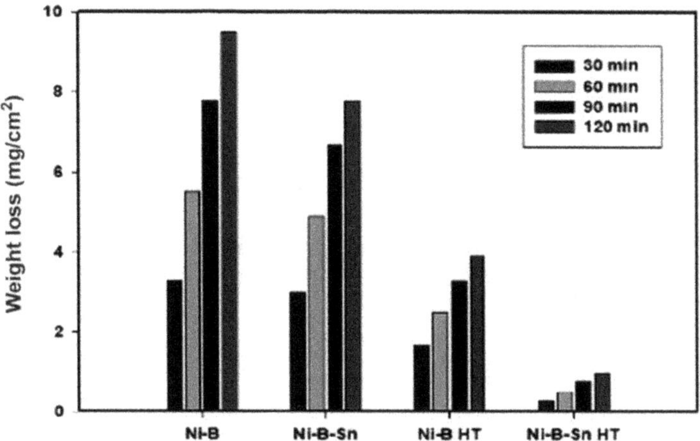

FIGURE 6.7 Wear resistance of electroless nickel-boron-tin coatings with and without heat treatment. (From Shimauchi, H. et al., *J. Electrochem. Soc.*, 141, 6–11, 1994.)

but that hardness is much lower than that reported by Bonin et al. (2019). Heat treating at 400°C led to an increase in hardness of similar amplitude to what is usually observed in electroless NiB coatings. They also observed a significant improvement in wear behavior for Ni-Sn-B that was still greater after heat treatment, as shown in Figure 6.7, as well as improved corrosion resistance, but heat treatment had a negative effect on this last parameter.

6.2.3 Electroless Ni-P-B

It is possible to produce ternary electroless Ni-P-B alloys by using two reducing agents in a single plating bath. The method was first developed in the late 1980s (El-Mallah et al. 1989a, 1989b; El-Mallah et al. 1993; El-Rehim et al. 1996; Lee and Chen 1999, 2001) but it has known renewed interest in the 2010s (Jiang et al. 2009; Omidvar et al. 2015; Shao et al. 2016; Venkatakrishnan et al. 2012). The boron source is usually borohydride but some plating solutions are operated in neutral to acid solutions, which is far from standard for the use of this compound (El-Mallah et al. 1989a). Using this kind of process, El Mallah et al. (1989b) were able to observe that an increase in boron content led to an increase in hardness but a decrease in corrosion resistance in alcohols.

There have been few studies of the plating process but Jiang et al. (2009) carried out a very in-depth study of the influence of plating parameters on the properties of coatings formed on optical fibers, using sophisticated statistical analysis and design of experiment. They were able to form very smooth layers with a low boron and phosphorus content (less than 1.5 wt.% of each) that presented good adhesion, low resistivity, and good solderability.

The most recent in-depth studies were carried out by Venkatakrishnan et al. (2012, 2014) on a very alkaline bath (pH = 14). They observed an increase in plating rate with borohydride concentration, linked to higher reducing efficiency of that

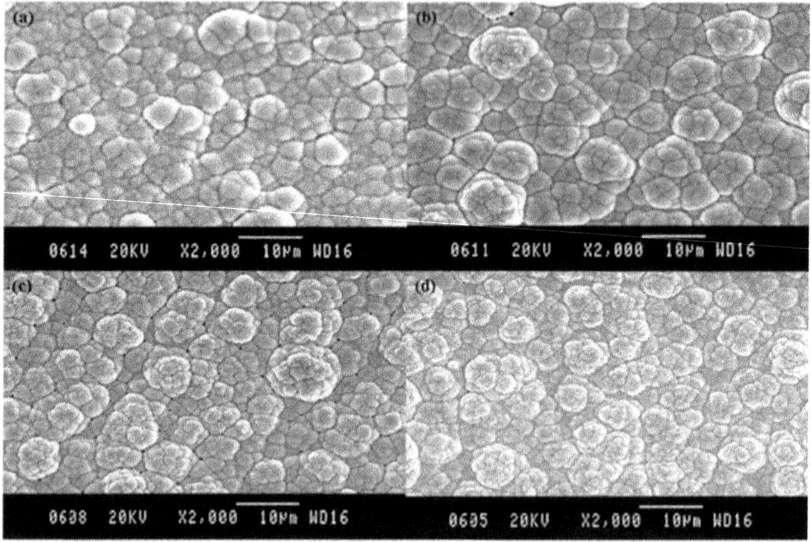

FIGURE 6.8 Scanning electron micrograph of Ni-P-B alloy coating deposited using alkaline bath containing 12 g/L NaH_2PO_2 and $NaBH_4$ concentration of (a) 0.2 g/L, (b) 0.4 g/L, (c) 0.6 g/L, and (d) 0.8 g/L. (From Venkatakrishnan, P.G. et al., *Int. J. Eng. Technol.*, 6, 1059–1064, 2014.)

compound. Borohydride seemed to be the dominant factor for the composition of the coating. They observed that the size of the features in the cauliflower-like texture decreased with increasing borohydride content in the plating bath, as shown in Figure 6.8. The increase in phosphorus content led to a smoothening of the coating. The structure of the coating is also strongly influenced by its composition: a higher boron content favors amorphous coatings while a higher phosphorus content (in the investigated range) leads to more crystalline coatings.

Shao et al. (2016) investigated the magnetic properties of electroless Ni-P-B coatings formed with dimethylamine borane as a boron source on polymer fibers. The amorphous material presented a weak paramagnetic behavior, which they attribute to the co-deposition of high amounts of phosphorus and boron.

Ternary electroless Ni-P-B alloys have also been investigated for their catalytic properties (Lee and Chen 1999, 2001).

Finally, the feasibility of depositing electroless Ni-P-B by dynamic chemical plating, a modified electroless plating method in which the material is not immersed in the solution but precursor solutions are sprayed on the surface, has been recently demonstrated (Omidvar et al. 2015).

6.3 QUATERNARY ELECTROLESS NICKEL ALLOYS

Owing to the excellent properties shown by some ternary alloys, some studies have investigated the formation and properties of quaternary alloys prepared by electroless plating.

TABLE 6.3
Quaternary Electroless Nickel Alloys that have been Investigated

NiB NiP→ ↓	Co	Cr	Cu	Fe	Mo	Pt	Sn	W	Zn	B	Others
Co			X								
Cr								X			
Cu								X	X		
Fe										X	
Mo								X			
Pt											
Sn								X			
W											Re
Zn											
P				X							
Others											

All of those studies have been carried out in phosphorus-based systems, as shown in Table 6.3: the only quaternary alloy that contains boron is based on a Ni-P-B system and thus also contains phosphorus (Wang et al. 2000). The most popular alloying elements for quaternary alloys are tungsten (Balaraju and Rajam 2005b; Balaraju et al. 2006a, 2006b; Balaraju et al. 2007, 2014; Che et al. 2013; Sha et al. 2010; Valova et al. 2001a) and copper (Balaraju and Rajam 2005b; Balaraju et al. 2006a, 2006b; Balaraju et al. 2007; Zaimi and Noda 2013; Zaimi et al. 2016), but iron (Wang et al. 2000), cobalt (Zaimi et al. 2016), chromium (Che et al. 2013), molybdenum (Balaraju et al. 2014), tin (Balaraju et al. 2007; Sha et al. 2010), and zinc (Zaimi and Noda 2013; Zaimi et al. 2015), as well as rhenium (Valova et al. 2001a), are also reported to be used.

The effect of the addition of a third element on electroless nickel is complex and depends greatly on the nature and content of phosphorus and the second alloying element.

It seems, however, that tungsten appears to increase hardness and thermal stability, while tin and copper improve corrosion resistance but not hardness. Molybdenum also improves thermal stability. Some elements inhibit phosphorus incorporation in the coating, which leads to the formation of crystalline deposits (Balaraju et al. 2006a, 2007, 2014). Zinc is reported to nullify the positive effect of copper incorporation on corrosion resistance (Zaimi and Noda 2013, 2014; Zaimi et al. 2015).

6.4 CONCLUSIONS

In the last 40 years, several research groups have aimed to improve the mechanical properties, corrosion resistance, thermal stability, catalytic activity, and magnetic properties of electroless nickel deposits by the addition of a second (and sometimes third) alloying element.

The effect of the addition of a second alloying element on the properties of electroless nickel is summarized in Table 6.4. Some, such as cobalt or iron, have the

TABLE 6.4
Effect of Second Allying Element on the Properties of Electroless Nickel Deposits

Alloy		Hardness	Corrosion Resistance	Magnetism	Thermal Stability	Catalytic Activity
Ni-X-P	**Ni-Co-P**			Soft magnetic properties, improved by increasing Co content	Possesses diffusion barrier properties	
	Ni-Cr-P	Not improved	Improved		Improved	
	Ni-Cu-P		Improved	Decreases magnetic behavior of phases formed during heat treatment	Improved	
	Ni-Fe-P	Improved	Improved			
	Ni-Mo-P	Improved for heat-treated coatings	Similar to NiP			
	Ni-Pt-P					
	Ni-Sn-P	Decreased	Improved		Not improved: Crystallization temperature decreases	
	Ni-W-P	Improved	Improved		Improved	
	Ni-Zn-P		Similar to NiP		Improved	
Ni-X-B	Ni-Co-B	Not improved	Good	Excellent, increases with Co content and heat treatment		Excellent
	Ni-Cr-B				Excellent	
	Ni-Cu-B	Not improved		Constant coercivity but tunable anisotropy		
	Ni-Fe-B			Soft magnetic properties		Best for Fe/(Fe+Ni) ratio of 0.3
	Ni-Mo-B				Excellent	Tested
	Ni-Sn-B	Improved	Improved			
	Ni-W-B	Improved	Improved		Excellent	

Note: Alloys in bold were also investigated in quaternary form.

most notable effect on the magnetic properties. Others, such as Mo, Cr, and W, are used mainly to increase thermal stability. Many elements have beneficial effects on the corrosion resistance (Cu, Sn, Fe, and W). Cobalt and iron seem to significantly improve the catalytic properties of NiB-based catalysts.

REFERENCES

Abdel-Gawad, S. A., M. A. Sadik, and M. A. Shoeib. 2019. "Preparation and Properties of a Novel Nano Ni-B-Sn by Electroless Deposition on 7075-T6 Aluminum Alloy for Aerospace Application." *Journal of Alloys and Compounds* 785: 1284–1292.

Abdel Hameed, R.M., and A.M. Fekry. 2010. "Electrochemical Impedance Studies of Modified Ni-P and Ni-Cu-P Deposits in Alkaline Medium." *Electrochimica Acta.* doi:10.1016/j.electacta.2010.05.046.

Abdullah, A., A. Talal, and A.-M. Abdullah. 2009. "Effect of Accelerators and Stabilisers on Crystallinity of Binary and Ternary Electroless Nickel Alloys." *International Journal of Nano and Biomaterials* 2: 42–51.

Aoki, K., O. Takano, and D. -H. Kim. 1995. "Electroless Nickel-Platinum-Phosphorus Alloy Plating." *Journal of the Surface Finishing Society of Japan* 46(1): 79–80. doi:10.4139/sfj.46.79.

Aoki, K., O. Takano, and Y. Sasaki. 1993. "Electroless Nickel-Tin-Boron Alloy Deposits." *Journal of the Surface Finishing Society of Japan* 44(10): 821.

Armyanov, S.A, and G.S. Sotirova. 1989. "Electroless Co-Ni-P Thin Films for Magnetic Recording." *Journal of the Electrochemical Society* 136(5): 134–137.

Aydeniz, A.I., A. Göksenli, G. Dil, and F. Muhaffel. 2013. "Electroless Ni-B-W Coatings for Improving Hardness, Wear and Corrosion Resistance." *Materialia* 47(6): 803–806.

Bai, C. -Y., Y.-H. Chou, C.-L. Chao, S.-J. Lee, and M.-D. Ger. 2008. "Surface Modifications of Aluminum Alloy 5052 for Bipolar Plates Using an Electroless Deposition Process." *Journal of Power Sources* 183(1): 174–181. doi:10.1016/J.JPOWSOUR.2008.04.082.

Balaraju, J.N., and K.S. Rajam. 2005a. "Electroless Deposition of Ni-Cu-P, Ni-W-P and Ni-W-Cu-P Alloys." *Surface and Coatings Technology.* doi:10.1016/j.surfcoat.2004.07.068.

Balaraju, J.N., N. Raman, and N.T. Manikandanath. 2014. "Nanocrystalline Electroless Nickel Poly-Alloy Deposition: Incorporation of W and Mo." *Transactions of the IMF* 92(3). Taylor & Francis: 169–176. doi:10.1179/0020296713Z.000000000123.

Balaraju, J.N., S.M. Jahan, A. Jain, and K.S. Rajam. 2007. "Structure and Phase Transformation Behavior of Electroless Ni–P Alloys Containing Tin and Tungsten." *Journal of Alloys and Compounds* 436(1–2): 319–327. doi:10.1016/J.JALLCOM.2006.07.045.

Balaraju, J.N., S.M. Jahan, C. Anandan, and K.S. Rajam. 2006a. "Studies on Electroless Ni–W–P and Ni–W–Cu–P Alloy Coatings Using Chloride-Based Bath." *Surface and Coatings Technology* 200(16–17): 4885–4890. doi:10.1016/J.SURFCOAT.2005.04.053.

Balaraju, J.N., and K.S. Rajam. 2005b. "Electroless Deposition of Ni–Cu–P, Ni–W–P and Ni–W–Cu–P Alloys." *Surface and Coatings Technology* 195(2–3): 154–161.

Balaraju, J.N., V.E. Selvi, V.K.W. Grips, and K.S. Rajam. 2006. "Electrochemical Studies on Electroless Ternary and Quaternary Ni–P Based Alloys." *Electrochimica Acta* 52(3): 1064–1074. doi:10.1016/J.ELECTACTA.2006.07.001.

Bangwei, Z., H. Wangyu, Q. Xuanyuan, Z. Qinglong, Z. Heng, and T. Zhaosheng. 1996. "Preparation, Formation and Corrosion for Ni-W-P Amorphous Alloys by Electroless Plating." *Transactions of the IMF* 74(2): 69–71. doi:10.1080/00202967.1996.11871098.

Bonin, L., V. Vitry, and F. Delaunois. 2019. "The Tin Stabilization Effect on the Microstructure, Corrosion and Wear Resistance of Electroless NiB Coatings." *Surface and Coatings Technology* 357: 353–363. doi:10.1016/j.surfcoat.2018.10.011.

Bouanani, M., F. Cherkaoui, M. Cherkaoui, S. Belcadi, R. Fratesi, and G. Roventi. 1999a. "Ni–Zn–P Alloy Deposition from Sulfate Bath: Inhibitory Effect of Zinc." *Journal of Applied Electrochemistry* 29(10): 1171–1176. doi:10.1023/A:1003426904395.

Bouanani, M., F. Cherkaoui, R. Fratesi, G. Roventi, and G. Barucca. 1999b. "Microstructural Characterization and Corrosion Resistance of Ni–Zn–P Alloys Electrolessly Deposited from a Sulphate Bath." *Journal of Applied Electrochemistry* 29(5): 637–645. doi:10.1023/A:1026441403282.

Campillo, B., P.J. Sebastian, S.A. Gamboa, J.L. Albarran, and L.X. Caballero. 2002. "Electrodeposited Ni–Co–B Alloy: Application in Water Electrolysis." *Materials Science and Engineering: C* 19(1–2): 115–18. doi:10.1016/S0928-4931(01)00457-X.

Che, L., M. Xiao, H. Xu, B. Wang, and Y. Jin. 2013. "Enhanced Corrosion Resistance and Micro-Hardness of Titanium with Electroless Deposition Ni-W-Cr-P Coating." *Materials and Manufacturing Processes*. doi:10.1080/10426914.2013.792412.

Chen, C.-H., B.-H. Chen, and L. Hong. 2006a. "Role of Cu^{2+} as an Additive in an Electroless Nickel–Phosphorus Plating System: A Stabilizer or a Codeposit?" *American Chemical Society*. doi:10.1021/CM0527571.

Chen, K., C. Liu, D.C. Whalley, D.A. Hutt, J.F. Li, and S.H. Mannan. 2006b. "Electroless Ni-W-P Alloys as Barrier Coatings for Liquid Solder Interconnects." *2006 1st Electronic System Integration Technology Conference* 421–427. IEEE. doi:10.1109/ESTC.2006.280037.

Chen, X.-M., G.-Y. Li, and J.-S. Lian. 2008. "Deposition of Electroless Ni-P/Ni-W-P Duplex Coatings on AZ91D Magnesium Alloy." *Transactions of Nonferrous Metals Society of China* 18: s323–s328. doi:10.1016/S1003-6326(10)60225-7.

Delaunois, F., J.P. Petitjean, M. Jacob-Dulière, and P. Liénard. 2000. "Autocatalytic Electroless Nickel-Boron Plating on Light Alloys." *Surface and Coatings Technology* 124: 201–209.

Drovosekov, A.B., M.V. Ivanov, V.M. Krutskikh, E.N. Lubnin, and Y.M. Polukarov. 2005. "Chemically Deposited Ni-W-B Coatings: Composition, Structure, and Properties." *Protection of Metals* 41(1): 55–62. doi:10.1007/s11124-005-0008-1.

Du, N., and M. Pritzker. 2003. "Investigation of Electroless Plating of Ni-W-P Alloy Films." *Journal of Applied Electrochemistry*. doi:10.1023/A:1026231532006.

El-Mallah, A.T., M.H. Abbas, M. E.-S. Aboul-Hassan, and I. Nagi. 1989a. "Deposition of Electroless Nickel-Phosphoros-Boron Alloys." *Plating and Surface Finishing* 12: 39–43.

El-Mallah, A.T., M.H. Abbas, I.H. Aly, and M.M. Younan. 1993. "Autocatalytic (Electroless) Deposition of Nickel-Phosphorus-Boron Alloys: Part 111. Deposit Properties." *Metal Finishing* 8: 23–28.

El-Mallah, A.T., M.H. Abbas, M.F. Shafei, M. E.-S. Aboul-Hassan, and I. Nagi. 1989b. "Structure of Electroless Nickel Deposits from Bath Containing Sodium Hypophosphite and Potassium Borohydride." *Plating and Surface Finishing* 5: 124–28.

El-Rehim, S.S.A., M. Shaffei, N. El-Ibiari, and S.A. Halem. 1996. "Effect of Additives on Plating Rate and Bath Stability of Electroless Deposition of Nickel-Phosphorus-Boron on Aluminum." *Metal Finishing* 94(12): 29–33. doi:10.1016/S0026-0576(96)80087-0.

Eraslan, S., and M. Ürgen. 2015. "Oxidation Behavior of Electroless Ni–P, Ni–B and Ni–W–B Coatings Deposited on Steel Substrates." *Surface and Coatings Technology* 265: 46–52. doi:10.1016/j.surfcoat.2015.01.064.

Gamboa, S.A., J.G. Gonzalez-rodriguez, E. Valenzuela, B. Campillo, P.J. Sebastian, and A. Reyes-Rojas. 2006. "Evaluation of the Corrosion Resistance of Ni–Co–B Coatings in Simulated PEMFC Environment." *Electrochimica Acta* 51 (19): 4045–4051. doi:10.1016/j.electacta.2005.11.021.

Georgieva, J., and S. Armyanov. 2007. "Electroless Deposition and Some Properties of Ni–Cu–P and Ni–Sn–P Coatings." *Journal of Solid State Electrochemistry*, 869–876. doi:10.1007/s10008-007-0276-6.

Hameed, R.A., and K.M. El-khatib. 2010. "Ni-P and Ni-Cu-P Modified Carbon Catalysts for Methanol Electro-Oxidation in KOH Solution." *International Journal of Hydrogen Energy* 35(6): 2517–2529. doi:10.1016/j.ijhydene.2009.12.145.

Hamid, Z.A., F.A. Mouez, and F.A. Morsy. 2013. "Electroless Ni–Cr–B on Diamond Particles for Fabricated Copper/Diamond Composites as Heat Sink Materials." *International Journal of Engineering Practical Research* 2(3): 112–117.

Haowen, X., Z. Bangwei, and Y. Qiaoqin. 1999. "Preparation, Structure and Corrosion Properties of Electroless Amorphous Ni-Sn-P Alloys." *Transactions of the IMF* 77(3): 99–102. doi:10.1080/00202967.1999.11871258.

Hu, W., L. Wu, L. Wang, B. Zhang, and H. Guan. 1995. "Crystallization of Amorphous Ni–Cu–B Alloys Obtained by Electroless Plating." *Physica B: Condensed Matter* 212(2): 195–200. doi:10.1016/0921-4526(94)00937-Q.

Huang, G.-F., W.-Q. Huang, L.-L. Wang, B.-S. Zou, D.-P. Chen, D.-Y. Li, J.-M. Wei, and J.-H. Zhang. 2007a. "Effects of Complexing Agents on the Corrosion Resistance of Electroless Ni–Fe–P Alloys." *International Journal of Electrochemical Science* 2: 321–328.

Huang, W.-Q., G.-F. Huang, L.-L. Wang, and X.-G. Shi. 2008. "Effects of Ligands on Electroless Ni-Fe-P Thin Films from Sulphate Bath." *International Journal of Electrochemical Science* 3: 1316–1324.

Huang, Y., K. Shi, Z. Liao, Y. Wang, L. Wang, and F. Zhu. 2007b. "Studies of Electroless Ni–Co–P Ternary Alloy on Glass Fibers." *Materials Letters* 61(8–9): 1742–1746. doi:10.1016/j.matlet.2006.07.122.

Itakura, K., T. Homma, and T. Osaka. 1999. "Effect of Deposition Site Condition on the Initial Growth Process of Electroless CoNiP® films." *Electrochimica Acta* 44: 3–4.

Jiang, B., L. Xiao, S. Hu, J. Peng, H. Zhang, and M. Wang. 2009. "Optimization and Kinetics of Electroless Ni-P-B Plating of Quartz Optical Fiber." *Optical Materials* 31(10): 1532–39. doi:10.1016/j.optmat.2009.02.016.

Kim, D.-H., K. Aoki, O. Takano, D.-H. Kim, K. Aoki, and O. Takano. 1995. "Soft Magnetic Films by Electroless Ni-Co-P Plating." *Journal of the Electrochemical Society* 142(11): 3763–3767. doi:10.1149/1.2048410.

Koiwa, I., M. Usuda, and T. Osaka. 1988. "Effect of Heat-Treatment on the Structure and Resistivity of Electroless Ni-W-P Alloy Films." *Journal of the Electrochemical Society* 135(5): 1222. doi:10.1149/1.2095931.

Krasteva, N., V. Fotty, and S. Armyanov. 1994. "Thermal Stability of Ni-P and Ni-Cu-P Amorphous Alloys." *Journal of the Electrochemical Society* 141: 2864–2867.

Krutskikh, V.M., M.V. Ivanov, A.B. Drovosekov, E.N. Lubnin, B.F. Lyakhov, and Y.M. Polukarov. 2007. "Structural Characteristics and Catalytic Activities of Nanocrystalline Ni-Mo-B Coatings Obtained by Catalytic Electroless Reduction." *Protection of Metals* 43(6): 560–566. doi:10.1134/S0033173207060070.

Kumar, A., M. Kumar, and D. Kumar. 2012. "Effect of Composition on Electroless Deposited Ni–Co–P Alloy Thin Films as a Diffusion Barrier for Copper Metallization." *Applied Surface Science* 258(20): 7962–7967. doi:10.1016/J.APSUSC.2012.04.145.

Lee, S.-P., and Y.-W. Chen. 1999. "Selective Hydrogenation of Furfural on Ni–P, Ni–B, and Ni–P–B Ultrafine Materials." *Industrial and Engineering Chemistry Research* 38(7): 2548–2566. doi:10.1021/IE990071A.

Lee, S.-P., and Y.-W. Chen. 2001. "Effects of Preparation on the Catalytic Properties of Ni–P–B Ultrafine Materials." *Industrial and Engineering Chemistry Research* 40(6): 1495–1499. doi:10.1021/IE000345Q.

Li, X., C. Wang, X. Han, and Y. Wu. 2013. "Surfactant-Free Synthesis and Electromagnetic Properties of Co–Ni–B Composite Particles." *Materials Science and Engineering: B* 178(3): 211–217. doi:10.1016/J.MSEB.2012.10.039.

Liu, H., P. Wang, Z. Liu, G. Harrison, and G.E. Thompson. 2016. "Comparison of the Corrosion Behaviour of Laser-Annealed Ni–P and Ni–Mo–P Deposits in H_2SO_4 and NaCl Solutions." *Transactions of the IMF* 94(2): 76–85. doi:10.1080/00202967.2015.1124639.

Liu, H., R.-X. Guo, Y. Zong, B.-Q. He, and Z. Liu. 2010. "Comparative Study of Microstructure and Corrosion Resistance of Electroless Ni-W-P Coatings Treated by Laser and Furnace-Annealing." *Transactions of Nonferrous Metals Society of China* 20(6): 1024–1031. doi:10.1016/S1003-6326(09)60252-1.

Liu, Y., and Q. Zhao. 2004. "Study of Electroless Ni–Cu–P Coatings and Their Anti-Corrosion Properties." *Applied Surface Science* 228(1–4): 57–62. doi:10.1016/j.apsusc.2003.12.031.

Lu, G., and G. Zangari. 2003. "Study of the Electroless Deposition Process of Ni-P-Based Ternary Alloys." *Journal of the Electrochemical Society* 150(11): C777. doi:10.1149/1.1614799.

Luo, H., H. Li, and L. Zhuang. 2001. "Furfural Hydrogenation to Furfuryl Alcohol over a Novel Ni–Co–B Amorphous Alloy Catalyst." *Chemistry Letters* 30(5): 404–5. doi:10.1246/cl.2001.404.

Mallory, G.O. 1974. "Ternary and Quaternary Electroless Nickel Alloys." *Transactions of the Institute of Metal Finishing* 52(15): 156–161.

Mallory, G.O, and T.R. Horn. 1979. "Electroless Deposition of Ternary Alloys." *Plating and Surface Finishing* 66(4): 40–46.

Matsubara, H., and A. Yamada. 1994. "Control of Magnetic Properties of Chemically Deposited Cobalt Nickel Phosphorus Films by Electrolysis." *Journal of the Electrochemical Society* 141(9): 3–7.

Matsuda, H., G.A. Jones, O. Takano, and P.J. Grundy. 1993. "Room-Temperature Electroless Deposition of High-Coercivity Co-Ni-P Films." *Journal of Magnetism and Magnetic Materials* 120(1–3): 338–341. doi:10.1016/0304-8853(93)91356-C.

Mendoza, L. Vargas, A. Barba, A. Bolarín, and F. Sánchez. 2006. "Age Hardening of Ni–P–Mo Electroless Deposit." *Surface Engineering.* doi:10.1179/174329406X84976.

Mukhopadhyay, A., T.K. Barman, and P. Sahoo. 2018. "Effect of Heat Treatment on the Characteristics of Electroless Ni-B, Ni-B-W and Ni-B-Mo Coatings." *Materials Today: Proceedings* 5(2): 3306–3315. doi:10.1016/j.matpr.2017.11.573.

Nie, M., Y.C. Zou, Y.M. Huang, and J.Q. Wang. 2012. "Ni–Fe–B Catalysts for $NaBH_4$ Hydrolysis." *International Journal of Hydrogen Energy* 37(2): 1568–1576. doi:10.1016/J.IJHYDENE.2011.10.006.

Omidvar, H., M. Sajjadnejad, G. Stremsdoerfer, Y. Meas, and A. Mozafari. 2015. "Characterization of NiBP-Graphite Composite Coatings Deposited by Dynamic Chemical Plating." *Anti-Corrosion Methods and Materials* 62(2): 116–122. doi:10.1108/ACMM-11-2013-1320.

Onoda, M., K. Shimizu, T. Tsuchiya, and T. Watanabe. 1993. "Preparation of Amorphous/Crystalloid Soft Magnetic Multilayer Ni-Co-B Alloy Films by Electrodeposition." *Journal of Magnetism and Magnetic Materials* 126(1–3): 595–598. doi:10.1016/0304-8853(93)90697-Z.

Osaka, T., I. Koiwa, K. Yamada, M. Nishikawa, and M. Usuda. 1987. "Effect of Molybdenum Codeposition on the Thermal Properties of Electroless Ni–B Alloy Plating Films." *Bulletin of the Chemical Society of Japan* 60(9): 3117–3124. doi:10.1246/bcsj.60.3117.

Oulladj, M., D. Saidi, E. Chassaing, and S. Lebaili. 1999. "Preparation and Properties of Electroless Ni–Zn–P Alloy Films." *Journal of Materials Science* 34(10): 2437–2439. doi:10.1023/A:1004566817439.

Pang, J., Q. Li, W. Wang, X. Xu, and J. Zhai. 2011. "Surface & Coatings Technology Preparation and Characterization of Electroless Ni–Co–P Ternary Alloy on Fly Ash Cenospheres." *Surface & Coatings Technology* 205(17–18): 4237–4242. doi:10.1016/j.surfcoat.2011.03.020.

Pearlstein, B.F. 1990. "Chapter 10 Electroless Deposition of Alloys." *Electroless Plating—Fundamentals and Applications*: 261–268. doi:10.1149/1.2403514.

Popov, K.I., S. Djokic, and B. Grgur. 2005. "Metal Deposition without an External Current." *Fundamental Aspects of Electrometallurgy*, 249–270. doi:10.1007/0-306-47564-2_10.

Ranganatha, S., T.V. Venkatesha, and K. Vathsala. 2010. "Development of Electroless Ni–Zn–P/Nano-TiO_2 Composite Coatings and Their Properties." *Applied Surface Science* 256(24): 7377–7383. doi:10.1016/J.APSUSC.2010.05.076.

Ranganatha, S., T.V. Venkatesha, and K. Vathsala. 2012. "Electroless Ni–W–P Coating and Its Nano-WS_2 Composite: Preparation and Properties." *Industrial & Engineering Chemistry Research* 51(23): 7932–7940. doi:10.1021/ie300104w.

Richardson, D., S. Kingston, and F.M.F. Rhen. 2015. "Synthesis and Characterization of Ni–Fe–B Nanotubes." *IEEE Transactions on Magnetics* 51(11): 1–4. doi:10.1109/TMAG.2015.2431552.

Richardson, D., and F.M.F. Rhen. 2014. "Magnetic Properties of Electroless Deposited Ni–Cu–B Nanotube Arrays." *IEEE Transactions on Magnetics* 50(11): 1–4. doi:10.1109/TMAG.2014.2322633.

Saito, T., E. Sato, M. Matsuoka, and C. Iwakura. 1998. "Electroless Deposition of Ni-B, Co-B and Ni-Co-B Alloys Using Dimethylamineborane as a Reducing Agent." *Journal of Applied Electrochemistry*. doi:10.1023/A:1003233715362.

Sankara Narayanan, T.S.N., S. Selvakumar, and A. Stephen. 2003. "Electroless Ni–Co–P Ternary Alloy Deposits: Preparation and Characteristics." *Surface and Coatings Technology* 172 (2–3): 298–307. doi:10.1016/S0257-8972(03)00315-3.

Sankara Narayanan, T.S.N., A. Stephan, and S. Guruskanthan. 2004. "Electroless Ni-Co-B Ternary Alloy Deposits: Preparation and Characteristics." *Surface and Coatings Technology*. doi:10.1016/S0257-8972(03)00788-6.

Schlesinger, M, and X. Meng. 1990. "Electroless Ni-Zn-P Films Electroless Copper Plating in the Presence of Excess Triethanol Amine." *Journal of the Electrochemical Society* 137(6): 1858–1859.

Serin, I.G., A. Göksenli, B. Yüksel, and R.A. Yildiz. 2015. "Effect of Annealing Temperature on the Corrosion Resistance of Electroless Ni-B-Mo Coatings." *Journal of Materials Engineering and Performance* 24(8): 3032–3037. doi:10.1007/s11665-015-1568-0.

Sha, W., N.H.J. Mohd Zairin, and X. Wu. 2010. "SEM-EDX of Morphology of Electroless Nickel Coatings with Tin and Tungsten." *Microscopy and Analysis-UK* 7: 11–14.

Shao, Q.-S., R.-C. Bai, Z.-Y. Tang, Y.-F. Gao, J.-L. Sun, and M.-S. Ren. 2016. "Durable Electroless Ni and Ni-P-B Plating on Aromatic Polysulfonamide (PSA) Fibers with Different Performances via Chlorine-Aided Silver Activation Strategy." *Surface and Coatings Technology* 302: 185–194. doi:10.1016/J.SURFCOAT.2016.05.087.

Shashikala, A.R., S.M. Mayanna, and A.K. Sharma. 2007. "Studies and Characterisation of Electroless Ni–Cr–P Alloy Coating." *Transactions of the IMF* 85(6): 320–324. doi:10.1179/174591907X246483.

Shimauchi, H., S. Ozawa, K. Tamura, and T. Osaka. 1994. "Preparation of Ni-Sn Alloys by an Electroless-Deposition Method." *Journal of the Electrochemical Society* 141(6): 6–11.

Song, G.-S., S. Sun, Z.-C. Wang, C.-Z. Luo, and C.-X. Pan. 2017. "Synthesis and Characterization of Electroless Ni–P/Ni–Mo–P Duplex Coating with Different Thickness Combinations." *Acta Metallurgica Sinica (English Letters)* 30(10): 1008–1016. doi:10.1007/s40195-017-0603-6.

Tai, F.C., K.J. Wang, and J.G. Duh. 2009. "Application of Electroless Ni–Zn–P Film for Under-Bump Metallization on Solder Joint." *Scripta Materialia* 61(7): 748–751. doi:10.1016/J.SCRIPTAMAT.2009.06.024.

Takai, M., K. Kageyama, and S. Takefusa. 1995. "Magnetic Properties of Electroless-Deposited NiFeB and Electrodeposited NiFe Alloy Thin Films." *IEICE Transactions on Magnetism* 78(11): 1530–1535.

Tarozaitė, R, and A. Selskis. 2006. "Electroless Nickel Plating with Cu^{2+} and Dicarboxylic Acids Additives." *Transactions of the IMF* 84(2): 105–112.

Toda, A., P. Chivavibul, and M. Enoki. 2013. "Effects of Plating Conditions on Electroless NiCoP Coating Prepared from Lactate-Citrate-Ammonia Solution Coating Cu Substrate." *Materials Transactions* 54(3): 337–343.

Ueda, M., H. Hayakawa, M. Mukaida, and Y. Imai. 2004. "Seebeck Coefficients of Iron Group Elements Borides." *Intermetallics* 12(1): 55–58. doi:10.1016/j.intermet.2003.07.005.

Valova, E., S. Armyanov, A. Franquet, A. Hubin, O. Steenhaut, J.-L. Delplancke, and J. Vereecken. 2001a. "Electroless Deposited Ni–Re–P, Ni–W–P and Ni–Re–W–P Alloys." *Journal of Applied Electrochemistry* 31(12): 1367–1372. doi:10.1023/A:1013862729960.

Valova, E., S. Armyanov, A. Franquet, K. Petrov, D. Kovacheva, J. Dille, J.-L. Delplancke, A. Hubin, O. Steenhaut, and J. Vereecken. 2004. "Comparison of the Structure and Chemical Composition of Crystalline and Amorphous Electroless Ni-W-P Coatings." *Journal of the Electrochemical Society* 151(6): C385. doi:10.1149/1.1705661.

Valova, E., I. Georgiev, S. Armyanov, J.-L. Delplancke, D. Tachev, T. Tsacheva, and J. Dille. 2001b. "Incorporation of Zinc in Electroless Deposited Nickel-Phosphorus Alloys I. A Comparative Study of Ni-P and Ni-Zn-P Coatings Deposition, Structure, and Composition." *Journal of the Electrochemical Society* 148(4): C266. doi:10.1149/1.1354598.

Veeraraghavan, B., H. Kim, and B. Popov. 2004. "Optimization of Electroless Ni-Zn-P Deposition Process: Experimental Study and Mathematical Modeling." *Electrochimica Acta*. doi:10.1016/j.electacta.2004.01.035.

Venkatakrishnan, P.G., S.S.M. Nazirudeen, and T.S.N.S. Narayanan. 2012. "Electroless Ni–B–P Ternary Alloy Coatings: Preparation and Evaluation of Characteristic Properties." *European Journal of Scientific Research* 82(4): 506–514.

Venkatakrishnan, P.G., S.S.M. Nazirudeen, and T.S.N.S. Narayanan. 2014. "Structural Characterization and Mechanical Properties of As-Plated and Heat Treated Electroless Ni-P-B Alloy Coatings." *International Journal of Engineering and Technology* 6(2): 1059–1064.

Wang, H., Z. Yu, H. Chen, J. Yang, and J. Deng. 1995. "High Activity Ultrafine Ni-Co-B Amorphous Alloy Powder for the Hydrogenation of Benzene." *Applied Catalysis A: General* 129(2): 143–149. doi:10.1016/0926-860X(95)00111-5.

Wang, L., L. Zhao, G. Huang, X. Yuan, B. Zhang, and J. Zhang. 2000. "Composition, Structure and Corrosion Characteristics of Ni–Fe–P and Ni–Fe–P–B Alloy Deposits Prepared by Electroless Plating." *Surface and Coatings Technology* 126(2–3): 272–278. doi:10.1016/S0257-8972(00)00545-4.

Wang, S.-L. 2004. "Studies of Electroless Plating of Ni–Fe–P Alloys and the Influences of Some Deposition Parameters on the Properties of the Deposits." *Surface and Coatings Technology* 186(3): 372–376. doi:10.1016/j.surfcoat.2004.01.017.

Wang, S.-L. 2007. "Electroless Deposition of Ni–Co–B Alloy Films and Influence of Heat Treatment on the Structure and the Magnetic Performances of the Film." *Thin Solid Films* 515(23): 8419–8423. doi:10.1016/j.tsf.2007.05.066.

Wang, W.-Y., Y.-Q. Yang, J.-G. Bao, and Z. Chen. 2009a. "Influence of Ultrasonic on the Preparation of Ni–Mo–B Amorphous Catalyst and Its Performance in Phenol Hydrodeoxygenation." *Journal of Fuel Chemistry and Technology* 37(6): 701–706. doi:10.1016/S1872-5813(10)60016-3.

Wang, W.-Y., Y.-Q. Yang, J.-G. Bao, and H.-A. Luo. 2009b. "Characterization and Catalytic Properties of Ni–Mo–B Amorphous Catalysts for Phenol Hydrodeoxygenation." *Catalysis Communications* 11(2): 100–105. doi:10.1016/j.catcom.2009.09.003.

Wang, Y., X. Chen, W. Ma, Y. Shang, Z. Lei, and F. Xiang. 2017. "Electroless Deposition of NiCrB Diffusion Barrier Layer Film for ULSI-Cu Metallization." *Applied Surface Science* 396: 333–338. doi:10.1016/J.APSUSC.2016.10.150.

Wangyu, H., W. Lingling, W. Lijun, Z. Bangwei, and G. Hengrong. 1994. "Preparation and Properties of Electroless Ni-Cu-B Alloy Deposits." *Transactions of the IMF* 72(4): 141–145. doi:10.1080/00202967.1994.11871041.

Wei, Y., W. Meng, Y. Wang, Y. Gao, K. Qi, and K. Zhang. 2017. "Fast Hydrogen Generation from $NaBH_4$ Hydrolysis Catalyzed by Nanostructured Co–Ni–B Catalysts." *International Journal of Hydrogen Energy* 42(9): 6072–6079. doi:10.1016/J.IJHYDENE.2016.11.134.

Wu, M.M., and B.Y. Lou. 2011. "Preparation and Corrosion Resistance of Electroless Plating of Ni-Cr-P/Ni-P Composite Coating on Sintered Nd-Fe-B Permanent Magnet." *Advanced Materials Research* 286: 2187–2190. doi:10.4028/www.scientific.net/AMR.284–286.2187.

Xie, G., W. Sun, and W. Li. 2008. "Synthesis and Catalytic Properties of Amorphous Ni–Co–B Alloy Supported on Carbon Nanofibers." *Catalysis Communications* 10(3): 333–335. doi:10.1016/J.CATCOM.2008.09.013.

Yamauchi, Y., T. Yokoshima, T. Momma, T. Osaka, and K. Kuroda. 2004. "Fabrication of Magnetic Mesostructured Nickel–cobalt Alloys from Lyotropic Liquid Crystalline Media by Electroless Deposition." *Journal of Materials Chemistry* 14(19): 2935–2940. doi:10.1039/B406265E.

Yildiz, R.A., K. Genel, and T. Gulmez. 2017. "Effect of Heat Treatments for Electroless Deposited Ni-B and Ni-WB Coatings on 7075 Al Alloy." *International Journal of Materials, Mechanics and Manufacturing* 5(2): 83–86. doi:10.18178/ijmmm.2017.5.2.295.

Yoshino, M., T. Masuda, S. Wakatsuki, J. Sasano, I. Matsuda, Y. Shacham-Diamand, and T. Osaka. 2006. "Fabrication of the Electroless NiMoB Films as a Diffusion Barrier Layer on the Low-k Substrate." *ECS Transactions* 1: 57–67. doi:10.1149/1.2218478.

Yu, Z.B., M.H. Qiao, H.X. Li, and J.F. Deng. 1997. "Preparation of Amorphous Ni-Co-B Alloys and the Effect of Cobalt on Their Hydrogenation Activity." *Applied Catalysis A: General* 163: 1–13. doi:10.1016/S0926-860X(96)00419-X.

Zaimi, M., and K. Noda. 2013. "Quaternary Alloy of Ni-Zn-Cu-P from Hypophosphite Based Electroless Deposition Method." *ECS Transactions* 45(19): 3–16. doi:10.1149/04519.0003ecst.

Zaimi, M., and K. Noda. 2014. "Effect of Coating Thickness on Corrosion Behavior of Electroless Quaternary Nickel Alloy Deposit in 3.5 wt% NaCl Solutions." *Journal of Advanced Manufacturing Technology (JAMT)* 8: 83–92.

Zaimi, M., M.A. Azam, A.H. Sofian, and K. Noda. 2015. "Electrochemical Impedance Behavior of Various Composition Quaternary Ni Alloy in 3.5 wt% NaCl." *Applied Mechanics and Materials* 761: 407–411. doi:10.4028/www.scientific.net/AMM.761.407.

Zaimi, M., M.N. Azran, M.K. Ahmad, M. Alif, M.S.A. Aziz, M.A. Azam, M.S. Kasim, R.F. Munawar, and M.E. Abd Manaf. 2016. "Electroless Ni-Co-Cu-P Alloy Deposition in Alkaline Hypophosphite Based Bath." *Key Engineering Materials* 694: 151–154. doi:10.4028/www.scientific.net/KEM.694.151.

Zhang, H., J. Zou, N. Lin, and B. Tang. 2014. "Review on Electroless Plating Ni–P Coatings For Improving Surface Performance of Steel." *Surface Review and Letters* 21(4): 1430002. doi:10.1142/S0218625X14300020.

Zhang, W.X., Z.H. Jiang, G.Y. Li, Q. Jiang, and J.S. Lian. 2008. "Electroless Ni–Sn–P Coating on AZ91D Magnesium Alloy and Its Corrosion Resistance." *Surface and Coatings Technology* 202(12): 2570–2576.

Zhao, L., G. Huang, X. Yuan, B. Zhang, and J. Zhang. 2000. "Composition, Structure and Corrosion Characteristics of Ni-Fe-P and Ni-Fe-P-B Alloy Deposits Prepared by Electroless Plating." *Surface and Coatings Technology* 126: 272–278.

Zhao, Q., Y. Liu, and E.W. Abel. 2004. "Effect of Cu Content in Electroless Ni–Cu–P–PTFE Composite Coatings on Their Anti-Corrosion Properties." *Materials Chemistry and Physics* 87(2–3): 332–335. doi:10.1016/j.matchemphys.2004.05.028.

Zheng, Y.X., S.B. Yao, and S.M. Zhou. 2004. "Study on Antioxidation of Nanosize Ni-Mo-B Amorphous Alloy." *Acta Physico-Chimica Sinica* 20: 1352–1356.

Zou, Y., Y.Y.H. Cheng, L. Cheng, and W. Liu. 2010. "Effect of Tin Addition on the Properties of Electroless Ni-P-Sn Ternary Deposits." *Materials Transactions* 51(2): 277–281. doi:10.2320/matertrans.MC200917.

7 Electroless Nickel-Based Multilayers

Asier Salicio-Paz and Eva García-Lecina

CONTENTS

7.1 INTRODUCTION TO ELECTROLESS NICKEL MULTILAYER COATINGS

Multilayered coatings are characterized by the combination of several layers ($n \geq 2$) into the same coating structure. The design of these multi-stacked coatings usually involve the combination of layers of different nature, as the synergistic effect of their mixture can boost the overall performance of this type of coatings, especially in terms of hardness, wear and corrosion resistance (Bull and Jones 1996; Clemens et al. 1999). On the other hand, there is also the possibility to combine layers of the same nature, creating artificial interphases along the coating thickness. In this case, the presence of these interphases, parallel to the substrate surface, can act as stoppers for the advance of defects and block dislocation motion, increasing the mechanical and protective properties beyond the limit offered by single-layered coatings. Thus, the multilayer approach fits perfectly on those technological fields in which superior mechanical, corrosion and tribological properties are demanded, such as aeronautics, automotive, military or cutting tools. Nowadays, multilayer systems are also finding new application niches in some current cutting-edge technological fields such as optoelectronics, electromagnetic interference (EMI) shielding and microelectromechanical systems (MEMS) in which optimized and tuned properties are required.

Multilayer coating architecture has been extensively investigated by sputtering and other physical vapor deposition techniques. However, these techniques often have high operational costs, sample size limitations and low throughput (Pogrebnjak et al. 2018). Because of these constraints, wet chemical technologies such as electrodeposition (ED) and electroless plating (EP) have emerged as suitable routes for the production of multilayered metallic coatings. The major advantages of wet chemical technologies over the other techniques rely on the simplicity to control, the possibility to process complicated parts and the low initial capital investment. In particular, EP stands out from electroplating for producing more homogeneous and harder coatings with enhanced wear and corrosion resistance. The other interesting features of EP, as its ability to provide a uniform coating even in parts with complex geometries and to coat dielectric substrates without the assistance of any electrical field, make electroless deposition one of the preferred surface-finishing treatments in many industrial sectors such as automotive, aeronautics and petrochemical. Among electrolessly plated metals, nickel and its alloys represent 95% of all the coatings obtained at the industrial scale, being hypophosphite the most used reducing agent. The excellent mechanical properties and superior corrosion resistance provided by the NiP coatings, along with its favorable processing conditions, make the electroless nickel (EN) plating as one of the most demanded solutions in the field of surface engineering. The in-depth review by Sudagar et al. (2013) on EP gathers the different plating modes of EN. Thus, suitable reducing agents, experimental parameters, chemistry of the electrolyte and a proper selection of reinforcing materials for composite production can give rise to an extensive variety of monolayered metallic coatings with a wide range of functional properties (Figure 7.1). Additionally, the combination of monolayered EN coatings into a single coating's architecture offers the possibility to enhance the performance and thus broaden the functionalities beyond those of single-layered materials.

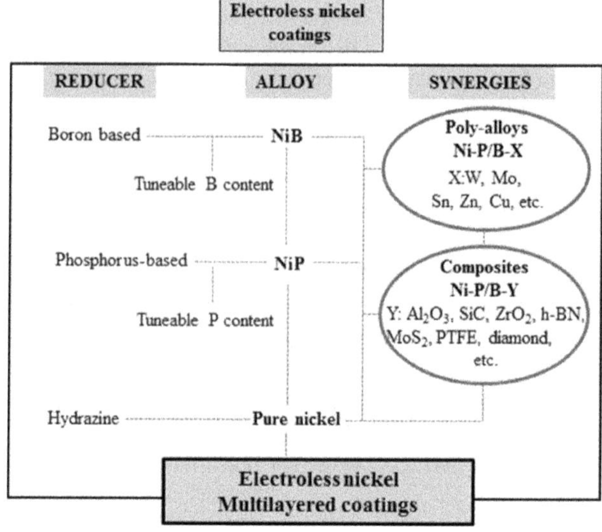

FIGURE 7.1 Electroless nickel plating modes.

In the next sections of this chapter, EN-based multilayer production methodologies along with the effect on their properties derived from the plating approach, before and after annealing, will be covered. Future trends on the field will be also discussed along the chapter.

7.2 DESIGN, PROCESSING AND FABRICATION

The design of multilayered coatings greatly depends on the performance requirements of the final application of the coated component. In general, a multilayered coating is expressed in terms of its constituent layers, microstructural features and properties, but it must also consider the integrity and coherence of the final system as a whole. Once the design of the multilayered coatings has been completed, a proper production strategy needs to be followed depending on the nature of the layers involved in the multilayered coating. There are two main plating approaches to produce EN-based multilayers, namely dual/multiple bath and single bath procedures.

7.2.1 Dual/Multiple Bath Approach

Dual bath approach is the most widely used strategy for the production of EN multilayers. In such procedure and once the proper substrate preparation has been carried out, the sample is placed in the first electrolyte, and after a given plating time, it is transferred to a second plating solution for plating a new layer over the first one. It is very important to maintain transfer times at a minimum to avoid surface oxidation, which can passivate the surface to be coated affecting the adhesion between subsequent layers (Anvari et al. 2015b). This dual/multiple bath allows varying the composition of the layer along the coating thickness, thus giving rise to the so-called compositionally modulated multilayer coatings (CMMC) (Figure 7.2).

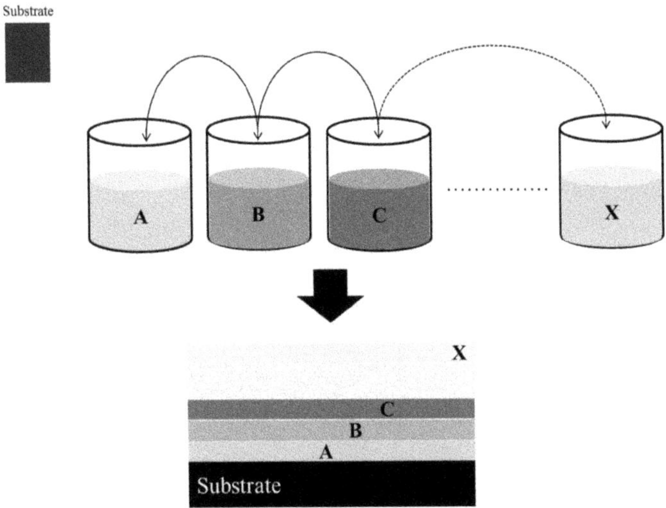

FIGURE 7.2 Dual/multiple bath approach.

In the dual/multiple bath approach, it is also possible to add particulate matter as a second phase to some of the intermediate plating baths, if not to all of them, for producing metallic matrix composite (MMC) coatings. The idea behind the MMC coatings is to combine the properties of the metallic matrix and the reinforcing material in a synergistic way, thus giving rise to a different class of materials with enhanced features compared to their individual constituents.

Using this plating strategy, one can combine different types of reinforcing materials (i.e. hard particles, lubricant materials, corrosion inhibitors, etc.) into a single coating. The multiple bath strategy also allows creation of coatings with a compositional thickness-dependent profile in which the functionalities of the composite coating change along the direction perpendicular to the substrate surface. These kinds of coatings are known as functionally graded materials (FGMs). This experimental approach allows to easily obtaining duplex and multilayered coatings by the EP technology.

7.2.2 Single Bath Approach

On the other hand, the single bath approach is a much simpler experimental setup, as just one single electrolyte is used for producing multilayered coatings. This approach is especially useful in the production of multilayered coatings in which all the single layers involved in the multi-layered coatings are of the same nature. The production of FGMs and/or CMMC coatings can be carried out by suitably selecting the plating parameters (solution pH, stirring, temperature, etc.). However, this plating strategy offers limited control over the final coating composition, as the time needed for reaching the equilibrium state after parameter adjustment widens the transitional region between layers (Figure 7.3).

In the recent years, another plating approach has emerged for the production of EN multilayers, the so-called hybrid electro-electroless deposited coatings (HEEDs), which can be considered a particular case of the single bath approach. This plating mode consists of the combination of ED and EP from a single plating bath, enabling the deposition of different metals present in the same electrolyte (Petro 2017; Petro and Schlesinger 2014). Usually, in the HEED plating mode, the electroless reduced metal or alloy is primary deposited, as it should possess a more noble character

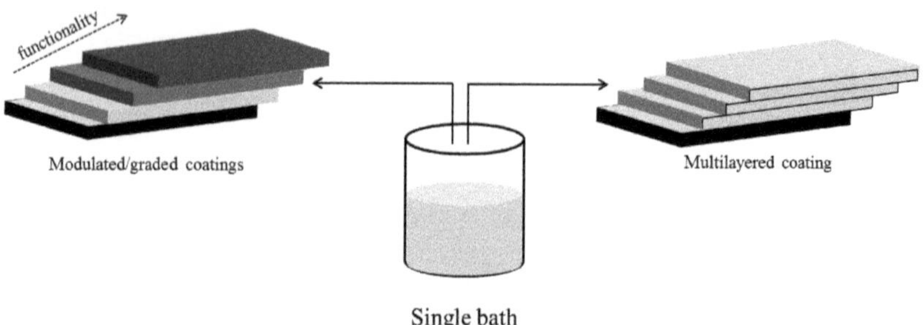

FIGURE 7.3 Single bath approach.

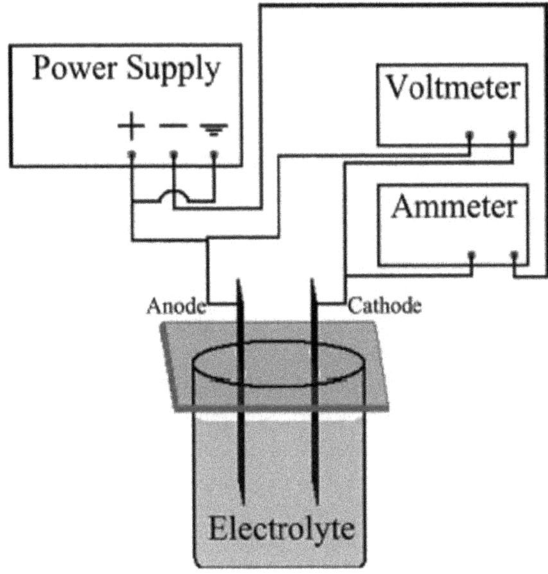

FIGURE 7.4 HEED experimental setup. (From Petro, R. and Schlesinger, M., *J. Electrochem. Soc.*, 161, D470–D475, 2014.)

than the metal or alloy to be electroplated. It must be noted that a proper selection of metallic constituents needs to be performed; thus, the electroplated metal does not inhibit the electroless process (Figure 7.4).

7.3 CHARACTERIZATION OF PROPERTIES

Mechanical properties, corrosion resistance and magnetic properties of the coatings can be tuned when the coatings are obtained by the multilayer approach. Addition of a second phase (i.e. particulate matter) at any level into the multilayer system can also modify the natural path for the advancement of some degradation mechanisms such as corrosion and/or crack propagation under mechanical stress. Stacking consecutive layers can also help in preventing the development of porosity along the thickness of the coatings due to the sealing effect of interlayers. The effect of annealing on the properties of the coatings can also differ from monolayer to multilayer coatings, boosting the performance of the latter. The possibility of combining the same or dissimilar NiP-based layers, the addition of nano/microparticles, and the combination of different plating approaches broadens the range of potential novel coatings to be produced by this technology, capable of fulfilling the most demanding requirements of the industry in the near future.

The properties provided by the multilayer approach are also dependent on the number of layers involved in the coating. Thus, these characterization sections have been divided according to the architecture of the coatings in terms of the number of layers involved in their design.

7.3.1 Duplex Coatings

Duplex coatings can be defined as the simplest multilayered type of coating and are easily produced by the dual bath approach. When it comes to blending of different electroless coatings, the most obvious choice is the combination of the most commonly used electroless coatings, namely NiP and NiB layers. Usually, these duplex coatings seek for pairing the excellent mechanical properties of the NiB coatings along with the superior corrosion resistance of the NiP coatings (Mallory and Hajdu 1990; Okinaka and Osaka 2008). Thus, Narayanan et al. produced different NiP/NiB coatings by alternating the order of the layers to a maximum thickness of 20 μm on mild steel substrates. The performance of duplex coatings was also compared to single NiP and NiB monolayers. The obtained coatings were dense and homogeneous and showed a very good compatibility between layers (Narayanan et al. 2003). The proper selection of outer layer made it possible to reach a hardness value of 652 ± 27 HV_{100} for the NiP/NiB as-deposited system in comparison to the 596 ± 24 HV_{100}, when NiP was placed as outer layer. Similarly, NiP/NiB showed a lower wear rate in comparison to the NiB/NiP and also compared to NiP and NiB single coatings on pin-on-disc tests. Single and duplex coatings exhibited adhesive wear failure; thus, no difference in terms of wear mechanism was found depending on the architecture of the coating. Despite the fact that the presence of a NiB outer layer exhibits lower wear rate, it should be considered the type of NiP coating used for comparison, as different P contents lead to different mechanical properties and corrosion resistance performance. Corrosion resistance showed the expected trend for NiP and NiB coatings, as NiP coatings have better protection ability against corrosion than NiB, mainly because of the not fully amorphous structure of NiB layer and the heterogeneous distribution of boron and co-deposited stabilizers.

More recently, Vitry et al. used a similar experimental approach for plating both mid NiP (7–9 wt.%) and NiB (5–7 wt.%) monolayers and duplex coatings on mild steel, the latter alternating the outer layer between NiP and NiB (Vitry et al. 2017). The composition analysis by Glow-discharge optical emission spectroscopy (GDOES) clearly determined the NiB/NiB interphase whereas, for N-B/NiB coatings this interphase is not clearly observed, masked by the own roughness of the coatings. On the other hand, the microstructural evaluation of the coatings undoubtedly revealed the columnar nature of the NiB coatings in all the coatings in which it was present, whereas no microstructural features could be spotted for NiP coatings without proper etching of the coatings.

The same coating configuration was characterized in terms of corrosion resistance in different media by Bonin et al. (2018). In chloride (Cl^-) media, NiP coatings showed the best corrosion performance. This behavior is mainly attributed to their tendency to passivation by forming a hydrated phosphate layer that prevents, or delays, the advance of the corrosive attack. This mechanism has been depicted elsewhere in the literature for single NiP coatings (Elsener et al. 2008). Only NiB coatings showed an improvement in the corrosion resistance when coated as a bilayer. This behavior was associated with its columnar structure, as the column boundaries act as paths for the progress of the corrosive attack. The interphases generated in the duplex coatings block these paths, thus improving the overall corrosion performance in this media. (Contreras et al. 2006). According to the research, the corrosion performance of the coating is driven by the nature of the selected top layer. Selecting NiP as outer layer provides the greatest protection ability in Cl^- media. Whereas, in the presence of sulfate anions (SO_4^{2-}), the duplex coatings do not alter the natural behavior of the coatings, when exposed to H_2SO_4 media, the duplex coatings perform better than single coatings, especially for NiB-based coatings. Experiments on alkaline media showed no improvement in the corrosion resistance of the coatings compared to the monolayer coatings, revealing NiP coatings as the best performers in this media.

Due to their excellent corrosion resistance and mechanical properties, duplex coatings are especially suited for conferring superior features to light alloys such as aluminum or magnesium. With regard to magnesium, despite being considered as one of the most interesting engineering materials, it suffers from a severe chemical and electrochemical reactivity, making it more prone to corrosion in almost any media they are in contact with (Li et al. 2006; Sridhar and Udaya Bhat 2013). In this context, the high-performing, and defect-free, surfaces obtained by EP have emerged as one of the most suitable technologies for protecting these kinds of materials. Georgiza et al. (2013) applied a mid-phosphorus/high-phosphorus duplex NiP coating on AZ-31 magnesium substrates.

The research showed the beneficial effect of duplex coatings in terms of corrosion resistance compared to monolayered NiP coatings; nevertheless, the use of coatings with dissimilar phosphorus content hid the real effect of the duplex approach. Following the protection of light alloys, Vitry et al. (2012) studied duplex coatings with NiB outer layer applied on 2024 aluminum alloy substrates. In terms of wear resistance, similar Taber wear index (TWI) was found for duplex coatings and the values were comparable to those measured on NiB-coated steel substrates. On the other hand, the corrosion resistance of the duplex NiP/NiB coating was found to be higher in comparison to plain NiB coatings in Cl^- media; the trend was maintained also on worn surfaces revealing the protective nature of the duplex coatings and its suitability for effectively protect aluminum substrates.

Depending on the final application of the duplex coatings, sometimes bare electroless coatings do not effectively preserve the substrates they are applied on. In this context, the versatility of the electroless coatings enables the production of more resistant coatings by including in the duplex structure electroless deposited ternary alloys and/or composite layers. With regard to ternary alloys, NiP coatings have been more intensively investigated for the production of these types of coatings in the form Ni-M-P (M: W, Cu, Sn, Cr, Co, Mo, etc.) (Balaraju et al. 2006; Lu and Zangari 2003; Mallory 1974; Wang et al. 2015a). Ni-W-P ternary alloys stand out over other known ternary electroless systems due to their superior corrosion resistance, wear resistance and thermal stability (Antonelli et al. 2006; Li et al. 2015). These features have made them suitable candidates for their inclusion in duplex coatings architecture for protecting both mild steel and magnesium substrates. In the case of steel substrates, a combination of NiP (9.3 wt.%P)/Ni-W-P (3.9 wt.%W, 13.3 wt.%P) showed very good chemical compatibility reaching a total thickness of 30 microns (Liu et al. 2012). Ni-W-P showed an increased hardness associated with strengthening via solid solution of tungsten in the alloy, this feature has a direct impact also on its wear behavior, thus making the duplex coating more wear-resistant than single

NiP coatings in as-deposited state. Despite the presence of tungsten in the outer layer, the wear mechanism remains as adhesive as has been already pointed out for other NiP duplex systems. Chen et al. (2008) have also tested the duplex NiP/Ni-W-P system on magnesium substrates. In this case, NiP (3.7 wt.%P)/Ni-W-P (0.65 wt.%W, 8.18 wt.%P) were used for protecting the AZ91D substrates. According to the results, the duplex NiP/Ni-W-P coatings provide better protection ability than NiP monolayers, with E_{corr} values of −0.45 and 0.26 V versus Saturated calomel electrode (SCE), respectively, and j_{corr} values of 0.001 and 0.002 A·cm^{-2}, respectively.

On the other hand, the presence of a composite based coating as inner or outer layer in duplex coatings can also benefit the overall performance of the coatings. As in the case of the ternary alloys described above, for composite materials based on EN, there have been intensive research works on suitable reinforcing materials for EN matrixes. Similarly to ternary alloys, successful co-deposition of the reinforcing material leads to the formation of Ni-X-P/B alloys (X: Al_2O_3, SiC, ZrO_2, diamond particles, WS_2, B_4C, PTFE, Fe_2O_3, h-BN, etc.) (De Hazan et al. 2012; Reddy et al. 2000; Ranganatha et al. 2012; Sharma and Singh 2012; Suiyuan et al. 2012). One of the preferred composite systems in duplex electroless coatings are those containing ZrO_2 as reinforcing material, as the proper nature of the particulate material can improve the mechanical properties of the coatings, promoting at the same time better electrochemical response in aggressive media (Szczygieł and Turkiewicz 2008, 2009). In the more recent research papers for duplex coating involving a NiP-ZrO_2 layer, this is usually placed as inner layer to take better advantage of their functionalities. Thus, Wang et al. deposited on steel substrates a sol-enhanced NiP-ZrO_2/NiP duplex coating in which the plain high-phosphorus coating acted as inner layer, whereas the outer layer was based on a low-phosphorus coating containing the sol-enhanced ZrO_2 nanoparticles (Wang et al. 2015b). This coating configuration was aimed to obtain coatings with very good wear resistance, while maintaining the corrosion resistance granted by the high-phosphorus layer. As can be seen in Figure 7.5, the coating has very good chemical compatibility for a total thickness of 22 μm, being the external NiP-ZrO_2 layer 16 μm thick. When compared to a duplex low-phosphorus NiP/NiP coating, which shows an average hardness of 648 HV_{100}, the incorporation of the sol-enhanced ZrO_2 nano-particles into the NiP matrix in the outer layer allows enhancing the hardness of the system to 752 HV_{100}. In this case, the grain refinement strengthening due to ZrO_2 nanoparticles co-deposition following the Hall-Petch relationship is attributed to main hardening mechanism along with dispersion hardening promoted by the ZrO_2 nanoparticles according to Orowan mechanism (Chang et al. 2007; Chen et al. 2010b). The presence of the sol-enhanced ZrO_2 nanoparticles promotes an increased wear resistance in comparison with the duplex NiP/NiP coating. Moreover, the difference in corrosion potential, approximately −0.24 and −0.18 V for outer and inner layers, respectively, enables the low-phosphorus layer to act as a sacrificial anode, while the NiP-ZrO_2 behaves as the cathode, decreasing the corrosion rate due to the small exposed area of the latter in salt spray corrosion tests. Duplex coatings allow a horizontal displacement of the corrosive attack, whereas it is usually developed vertically in monolayered coatings, progressing through the pores of the coatings.

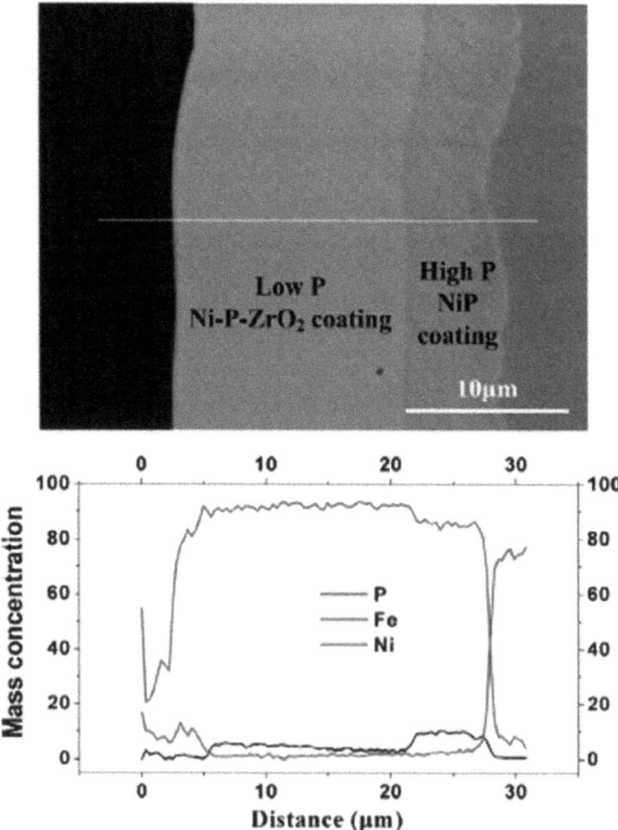

FIGURE 7.5 Cross-section and compositional profile for the duplex NiP/NiP-ZrO_2 coating. (Reprinted from *J. Alloy. Compd.*, 630, Wang, Y. et al., Duplex NiP-ZrO_2/NiP electroless coating on stainless steel, 189–194, Copyright 2015b, with permission from Elsevier B.V.)

Following this approach, Luo et al. obtained duplex coatings based on NiP-yttria-stabilized zirconia (YSZ)/NiP as inner and outer layers, respectively. The metallic matrix containing the YSZ nanoparticles was a low-phosphorus nickel coating, whereas a high-phosphorus nickel deposit acted as outer layer. The addition of YSZ nanoparticles was studied up to a concentration of 18 $g{\cdot}L^{-1}$ (Luo et al. 2017). It was confirmed that duplex coatings were harder than monolayer coatings, showing a sharp increase in the hardness of the coatings in the range of 2–7 $g{\cdot}L^{-1}$ of YSZ nanoparticle concentration. Beyond this limit, the relationship of hardness with regard to particle concentration increases moderately mainly due to particle agglomeration-related issues. The presence of YSZ nanoparticles also enhances the corrosion resistance of the coatings. The curved shape in the polarization studies changes, showing an increase in the corrosion potential of the coatings. When [YSZ] $\geq$ 7 $g{\cdot}L^{-1}$, a plateau region is observed in the anodic branch indicating that the presence of the YSZ nanoparticles promotes the formation of a passive film providing better corrosion protection in comparison to monolayered NiP coatings in Cl^- media. Again, the difference in corrosion potential between the inner and outer layers allows decreasing of the corrosion rate of the NiP-YSZ/NiP coating. Moreover, the presence of the YSZ nanoparticles acts as nucleation sites during the electroless process, decreasing the nodule size of the NiP coatings blocking the nodule boundary, which are the main path for the advance of the corrosive attack as schematically depicted in Figure 7.6. A similar mechanism has been proposed for composites based on NiP/nano-diamond (Ashassi-Sorkhabi and Es′haghi 2013).

Finally, the performance of this kind of coatings can be further increased by the addition of an alloying element to the NiP inner layer. As mentioned earlier in this chapter, tungsten-alloyed NiP coatings exhibit better corrosion resistance and enhanced mechanical properties. Thus, Luo et al. have developed novel electroless coatings combining Ni-W-P/NiP-nanoZrO_2 on low-carbon steel substrates, in which a high-phosphorus Ni-W-P coating was placed as an inner layer (Luo et al. 2018). The coating showed very good chemical compatibility between layers with a co-deposited ZrO_2 content of 1.72 wt.%. The presence of the ZrO_2 nanoparticles, even though not completely dispersed into the metallic matrix, provides a better response in 5wt.% NaCl media saturated in H_2S and CO_2. In this media, no passivated areas are present and the dissolution reaction occurs. The preferential dissolution of nickel leads to a phosphorus enrichment at the coating–solution interface that promotes the formation of an adsorbed $H_2PO_2^-$ layer, which blocks the advance of the corrosive attack. Moreover, the presence of ZrO_2 nano-particles decreases the nodule size, thus blocking the nodule boundary and so decreasing the overall corrosion rate. In addition to this feature, duplex coatings avoid the progress of the corrosive attack because of the better coverage of coatings defects by the duplex approach.

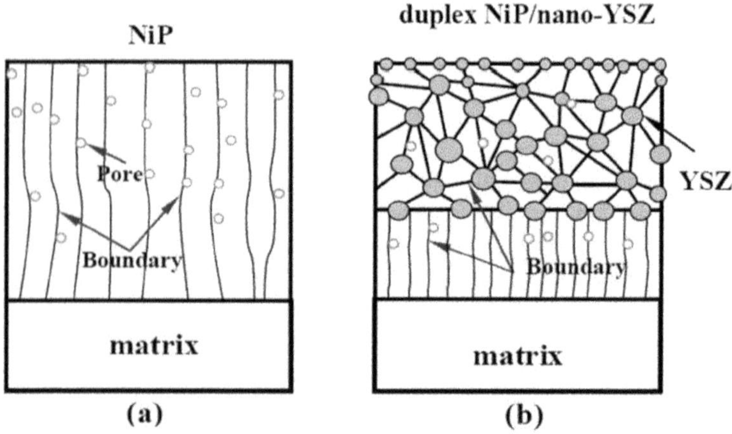

FIGURE 7.6 Corrosion protection mechanism for (a) NiP and (b) NiP-YSZ. (Reprinted from *Surf. Coat. Technol.*, 311, Luo, H. et al., Synthesis of a duplex NiP-YSZ/NiP nanocomposite coating and investigation of its performance, 70–79, Copyright 2017, with permission from Elsevier B.V.)

7.3.2 Multilayer Coatings

Multilayer coatings offer a wider range of possible layer arrangements, enabling the production of innovative systems with dissimilar functionalities. The evaluation of the effect of the multilayer approach needs to be performed firstly using multilayered coatings with the same composition along the coating thickness. Thus, Salicio-Paz et al. (2019) produced tri-layered coatings from a low-phosphorus electrolyte following the single bath approach. In this plating approach, the plating process was interrupted accordingly by removing the sample from the bath at different intervals. The multilayered coatings showed the ability to effectively block the advance of defects through the coatings' thickness as shown in Figure 7.7. The multilayered coatings showed a slightly nobler E_{corr} and lower j_{corr} values in comparison with monolayered coatings in aerated Cl^- media.

When it comes to combining layers of different nature, the most logical selection of layers would be those containing the three different types of phosphorus-based electroless coatings; namely low-, medium- and high-phosphorus NiP coatings. Thus, Narayanan et al. coated mild steel substrates with low-P/medium-P/high-P multilayer coating to achieve a phosphorus gradient along the coatings' thickness, thus creating an FGM structure (Sankara Narayanan et al. 2006). The graded coating had a thickness of 24 ± 1 µm; single NiP coatings, at the same thickness range, with low, medium and high P content were also analyzed for the sake of comparison. The chemical compatibility among layers was claimed as good enough to perform the characterization studies.

As described elsewhere (Gool et al. 1987) and shown in Figure 7.8, corrosion resistance experiments in chloride media showed that high-P monolayer coatings exhibited more positive E_{corr} and lower I_{corr} values. This trend was also maintained in the graded coatings when high P is placed as outer layer. The corrosion performance of the high-P, for both single and graded coatings, is promoted via $H_2PO_2^-$ layer development on the coating surface, which blocks the water supply toward the surface of the coating preventing nickel dissolution. This mechanism is not that effective for low and medium phosphorus coatings because of the different nature of the protective film, which decreases its performance against corrosion. However, the graded structure with low-phosphorus outer layer showed better corrosion response because of the barrier effect provided by the underlying nickel coatings in comparison to

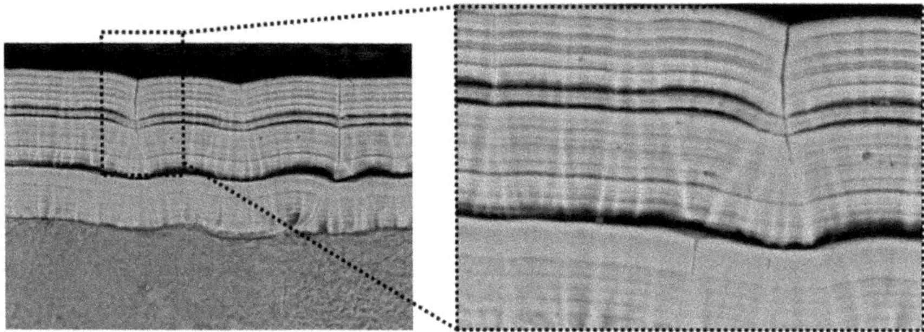

FIGURE 7.7 Cross-section of NiP multilayer coating at different magnifications. (Extracted from Salicio, A. et al., *Surf. Coat. Technol.*, 368, 138–146, 2019.)

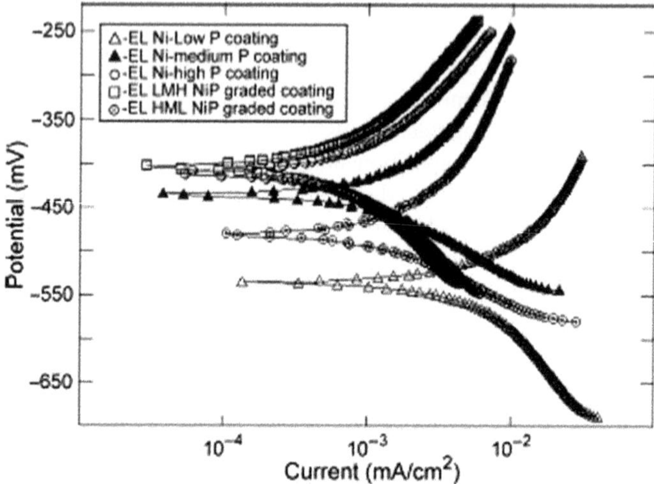

FIGURE 7.8 Potentiodynamic studies for NiP-graded and single coatings in chloride media. (From Sankara Narayanan, T.S.N. et al., *Surf. Coat. Technol.*, 200, 3438–3445, 2006.)

low-phosphorus monolayer coatings. More recently, Anvari et al. produced similar EN FGM coatings but using the single bath approach in comparison with the multiple bath strategy carried out by Narayanan. The characterization of the coatings by nano-indentation revealed a graded hardness profile depending on the selected layer configuration as shown in Figure 7.9. L-H refers to low-phosphorus as outer layer, whereas H-L refers to a high-phosphorus outer layer, a medium-phosphorus interlayer was used in both cases (Anvari et al. 2015b). The observed trends match those observed for single layers as a function of the phosphorus content in the layer (Hamada et al. 2015a). The wear behavior expressed as the ratio of hardness (H) to elastic modulus (E), obtained from nano-indentation measurements showed that

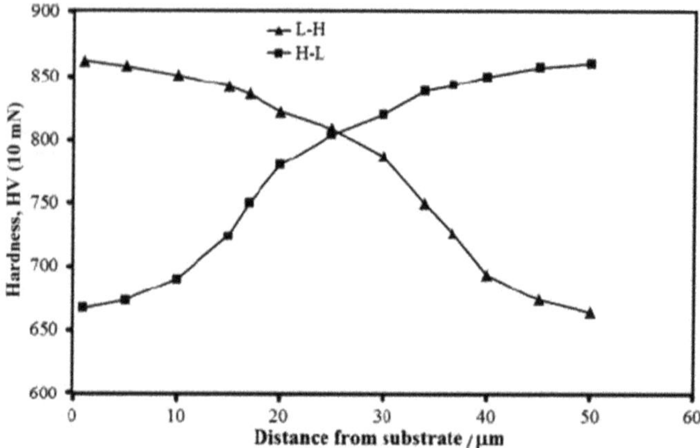

FIGURE 7.9 Hardness profile for graded coatings along the coating's thickness. (From Anvari, S.R. et al., *Surf. Eng.*, 31, 693–700, 2015b.)

the FGM coatings were more wear-resistant in comparison to monolayer coatings. This behavior can be explained by the effectiveness in blocking surface defects under wear conditions provided by the multilayer coatings. Among FGM coatings, the H-L ones showed better wear resistance as indicated by their higher hardness and H/E ratio, which prevents crack propagation and mass loss.

Further research in the field carried out by Anvari et al. compared FGM NiP coatings obtained from single bath approach with those obtained by the so-called stepwise multilayer coatings (SMCs) (Anvari et al. 2015a). As can be seen in Figure 7.10, the obtained deposits were dense and homogeneous, the thickness of the plated layers was 50 ± 5 µm. FGM coating showed a phosphorus-graded content ranging from 12.3 to 3 wt.% from the substrate–coating interface toward the coating–air interface. In the case of the SMC coatings, the measured phosphorus content, obtained from the individual plated coatings, was 3.03, 6.40 and 12.21 wt.% for the low, medium and high phosphorus layers, respectively.

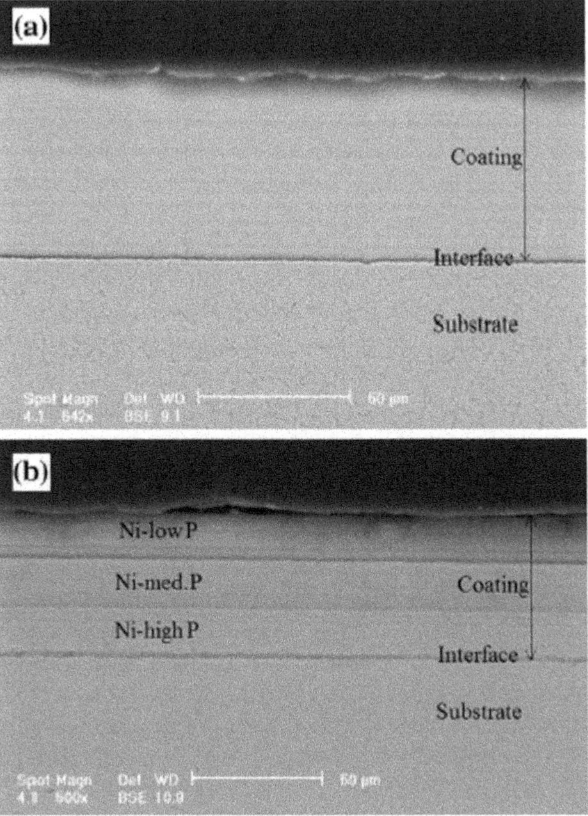

FIGURE 7.10 Cross-section of (a) FGM and (b) SMC coatings. (With kind permission from Springer Science+Business Media: *J. Mater. Eng. Perform.*, Novel investigation on nanostructured multilayer and functionally graded NiP electroless coatings on stainless steel, 24, 2015a, 2373–2381, Anvari, S.R. et al.)

The analysis of the H/E ratio carried out by nano-indentation confirmed the hardness profile along the thickness of the coating for FGMs obtained from the single bath approach. This hardness profile was related to changes in the microstructure of the coatings because of the graded phosphorus content. On the other hand, the SMC coatings showed a more defined hardness profile due to the stepwise changes in phosphorus content. Both methods allow the production of coatings with gradual changes in hardness, which help to reduce the stress accumulation at the substrate-coating interphase, by the presence of the high-phosphorus layer, whereas wear and hardness are maximized at the top surface of the coating by the low-phosphorus coating.

The single bath approach has also been investigated to produce nano-sized NiP multilayers coating for its application in magnetic heads for writing and reading magnetic-recorded high-density data (Chen et al. 2010a). Thus, nano-multilayer structures were deposited on silicon and Fe-Si substrates by appropriately changing the pH of the bath during the process. Alternating pH during the plating bath provides changes in the crystallinity of the layers due to the phosphorus content. Varying the phosphorus content promotes changes in the coercivity and squareness ratio for as-deposited coatings. The application of nickel multilayer on the iron film promotes the enhancement of its coercivity and squareness ratio. This effect appears to be independent of the number of periodic duplex layers applied.

Another interesting choice for electroless multilayer production arises from the combination of NiP and NiB coatings aiming to benefit from their complementary features in terms of mechanical properties and corrosion resistance. Vitry et al. investigated the effect of combining NiP and NiB coatings in multilayered coatings composed of 10 consecutive layers (Vitry and Bonin 2017). The microstructure of obtained coatings is shown in Figure 7.11. Interphases among layers can be clearly observed in the multilayered coatings and the columnar growth of the NiB coatings can be observed in both mono and multilayer coatings.

GDOES analysis showed an increased phosphorus content during the plating time for NiP layers in the range 7–8 wt.%, whereas the boron content remains at 6.5 wt.% during the plating process. The hardness of the multilayer coatings is influenced by the top layer, showing similar values for NiP in both the monolayer and multilayer approach. On the other hand, outer NiB layer is harder in comparison to NiP-based coatings as expected. Regarding wear resistance, multilayer coatings showed a very similar behavior between them but interestingly they exhibit a TWI 2–3 times lower than monolayer coatings. This different behavior of the multilayered coatings is probably connected to the alternating layers structure and the presence of the interfaces along the coating thickness. The corrosion resistance of the coatings is also sensitive to the outer layer. These coatings with a NiP outer layer (monolayer and multilayer systems) have a tendency to passivation at 0.2 V versus Ag/AgCl (KCl saturated), whereas NiB monolayers showed passivation at −0.2 V versus Ag/AgCl and the NiP/NiB multilayer at 0 V versus Ag/AgCl. Moreover, the salt spray chamber test revealed slightly better corrosion resistance for the NiP/NiB system than that for the Ni-B/NiP multilayer. The presence of the interfaces between stacked layers favors the sealing of the developed pores, especially in the case of the NiB coatings.

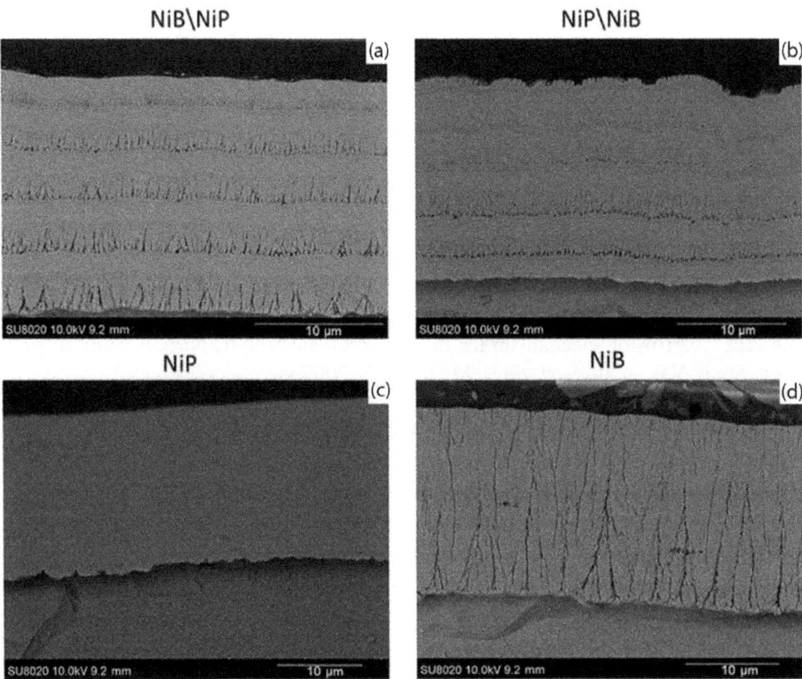

FIGURE 7.11 Cross-section of obtained coatings. (a) NiB/NiP multilayer, (b) NiP/NiB multilayer, (c) NiP monolayer and (d) NiB monolayer. (From Vitry, V. and Bonin, L., *Electrochim. Acta*, 243, 7–17, 2017.)

As in the case of duplex coatings, multilayered coatings are of great interest for protecting light alloys such as aluminum and/or magnesium, which show poor corrosion resistance, while enhancing their mechanical properties. Gu et al. plated a multilayer system composed of 2.5 wt.% P protective layer followed by 9.2 wt.% P and 5.4 wt.% P as outer layers on AZ91D substrates (Gu et al. 2005a). Polarization studies on 3wt.% NaCl media revealed the corrosion mechanism provided by the different layers as a result of the difference in corrosion potential showed by the layers with different phosphorus content. In this case, when the multilayer coating is exposed to the corrosive media, pitting occurred at the surface layer reaching the intermediate layer. The more noble character of the 9 wt.% P layer promotes the preferential dissolution of the outer layer, thus acting as a sacrificial coating with regard to the intermediate layer. This mechanism was confirmed by etching of the duplex intermediate/outer layers with proper oxidant solution. Another plating strategy for protecting light alloys is the combination of electroless and electrodeposited coatings. Moreover, according to the better protection ability of the ZrO_2-based NiP composites, this seems to be a promising plating approach for light alloys. Song et al. tested different combinations of NiP, electrodeposited nickel and NiP-ZrO_2 layers for protecting AZ91D substrates from corrosion (Song et al. 2007). In the salt spray chamber test, the multilayered coating formed by NiP-ZrO_2/electrolytic Ni/NiP showed the best performance. The proposed mechanism is shown in Figure 7.12.

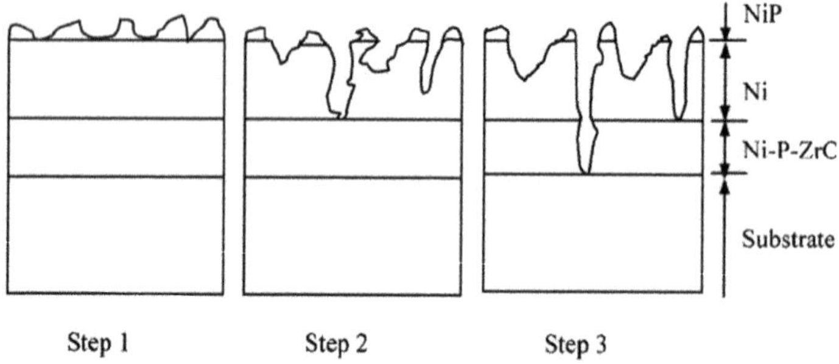

FIGURE 7.12 Corrosion mechanism for the NiP-ZrO_2/electrolytic Ni/NiP multilayer coating. (From Song, Y.W. et al., *Surf. Eng.*, 23, 329–333, 2007.)

In this coating configuration, when the outer NiP coating is exposed to the corrosive media, the aggressive agent can penetrate the NiP layer through its pores reaching the underlying electroplated nickel. Because of the lower E_{corr} value of the NiP coating, in comparison with the electroplated layer, the former acts as a sacrificial anode dissolving preferentially and preventing the attack on the electroplated nickel layer (step 1). Once the NiP layer is completely corroded, the electroplated nickel layer is exposed to the corrosive media, protecting the bottom layer (step 2). When the corrosive agent breaks down the electroplated nickel, the NiP-ZrO_2 plays the role of protecting the substrate (step 3).

A similar approach was used earlier by Gu et al. for protecting steel substrates by intercalating an electroplated nickel layer between inner and outer layers based on electroless NiP 9.5 wt.% (Gu et al. 2005b). In this case, the electrolytic nickel exhibited lower E_{corr} values among the single layers; thus, at the potential value of −0.82V versus Ag/AgCl, it started to corrode after being reached by the pores of the upper NiP layer. Increasing the potential until −0.77V versus Ag/AgCl showed the dissolution of both the outer NiP layer and pure nickel intermediate. Subsequently, it can be clearly noted that the formation of a passive region in the potential range from −0.77 to −0.54 V versus Ag/AgCl after which the corrosion of the inner layer begins (Figure 7.13).

Copper has also been proposed as suitable interlayer in tri-layered NiP-based coatings as reported by Zhao et al. (2015). The aim of this approach was to block the existing pores of the electroless coatings by alternating an electrolytic copper layer between two high-phosphorus NiP coatings. The thickness of inner layer was varied during the study. Porosity analysis revealed a lower porosity degree for the multilayered coatings (samples 7–11) than that for a NiP monolayer with the same coating thickness as depicted in Figure 7.14.

Porosity is also related to the thickness of the inner NiP layer, here expressed as plating time. The results showed that 20–40 min of plating time provide the maximum porosity decrease. Polarization studies showed the improved corrosion performance of the tri-layer coatings, in comparison to the monolayer coating, revealing the positive effect of the copper interlayer on blocking porosity. As shown in Figure 7.15, the tri-layer coating exhibits two passivation regions associated with NiP outer layer and Cu interlayer dissolution in the range −0.4 to −0.3 V versus SCE. Once the copper layer dissolves, the second passivation region between −0.15 and −0.1 V versus SCE corresponds to the inner NiP layer.

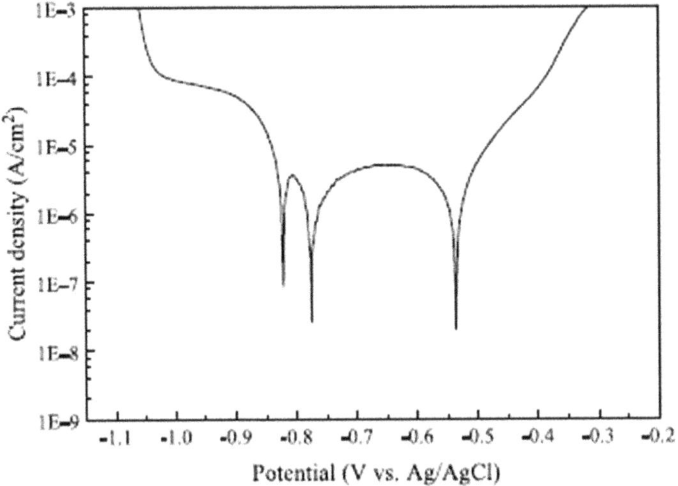

FIGURE 7.13 Polarization curve for the NiP/Ni/NiP coating. (From Gu, C. et al., *Surf. Coat. Technol.*, 197, 61–67, 2005b.)

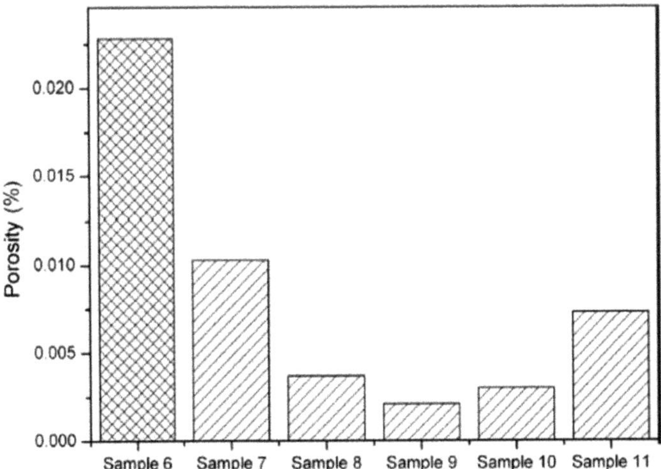

FIGURE 7.14 Porosity ratio for multilayer NiP/Cu/NiP and single NiP coatings. (From Zhao, G.L. et al., *Arch. Metall. Mater.*, 60, 1003–1008, 2015.)

ZnO nanoparticles have also been used for improving the anti-corrosion properties of hybrid electro-electroless deposition of nickel multilayers by Varmazyar et al. (2018). The multilayer arrangement of NiP-ZnO/Ni/NiP showed lower corrosion rates and higher charge transfer resistance than the best performing monolayer. The layers arrangement is based on the differences in corrosion potential between the different layers present in the multilayer coating. Thus, the Ni/NiP pair is a strong galvanic cell that favors the dissolution of the outer NiP layer acting as sacrificial

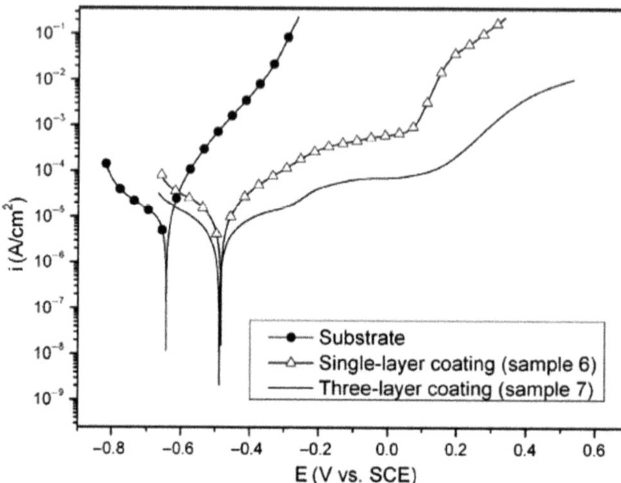

FIGURE 7.15 Potentiodynamic studies for monolayer, tri-layer and substrate. (From Zhao, G.L. et al., *Arch. Metall. Mater.*, 60, 1003–1008, 2015.)

anode under corrosive conditions. Zhao et al. (2005) proposed a graded multilayered system based on NiP/Ni-Cu-P/Ni-Cu-P-PTFE for antifouling applications. The multilayer coating emerged as the most effective combination of layers by improving the adhesion of the whole coating system, whereas copper and PTFE nanoparticles provide better corrosion resistance and anti-sticking properties to the intermediate and outer layers, respectively.

The EP technology also allows the creation of more complex multilayer coatings as the so-called "3D-latticed compositional modulated multilayer," firstly, introduced by Liu et al. (2013b). These classes of coatings are composed of alternated Zn-Ni/NiP layers obtained from the dual bath approach in which the former are applied by ED, whereas the NiP layer is electrolessly plated. Thus, the immersion of the Zn-Ni electroplated layer in the NiP electrolyte promotes the generation of a pattern of ditches, both parallel and transverse to the plated surface. These ditches are filled with the new NiP coating that is being formed, interconnecting in this way both layers in the X and Y direction. At longer plating times, and once the ditches are filled, the NiP coating grows until covering the entire surface. The repetition of this pattern gives rise to a 3D-lattice structure in which the NiP inserts in the Zn-Ni sublayer. The architecture of the 3D-latticed multilayer coatings is shown in Figure 7.16.

When compared to traditional compositionally modulated Zn-Ni/Ni coatings obtained by ED, the 3D lattice coatings provide a superior corrosion resistance. The traditional modulated coatings can only block the advance of the corrosion in the Y-axis direction by the interphases parallel to the substrate direction. 3D lattice coatings can block the advance of the corrosive attack in the X- and Z-axis directions by means of the NiP inserts into Zn-Ni inner layer as depicted in Figure 7.17, providing enhanced corrosion resistance to the multilayer system.

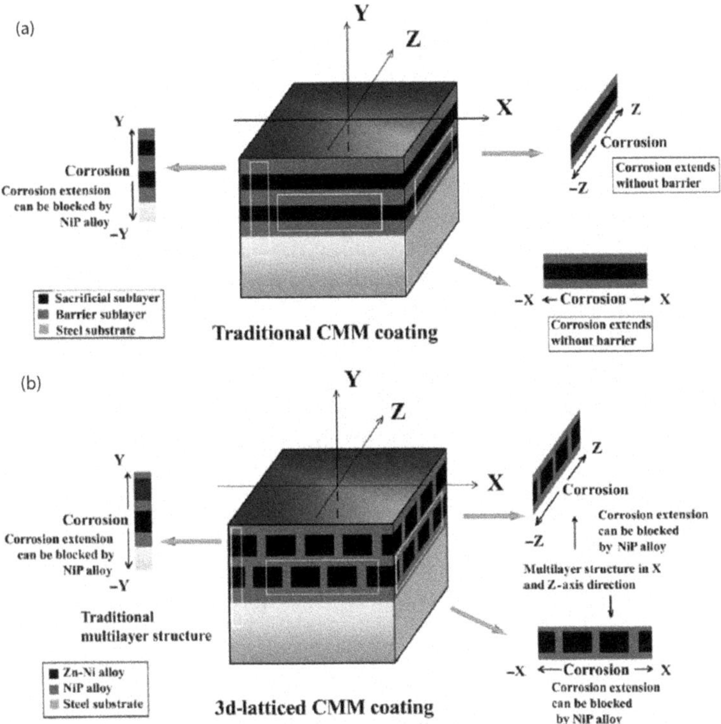

FIGURE 7.16 Multilayered coatings: (a) traditional and (b) 3D-latticed architecture. (From Liu, J.H. et al., *Mater. Corros.*, 64, 335–340, 2013b.)

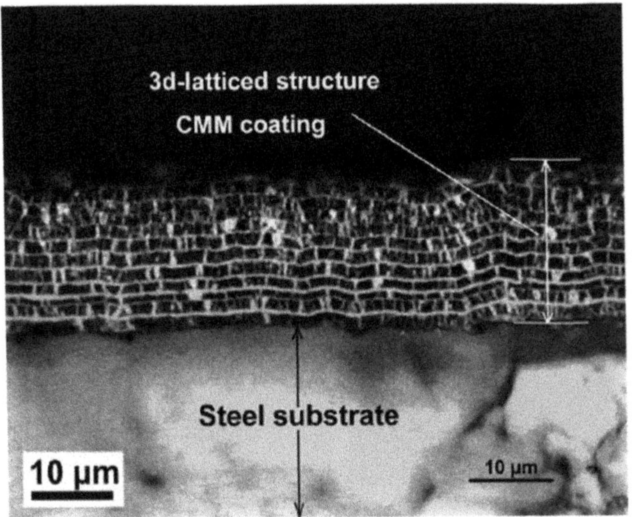

FIGURE 7.17 Optical Microscope (OM) image of a 3D-latticed multilayer (22 layers). (From Liu, J.H. et al., *Mater. Corros.*, 64, 335–340, 2013b.)

7.4 ANNEALING EFFECT

As already described in various chapters of this book, annealing of electroless coatings, either on phosphorus or boron-based coatings, has a profound effect on their crystalline structure and so in their properties. The annealing of electroless coatings can pursue different purposes such as elimination of hydrogen embrittlement, increasing the mechanical properties of the coating and enhancing the adhesion between the electroless coating and the substrate. Usually, the annealing of EN coatings is carried out in the range of 320°C–500°C for annealing times ranging from 1 to up to 8 h, being 400°C for 1 h time the most commonly used annealing conditions for improving the performance of the coatings (Mallory and Hajdu 1990; Riedel 1991).

In the case of the multilayer coatings, the microstructural changes promoted by the thermal treatment have a synergistic effect with the interfaces present at the multilayered coatings, thus further improving the performance, especially in terms of corrosion resistance and adhesion, of this kind of coating. Nevertheless, higher temperatures, or long annealing times, can lead to the formation of a diffusion layer between the coating and the substrate, giving rise to a bi-layered structure as shown in Figure 7.18.

Annealing of the coating promoted an increase in the hardness values up to 11.4 GPa with a small plastic indentation depth of 1.4 μm (Hamada et al. 2015b).

FIGURE 7.18 The annealed NiP coating at 700°C for 1 h on TWI steel. (a, b) Ni-Ni_3P grains, (c) cross-section of the coating showing the diffusion layer and (d) cross-section of the coated TWI steel after tapered. (From Hamada, A.S. et al., *Appl. Surf. Sci.*, 356, 1–8, 2015b.)

High-temperature ranges lead to grain coarsening of the Ni_3P phase, thus having a detrimental effect on the final properties of the annealed coatings. High annealing temperatures can also promote diffusion of some elements from both the substrate and the electroless coatings. Nevertheless, on steel substrates, even at high temperatures, phosphorus distribution along the coating usually remains unchanged due to its low diffusivity into this material (Hamada et al. 2015b).

On the other hand, the properties of duplex coatings can also be optimized by proper thermal annealing. However, thermal annealing below the optimal temperature range for nickel matrix recrystallization has also confirmed the suitability of duplex layers for working on high-temperature environments. Thus, annealing at 200°C for 2 h in NiP(mid)/NiP-ZrO_2 coatings did not lead to any microstructural changes but could increase the internal stress accumulation in the coating. The presence of the outer layer of composite material helps to block the rack propagation originated during internal stress development (Georgiza et al. 2013).

Annealing of NiP/NiP-YSZ duplex coatings confirmed the expected changes in the microstructure of the coating as a result of matrix recrystallization and precipitation of Ni_3P phase. Due to the penetration depth of the X-ray beam during X-ray diffraction (XRD) experiments, the results obtained with this technique are completely dependent on the selected order of the single layers used in the production of the duplex coatings. In the case of NiP/NiP-YSZ, the outer layer is based on a low-phosphorus NiP coating with YSZ as reinforcing material. The XRD spectra shown in Figure 7.19 depict the diffractogram of as-deposited NiP single layer exhibiting a broad diffraction peak at 45°, ascribed to the (111) diffraction plane of the fcc nickel phase. After annealing, sharpening of the (111) plane along with Ni_3P phase reflections can be clearly described. YSZ cannot be detected but influence the (111) reflection of the fcc nickel (Luo et al. 2017).

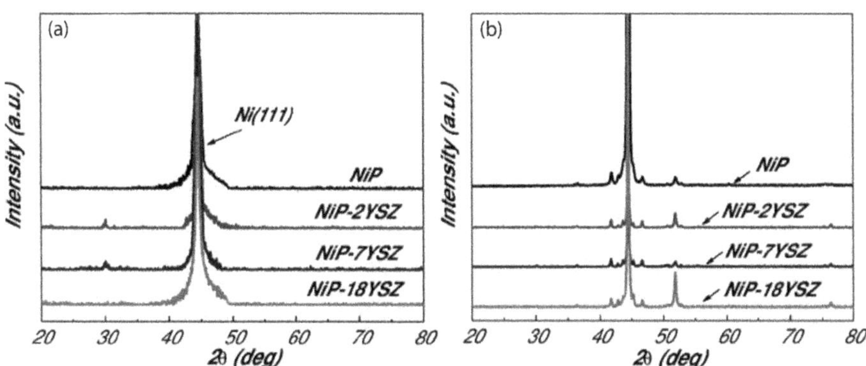

FIGURE 7.19 XRD spectra of single NiP coating and NiP/NiP-YSZ duplex coating (a) before and (b) after thermal annealing at 400°C for 1 h. (Reprinted from *Surf. Coat. Technol.*, 311, Luo, H. et al., Synthesis of a duplex NiP-YSZ/NiP nanocomposite coating and investigation of its performance, 70–79, Copyright 2017, with permission from Elsevier B.V.)

The Increment of hardness on duplex coatings including composite layers is usually related to the formation of hard precipitates, boron- or phosphorus-based, depending on the reducing agent used, along with the strengthening effect of particle co-deposition that blocks dislocation at the grain boundaries. In the case of NiP/NiB coatings, the temperature required for starting phase transformation during annealing is in the range of 300°C–350°C; a further increase in the annealing temperature leads to higher degrees of crystallinity for both NiP and NiB coatings (Baibordi et al. 2012). High hardness and low wear rate of annealed duplex coatings are also dependent on the selected outer layer. Additionally, NiP/NiB coatings are also suitable surface-finishing treatments for improving the properties of light materials such as magnesium and aluminum. However, the usual temperature range for thermal annealing cannot be applied to these materials, thus making it necessary to carry out annealing treatments at lower temperatures and longer periods. In the case of aluminum substrates, duplex NiP/NiB layers exhibit increased corrosion resistance and lower wear rates in comparison with as-deposited duplex coatings (Vitry et al. 2012). This improvement is ascribed to the initial crystallization of the NiB outer layer coating caused by the low-temperature annealing. For NiP/NiB duplex coatings on steel substrates, annealing can have a substantial impact on alloying element distribution along the coating thickness, whereas the total alloying element concentration remains constant. Under annealing, phosphorus migrates outward, whereas boron migrates mainly inward the coating, being this effect more pronounced in NiP bilayers than in NiB bi-layered coatings (Vitry et al. 2017). Laser can also be used for annealing duplex coatings. In the case of NiP/Ni-W-P(outer), the laser annealing treatments can lead to incomplete crystallization or different degrees of crystallization between the inner and outer layers. This fact is related to the temperature gradient of the laser annealing process along the thickness of the coating. Even in not fully crystallized duplex coatings, hardness and wear are remarkably increased due to Ni_3P phase precipitation as confirmed by XRD. Laser-annealed bilayer coatings revealed the suitability of a NiP under layer for improving the wear resistance of duplex coatings in comparison with traditional Ni-W-P monolayered coatings (Liu et al. 2012; Liu et al. 2013a).

In the case of multilayer coatings, few reports can be found on the effect of annealing on the final properties of the coatings. Usually, the improvement of the coating performance is mainly associated with the nature of the single layers, which compose the multilayer coatings. In order to determine the effect of the annealing process on multilayer structures, stacking layers of the same material is the only way to prove the sole effect of the multilayer approach. The annealing of low-phosphorus multilayer coatings (three layers) showed an increased corrosion protection after annealing in comparison with as-deposited multilayered and monolayer coatings. In this case, as there are no differences in terms of E_{corr} values for the different layers, both grain coarsening and the presence of the interfaces are the main contributors to the corrosion resistance. At the same polarization conditions, the multilayer coating was still protecting the substrate from the corrosive media, whereas monolayered coatings suffer severe corrosion. The enhanced interfacial area, promoted by the multilayer architecture, along with the recrystallization of the metallic matrix allows a change in the corrosion mechanism. Thus, in the multilayer coating, when the Cl^- anion reaches the surface of the coating, it can cross the first layer through the defects or porosity of the layer. At the interface, the corrosion spreads laterally where the corrosion products accumulate. At some point, the amount of corrosion products promotes the exfoliation of the outer layer exposing the next layer. This mechanism observed for the annealed samples was not obvious for as-deposited coatings, revealing the importance of the annealing process on the performance of multilayered coatings (Salicio-Paz et al. 2019). On the other hand, the detrimental effect of annealing treatment at 400°C for 1 h has also been described for graded coatings (Sankara Narayanan et al. 2006). The lower corrosion protection of a low-P/med-P/high-P-graded multilayer coating is related to phosphorus migration among the layers along with crystallization of the metallic matrix, which enhances the number of grain boundaries that act as new paths for the corrosive attack. However, in terms of mechanical properties, this coating configuration allows obtaining harder coatings with improved wear resistance. Thus, for low-P/mid-P/high-P (outer) multilayers the presence of a hardness profile after annealing along the coating thickness may have a positive effect on the wear behavior of the whole system (Hadipour et al. 2015) (Figure 7.20).

The number of layers influences the wear behavior of the system, something associated also with the graded hardness profile and with the gradual change in phases composition along the thickness achieved after annealing. These facts allow the inhibition of crack propagation during the wear test due to the graded content of brittle phases (Ni_3P, Ni_2P) along the coating thickness. Similar results can be found for the comparison between SMCs and functionally graded coatings obtained by the dual and single bath approach, respectively.

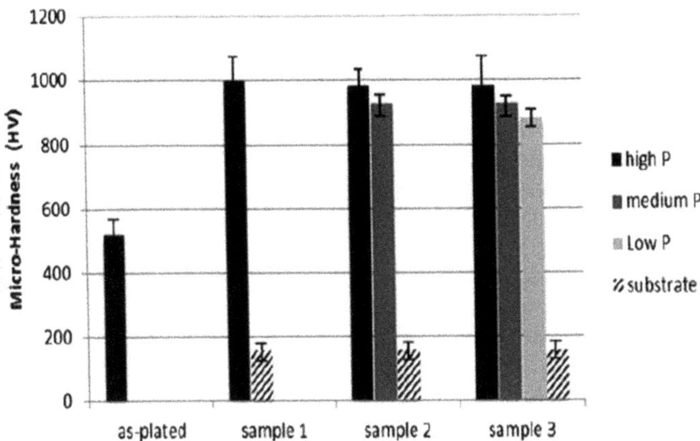

FIGURE 7.20 Hardness profile for as-plated and annealed coatings. Sample 3 refers to the high-P/med-P/low-P multilayer. (From Hadipour, A. et al., *Surf. Eng.*, 31, 399–405, 2015.)

7.5 FUTURE TRENDS

As it has been described along this section, electroless multilayered coatings possess a high potential for improving the characteristics of traditionally obtained electroless monolayer coatings. It is clear that the multilayer approach has a great benefit on the final performance of the coatings compared to single coatings. Moreover, the vast number of possible combination of layers of different nature allows tailor-made solutions depending on the final application of the coated components. This process versatility along with the known features of the electroless process; namely even coatings on complex 3D geometries, absence of electrical current, ability to plate on dielectric substrates and alloying element content dependent properties, highlight the role of the electroless multilayer approach for the development of novel metallic coatings. Moreover, it is worthy to highlight that the electroless multilayer approach is a cost-effective technology especially in comparison with vacuum-based technologies usually employed for the production of multilayered coatings.

In the near future, it is expected that novel combination of single layers, polyalloys and composite layers with new reinforcing materials could open new venues in the field of surface engineering. Many of the most important technological fields at present (i.e. electric vehicle, electronics, aerospace, etc.) are steadily demanding technical coatings with enhanced features (hardness, wear resistance, magnetic properties, corrosion resistance, etc.), and it is expected that multilayer coatings will be able to fulfill these requirements in a cost-effective manner. In this way, it will be needed to find new application market niches in which the electroless multilayer coatings could fit, going further than the traditional sectors in which the EN coatings have been applied. Thus, the current understanding of the electroless process will favor the development of new electroless-based coatings suitable of being incorporated into multilayered coatings, thus enhancing the potential applications fields in which they can be applied.

REFERENCES

Antonelli, S. B., T. L. Allen, D. C. Johnson, and V. M. Dubin. 2006. "Determining the Role of W in Suppressing Crystallization of Electroless Ni–W–P Films." *Journal of The Electrochemical Society* 153 (6): J46–J49. doi:10.1149/1.2193332.

Anvari, S. R., S. M. Monirvaghefi, and M. H. Enayati. 2015a. "Novel Investigation on Nanostructured Multilayer and Functionally Graded NiP Electroless Coatings on Stainless Steel." *Journal of Materials Engineering and Performance* 24 (6). Springer US: 2373–2381. doi:10.1007/s11665-015-1533-y.

Anvari, S. R., S. M. Monirvaghefi, and M. H. Enayati. 2015b. "Wear Characteristics of Functionally Graded Nanocrystalline Ni–P Coatings." *Surface Engineering* 31 (9): 693–700. doi:10.1179/1743294415Y.0000000023.

Ashassi-Sorkhabi, H., and M. Es'haghi. 2013. "Corrosion Resistance Enhancement of Electroless Ni–P Coating by Incorporation of Ultrasonically Dispersed Diamond Nanoparticles." *Corrosion Science* 77: 185–193. doi:10.1016/j.corsci.2013.07.046.

Baibordi, A., M. H. Bina, K. Amini, and A. Dehghan. 2012. "Investigating Tribological Characteristics of NiP and Double—Layered NiP/NiB Electroless Coatings Applied to the Carbon Mild Steel Ck45." *International Journal of Iron & Steel Society of Iran* 9 (2): 1–4.

Balaraju, J. N., V. E. Selvi, V. K. W. Grips, and K. S. Rajam. 2006. "Electrochemical Studies on Electroless Ternary and Quaternary NiP Based Alloys." *Electrochimica Acta* 52 (3): 1064–1074. doi:10.1016/j.electacta.2006.07.001.

Bonin, L., V. Vitry, and F. Delaunois. 2018. "Corrosion Behaviour of Electroless High Boron-Mid Phosphorous Nickel Duplex Coatings in the as-Plated and Heat-Treated States in NaCl, H_2SO_4, NaOH and Na_2SO_4 Media." *Materials Chemistry and Physics* 208 (C): 77–84. doi:10.1016/j.matchemphys.2017.12.030.

Bull, S. J., and A. M. Jones. 1996. "Multilayer Coatings for Improved Performance." *Surface and Coatings Technology* 78 (1–3): 173–184. doi:10.1016/0257-8972(94)02407-3.

Chang, L., P. W. Kao, and C. H. Chen. 2007. "Strengthening Mechanisms in Electrodeposited NiP Alloys with Nanocrystalline Grains." *Scripta Materialia* 56 (8): 713–716. doi:10.1016/j.scriptamat.2006.12.036.

Chen, W. J., S. H. Hsieh, W. L. Liu, and W. L. Chen. 2010a. "Structures and Magnetic Properties of Electroless Ni Multilayers on Fe/Si Substrate." *Journal of Nanoscience and Nanotechnology* 10 (7): 4706–4710. doi:10.1166/jnn.2010.1704.

Chen, W. W., Y. D. He, and W. Gao. 2010b. "Synthesis of Nanostructured Ni-TiO_2 Composite Coatings by Sol-Enhanced Electroplating." *Journal of the Electrochemical Society* 157 (8): E122–E128. doi:10.1149/1.3442366.

Chen, X.-M., G.-Y. Li, and J.-S. Lian. 2008. "Deposition of Electroless NiP/Ni-W-P Duplex Coatings on AZ91D Magnesium Alloy." *Transactions of Nonferrous Metals Society of China* 18. Elsevier: s323–s328. doi:10.1016/S1003-6326(10)60225-7.

Clemens, B. M., H. Kung, and S. A. Barnet. 1999. "Structure and Strength of Multilayers." *MRS Bulletin*: 20–26. doi:10.1086/651387.

Contreras, A., C. León, O. Jimenez, E. Sosa, and R. Pérez. 2006. "Electrochemical Behavior and Microstructural Characterization of 1026 Ni–B Coated Steel." *Applied Surface Science* 253 (2): 592–599. doi:10.1016/j.apsusc.2005.12.161.

De Hazan, Y., F. Knies, D. Burnat, T. Graule, Y. Yamada-Pittini, C. Aneziris, and M. Kraak. 2012. "Homogeneous Functional Ni–P/Ceramic Nanocomposite Coatings via Stable Dispersions in Electroless Nickel Electrolytes." *Journal of Colloid and Interface Science* 365 (1). Academic Press: 163–171.

Elsener, B., M. Crobu, M. A. Scorciapino, and A. Rossi. 2008. "Electroless Deposited NiP Alloys: Corrosion Resistance Mechanism." *Journal of Applied Electrochemistry* 38 (7): 1053–1060. doi:10.1007/s10800-008-9573-8.

Georgiza, E., J. Novakovic, and P. Vassiliou. 2013. "Characterization and Corrosion Resistance of Duplex Electroless NiP Composite Coatings on Magnesium Alloy." *Surface and Coatings Technology* 232. Elsevier B.V.: 432–439. doi:10.1016/j.surfcoat.2013.05.047.

Gool, A. P. van, P. J. Boden, and S. J. Harris. 1987. "Corrosion Behaviour of Some Electroless Nickel–Phosphorus Coatings." *Transactions of the IMF* 65 (1): 108–114. doi:10.1080/00202967.1987.11870782.

Gu, C., J. Lian, and Z. Jiang. 2005a. "Multilayer NiP Coating for Improving the Corrosion Resistance of AZ91D Magnesium Alloy." *Advanced Engineering Materials* 7 (11): 1032–1036. doi:10.1002/adem.200500136.

Gu, C., J. Lian, G. Li, L. Niu, and Z. Jiang. 2005b. "High Corrosion-Resistant NiP/Ni/NiP Multilayer Coatings on Steel." *Surface and Coatings Technology* 197 (1): 61–67. doi:10.1016/j.surfcoat.2004.11.004.

Hadipour, A., S. M. Monirvaghefi, and M. E. Bahrololoom. 2015. "Electroless Deposition of Graded Ni–P Coatings." *Surface Engineering* 31 (6): 399–405. doi:10.1179/1743294414Y.0000000430.

Hamada, A. S., P. Sahu, and D. A. Porter. 2015a. "Indentation Property and Corrosion Resistance of Electroless Nickel-Phosphorus Coatings Deposited on Austenitic High-Mn TWIP Steel." *Applied Surface Science* 356. Elsevier B.V.: 1–8. doi:10.1016/j.apsusc.2015.07.153.

Hamada, A. S., P. Sahu, and D. A. Porter. 2015b. "Indentation Property and Corrosion Resistance of Electroless Nickel-Phosphorus Coatings Deposited on Austenitic High-Mn TWIP Steel." *Applied Surface Science* 356: 1–8. doi:10.1016/j.apsusc.2015.07.153.

Li, J., Y. Tian, Z. Huang, and X. Zhang. 2006. "Studies of the Porosity in Electroless Nickel Deposits on Magnesium Alloy." *Applied Surface Science* 252: 2839–2846.

Li, J., D. Wang, H. Cai, A. Wang, and J. Zhang. 2015. "Competitive Deposition of Electroless Ni-W-P Coatings on Mild Steel via a Dual-Complexant Plating Bath Composed of Sodium Citrate and Lactic Acid." *Surface and Coatings Technology* 279 (5). Elsevier B.V.: 9–15. doi:10.1016/j.surfcoat.2015.08.017.

Liu, H., R. X. Guo, J. S. Bian, and Z. Liu. 2013a. "Effect of Laser-Induced Nanocrystallisation on the Properties of Electroless NiP/Ni-W-P Duplex Coatings." *Crystal Research and Technology* 48 (2): 100–109. doi:10.1002/crat.201200427.

Liu, H., R. X. Guo, and Z. Liu. 2012. "Characteristics of Microstructure and Performance of Laser-Treated Electroless NiP/Ni-W-P Duplex Coatings." *Transactions of Nonferrous Metals Society of China (English Edition)* 22 (12). The Nonferrous Metals Society of China: 3012–320. doi:10.1016/S1003-6326(11)61564-1.

Liu, J. H., J. L. Chen, Z. Liu, M. Yu, and S. M. Li. 2013b. "Fabrication of Zn-Ni/NiP Compositionally Modulated Multilayer Coatings." *Materials and Corrosion* 64 (4): 335–340. doi:10.1002/maco.201106140.

Lu, G., and G. Zangari. 2003. "Study of the Electroless Deposition Process of NiP-Based Ternary Alloys." *Journal of The Electrochemical Society* 150 (11). The Electrochemical Society: C777. doi:10.1149/1.1614799.

Luo, H., M. Leitch, H. Zeng, and J.-L. Luo. 2018. "Characterization of Microstructure and Properties of Electroless Duplex Ni-W-P/NiP Nano-ZrO_2 Composite Coating." *Materials Today Physics* 4. Elsevier Ltd: 36–42. doi:10.1016/j.mtphys.2018.03.001.

Luo, H., X. Wang, S. Gao, C. Dong, and X. Li. 2017. "Synthesis of a Duplex NiP-YSZ/NiP Nanocomposite Coating and Investigation of Its Performance." *Surface and Coatings Technology* 311: 70–79. doi:10.1016/j.surfcoat.2016.12.075.

Mallory, G. O. 1974. "Ternary and Quaternary Electroless Nickel Alloys." *Transactions of the IMF* 52 (1). Taylor & Francis: 156–161. doi:10.1080/00202967.1974.11870322.

Mallory, G. O., and J. B. Hajdu. 1990. *Electroless Plating: Fundamentals and Applications. American Electroplaters and Surface Finishing Society, Orlando.* Orlando, FL: American Electroplaters and Surface Finishers Society.

Narayanan, T. S., K. Krishnaveni, and S. K. Seshadri. 2003. "Electroless NiP/NiB Duplex Coatings: Preparation and Evaluation of Microhardness, Wear and Corrosion Resistance." *Materials Chemistry and Physics* 82: 771–779. doi:10.1016/S0254-0584(03)00390-0.

Okinaka, Y., and T. Osaka. 2008. "Electroless Deposition Processes: Fundamentals and Applications." *Advances in Electrochemical Science and Engineering* 3: 55–116. doi:10.1002/9783527616770.ch2.

Petro, R. A. 2017. "Developments in Co(NiP) Hybrid Electro-Electroless Deposited (HEED) Alloys and Composites." *ECS Transactions* 75 (34): 61–66.

Petro, R., and M. Schlesinger. 2014. "Development of Hybrid Electro-Electroless Deposit (HEED) Coatings and Applications." *Journal of the Electrochemical Society* 161 (10): D470–D475. doi:10.1149/2.0331410jes.

Pogrebnjak, A. D., Y. O. Kravchenko, O. V. Bondar, B. Zhollybekov, and A. I. Kupchishin. 2018. "Structural Features and Tribological Properties of Multilayer Coatings Based on Refractory Metals." *Protection of Metals and Physical Chemistry of Surfaces* 54 (2): 240–258. doi:10.1134/S2070205118020107.

Ranganatha, S., T. V. Venkatesha, and K. Vathsala. 2012. "Electroless Ni-W-P Coating and Its Nano-WS_2 Composite: Preparation and Properties." *Industrial and Engineering Chemistry Research* 51 (23): 7932–7940. doi:10.1021/ie300104w.

Reddy, V. V. N, B. Ramamoorthy, and P. K. Nair. 2000. "A Study on the Wear Resistance of Electroless Ni–P/Diamond Composite Coatings." *Wear* 239 (1): 111–116. doi:10.1016/S0043-1648(00)00330-6.

Riedel, W. 1991. *Electroless Nickel Plating.* ASM International and Finishing Publications, Stevenage, England.

Sankara Narayanan, T. S. N., I. Baskaran, K. Krishnaveni, and S. Parthiban. 2006. "Deposition of Electroless NiP Graded Coatings and Evaluation of Their Corrosion Resistance." *Surface and Coatings Technology* 200: 3438–3445. doi:10.1016/j.surfcoat.2004.10.014.

Salicio-Paz, A., Grande, H., Pellicer, E., Sort, J., Fornell, J., Offoiach, R., Lekka, M., García-Lecina, E. 2019. "Monolayered versus multilayered electroless NiP coatings: Impact of the plating approach on the microstructure, mechanical and corrosion properties of the coatings." *Surface and Coatings Technology* 368: 138–146, doi: 10.1016/j.surfcoat.2019.04.013.

Sharma, A., and A. K. Singh. 2012. "Electroless NiP and NiP-Al2O3 Nanocomposite Coatings and Their Corrosion and Wear Resistance." *Journal of Materials Engineering and Performance* 22 (1): 176–183. doi:10.1007/s11665-012-0224-1.

Song, Y. W., D. Y. Shan, and E. H. Han. 2007. "High Corrosion Resistance Multilayer Nickel Coatings on AZ91D Magnesium Alloys." *Surface Engineering* 23 (5): 329–333. doi:10.1179/174329407X260528.

Sridhar, N., and K. Udaya Bhat. 2013. "Effect of Deposition Time on the Morphological Features and Corrosion Resistance of Electroless Ni-High P Coatings on Aluminium." *Journal of Materials* 2013: 1–7. doi:10.1155/2013/985763.

Sudagar, J., J. Lian, and W. Sha. 2013. "Electroless Nickel, Alloy, Composite and Nano Coatings—A Critical Review." *Journal of Alloys and Compounds* 571. Elsevier B.V.: 183–204. doi:10.1016/j.jallcom.2013.03.107.

Suiyuan, C., S. Ying, F. Hong, L. Jing, L. Changsheng, and S. Kai. 2012. "Synthesis of NiP-PTFE-Nano-Al_2O_3 Composite Plating Coating on 45 Steel by Electroless Plating." *Journal of Composite Materials* 46 (12): 1405–1416. doi:10.1177/0021998311420312.

Szczygieł, B., and A. Turkiewicz. 2008. "The Effect of Suspension Bath Composition on the Composition, Topography and Structure of Electrolessly Deposited Composite Four-Component Ni–W–P–ZrO_2 Coatings." *Applied Surface Science* 254 (22). North-Holland: 7410–7416. doi:10.1016/J.APSUSC.2008.05.342.

Szczygieł, B., and A. Turkiewicz. 2009. "The Rate of Electroless Deposition of a Four-Component Ni-W-P-ZrO_2 Composite Coating from a Glycine Bath." *Applied Surface Science* 255 (2008): 8414–8418. doi:10.1016/j.apsusc.2009.05.145.

Varmazyar, A., S. R. Allahkaram, and S. Mahdavi. 2018. "Deposition, Characterization and Evaluation of Monolayer and Multilayer Ni, Ni–P and Ni–P–Nano ZnO_p Coatings." *Transactions of the Indian Institute of Metals* 71 (6). Springer India: 1301–1309. doi:10.1007/s12666-018-1279-y.

Vitry, V., and L. Bonin. 2017. "Formation and Characterization of Multilayers Borohydride and Hypophosphite Reduced Electroless Nickel Deposits." *Electrochimica Acta* 243. Pergamon: 7–17. doi:10.1016/j.electacta.2017.04.152.

Vitry, V., A. Sens, A.-F. F. Kanta, and F. Delaunois. 2012. "Wear and Corrosion Resistance of Heat Treated and As-Plated Duplex NiP/NiB Coatings on 2024 Aluminum Alloys." *Surface and Coatings Technology* 206 (16): 3421–3427. doi:10.1016/j.surfcoat.2012.01.049.

Vitry, V., L. Bonin, and L. Malet. 2017. "Chemical, Morphological and Structural Characterisation of Electroless Duplex NiP/NiB Coatings on Steel." *Surface Engineering* 34 (6). Taylor & Francis: 1–10. doi:10.1080/02670844.2017.1320032.

Wang, H., M. Xie, Q. Zong, Z. Liu, X. Huang, and Y. Jin. 2015a. "Electroless Ni–W–Cr–P Alloy Coating with Improved Electrocatalytic Hydrogen Evolution Performance." *Surface Engineering* 31 (3): 226–231. doi:10.1179/1743294414Y.0000000373.

Wang, Y., X. Shu, S. Wei, C. Liu, W. Gao, R. A. Shakoor, and R. Kahraman. 2015b. "Duplex NiP-ZrO_2/NiP Electroless Coating on Stainless Steel." *Journal of Alloys and Compounds* 630. Elsevier B.V.: 189–194. doi:10.1016/j.jallcom.2015.01.064.

Zhao, G. L., Y. Zou, Y. L. Hao, and Z. D. Zou. 2015. "Corrosion Resistance of Electroless NiP/Cu/NiP Multilayer Coatings." *Archives of Metallurgy and Materials* 60 (2A): 1003–1008. doi:10.1515/amm-2015-0250.

Zhao, Q., Y. Liu, C. Wang, S. Wang, and H. Müller-Steinhagen. 2005. "Effect of Surface Free Energy on the Adhesion of Biofouling and Crystalline Fouling." *Chemical Engineering Science* 60 (17): 4858–4865. doi:10.1016/j.ces.2005.04.006.

8 Electroless Composite Coatings

Sudagar Jothi, R. Muraliraja, T. R. Tamilarasan, Sanjith Udayakumar, and A. Selvakumar

CONTENTS

8.1 HISTORY AND EVOLUTION OF ELECTROLESS COMPOSITE COATING

At first, electroless composite coatings were unsuccessful and frequently resulted in the decomposition of the bath due to the dispersion of fine particles that increased the surface area of the electroless bath by nearly 700–800 times. However, the problem was solved with the help of appropriate stabilizers.

The research pertaining to incorporation of second phase particles in the electroless nickel bath began as early as 1960s. Odekerken (1968) interposed the structure with an intermediate layer containing finely divided particles distributed within the metallic matrix. During the 1970–1980 period, the research on electroless composite coatings was in the beginning stage. There are approximately 200 papers since this period in the Scopus database and the earliest one was entitled "Recent advances in electroless nickel deposits" (Parker 1972). Metzger and Florian (1976) developed electroless nickel coatings containing micron-sized alumina (Al_2O_3) particles in 1976. The first commercial application of electroless SiC composite coatings was used for Wankel internal combustion engine, and another commercial composite incorporating nickel-phosphorus-polytetrafluoroethylene (NiP-PTFE) was co-deposited in 1981. Nevertheless, the co-deposition of diamond and PTFE particles was more challenging than the synthesis of composites incorporating Al_2O_3 or SiC (Feldstein et al. 1983). The possibility to incorporate fine second-phase particles of submicron to nanosize within a metal/alloy matrix has introduced a new generation of composite coatings.

8.2 PROCESSING OF ELECTROLESS NICKEL COMPOSITE COATING

8.2.1 Electroless Nickel Composite Coatings

Composite electroless nickel coatings are defined as those that incorporate distinct particles into the deposit to impart a specific property. The functional particles are evenly and thoroughly distributed in the Electroless Ni (EN) matrix, which is firmly bonded to the substrate. This unique combination of distribution and bond strength makes composite EN coatings extremely long lasting and durable compared with many other wear and lubrication alternatives. Theoretically, almost any type of particle can be co-deposited, as long as it can withstand the conditions within an EN bath and its size is appropriate.

The advantages of electroless composite plating technique over the conventional composite plating techniques include the quality of the deposit, uniformity and excellent tribological properties (Baudrand 1978; Mallory and Hajdu 1991). The process operates by impact (impingement) and deposition of particles on the surface of the workpiece and the subsequent entrapment of these particles by the electroless matrix. There is no molecular bonding between the metal matrix and the incorporated hard particles.

8.2.2 Bath Composition for Electroless Nickel Composite Coating

The chief constituents of the electroless bath solution are source of nickel ions, reducing agent, hard particles and other components for supporting the electroless coating process as shown in Figure 8.2, where surfactants are added for the composite electroless plating. The idea of preparing electroless NiP/B composite coatings is to co-deposit secondary phase of solid particles in electroless nickel (phosphorus or boron) coatings and to take benefits of their uniformity, hardness, corrosion resistance and wear resistance. The process is similar to the conventional reduction technique but with suspended particles. The effectiveness of co-deposition depends on bath chemistry, particles nature and operating conditions. The incorporation of particles in the coating is linked to the growth of the surrounding deposit: no molecular bonding is present between the NiP/B matrix and the co-deposited particles. The concentration of the particles in the plating bath has an influence on the level of incorporation. The co-deposition improves the properties of these composite coatings such as wear resistance, lubrication property and oxidation resistance.

Particles and surfactants addition are the main specific components of electroless composite coatings (Figure 8.1). One or more types of particles are selected from the following groups: carbides, nitrides, oxides and carbon-based elements. Particles

Component	Function
Metal ion	Source of metal
Reducing agent	Source of electrons to reduce the metal ions
Complexants	Prevents Ni-phosphate precipitation
Accelerators	Activates the reducing agent and increases the deposition
Buffers	Controls the pH
Stabilizer	Prevents the bath decomposition
Hard Particles	**Desired composite coating**
Wetting agents (surfactants)	**Increases the surface wettability (preferred for composite coating)**

FIGURE 8.1 Electroless composite bath components and their functions.

selection is based on the type of surface to be plated and the desired properties. Surfactants are generally used to favor uniform distribution of hard particles in the NiP/nickel-boron (NiB) matrix and can be anionic, cationic or nonionic surfactants. Their influence will be described in the next section.

8.2.3 Stability of Particles in Suspension in Electroless Plating Baths

Successful co-deposition of particles in electroless nickel plating requires that the particles are suspended in the plating bath and are free of agglomeration. The dispersion of particles in the plating bath is thus a major concern in the electroless plating of composite, even more so for nano-sized particles (Necula et al. 2007). Particles that are several microns in size can be kept in suspension by diverse mixing methods such as circulation, magnetic or mechanical stirring, ultrasound agitation, (Liu and Zhu 2015; Xiang et al. 2001) addition of baffles in the plating system (Zhu and Zhu 2013) and so on. However, this is more difficult to achieve when the size of particles reaches the nanometer range. In that case, mechanical agitation, and the directional flows it creates, leads to lesser incorporation of particles (Xiang et al. 2001). Similar problems are encountered in composite electroplating (Kuo et al. 2004), and ultrasound and surfactants have been proposed as solutions to the agglomeration problem. However, this can result in the lower incorporation of particles in the coating (Liu and Zhu 2015).

In some cases (for electroless plating), measures taken to avoid agglomeration of particles can be very sophisticated, for example, the deposition of a layer of aluminum oxide on diamond particles (Liu et al. 2019).

The control of dispersion by zeta potential (particles with a zeta potential higher than 30 mV or lower than –30 mV are considered as stable in solution) has also been investigated for electroless nickel plating (Necula et al. 2007) but as the plating process is extremely sensitive to pH, it is not possible to use this method as a way to improve the dispersion of particles (Liu and Zhu 2015). It was nevertheless possible to assess the stability of some particles in electroless nickel baths and even to formulate an analytical model for particles dispersion (Liu and Zhu 2015; Necula et al. 2007).

The use of surfactant is thus widely reported to improve particles dispersion, to supplement agitation (excessive agitation leads to some issues) and zeta potential that cannot be modified due to fixed pH.

8.2.3.1 Surfactants

Surfactants are surface active agents used to lower the surface tension of fluid, thus allowing easier dispersion and lowering the interfacial tension between two fluids or at a solid–liquid interface. Generally, in an electroless bath, surfactants are added to promote the plating reaction between the solution and the substrate. These surfactants are usually organic compounds that are amphiphilic in nature, i.e., they contain both hydrophobic tail and hydrophilic head, as a result of which they are soluble in both organic solvents and water. Surfactants reduce the surface tension of water by absorption onto liquid–gas or solid–liquid surfaces. Surfactants are present as isolated molecules in low concentration, and they assemble to form micelles at higher concentration (Schaller et al. 1992).

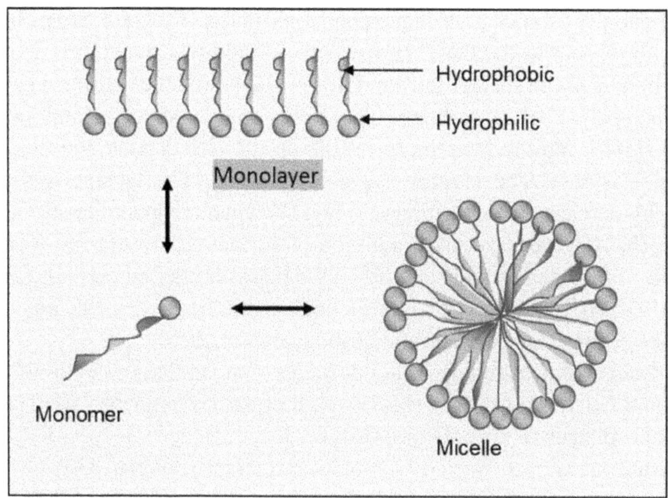

FIGURE 8.2 Monolayer and micelle structure as a function of surfactant concentration.

Surfactants can be classified by the presence of formally charged groups in their head (Elansezhian et al. 2008) (Figure 8.2). *Anionic surfactant:* In this type, the hydrophilic group bears a negative charge, for example, sulfonate salts, alcohol, sulfates, alkyl benzene, phosphoric esters and carboxylic acid salts. *Cationic surfactant:* In this type, the hydrophilic group bears a positive charge, for example, polyamide and its salt, quaternary ammonium salts and amine oxides and salt of long chain amine. *Nonionic surfactant:* In this type, the hydrophilic group has no charge, for example, polyethylenated alkyl phenols, alcohol ethoxylates and alkanolamides. *Zwitterionic Surfactant:* In this type, both positive and negative charges may be present on the hydrophilic group, for example, long chain amino acid, betaines and sulfobetaines.

Surfactants are responsible for enhancing the stability of suspension and have a strong influence on wetting, absorption and adhesion behavior in the electroless composite coatings. Hence, surfactants are used to achieve uniform distribution of particles in the coatings, which will enhance the properties of the deposits. The critical micelle concentration (CMC), an important characteristic of a surfactant, is defined as the concentration of surfactants above which micelles form and all additional surfactant molecules added to the system move to micelles. When surfactant concentration is below CMC, the surfactant molecules are loosely integrated into the water structure, and they are termed as monomer. At CMC, the structure of the surfactant-water solution becomes modified such that the surfactant molecules start building up the structure of micelles as shown in Figure 8.2. The CMC value of surfactants may be significantly modified with appropriate room temperature and ionic liquids, and it varies for specific applications (Modaressi et al. 2007). When the concentration of surfactant reaches the CMC, the physical properties of the electrolytic solution will be affected to a considerable extent. The CMC of the surfactant is generally determined by the conductance method. Henceforth, CMC is used to distinguish and study the surface activity of surfactants (Schaller et al. 1992). The presence of surfactants in electroless plating baths modifies the properties of coatings: hardness, quality of the deposits, abrasion resistance and dispersion of particles. They are added in definite quantities with respect to the CMC of the corresponding surfactant to decrease the vertical component of the surface tension forces, which bind the hydrogen gas bubbles produced during the coating reaction. The incorporation of surfactants in the bath solution can significantly reduce micro-pitting on the NiP deposit and thus improve the corrosion resistance. The addition of a small quantity of surfactants to the electroless bath could increase the deposition rate by 25% when compared to surfactant-free electroless bath. However, excessive addition of surfactants would result in inferior surface finish of the coating and slow down the deposition (Elansezhian et al. 2008; Sudagar et al. 2013): the influence of surfactants is maximum at CMC and stabilizes or deteriorates beyond this concentration (Sudagar et al. 2012).

In the absence of surfactants, nanoparticles can agglomerate in the coating (Rabizadeh and Allahkaram 2011), this is why they are so frequently used in electroless nickel (phosphorus or boron) composites.

The reported surfactants include cationic surfactant (Aliquat 336) (Ansari and Thakur 2017)—cetyl trimethyl ammonium bromide (CTAB) (Chen et al. 2017; Czagány et al. 2017; Ger 2004; Zielińska et al. 2012); anionic surfactant (sodium dodecyl sulfate, SDS) (Chen et al. 2017; Nwosu et al. 2013; Zielińska et al. 2012)—sodium dodecyl benzene sulfonate (Chen et al. 2017; Czagány and Baumli 2019), sodium lauryl sulfate (Chen et al. 2017) and nonionic surfactants (polyvinylpyrrolidone) (Chintada and Koona 2018a; Czagány and Baumli 2019)—comb polyelectrolytes (de Hazan et al. 2008a, 2008b)—tetraethylene glycol dodecyl ether (Brij 30) (Zielińska et al. 2012).

Sodium lauryl sulfate and CTAB are popular surfactants and present a direct effect on the surface properties of coatings on magnesium alloy that lead to considerable improvement in the deposition rate, smoothness and microhardness of coatings. The study of the effects of SDS and CTAB reported that the surfactants significantly reduced the surface tension and enhanced the plating efficiency from 32% to 95% (Elansezhian et al. 2009; Liu et al. 2009). Furthermore, SDS and CTAB present positive effects on wastewater treatment, as they significantly reduced the surface tension of aqueous solution (Sineva et al. 2007). The impact of CTAB surfactant was evaluated particularly on the corrosion resistance of NiP-Al_2O_3 coatings and its presence improved the amount of co-deposited Al_2O_3 particles. In addition, the corrosion resistance of the composite coatings increased with CTAB concentration up to the optimal concentration, above which corrosion resistance decreased as observed from the salt spray test (Liu et al. 2009). Surfactants have also been used to reduce the effect of surface free energy of graded NiP-PTFE electroless coatings on bacterial adhesion (Zhao 2004).

The other surfactants used in coating technologies are dodecyl trimethyl ammonium bromide (DTAB)—cationic, Brij 30—non-ionic and SDS—anionic. The properties of NiP-nano-ZrO_2 composite coatings in the presence of these surfactants showed their addition had greatly increased the weight percentage of zirconia (22.10–21.88 wt%) for the bath containing a concentration of DTAB higher or equal to CMC (Zielińska et al. 2012). The behavior of electroless NiP-nano-TiO_2 composite coatings on low-carbon steel in the presence of SDS and DTAB surfactants also shows the promising contribution in composite coatings. At an optimum concentration of DTAB surfactant, uniform distribution of TiO_2 particles with no defects was observed. The corrosion and tribological properties were improved by the incorporation of TiO_2 particles in the NiP matrix. The increase in TiO_2 content of the coatings significantly depends on the surfactant and its concentration (Tamilarasan et al. 2015, 2016).

8.2.4 Incorporation of Particles in Electroless Nickel (Phosphorus or Boron) Coatings

The entrapment of particles in electroless nickel matrix composites does not include chemical reactions between the particles and the growing coating (Balaraju et al. 2003). As such, the particles are usually considered as inert with regard to the plating reaction. It is thus the growth of electroless nickel itself, next to and around the particles, that leads to their entrapment in the material. Xiang et al. (2001) have described the growth of electroless nickel around the entrapped particles, as shown in Figure 8.3. They observed that the process began with the absorption of particles on the surface that were progressively entrapped (Figure 8.3a). The number of entrapped particles increases with time (Figure 8.3b). The co-deposition of a large number of particles for the roughening of the coating (Figure 8.3c). The presence of particles, creating a rougher surface and thus an increased reaction surface, is reported to lead to an increase of plating rate.

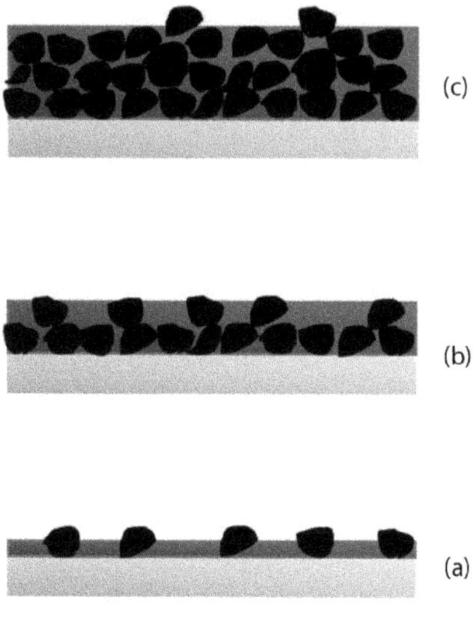

FIGURE 8.3 Growth process of electroless composite coating. (a) adsorption and entrapment of particles (b) growth of coating and entrapment of a second layer of particles (c) roughening of coating due to presence of particles (Adapted from Xiang, Y. et al., *Plating and Surface Finishing,* 2, 64–68, 2001.)

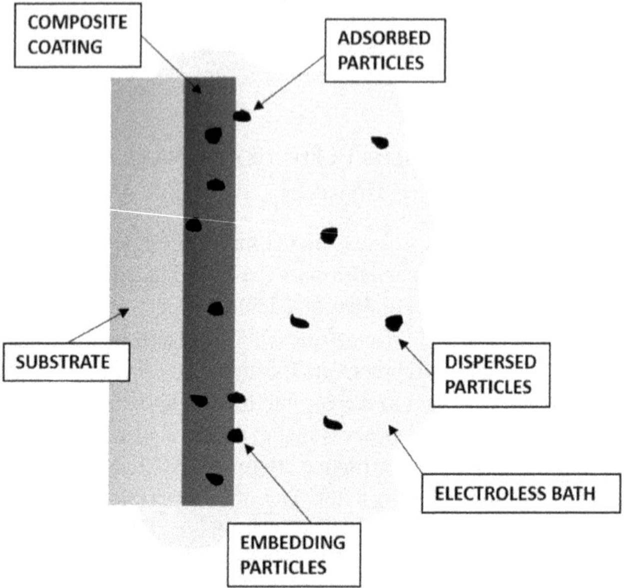

FIGURE 8.4 The adsorption of particles during electrolytic composite plating processes. (Adapted from Guglielmi, N., *J. Electrochem. Soc.*, 119, 1009, 1972.)

Guglielmi (1972) explained the peculiarities of co-deposition of inert particles from electrolytic baths by a mechanism based on two successive adsorption steps. In the first step, the particles are loosely adsorbed so that they are in equilibrium with those remaining in suspension. They are irreversibly adsorbed during the second step. The first step corresponds to a stage when particles are still coated with adsorbed ions and water molecules and the second one to a stage when those ions have desorbed and the particle is directly adsorbed on the surface, as shown in Figure 8.4. This made possible to deduce a relationship between the concentration of the embedded particles and that of particles in suspension that fits experimental results. This model was successfully adapted for electroless nickel plating by introducing an equivalent current density corresponding to the global reaction of the electroless system (Grosjean et al. 1998).

8.2.5 Electroless Nano-composite Coatings

Many researchers are showing major interest to study the effect of particles size on electroless nickel (phosphorus or boron) composite coatings, and mostly of the effect of nanoparticles addition. The addition of nanoparticles into the matrix results in ultra-thin and pore-free deposition. This technology helped to obtain nanostructured composite materials and coatings with specified properties. This method of coating increased the functional capabilities of the substrates being coated and was found to be more advantageous than other methods of synthesizing nano-sized system (Larson et al. 2013).

It has on the other hand been shown that nano-sized particles had a bigger influence on the growth of electroless nickel than micro-sized ones (Sarret et al. 2006).

The addition of nano-sized solid particles to electroless coatings is a significant approach to widen the opportunities of the coatings in engineering industries, because several types of nanoparticles have distinctive properties, which are much different from their bulk counterparts, and also endow the coating with superior functionality (Li et al. 2006a). Nanoparticles greatly modify the growth mechanism of metal matrix compared to micron-sized particles and the co-deposition depends on the particle size and not on the particle nature (Sarret et al. 2006). It was shown for example that SiC nanoparticles allowed to provide higher cavitation erosion resistance than microparticles but the system still required a post-heat treatment to reach optimal cavitation erosion resistance (Ranganatha et al. 2010). Likewise, the use of nano-zinc particles in electroless coatings resulted in surface uniformity, homogeneity and also improvement in hardness and corrosion characteristics of the nanocomposite coatings (Shibli et al. 2006b); Electroless NiP-nano-TiN coatings on magnesium AZ31 alloy evolved with the introduction of secondary phase particles showed dense fine nodules uniform in size without any pores and cracks and highly improved the wear resistance (Yu et al. 2011).

8.2.6 Complex Composite Coatings

It is shown in Chapter 6 of this book that it is possible to prepare complex alloys (with three or more components) by electroless plating and that this method allows to obtain interesting properties. It is possible to add inert particles to such alloys to further modify the properties of the coating. The following ternary composites have been reported: Ni-B-P-PTFE (Omidvar et al. 2016b), Ni-B-P-graphite (Omidvar et al. 2015), Ni-Co-P-SiO_2 nanocomposite (Seifzadeh and Hollagh 2014), Ni-Co-P-TiN (Shi et al. 2016), Ni-Cu-P/nano-graphite composite (Lee 2012), Ni-Cu-P/CNT (Yucheng et al. 2008), Ni-W-P-cerium oxide (Sun et al. 2017), Ni-P-W-Nb_2O_5 (Chen et al. 2015), Ni–W–P–ZrO_2 (Szczygieł and Turkiewicz 2009), Ni-W-P-Al_2O_3 (Balaraju et al. 2010a), Ni-W-P/nano-WS_2 (Ranganatha et al. 2012), Ni-W-P/nano-MoS_2 (Ranganatha and Venkatesha 2012) and Ni-Zn-P/nano-TiO_2 (Ranganatha et al. 2010).

Tungsten is specifically popular as an alloying element in electroless nickel composites, like it was in alloy coatings. Similarly, most composites are developed on the basis of NiP coatings rather than NiB.

8.2.7 Post Processing/Treatments

Post-treatments (mostly heat treatment) that improve the properties of electroless nickel coatings can also be applied to composite coatings. Although several studies have been conducted to explore the potential of these process for composite coatings, they are not commercially viable on an industrial scale yet. Therefore, a sustainable method of posttreatment that can cater specifically to particular properties depending upon the nature of application needs to be established. Investigations have been carried out on the posttreatment of composites containing various types of particles (Kaczmar et al. 2017; Sahoo and Das 2011; Srinivasan and John 2005; Wu et al. 2006b).

8.3 COMMONLY USED PARTICLES AND PROPERTIES OBTAINED WITH THEM

Several types of particles can be incorporated in electroless nickel coatings and their addition will modify the properties of coatings in various ways depending on the nature and size of the added particles. The most sought-after modifications are hardness and friction. The first is obtained by the addition of hard particles (e.g. diamond, silicon carbide, and titanium nitride) while the second relies on the addition of lubricating materials such as PTFE, graphite MoS_2 or hexagonal boron nitride.

In this section, the effect of added particles will be examined according to their chemical nature. The effect of size will also be discussed, when possible, in a group of particles.

8.3.1 Carbon-Based Particles

Carbon-based particles are the widest group of materials added to electroless coatings. These particles can take nearly any of the many forms of carbon, from graphite to nanotubes and of carbon compounds, from carbides to polymers. As carbon-based materials have an extremely wide range of properties due to their very different nature and structure, each group will be discussed separately.

8.3.1.1 Pure Carbon

Pure carbon covers all materials that are made of only carbon. They thus vary only in shape and structure and not in chemical composition. The three types of pure-carbon materials are graphite, diamond and graphene-based materials, which include nanotubes, fullerenes and reduced graphene oxide (rGO).

8.3.1.1.1 Graphite

The first paper reporting graphite incorporation into electroless nickel was published in 1987 (Izzard and Dennis 1987) and reported the addition of 2 and 6 μm particles in a commercial electroless nickel solution. A low coefficient of friction and wear rate were obtained for these coatings. The plating rate of electroless nickel-phosphorus-graphite (Ni-P-Cg) coatings is relatively limited (Mohtar et al. 2013b) but films are adherent and heat treatment modifies the electroless nickel coating similarly to coatings free of added particles (Mohtar et al. 2013a), with peak hardness obtained for heat treatment at 400°C for 1 h (Liu et al. 2015; Wu et al. 2006d). Graphite particles are usually well spread inside the coating (Wu et al. 2006b).

Similar results have been reproduced by several authors: low and stable coefficient of friction (Han et al. 2008; Liu et al. 2015) (approximately 0.1 or less against GCr5 for a load of 5 N (Liu et al. 2015; Wu et al. 2006a) and 0.4 against carbon steel (Wu et al. 2006c) and limited wear rate that is improved by heat treatment (Wu et al. 2006d).

Heat-treated electroless nickel-phosphorus-graphite composite coatings presented significantly reduced wear rate and friction coefficient compared to NiP (Chen et al. 2002), as shown in Table 8.1. The modifications of structure encountered due to heat treatment are similar to those observed in particle-free coatings: the initially

TABLE 8.1
Wear and Friction Properties

	Treated at 473 K for 1 h		Treated at 473 K for 2 h	
Coating	**Mass Loss (mg)**	**Friction Coefficient**	**Mass Loss (mg)**	**Friction Coefficient**
NiP	32.4	0.102	15.6	0.090
NiP-SiC	8.5	0.124	3.5	0.092
NiP-graphite	10.4	0.077	4.3	0.067
NiP-CNTs	6.2	0.063	2.6	0.061

Source: Chen, W.X. et al., *Surf. Coat. Technol.*, 160, 68–73, 2002.

amorphous nickel matrix crystallizes during heat treatment, without visible interaction with the embedded particles (Wu et al. 2006b).

The improved wear behavior of electroless nickel-phosphorus-graphite composite coatings has been attributed to the formation of a graphite-rich film at the surface of the material, from reaction of the material and wear counterpart (Wu et al. 2006d). It is important to note that the hardness of electroless NiP coatings is decreased by the addition of graphite particles (Wu et al. 2006b).

There are few studies on the corrosion resistance of electroless nickel-phosphorus-graphite coatings but they present results similar to those obtained for standard electroless nickel coatings (Liu et al. 2015).

Electroless nickel-boron-graphite composites have also been investigated (Omidvar et al. 2016a). Apart from modifying the morphology, leading to the formation of finer globular features, additions of graphite improved the friction coefficient (with an optimum reached for 12–13% Cg), without affecting too much the deposition rate (those coatings were deposited by dynamic chemical plating technique, which is a variant of electroless plating).

8.3.1.1.2 Diamond

Diamond has been incorporated in electroless NiP as early as the late 1970s (Lukschandel 1978) and has been the object of several publications (Hung et al. 2008; Jappes et al. 2009; Lee et al. 2015b; Petrova et al. 2016). Its incorporation in electroless nickel brings modifications of the coating morphology, with diamond particles often protruding from the coating (Petrova et al. 2011), leading to an aspect that can be described as foggy and rough (Xiang et al. 2001). The most significant effect of the addition of diamond in electroless NiP is the increase in hardness (Bozzini and Boniardi 2001; Bozzini et al. 2001; Huang 2011; Mazaheri and Allahkaram 2012; Petrova et al. 2011; Sheela and Pushpavanam 2002; Wang et al. 2018; Xiang et al. 2001; Xu et al. 2005) that is followed by an increased wear resistance (Bozzini et al. 2001; Huang 2011; Mazaheri and Allahkaram 2012; Petrova et al. 2011; Sheela and Pushpavanam 2002; Xu et al. 2005), and also by an increase in elastic modulus (Bozzini et al. 2001). However, the structure of the nickel matrix is not modified by the presence of particles (Mazaheri and Allahkaram 2012; Xiang et al. 2001).

The toughness of coatings is also improved by the presence of dispersed diamonds. The corrosion resistance of electroless NiP appears to be improved by diamond additions (Ashassi-Sorkhabi and Es′haghi 2013; Mazaheri and Allahkaram 2012; Xu et al. 2005); this phenomenon is explained by the disturbance of the natural corrosion paths inside the electroless coating and the decrease in the porosity due to the presence of diamond particles.

Diamonds are difficult to disperse in electroless plating solutions (Liu and Zhu 2015; Liu et al. 2019). For this reason, authors have investigated the effects of surface modification (Liu et al. 2019) and surfactants (Petrova et al. 2016) on their incorporation. However, the plating rate of electroless NiP increases when diamonds are added to the solution, up to a certain concentration (Hou and Gao 2016; Petrova et al. 2011).

The size of diamond particles that are incorporated in electroless nickel also influences the properties of coatings: higher deposition rate for smaller particles (Petrova et al. 2011), linked with transport limitations for larger particles, more significant increase in hardness for smaller particles (Petrova et al. 2011), and improved wear resistance (Reddy et al. 2000), probably linked with an higher incorporation rate for finer particles (Reddy et al. 2000). However, for very small particles, the tendency is reversed (Wang et al. 2018). The optimal size of particles is not clearly defined.

Like particle-free electroless nickel, electroless NiP/diamond composites are positively influenced by heat treatment, which leads to improved hardness and wear resistance, as well as increased elastic modulus (Bozzini et al. 2001; Reddy et al. 2000; Xu et al. 2005).

Some authors also investigated the incorporation of diamond in electroless NiB coatings (Kaya et al. 2009; Ogihara et al. 2010) and reported a significant increase in hardness and excellent wear resistance (up to 14 times better than that of plain electroless NiB).

8.3.1.1.3 Graphene-Based Materials

Graphene-based materials are all materials whose crystalline structure is based on graphene sheets, except graphite: graphene, rGO and carbon nanotubes (CNTs). Their addition to electroless nickel coatings is a recent development but the properties of electroless nickel composites containing these materials are nevertheless very interesting.

For example, Tamilarasan et al. studied electroless NiP-rGO coatings on steel substrates and reported that there was even dispersion of rGO particles for coatings deposited with 50 mg/L of rGO bath concentration. The increase in the rGO concentration in the bath led to changes in the surface heterogeneity and enlarged nodule sizes in the coating matrix. Nodular structures are distinguished on the surface, enlarging with an increase in the rGO concentration in the plating bath. Energy-dispersive X-ray spectroscopy and X-Ray diffraction results confirm the existence of incorporated rGO particles in the coatings and the amount of incorporated rGO particles in the coatings increased up to a concentration of 50 mg/L of rGO in the bath and tended to decrease on further addition. They reported that the corrosion, erosion-corrosion, wear resistance and friction coefficient of the coating are strongly influenced by the content of the rGO particles at the optimum concentration (Tamilarasan et al. 2017).

Other authors investigated the additions of graphene or rGO in electroless NiP coatings (Lee et al. 2015b) and observed that the composite coating were rougher (Sadhir et al. 2014; Wu et al. 2015a; Yu et al. 2018) but had similar crystalline structure than electroless nickel (Jiang et al. 2016; Kumari et al. 2018; Sadhir et al. 2014; Wu et al. 2015a; Yu et al. 2018). The effect of graphene and rGO on mechanical properties seemed to evolve with content: hardness was increased for low level of particles addition but decreased for higher particles content (Wu et al. 2015). Corrosion and wear resistance were improved by the addition of particles (Kumari et al. 2018; Sadhir et al. 2014; Uysal 2019; Wu et al. 2015a) and the composite coatings responded to heat treatment in a similar manner to plain NiP coatings (Jiang et al. 2016; Sadhir et al. 2014; Yu et al. 2018). However, the additions of graphene allowed to keep the level of corrosion resistance after heat treatment (Sadhir et al. 2014), which is not the case for graphene-free electroless nickel coatings. The use of ultrasound leads to further improved properties (Jiang et al. 2019).

Carbon nanotubes have also been added to electroless nickel, as shown in Figure 8.5. All samples were annealed at 473 K for 1 h and 673 K for 2 h to avoid the hydrogen brittleness and for improving microhardness [69]. The friction and the wear

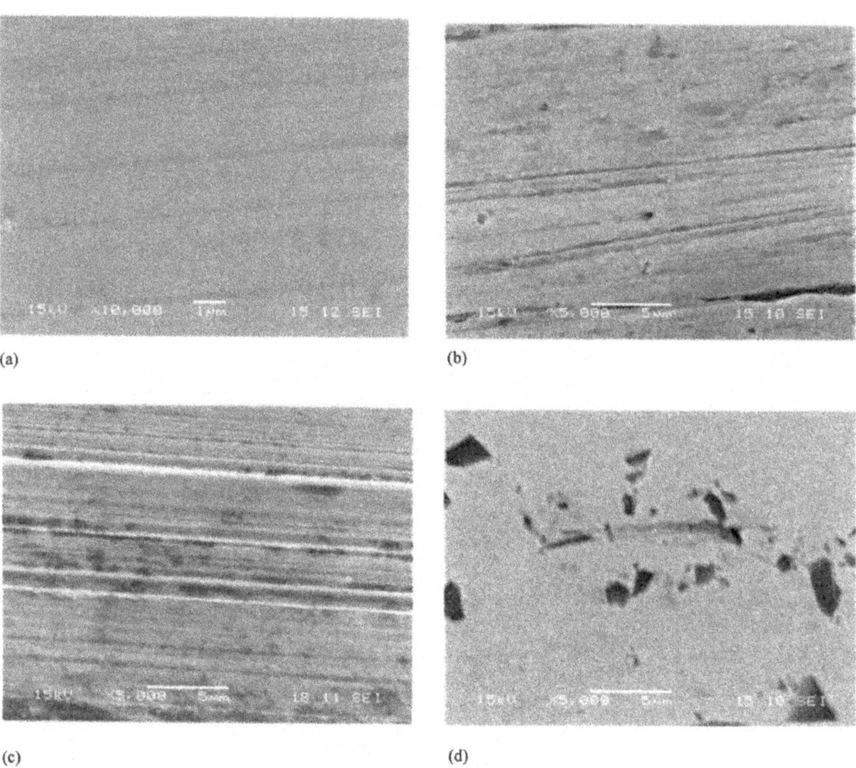

FIGURE 8.5 The worn surface morphology of (a) Ni-P-CNT, (b) NiP, (c) Ni-P-graphite and (d) Ni-P-SiC electroless coatings after wear testing for 6 h. The coatings were treated at 673 K for 2 h. (From Chen, W.X. et al., *Surf. Coat. Technol.*, 160, 68–73, 2002.)

of the NiP-CNTs composite have very lower mass loss (6.2 mg) and friction coefficient (0.063) than NiP and other composites as shown in Table 8.1. The improved mechanical properties are obtained due to the excellent lubrication and strength of CNTs [69].

Arai et al. (2010) investigated the magnetic properties of NiP-multiwalled carbon nanotube (MWCNT) composite and pure-nickel-coated MWCNT fabricated by the electroless deposition method. It was found that the microstructure of the pure nickel-coated MWCNTs deposit is comparatively less smooth than that of the NiP-coated MWCNT composite, and it has higher magnetization and coercivity. The heat treatment process significantly improves the surface smoothness of the pure-nickel coating.

Liu et al. (2007c) proposed a method to fabricate CNT-based field emitters by the electroless deposition process. In order to get a conductive layer, the MWCNT-reinforced nickel composite is deposited on a glass substrate with Ni film in the electroless nickel bath containing sulfate hexahydrate. For effective dispersion and co-deposition of MWCNT with nickel film, the raw MWCNTs were annealed and acid-treated for removing the surface impurities before the electroless plating process. It was found that the impure MWCNT-coated nickel film shows uneven deposition of MWCNTs on the nickel film. On the other hand, the acid- and heat-treated MWCNTs are uniformly distributed on to the nickel film in a significant manner due to the friction process of MWCNTs. The CNT/Ni film significantly improves the field-emitted image (Liu et al. 2007c) (Figure 8.6) with respect to light uniformity and the image quality in a field emission display systems compared with existing systems. It was concluded that the proposed coating method has significant advantages for producing uniform coating of MWCNTs on nickel matrix compared to other methods such as screen printing or the electrophoretic method.

Chen et al. (2003) proposed a method for the preparation of NiP-CNT composite through electroless plating and powder metallurgy technique. The NiP-CNT

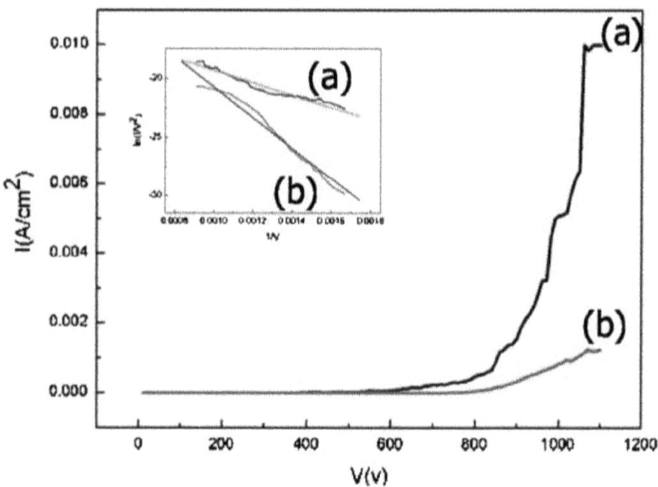

FIGURE 8.6 Field emission properties of MWCNT/Ni film plated.

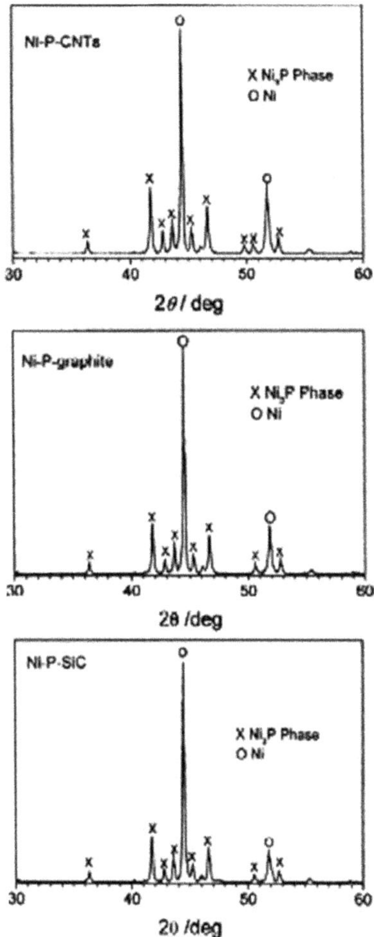

FIGURE 8.7 X-Ray diffraction patterns of Ni-P-CNT, Ni-P-graphite and Ni-P-SiC electroless composite coatings after heat treating at 673 K for 2 h. (From Chen, W.X. et al., *Carbon*, 41, 215–222, 2003.)

composite has very low friction coefficient and higher wear resistance in comparison with nickel-phosphorus-silicon carbide (NiP-SiC) and NiP-graphite. The heat treatment of the NiP-CNT composite was carried out at 476 K for 2 h to develop the crystalline structure and the NiP_3 forms in the nickel phase (Figure 8.7).

NiP-CNT composite material is compared with NiP composite in identical heat treatment state. The microhardness of the composite was comparatively studied before and after annealing process at 673 K for 1 h for all samples and the results are presented in Figure 8.8. The corrosion resistance of CNT-incorporated composite coatings is improved due low chemical reactivity of CNT and by the formation of passive film.

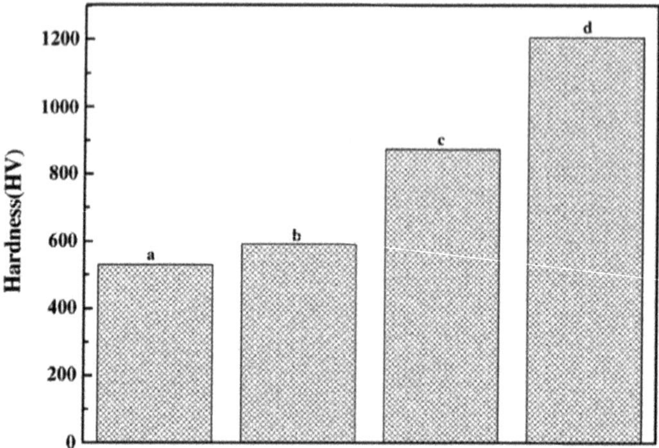

FIGURE 8.8 A bar diagram represents the microhardness values of the composite coatings: (a) as-prepared Ni-P coating; (b) as-prepared Ni-P-CNT composite coatings; (c) Ni-P coating after heat treatment; (d) Ni-P-CNT composite coatings after heat treatment. (From Yang, Z. et al., *Mater. Res. Bull.*, 40, 1001–1009, 2005.)

Yucheng et al. (2008) studied the properties of Ni-Cu-P/CNT quaternary composite fabricated on a carbon-steel substrate by electroless plating technique. It was found that the hardness, smoothness and corrosion resistance of the Ni-Cu-P/CNTs composite significantly increased after heat treatment process at 400°C/h (at hydrogen-free atmosphere) compared with the as-plated Ni-Cu-P composite (see Figure 8.9). The heat treatment process yields improved hardness due to the formation of intermetallic phases such as Ni_3P and Cu_3P phases. Conversely, the hardness has a decreasing trend beyond the CNTs concentration of 1.6 g/L, due to its conglomeration nature (Figure 8.9).

Other authors also reported an improvement in the wear resistance of electroless NiP coatings due to CNT additions (Alishahi et al. 2012; Chen et al. 2002, 2006; Li et al. 2006b; Meng et al. 2012; Tian et al. 2014; Yang et al. 2004; Xu et al. 2014).

Yazdani investigated CNT addition in electroless NiB (Yazdani et al. 2018). He observed increased crystallinity of the coating when CNTs were added that they attribute to the release of H^+ ions from carboxylic groups at the surface of nanotubes and the reduction of nickel ions by the remaining $HCOO^-$ ions. They also observed that they increased roughness and that nanotubes tended to agglomerate at high concentrations. The addition of CNTs leads to an increase in hardness for low concentrations but that effect decreased for higher concentrations. The friction coefficient and wear rate of the electroless NiB coatings were improved by CNTs additions.

8.3.1.2 Carbides

Carbides are known as hard materials. For this reason, they can be added to electroless nickel coatings, just like diamond, in order to increase the coating hardness. The most popular carbide for this use is silicon carbide but other forms of carbides, such as titanium carbide (Huang et al. 2019), boron carbide (Araghi and Paydar 2013;

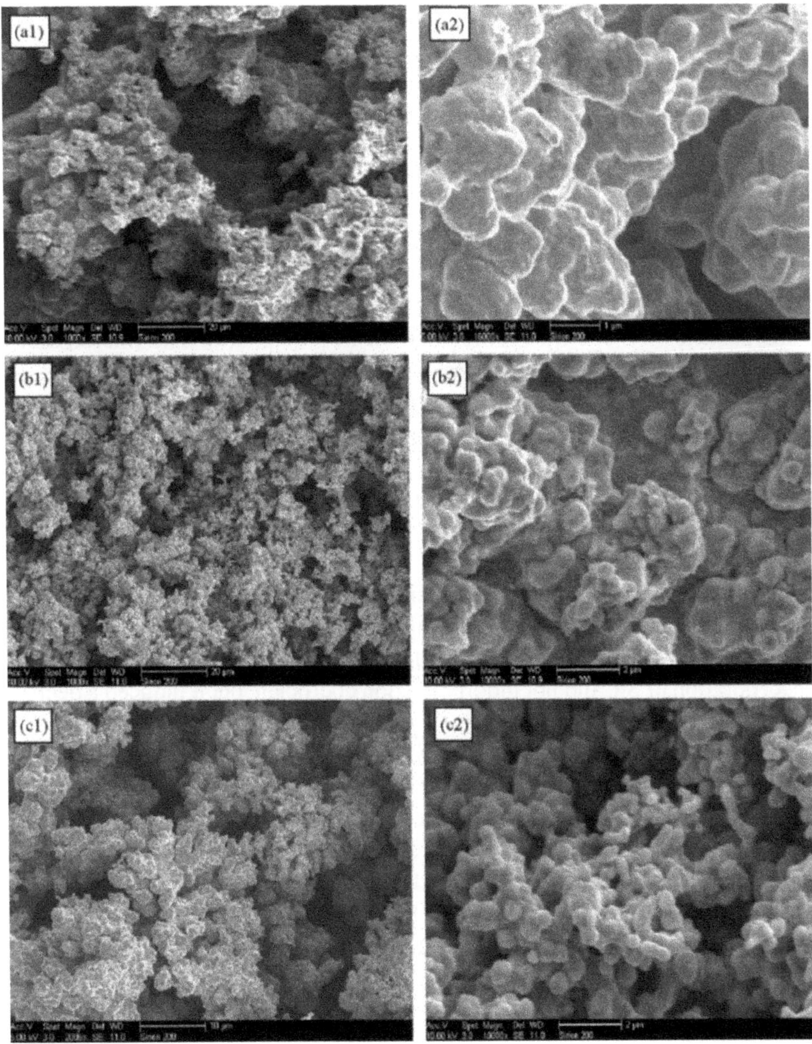

FIGURE 8.9 SEM morphology of as-deposited Ni-Cu-P/CNTs composite coatings: (a1,a2) CNTs concentration 0.8 g/L; (b1,b2) CNTs concentration 1.6 g/L; (c1,c2) CNTs concentration 2.4 g/L. (From Yucheng, W. et al., *Mater. Res. Bull.*, 43, 3425–3432, 2008.)

Ebrahimian-Hosseinabadi et al. 2006), tungsten carbide (Hamid et al. 2007b; Liu et al. 2007a; Luo et al. 2015) or chromium carbide (Sherkat et al. 2011) have also been used.

The study on the effect of reinforcement and heat treatment on elevated temperature sliding of electroless NiP-SiC composite coatings was investigated by Franco et al. The investigation is focused on the wear behavior at the elevated test temperature of composite NiP-SiC deposit, with varying concentration of the reinforcing SiC particles. On reinforcing with particles, reduction of wear rate is observed as compared to the particle-free coating. Heat treatment further increases the wear resistance due to the better bonding of the reinforcing particles and the matrix. The surface roughness measurement confirmed the effectiveness of ceramics in the heat-treated composite coating (Franco et al. 2016).

A similar research based on laser-induced surface treatment of electroless NiP-SiC coating on A1356 alloy was conducted by Hashemi et al. Laser heat treatment was performed with no major defects, melted area, protuberances and cracks. The laser-treated zone structure contained mainly of Ni-P crystal phases. The results of micro-structural characterization indicated that the laser treatment under different operating conditions produced composite coating contained nanocrystalline Ni-based matrix with SiC particles: Ni_3P, $Ni_{12}P_5$, Ni_5P_2 and Ni_8P_3 precipitates. The microhardness measurements showed that the hardness of the coating was increased up to 60%, due to laser heat treatment, without affecting the base metal (Hashemi and Shoja-Razavi 2016a).

Ma et al. (2014) studied the influence of heat treatment process on the electroless NiP-SiC nanocomposites. For this analysis, four levels of temperature range were chosen with constant treatment time. The grain sizes observed after heat treatment were still in the nanometer range.

Similar to the NiP-CNT and NiP-graphite composites, NiP-SiC composites were synthesized by Chen et al. (2002). All samples were annealed at 473 K for 1 h and 673 K for 2 h to avoid the hydrogen brittleness and for improving microhardness. The incorporation of SiC did not modify friction coefficient but it induced a decrease in wear rate, as shown in Table 8.1.

The other aspects of electroless NiP-SiC coatings are also reported in the literature: roughness (Chang et al. 2016), morphology (Grosjean et al. 2000; Hu et al. 2017; Islam et al. 2015), structure (Farzaneh et al. 2013; Hashemi and Shoja-Razavi 2016b; Hu et al. 2017), adhesion (Hashemi and Shoja-Razavi 2016b), corrosion resistance (Allahkaram and Bigdeli 2010; Bigdeli and Allahkaram 2009; Chintada and Koona 2018b; Farzaneh et al. 2013, 2016; Islam et al. 2015; Soleimani et al. 2015a, 2015b; Zarebidaki and Allahkaram 2012; Zhang et al. 2016), growth mechanism (Sarret et al. 2006), friction coefficient (Staia et al. 2002) and high temperature wear (Staia et al. 2002). The process parameters of coatings were even optimized by some authors (Su 2013).

Among other carbides incorporated in electroless NiP coatings, boron carbide (B_4C) have been reported to increase hardness, wear resistance, elastic modulus and fracture toughness of coatings in as-plated and heat-treated conditions. However, the coefficient of friction of the coatings increased due to the presence of particles (Araghi and Paydar 2013, 2014; Bozzini et al. 1999; Ebrahimian-Hosseinabadi et al. 2006; Ge et al. 1998; Monir Vaghefi et al. 2003; Zhu et al. 2008).

Titanium carbide has also been investigated as reinforcing particle in electroless NiP coatings (Afroukhteh et al. 2012; Huang et al. 2019). The composite coatings presented unmodified structure, increased hardness, significantly reduced wear and responded to heat treatment in a similar way to standard electroless nickel coatings. The impact on corrosion resistance was limited and the presence of particle allowed to decrease the impact of heat treatment on corrosion resistance. The properties of such composite coatings can be very much impacted by the use of surfactants during plating.

As for tungsten carbide, it has been reported to increase hardness and corrosion resistance without modifying the structure of electroless NiP coatings (Hamid et al. 2007a; Liu et al. 2006, 2007b; Luo et al. 2015).

Studies on incorporation of carbides in electroless NiB coatings are relatively less diverse compared to electroless NiP coatings, in which a rich range of composite coating methods have been developed. SiC particles were co-deposited with electroless NiB coatings using surfactant Tween 20 by steric stabilization and their addition tended to improve the corrosion and hardness behavior simultaneously. The cauliflower-like structure of noncomposite deposits, which is an indication of a nanocrystalline phase, becomes more nodular in the presence of SiC particles, because they can act as nucleation centers during coating development (Georgiza et al. 2017). The embedded SiC particles reduce the metallic area prone to corrosion, thus enhancing the corrosion barrier effect, provided by non-composite NiB coatings (Georgiza et al. 2017). One author also reported a significant improvement in the hardness, wear and corrosion resistance of electroless NiB coatings due to the addition of B_4C particles (Rezagholizadeh et al. 2015).

8.3.1.3 Polymers

The addition of polymers in electroless nickel coatings is mostly used for PTFE. The rationale for the addition of those particles is that they provide excellent friction properties, thus leading to a reduction of wear. Electroless NiP-PTFE coatings are actually one of the few electroless nickel composite coatings that are widely used in commercial applications.

The reduction of wear linked to the incorporation of PTFE in electroless NiP is very significant (of nearly one order of magnitude) (Ankita and Singh 2011; Jiang et al. 2013; Liew et al. 2013; Mohammadi and Ghorbani 2011; Ramalho and Miranda 2005; Rossi et al. 2003; Shaffer and Rogers 2007; Sharma and Singh 2013; Srinivasan and John 2005; Wu et al. 2006c) and the coatings containing PTFE present low friction coefficients (Liew et al. 2013; Ramalho and Miranda 2005; Shaffer and Rogers 2007; Srinivasan and John 2005; Wu et al. 2006c). Moreover, the corrosion resistance of electroless NiP is increased by the addition of PTFE (the corrosion potential is not modified much and thus the corrosion rate is significantly decreased) (Ankita and Singh 2011; Hou et al. 2013; Huang et al. 2005; Mafi and Dehghanian 2011a; Mohammadi and Ghorbani 2011; Rossi et al. 2003; Srinivasan and John 2005; Zhao et al. 2004; Zhao and Liu 2005).

There are also a few reports of PTFE additions in electroless NiB coatings, and in ternary Ni-P-B alloys (Omidvar et al. 2016a, 2016b). The morphology of electroless NiB is not really modified by the presence of PTFE but the particles can be seen on the surface. The friction coefficient of electroless NiB is significantly reduced by PTFE incorporation (Omidvar et al. 2016a). Similar results accompanied by a significant decrease in wear have been observed in the case of electroless Ni-P-B/PTFE composites (Omidvar et al. 2016b).

8.3.2 Nitrides

Nitrides can be added to electroless nickel coatings as reinforcing particles with an aim to increase hardness and wear resistance, with similar reinforcement mechanisms as additions of carbide particles.

Silicon nitride is by far the most popular material of this group that can be added to electroless nickel coatings. Wang et al. observed the behavior of NiP-Si_3N_4 nanowire composite coatings that were prepared by electroless technique on AZ31 alloy substrate, which confirmed the fact that a crystalline component could be embedded into the amorphous NiP matrix. The study reported that the incorporation of Si_3N_4 nanowire in the coating (1.5 g/L in bath) significantly modified the surface morphology with a fine nodular microstructure, compact and smooth without porosity and improved the mechanical properties (Wang et al. 2015).

The addition of Si_3N_4 increased the hardness (Balaraju and Seshadri 1999; Sarret et al. 2006) and wear (Sarret et al. 2006) resistance of coatings and did not modify the crystalline structure (Balaraju and Rajam 2008; Balaraju et al. 2006b; Sarret et al. 2006) or grain size (Balaraju and Rajam 2008) of electroless nickel. The effect on corrosion resistance is not clear as authors have reported contradictory findings (Balaraju and Seshadri 1998; Balaraju et al. 2010b). Heat treatment usually improves the hardness of electroless NiP/silicon nitride composite coatings (Balaraju et al. 2006b; Balaraju and Seshadri 1999).

Yu et al. (2011) studied the electroless NiP-nano-TiN coatings on magnesium AZ31 alloy and reported that the introduction of secondary phase particles showed dense fine nodules uniform in size without any pores and cracks and highly improved the wear resistance. The addition of TiN particles in electroless NiP coatings led to increased deposition rate, increased hardness but unmodified structure and mediocre corrosion properties (Mafi and Dehghanian 2011b).

Boron nitrides have also been used in electroless nickel-based composites. Most studies focus on the incorporation of hexagonal boron nitride, which presents lubricating properties similar to those of graphite (Hsu et al. 2015; León 2003; León et al. 1998) ones as compared to both cubic and hexagonal boron nitrides, but mostly focus on process (Chakarova et al. 2016). The additions of boron nitride particles in the plating bath will decrease the plating rate (León et al. 1998). The reduction of wear reported for h(BN)-loaded electroless NiP coatings is of two orders of magnitude (León et al. 1998) compared to standard coatings, and the friction coefficient of h(BN)-loaded electroless NiP coatings is low (Hsu et al. 2015; León et al. 1998). The coatings also present an interesting wear behavior at high temperature (León et al. 2003). The hexagonal boron nitride particles, while they confer solid-lubricant properties to the coating, are reported by some authors to also increase hardness and corrosion resistance (Ranganatha and Venkatesha 2017). However, other authors reported a slight decrease in hardness due to hexagonal boron nitride particles incorporation (Hsu et al. 2015).

Two-dimensional carbon nitride (C_3N_4) particles have also been successfully incorporated in electroless NiP coatings (Fayyad et al. 2018, 2019). This allowed to increase hardness and corrosion resistance of the coatings.

8.3.3 Oxides

Oxide particles can present several useful properties for the reinforcement of electroless nickel coatings: hardness, corrosion resistance, self-healing enabling properties, catalytic properties and so on. For this reason, several types of oxide particles have been incorporated in electroless nickel coatings over the years.

The incorporation of TiO_2 in coating significantly modified the surface morphology and increased the corrosion resistance (Tamilarasan et al. 2015). Studies on the effects of phosphorus and TiO_2 content on the electrocatalytic properties of NiP-TiO_2 coating revealed that the reinforcement of nano-TiO_2 further improved the number of electro-active sites and hydrogen adsorption ability and decreased the roughness. Furthermore, the improved electrocatalytic activity can be used for hydrogen evolution reaction and applied as a catalytic support, reinforcement and inert filler (Shibli and Dilimon 2007).

TiO_2 nanoparticles were also used in electroless Ni-Zn-P ternary alloy. The co-deposited particles did not influence structure and phase transformation behavior but the stable phases were identified in the deposits heated at 400°C temperature. The electrochemical measurements showed that the annealed Ni-Zn-P-TiO_2 coatings have good corrosion resistance compared to Ni-Zn-P coatings (Ranganatha et al. 2010).

Titanium dioxide particles have also been used in electroless NiP plating to increase hardness (Gadhari and Sahoo 2016; Novakovic and Vassiliou 2009; Wu et al. 2015b), wear (Gadhari and Sahoo 2016; Nad and Ehteshamzadeh 2014; Tamilarasan et al. 2016), stress corrosion cracking (Lee et al. 2015a) and high-temperature corrosion resistance (up to at least 150°C) (Kim et al. 2017). Heat treatment improved the mechanical properties of TiO_2-containing electroless NiP coatings (Gadhari and Sahoo 2016; Wu et al. 2015b) but deteriorated the corrosion resistance (Novakovic and Vassiliou 2009).

As titanium dioxide is known for its photocatalytic activity and NiP is also catalytic, several research groups have used electroless NiP-TiO_2 coatings for applications that require catalytic activity or chemical sensitivity (Chiu et al. 2019; de Hazan et al. 2012; Ding et al. 2019; Rattanawaleedirojn et al. 2016; Shibli and Dilimon 2007, 2008) and even to avoid bacterial adhesion (Liu and Zhao 2011).

Alumina particles can also easily be added to electroless nickel coatings. Karthikeyan et al. prepared the electroless NiP-Al_2O_3 coatings on steel substrates by varying reducing agent concentration and reported that the deposition rate and surface roughness of the coatings are highly influenced by sodium hypophosphite reducing agent. With an increase in sodium hypophosphite, Ni forms amorphous phase and as a result, the microhardness of the coating is reduced. The morphology showed a uniform distribution of nano-alumina particles without any pores and cracks and highly improved the wear resistance (Karthikeyan and Ramamoorthy 2014). Alirezaei et al. (2013) studied the tribological properties of electroless NiP-Ag-Al_2O_3 hybrid coatings at high temperature and reported that the wear resistance and friction coefficient of the hybrid coating are strongly influenced by the self-lubricating silver layers formed between mating surfaces during high-temperature sliding wear. In simpler coatings, additions of alumina particles brought the following benefits: improved wear rate (Hu et al. 2017, 2018; Karthikeyan and Ramamoorthy 2014; León-Patiño et al. 2019; Novák et al. 2009), increase in hardness after heat treatment (Alirezaei et al. 2004; Apachitei et al. 1998; Balaraju et al. 2006a), decrease in abrasion rate (Vojtěch 2009) and improved corrosion resistance (Hu et al. 2018).

Silicon dioxide particles had a observable effect on the wear behavior of NiP coatings. The effect of nano-SiO_2 on the electroless deposition and heat treatment for 1 h at various temperatures to determine the wear resistance of NiP-SiO_2 composite coatings was studied. Results showed that there was a significant improvement in the microhardness and wear resistance. Heat treatment at 400°C led to a change in the composite structure. This may be largely due to the reinforcing action of embedded SiO_2 nanoparticles and the strengthening action of hard Ni_3P phase in the composite coating annealed at 400°C (Dong et al. 2009).

Apart from these, refractory metal oxides were also introduced for coatings with superior hardness properties. ZrO_2 nanoparticles were embedded in the NiP matrix by adding ZrO_2 solution to electroless plating bath, and it evidenced increased microhardness of 1045 HV and wear resistance (Yang et al. 2011). Bostani et al. prepared the composite coatings by dispersing ZrO_2 and electroless nickel-coated ZrO_2 (NCZ) particle concentrations (continuously increased) in the bath. Functionally graded Ni-NCZ (FGN-NCZ) coating has a rough surface morphology with a smaller Ni crystallite size than functionally graded Ni-ZrO_2 (FGN-Z) due to the conductive properties of the NCZ particle, which causes the nucleation of electrodeposited Ni clusters on the surface of the particles. In addition, corrosion resistance of FGN-NCZ is higher than that of FGN-Z possibly due to the higher levels of co-deposited secondary particles, more change in Ni microstructure and stronger bonds between second phase particles and matrix attained for FGN-NCZ (Bostani et al. 2017; Szczygieł and Turkiewicz 2008; Zielińska et al. 2012).

Other oxides such as iron oxide (de Hazan et al. 2012; Zuleta et al. 2009), cerium oxide (de Hazan et al. 2012; Pancrecious et al. 2018), indium-tin-oxide (ITO) (de Hazan et al. 2012), zinc oxide (Shibli et al. 2006a, 2006b), niobium oxide (Chen et al. 2015) and even GO (Jiang et al. 2016, 2019; Wu et al. 2015a) have been used in electroless NiP coatings, with various results in terms of hardness, wear and so on.

Some oxides, including titanium dioxide have also been incorporated in electroless NiB coatings with the following benefits: increase in hardness and corrosion resistance in the as-plated state (Ekmekci and Bülbül 2015; Niksefat and Ghorbani 2015; Shu et al. 2015). However, these benefits were lost after heat treatment (Niksefat and Ghorbani 2015).

Dursun Ekmekc and Ferhat Bülbül (Casula et al. 2004) investigated the effect of the addition of other oxide particles (nano-SiO_2, Al_2O_3 and CuO) on the mechanical properties of electroless NiB coating. The nanoparticles act as a crystal in the deposit. When the load is applied, the matrix resists the force and the added reinforcement obstructs the particle dislocation. Thus, the resistance to the applied force is increased. The alloying element like P or B along with Ni has also impact on the microhardness of the deposit. In the case of NiB, the hardness increases with the increase in the percentage of B content (Vitry and Bonin 2017).

8.3.4 Natural Materials

Natural materials can be of two kinds: natural stones and bio-sourced materials. Both types have been used as added particles in electroless nickel composites.

First, for bio-sourced particles, a comparative analysis on the friction-wear property of as-plated, Nd:YAG laser-treated, and heat-treated electroless NiP-crab-shell particle (CSP) composite coatings on mild steel was conducted by Arulvel et al. Fabricated coatings for different concentrations of CSPs were post-heat-treated and

Nd:YAG laser treatment process and their characteristics were compared. Dendritic structures were observed for laser-treated coatings and its formation was affected by the incorporation of CSP in the NiP matrix. The $FeNi_2P$ phase was formed for laser-treated coatings in addition to the Ni_3P and $Ni_{12}P_5$ phases. The surface hardness increased for laser-treated NiP and NiP-CSP composite coatings compared to the heat treatment process and as-plated condition. A low friction coefficient was obtained for the heat treatment process followed by the as-plated and laser treatment process. The formation of $FeNi_2P$ phase caused an increase in the friction coefficient for laser-treated NiP and NiP-CSP composite coatings (Arulvel et al. 2019).

Second, some studies were carried out with talc as a reinforcing agent (Alexis et al. 2013; Etcheverry 2006; Petit and Celis 2006). Despite significantly increasing the roughness of coatings, talc additions provided a reduction of wear for heat-treated coatings.

8.3.5 Others

It is not possible to exhaustively enumerate all the particles that have been added to electroless nickel coatings. It can be noted that besides the families that have been previously described, boride (Huang et al. 2019), fluoride (Du et al. 2018) and sulfide (Du et al. 2018; Li et al. 2013; Liew et al. 2013; Ranganatha and Venkatesha 2012; Rossi et al. 2003; Zou et al. 2006c) particles have also been used in electroless nickel composites. The boride particles are usually used because of their hardness, while sulfides are preferred due to their self-lubricating properties.

Du et al. investigated, for example, the tribological behavior of composite electroless MoS_2 and CaF_2 coatings after heat treatment from room temperature to approximately 570°C. The results showed that the heat treatment of composite coatings leads to the changes in the microstructure of the NiP matrix from amorphous to crystalline structure and showed the influence of heat treatment temperature on the tribological behavior of the coatings (Du et al. 2018). The good self-lubricating property of the composite coatings over a relatively broad range of temperature was mainly attributed to the synergistic lubricating effects of the oxides from Ni and Mo, MoS_2, sulfate, phosphate and the small amounts of CaF_2 and $CaMoO_4$. The study revealed that the electroless NiP composite coating with MoS_2 and CaF_2 might be a potential candidate as a protective and functional coating under dry sliding conditions at elevated temperatures (Du et al. 2018). The incorporation of WS_2 nanoparticles to Ni-W-P alloy coating reduced the coefficient of friction from 0.16 to 0.11 and also successfully further improved the corrosion resistance of the coating (Ranganatha et al. 2012).

8.3.6 Composites with More Than One Type of Particles

Considering the two main effects of added particles—hardness increase and solid lubrication—some authors have combined two different types of particles in their coatings. Electroless nickel composite coatings containing SiC and graphite (Bahaaideen et al. 2010; Mohtar et al. 2013b; Wu et al. 2006b, 2006c), TiO_2 and PTFE (Liu and Zhao 2011) and TiO_2 and graphite (Liu et al. 2015) have been described in the literature and have proved to bring beneficial effects to electroless nickel coatings.

More rarely, particles having similar effects have been combined in electroless nickel coatings, such as PTFE and MoS_2 (Mohammadi and Ghorbani 2011), iron and titanium oxide (Shibli and Sebeelamol 2013), zirconium and aluminum oxide (Sharma et al. 2002).

8.4 SUMMARY OF THE EFFECT OF PARTICLES ON COATING PROPERTIES

Surface composite coatings can significantly influence the properties of structural components. Most of the aerospace materials are stressed by a combination of tribological, thermal and mechanical factors. In these conditions, the composite coating on the materials can provide better surface protection against the environment. Electroless composite coating can be achieved on any sort of materials such as metals, ceramics, glass, plastics and so on. The important properties of electroless coating are shown in Figure 8.10.

8.4.1 Physical Properties

8.4.1.1 Coating Thickness

The thickness of coatings is influenced by particles in two distinct ways: first by the increase in volume due to the presence of the particles themselves, and second by inhibiting or catalytic effects of the particles on the plating process. Even as the particles are considered as inert for deposition, some of them are not exactly inert. Diamond, for example, can increase the plating rate.

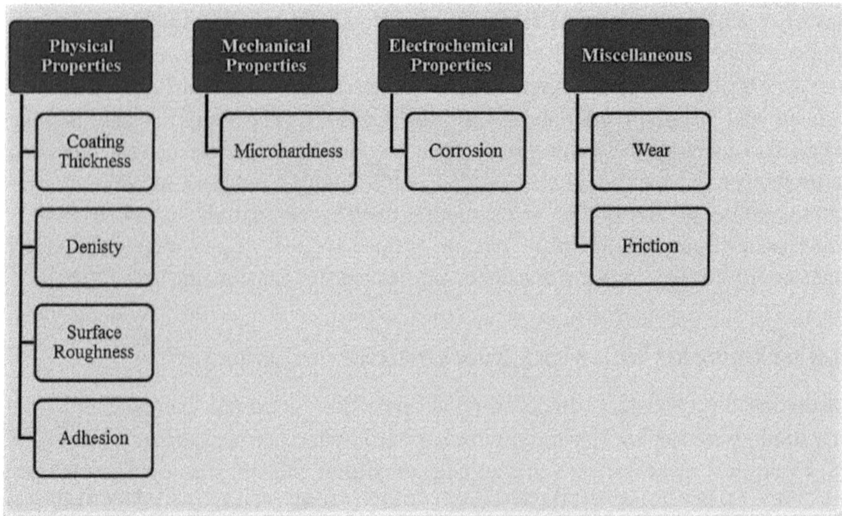

FIGURE 8.10 Properties of electroless composite coatings.

8.4.1.2 Density

In the case of composite coatings, the deposit density is decreased with an increase in the reinforcement percentage. As discussed, the average density of NiP and NiB is 8.3 g/cm^3, whereas the average density of the deposit that contains NiP-Si_3N_4 is 7.34 g/cm^3. According to the reinforcement density, the deposit density and the mechanical strength also vary (Matik 2016). Values of 6.67 and 5.8 g/cm^3 have been reported for electroless NiB coatings with 12% graphite and 28% PTFE, respectively (Omidvar et al. 2016a).

8.4.1.3 Surface Roughness

Surface roughness is one of the essential parameters to determine the wear resistance of the deposit because it contacts metal surface directly during the process, which in turn produces better resistance to friction (Balaraju et al. 2003). There are many factors that influence the surface finish of the electroless prepared samples. They are:

1. Incorporation of hard (or) soft particles,
2. Wetting agents,
3. Alloying elements,
4. Surface treatments,
5. Inhibitors and so on.

The addition of particles into the NiP/B coating alters the composition of the deposit. Thus, surface roughness and appearance are also affected. When the phosphorus content in the deposit increases, the roughness decreases and produces bright appearance (Alirezaei et al. 2007). Many research results concluded that the heat treatment process can alter the structure from amorphous to crystalline in the electroless coated deposits (Apachitei et al. 2002). Surface roughness is an adverse property for the deposits because it causes friction and wear. However, in the case of lubricated environment, it helps to store the oil and prevent from getting welded together. In general, electroless process produces very good surface finish because it follows the surface profile for deposition rather than filling the gaps between particles (Sahoo and Das 2011). To reduce the roughness further, duplex coatings with NiP and NiBlayers or the ternary alloy coating can be prepared (Balaraju et al. 2005). The relation between roughness, corrosion and wear are depicted in Figure 8.11. When the surface roughness of the deposit increases, the wear resistance is improved. The contact area between the two mating surface is reduced for the deposit which has high surface roughness. Consequently, a smooth surface finish provides more contact surface, and hence, the corrosion rate is influenced.

8.4.1.4 Structure

The structural characteristics and phase transformation behavior of electroless NiP, NiP-Si_3N_4, NiP-CeO_2 and NiP-TiO_2 coatings were comparatively analyzed and revealed that the incorporation of these particles in the NiP matrix did not have any significant influence on the structure and phase transformation behavior (Balaraju and Narayanan 2006). The heat treatment of the NiP-CNT composite was carried out at 476 K for 2 h to develop the crystalline structure and the NiP_3 forms in the nickel phase. Similar results have been presented in nearly all studies of electroless nickel composite coatings.

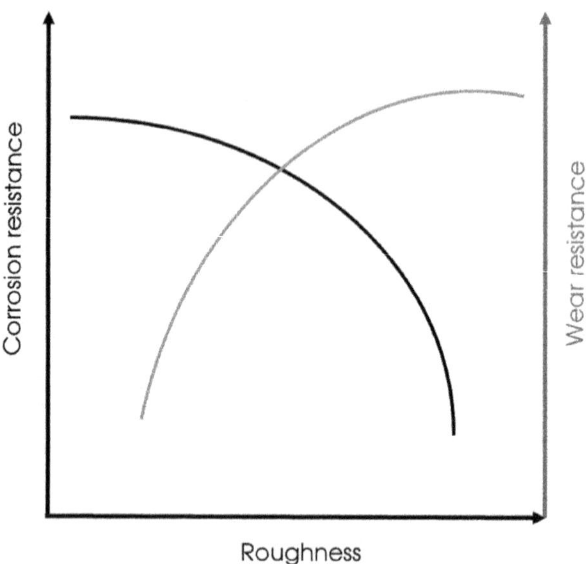

FIGURE 8.11 Relationship between corrosion and wear and surface roughness.

8.4.2 Mechanical Property

8.4.2.1 Microhardness

Microhardness of composite coatings is obviously affected by the presence of particles. This influence can appear in two different ways. First by a mixing law: when particles whose hardness is different from that of the matrix are added to electroless nickel, the hardness of the resulting composite lie between the hardness of the matrix and that of the particles. For example, hard particles such as diamond and silicon carbide will increase the hardness of electroless nickel coatings, while soft particles such as PTFE or graphite will lead to a decrease in hardness. The second way of action is observed for nanoparticles that will act as new sites for germination of the coating or as obstacles for the propagation of dislocation in the matrix. When these mechanisms are observed, the hardness of the composite coatings is increased whatever the hardness of the nanoparticles.

8.4.3 Electrochemical Property

8.4.3.1 Corrosion

The effect of particles on corrosion resistance of electroless nickel composite coatings cannot be easily described because it depends highly on the nature of the particles that were added. Some particles such as diamond, PTFE and carbides improve the corrosion resistance of the electroless nickel coating, while some others such as titanium nitride clearly decrease it. Some also do not really alter the electrochemical properties of coatings.

8.4.4 Tribological Properties

8.4.4.1 Friction

Friction is the force between two objects in contact that restricts the motion, and its direction is opposite to the motion of the object. The most difficult limitation in tribology is achieving and retaining less friction at elevated temperatures. There are some oxides, molybdates and fluorides that are promising as solid lubricants at high temperature (Peterson et al. 1959). However, no single solid lubricant is available for use for high-temperature applications. The effect of incorporating two or more solid lubricants in the composites has attracted attention. The nickel composite coating is prepared with solid lubricants such as graphite and MoS_2, and it acted as excellent self-lubricant and produced minimum friction even at 600°C (Peterson et al. 1959). Coefficient of friction is varied by altering the percentage of alloying elements such as phosphorus or boron, reinforced particles, heat treatment. (Tamilarasan et al. 2016) and (Uysal et al. 2016) studied the influence of surfactant on the frictional behavior of electroless plating. The frictional coefficient decreased with the increase in surfactant concentration in the plating bath. In addition, the surfactant helps to produce smoother, low porous and uniform surfaces. Without surfactant addition, the nanoparticles agglomerated and increased the friction between the surfaces whereas, in the presence of surfactant, the nanoparticles distributed homogenously without agglomeration and friction is reduced at the surfactants CMC.

8.4.4.2 Wear

Electroless coatings are preferred in many applications because of their excellent wear resistance. To increase the wear resistance and decrease the frictional coefficient, the particles such as Al_2O_3, SiC and PTFE are reinforced in NiP/B coatings. Most electroless nickel composite coatings are actually intended to reduce wear by one or more means of action. The different types of wear that can be encountered by electroless nickel composite coatings are as follows:

1. **Adhesive wear:** It occurs due to hard particles forced against a mating surface and move along the solid component. Electroless NiP is incorporated with diamond particles to increase the wear resistance, and it has many potential industrial applications. During sliding motion, initially, the hard particles protrude off from the substrate further increase in sliding distance smooth the surface of the substrate and reduce the wear rate. This mechanism is indicated in most of the earlier research studies (Radhika et al. 2018). This mechanism can be observed in most of the room temperature applications.
2. **Abrasive wear:** It is the transfer of material between two mating solid surface due to localized bonding between them. Wang et al. (2018) studied the diamond particle addition with NiP alloy coating by varying the particles size in micron level. The results showed that the friction and wear resistance improved significantly with the particle sizes increased. Mukhopadhyay et al. concluded that at 100°C, both adhesion and abrasion are predominant. Due to higher three-body abrasion, NiB coatings show the highest wear

rate (Mukhopadhyay et al. 2017). Kundu et al. (2016) observed both adhesion and abrasion phenomena in the coated substrate. The surface of the substrate exhibits ploughing effect, coating detachment and cracks.

3. **Surface fatigue wear:** Fatigue wear is caused by fracture arising due to material fatigue. Wear generated after repeated contact cycles is called fatigue wear. When the number of contact cycles is high, the high-cycle fatigue mechanism is expected to be the wear mechanism. When it is low, the low-cycle fatigue mechanism is expected. Zhou et al. (2016) observed the fatigue wear in the composite coating, and it is attributed due to the reciprocating action under high stress, or the internal stress resulting from the reinforcing of the hard particles.
4. **Oxidation wear:** The surface of any material has oxide layers over it. Depending upon the material characteristic, the layer thickness may get varied. This oxide layer has the tendency to prevent the surface from metal-to-metal contact. However, when the load is higher, the oxide layers removed and metal-to-metal contact produce either adhesive or abrasive wear (Alirezaei et al. 2007). At higher temperatures, the formation of a protective oxide glaze layers consisting of oxides of Ni and oxides of reinforced materials improves the wear resistance significantly (Mukhopadhyay et al. 2017).

The means of actions that are usually observed in electroless nickel composite coatings to reduce wear are increase of hardness (by addition of hard particles) or increase of lubrication by the addition of soft particles that can be considered as solid lubricant, such as graphite or PTFE. In some cases, the two modes of actions can be encountered in a single coating when both hard and lubricant particles are added.

8.4.5 Conclusion

Electroless nickel composite coatings pursue in most cases the reduction of wear, and there are two main ways to achieve that goal: addition of hard particles and addition of solid lubricant particles. The effect of particles on the properties of the composite coating will thus differ depending on the group they belong to (hard or solid lubricant).

Table 8.2 presents an overview of the effect of adding different types of reinforcing particles in electroless nickel (phosphorus or boron) coatings. In the table, the hard particles are presented in bold and the solid lubricants are presented in italics. Some effects that are similar for all types of particles are not shown in Table 8.2: all particles increase roughness and none of them modify the structure of the electroless nickel matrix, except CNTs that make electroless NiB coatings more crystalline; addition of particles, whatever their type, also increases wear resistance.

TABLE 8.2
Summary of the Effect of Additional Particles on Electroless Nickel Composite Coatings

Type of Electroless Nickel	Particle Type	Particle	Hardness	Friction Coefficient	Corrosion Resistance	Others
NiP	Carbon-based	**B_4C**	Increased			
NiP	Carbon-based	*Cg*	Decreased	Decreased	Tested—seems similar to standard NiP	
NiP	Carbon-based	*CNT*				Better field-emitted image; higher magnetization and coercivity
NiP	Carbon-based	**Diamond**	Increased		Improved	Increase of toughness and elastic modulus
NiP	Carbon-based	*Graphene*	Increased but decreases when limit content is reached		Improved	
NiP	Carbon-based	*PTFE*	Decreased	Decreased	Improved	
NiP	Carbon-based	**SiC**	Increased		Improved	
NiP	Carbon-based	**TiC**	Increased		Improved	
NiP	Carbon-based	**WC**	Increased		Improved	
NiP	Nitride	**Cubic BN**	Increased		Improved	
NiP	Nitride	*Hexagonal BN*			Improved	
NiP	Nitride	**Si_3N_4**	Increased		Disputed	
NiP	Nitride	**TiN**	Increased		Lowered	

(Continued)

TABLE 8.2 (*Continued*)
Summary of the Effect of Additional Particles on Electroless Nickel Composite Coatings

Type of Electroless Nickel	Particle Type	Particle	Hardness	Friction Coefficient	Corrosion Resistance	Others
NiP	Oxide	**Al_2O_3**	Improved for heat-treated coatings		Improved	
NiP	Oxide	**TiO_2**	Increased		Improved	Also photocatalytic properties
NiP	Oxide	**ZrO_2**	Increased		Improved	
NiP	Oxide	**SiO_2**	Increased			
NiP	**Boride**		Increased	Decreased		
NiP	*Sulfide*			Decreased		
NiB	Carbon-based	*Cg*		Decreased		
NiB	Carbon-based	*CNT*	Increased but decreases when limit content is reached			
NiB	Carbon-based	**Diamond**	Ultra-hard			
NiB	Carbon-based	*PTFE*	Decreased			
NiB	Carbon-based	**SiC**			Improved	

Note: Hard particles are indicated in bold. Solid lubricants are indicated in italics.

8.5 EFFECT OF HEAT/POSTTREATMENT ON PROPERTIES OF ELECTROLESS NICKEL (PHOSPHORUS OR BORON) COATINGS

Similarly to standard electroless nickel coatings, composite electroless nickel coatings present, in their as-deposited state, a matrix with an amorphous supersaturated structure. It is thus possible to modify the structure of this matrix by heat treatment and most electroless nickel composite answer to heat treating in a similar way to the particle-free coatings. Although quite several studies have been conducted to explore the potential of posttreatment process for improving the coating performance, they have not been commercially viable on an industrial scale. Therefore, a sustainable method of posttreatment that can cater specifically to particular properties depending upon the nature of application needs to be established.

8.6 VARIATIONS OF THE ELECTROLESS COMPOSITE DEPOSITION: ELECTROLESS NICKEL COATING OF POWDERS TO PREPARE METAL-MATRIX COMPOSITES

Next to composite electroless nickel coatings, similar methods have been used to prepare core-shell particles and composite powders that are used in the metal matrix composite industry. Tungsten carbide powders (Alhaji et al. 2019; Chuankrerkkul et al. 2013; Erfanmanesh et al. 2018; Guo et al. 2017; Jafari et al. 2013a, 2013b, 2014; Liu et al. 2006; Qui et al. 1999; Vélez et al. 1999; Wang et al. 2019) are among the most popular materials for this kind of treatment but numerous other types of particles have been used (Alavi et al. 2019; Dai et al. 2006; Guo et al. 2010; Gui et al. 2018; Heydari et al. 2018; Jang et al. 2007, 2014; Lin et al. 2013; Liu et al. 2017; Palaniappa et al. 2007; Park et al. 2017; Rattanawaleedirojn et al. 2014; Song et al. 2009; Ulicny and Mance 2004; Uysal et al. 2011, 2013; Xue et al. 2015; Zangeneh-Madar and Jafari 2012; Zou et al. 2006a, 2006b, 2011. The formed composite powders can be used for a variety of applications, from electrodes in modern batteries to feedstock for advanced powder metallurgy fabrication, including sensors, catalysts and metal hollow spheres.

8.7 CHALLENGES IN ELECTROLESS COMPOSITE COATING

- The study on the co-deposition mechanism of each type of ceramic particles is essential, because the reinforcement of nano-sized particles increases the chances of agglomeration of particles in the electrolyte (Gawad et al. 2013).
- Higher coating thickness of approximately 2 mg/cm^2.
- Coating is difficult to plate upon Inconel-600 electrode material (Sathiyamoorthy 2017).
- Dispersing PTFE in solution is a challenge because the particles tend to shed water (Guo et al. 1995).

8.8 RECOMMENDATIONS

1. Addition of any sort of surfactant in the electrolyte during coating process increases the life of the bath. Particularly, Zwitterionic surfactant increases the life of the bath in binary alloy coatings.
2. Surfactants are used in the electroless bath to avoid agglomeration of reinforcement nanoparticles.
3. The percentage of embedded ceramic particles in the deposit increases with increasing concentration of suspended particles in the electrolyte. However, after the certain limit, agglomeration appeared.
4. To achieve proper coating on the substrate with good adhesion, the surface preparation must be proper, should follow the activation procedure based on the base substrate, and surface must be free from contaminants such as oil.
5. Good plating rate can be obtained by the following simple ways: preparing the electrolyte at the time of plating, maintaining the nickel and reducing agent concentration, avoiding bath overheating, ensuring temperature homogeneity throughout the bath and keeping the optimal concentration of stabilizer.
6. According to the need, the pH level needs to be maintained at acidic, neutral or alkaline. If it exceeds a certain level, the electrolyte will become cloudy and precipitation will be observed in solution.
7. The pH in the electrolyte decreases rapidly in some cases; to avoid this problem, the buffering agent has to be added to control the pH.
8. The following reasons are the causes of the zero plating rate on the substrate: (a) improper solution preparation, (b) presence of unwanted materials, (c) too high pH level and (d) overheated electrolyte.
9. Peeling or flaking is usually due to improper cleaning or activation of substrates. The importance of pretreatment cannot be overemphasized. Matching the proper cleaning and activation processes for the type of metal and the manufacturing method are the prerequisites to having a robust process to handle variations in soils and oils left on the part.
10. The dark or dull deposit may be obtained due to the low concentration of nickel and reducing agent and low pH and temperature.
11. The decomposition of bath occurs due to (a) high temperature, (b) high pH, (c) local overheating, (d) plating tank not passivated properly, (e) airborne particles or (f) large additions to the bath.
12. The recommended electroless composite coating setup for electroless plating is shown in Figure 8.12.
13. The recommended bath conditions for electroless NiP and NiB composite coatings are presented in Tables 8.3 and 8.4.

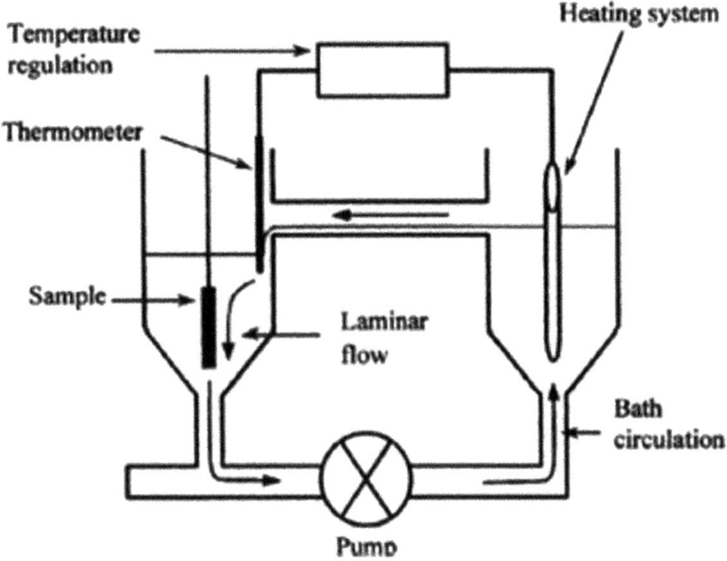

FIGURE 8.12 A schematic of recommended electroless composite coating setup.

TABLE 8.3
Recommended Conditions for Electroless Nickel-Phosphorus Composite Plating

NiP Bath Conditions	
Nickel sulfate	25–35 (g/L)
Sodium hypophosphite	35–45 (g/L)
Lactic acid	32–38 (mL/L)
Propionic acid	2–3 (mL/L)
Stabilizer	1 ppm
Hard particles	Desired level
Surfactants	(0.25–2) × CMC
pH	6
Temperature	88°C (±2°C)

TABLE 8.4
Recommended Conditions for Electroless Nickel-Boron Composite Plating

NiB Bath Conditions	
Nickel chloride	20–25 (g/L)
Sodium borohydride	1–2 (g/L)
Ethylenediamine	80–40 (mL/L)
Sodium hydroxide	80–40 (g/L)
Stabilizer	1 ppm
Hard particles	Desired level
Surfactants	(0.25–2) × CMC
pH	9
Temperature	88°C (±2°C)

8.9 SUMMARY

As the research on electroless nickel (phosphorus or boron) has almost saturated, since the last decade, more researchers have focused on the electroless nickel composite coatings. It has an optimistic future for wear-resistant or self-lubricating coatings and other special treatments. Since 2010, research publications have shown that many composite materials fulfill new benefits and property. With advancements in the automatic controllers and more advanced processing technology, new polyalloys with new composites developed the new era of electroless technology. The composite coatings are providing an excellent layer of coating containing nickel/boron and other components to be used for defense, textiles, heavy industries, and intelligent materials. In addition to composite coating, the electroless nickel nanocoating has revealed that the second phase nanoparticles (SiC, CeO_2, TiO_2, Al_2O_3, CNTs, Fe_3O_4, B_4C, nanometer-size diamond, etc.) can be incorporated successfully in the NiP/B matrix by the electroless plating technique.

In the machinery construction industry, components usually coated with electroless composite coatings are lathe beds, shafts, levers, calipers and fasteners. The applications of electroless composite in chemical and plastic industries involve coatings for autoclaves, filters, pipe works, heat exchangers, screw feeders, valves, pump components and so on. The aesthetic attribute and functionality of electroless composite are used in the automobile industry for brake pistons, radiator jackets, clutch bosses, steering gears and so on (Parkinson 1997; Parker 1972). Apart from these, electroless nickel-based coatings are also used in hydraulics, mining, offshore technology, oil and gas, electrical and electronics, aerospace and printing industries (Parkinson 1997).

REFERENCES

Afroukhteh, S., C. Dehghanian, and M. Emamy. 2012. "Corrosion Behavior of Ni–P/Nano-TiC Composite Coating Prepared in Electroless Baths Containing Different Types of Surfactant." *Progress in Natural Science: Materials International* 22 (5): 480–487. doi:10.1016/J.PNSC.2012.10.005.

Alavi, B., H. Aghajani, and A. Rasooli. 2019. "Electrophoretic Deposition of Electroless Nickel Coated YSZ Core-Shell Nanoparticles on a Nickel Based Superalloy." *Journal of the European Ceramic Society* 39 (7): 2526–2534. doi:10.1016/j.jeurceramsoc.2019.01.028.

Alexis, J., C. Gaussens, B. Etcheverry, and J. P. Bonino. 2013. "Development of Nickel-Phosphorus Coatings Containing Micro Particles of Talc Phyllosilicates." *Materials Chemistry and Physics* 137: 723–733. doi:10.1016/j.matchemphys.2012.09.049.

Alhaji, A., M. Shamanian, M. Salehi, S. M. Nahvi, and M. Erfanmanesh. 2019. "Electroless Nickel–phosphorus Plating on WC–Co Powders Using HVOF Feedstock." *Surface Engineering* 35 (2): 120–127. doi:10.1080/02670844.2018.1477558.

Alirezaei, S., S. M. Monirvaghefi, M. Salehi, and A. Saatchi. 2007. "Wear Behavior of Ni-P and Ni-P-Al_2O_3 Electroless Coatings." *Wear* 262 (7–8): 978–985. doi:10.1016/j.wear.2006.10.013.

Alirezaei, S., S. M. Monirvaghefi, A. Saatchi, M. Ürgen, and A. Motallebzadeh. 2013. "High Temperature Friction and Wear Behaviour of Ni–P–Ag–Al_2O_3 Hybrid Nanocomposite Coating." *Transactions of the IMF* 91 (4): 207–213. doi:10.1179/0020296713Z.000000000105.

Alirezaei, S., S. M. Monirvaghefi, M. Salehi, and A. Saatchi. 2004. "Effect of Alumina Content on Surface Morphology and Hardness of Ni–P–$Al_2O_3(\alpha)$ Electroless Composite Coatings." *Surface and Coatings Technology* 184: 170–1753.

Alishahi, M., S. M. Monirvaghefi, A. Saatchi, and S. M. Hosseini. 2012. "The Effect of Carbon Nanotubes on the Corrosion and Tribological Behavior of Electroless Ni–P–CNT Composite Coating." *Applied Surface Science* 258 (7): 2439–2446. doi:10.1016/j.apsusc.2011.10.067.

Allahkaram, S., and F. Bigdeli. 2010. "Influence of Particle Size on Corrosion Resistance of Electroless Ni-P-Sic Composite Coatings." *Iranian Journal of Science & Technology* 34: 231–234.

Ankita, S., and A. K. Singh. 2011. "Corrosion and Wear Resistance Study of Ni–P and Ni–P–PTFE Nanocomposite Coatings." *Central European Journal of Engineering* 1 (3): 234–243. doi:10.2478/s13531-011-0023-8.

Ansari, M. I., and D. G. Thakur. 2017. "Influence of Surfactant: Using Electroless Ternary Nanocomposite Coatings to Enhance the Surface Properties on AZ91 Magnesium Alloy." *Surfaces and Interfaces* 7: 20–28. doi:10.1016/J.SURFIN.2017.02.004.

Apachitei, I., F. D. Tichelaar, J. Duszczyk, and L. Katgerman. 2002. "The Effect of Heat Treatment on the Structure and Abrasive Wear Resistance of Autocatalytic NiP and NiP–SiC Coatings." *Surface and Coatings Technology* 149 (2–3): 263–278. doi:10.1016/S0257-8972(01)01492-X.

Apachitei, I., J. Duszczyk, L. Katgerman, and P. J. B. Overkamp. 1998. "Electroless Ni–P Composite Coatings: The Effect of Heat Treatment on The Microhardness of Substrate and Coating." *Scripta Materialia* 38: 1347–1353.

Araghi, A., and M. H. Paydar. 2013. "Electroless Deposition of Ni–W–P–B4C Nanocomposite Coating on AZ91D Magnesium Alloy and Investigation on Its Properties." *Vacuum* 89 (1): 67–70. doi:10.1016/j.vacuum.2012.09.011.

Araghi, A., and M. H. Paydar. 2014. "Wear and Corrosion Characteristics of Electroless Ni–W–P–B4C and Ni–P–B4C Coatings." *Surface and Interfaces* 8 (3): 146: 153.

Arai, S., M. Kobayashi, T. Yamamoto, and M. Endo. 2010. "Pure-Nickel-Coated Multiwalled Carbon Nanotubes Prepared by Electroless Deposition." *Electrochemical and Solid-State Letters* 13 (12): D94. doi:10.1149/1.3489535.

Arulvel, S., A. Elayaperumal, M. S. Jagatheeshwaran, and K. A. Satheesh. 2019. "Comparative Study on the Friction-Wear Property of As-Plated, Nd-YAG Laser Treated, and Heat Treated Electroless Nickel-Phosphorus/Crab Shell Particle Composite Coatings on Mild Steel." *Surface and Coatings Technology* 357: 543–558. doi:10.1016/J.SURFCOAT.2018.10.051.

Ashassi-Sorkhabi, H., and M. Es′haghi. 2013. "Corrosion Resistance Enhancement of Electroless Ni–P Coating by Incorporation of Ultrasonically Dispersed Diamond Nanoparticles." *Corrosion Science* 77: 185–193. doi:10.1016/j.corsci.2013.07.046.

Bahaaideen, F. B., Z. M. Ripin, and Z. A. Ahmad. 2010. "Electroless Ni-P-Cg(Graphite)-SiC Composite Coating and Its Application onto Piston Rings of a Small Two Stroke Utility Engine." *Journal of Scientific and Industrial Research JSIR*. Vol. 69, Nov. 2010, pp. 830–834.

Balaraju, J. N., C. Anandan, and K. S. Rajam. 2005. "Electroless Deposition of Ternary Ni–W–P Alloys from Sulphate and Chloride Based Baths." *Surface Engineering* 21 (3): 215–220. doi:10.1179/174329405X50046.

Balaraju, J. N., T. S. N. Sankara Narayanan, and S. K. Seshadri. 2003. "Electroless Ni-P Composite Coatings." *Journal of Applied Electrochemistry* 33 (9): 807–816. doi:10.1023/A:1025572410205.

Balaraju, J. N., T. S. N. Sankara Narayanan, and S. K. Seshadri. 2006. "Structure and Phase Transformation Behaviour of Electroless Ni–P Composite Coatings." *Materials Research Bulletin* 41 (4): 847–860. doi:10.1016/j.materresbull.2005.09.024.

Balaraju, J. N., and S. K. Seshadri. 1998. "Synthesis and Corrosion Behavior of Electroless Ni–P–Si_3N_4 Composite Coatings." *Journal of Materials Science Letters* 17 (1 5): 1297–1299. doi:10.1023/A:1006528229614.

Balaraju, J. N., N. Kalavati, and K. S. S. Rajam. 2010a. "Electroless Ternary Ni–W–P Alloys Containing Micron Size Al_2O_3 Particles." *Surface and Coatings Technology* 205 (2): 575–581. doi:10.1016/J.SURFCOAT.2010.07.047.

Balaraju, J. N., V. E. Selvi, and K. S. Rajam. 2010b. "Electrochemical Behavior of Low Phosphorus Electroless Ni–P–Si_3N_4 Composite Coatings." *Materials Chemistry and Physics* 120 (2–3): 546–551. doi:10.1016/j.matchemphys.2009.11.047.

Balaraju, J. N., N. Kalavati, and K. S. S. Rajam. 2006a. "Influence of Particle Size on the Microstructure, Hardness and Corrosion Resistance of Electroless Ni–P–Al_2O_3 Composite Coatings." *Surface and Coatings Technology* 200 (12–13): 3933–3941. doi:10.1016/j.surfcoat.2005.03.007.

Balaraju, J. N., T. S. N. Sankara Narayanan, and S. K. Seshadri. 2006b. "Structure and Phase Transformation Behaviour of Electroless Ni–P Composite Coatings." *Materials Research Bulletin* 41 (4): 847–860. doi:10.1016/j.materresbull.2005.09.024.

Balaraju, J. N., and S. K. Seshadri. 1999. "Preparation and Characterization of Electroless Ni–P and Ni–P–Si_3 N_4 Composite Coatings." *Transactions of the IMF* 77 (2): 84–86. doi:10.1080/00202967.1999.11871253.

Balaraju, J. N., and T. S. N. Sankara Narayanan. 2006. "Bulletin Structure and Phase Transformation Behaviour of Electroless Ni–P Composite Coatings." *Materials Research Bulletin* 41 (2006): 847–860. doi:10.1016/j.materresbull.2005.09.024.

Balaraju, J. N., and K. S. Rajam. 2008. "Preparation and Characterization of Autocatalytic Low Phosphorus Nickel Coatings Containing Submicron Silicon Nitride Particles." *Journal of Alloys and Compounds* 459 (1779): 311–319. doi:10.1016/j.jallcom.2007.04.228.

Baudrand, D. 1978. "Metals Handbook." *ASM Handbook* 8: 215–230.

Bigdeli, F., and S. R. Allahkaram. 2009. "An Investigation on Corrosion Resistance of As-Applied and Heat Treated Ni–P/NanoSiC Coatings." *Materials and Design* 30 (10): 4450–4453. doi:10.1016/j.matdes.2009.04.020.

Bostani, B., R. Arghavanian, N. P. Ahmadi, and S. Yazdani. 2017. "Fabrication and Properties Evaluation of Functionally Graded Ni-NCZ Composite Coating." *Surface and Coatings Technology* 328 (2017): 276–282. doi:10.1016/j.surfcoat.2017.09.004.

Bozzini, B., and M. Boniardi. 2001. "Measurement and Modelling of Some Mechanical Properties of Electroless NiP/Diamond Particulate Metal Matrix Composite Thin Films." *International Journal of Materials and Product Technology* 16 (6/7): 626. doi:10.1504/IJMPT.2001.001285.

Bozzini, B., M. Boniardi, A. Fanigliulo, and F. Bogani. 2001. "Tribological Properties of Electroless Ni–P/Diamond Composite Films." *Materials Research Bulletin* 36 (11): 1889–1902. doi:10.1016/S0025-5408(01)00672-9.

Bozzini, B., C. Martini, P. L. Cavallotti, and E. Lanzoni. 1999. "Relationships among Crystallographic Structure, Mechanical Properties and Tribological Behaviour of Electroless Ni–P(9%)/B4C Films." *Wear* 225–229: 806–813. doi:10.1016/S0043-1648(98)00389-5.

Casula, M. F., A. Falqui, G. Paschina, and S. Chimiche. 2004. "Preparation and Characterization of FeCo–Al_2 O_3 and Al_2 O_3 Aerogels." *Journal of Sol-Gel Science and Technology* 38 (3): 83–86. doi:10.1023/B:JSST.0000047965.22232.53.

Chakarova, V., M. Georgieva, M. Petrova, E. Dobreva, and D. Stoychev. 2016. "Electroless Deposition and Investigation of Composite Coatings Based on Nickel or Cobalt Matrix, Including Boron Nitrides as Dispersoids, on Polyethylene Terephthalate Substrate." *Transactions of the IMF* 94 (5): 259–264. doi:10.1080/00202967.2016.1208386.

Chang, C.-S., K.-H. Hou, M.-D. Ger, C.-K. Chung, and J.-F. Lin. 2016. "Effects of Annealing Temperature on Microstructure, Surface Roughness, Mechanical and Tribological Properties of Ni–P and Ni–P/SiC Films." *Surface and Coatings Technology* 288: 135–143. doi:10.1016/j.surfcoat.2016.01.020.

Chen, H., Q. Wang, H. Dong, L. Xi, X. Lin, F. Pan, Z. Ma. 2015. "Electroless Plating of Ni-P-W Coatings Containing Scattered Nb2O5 on Sintered NdFeB Substrate." *Materials Research* 18 (5): 1089–1096. doi:10.1590/1516-1439.032015.

Chen, W. X., J. P. Tu, H. Y. Gan, Z. D. Xu, Q. G. Wang, J. Y. Lee, Z. L. Liu, and X. B. Zhang. 2002. "Electroless Preparation and Tribological Properties of {Ni–P-Carbon} Nanotube Composite Coatings under Lubricated Condition." *Surface and Coatings Technology* 160 (1): 68–73. doi:10.1016/S0257-8972(02)00408-5.

Chen, W. X., J. P. Tu, L. Y. Wang, H. Y. Gan, Z. D. Xu, and X. B. Zhang. 2003. "Tribological Application of Carbon Nanotubes in a Metal-Based Composite Coating and Composites." *Carbon* 41: 215–222.

Chen, X. H., C. S. Chen, H. N. Xiao, H. B. Liu, L. P. Zhou, S. L. Li, and G. Zhang. 2006. "Dry Friction and Wear Characteristics of Nickel/Carbon Nanotube Electroless Composite Deposits." *Tribology International* 39: 22–28. doi:10.1016/j.triboint.2004.11.008.

Chen, Y., Y. Hao, W. Huang, Y. Ji, W. Yang, X. Yin, Y. Liu, and X. Ling. 2017. "Corrosion Behavior of Ni–P-Nano-Al_2 O_3 Composite Coating in the Presence of Anionic and Cationic Surfactants." *Surface and Coatings Technology* 310: 122–128. doi:10.1016/j.surfcoat.2016.12.089.

Chintada, V. B., and R. Koona. 2018a. "Influence of Surfactant on the Properties of Ni–P-Nano ZnO Composite Coating." *Materials Research Express* 6 (2): 025030. doi:10.1088/2053-1591/aaeee8.

Chintada, V. B., and R. Koona. 2018b. "Influence of SiC Nano Particles on Microhardness and Corrosion Resistance of Electroless Ni–P Coatings." *Journal of Bio- and Tribo-Corrosion* 4 (4): 68. doi:10.1007/s40735-018-0186-4.

Chiu, W.-T., C.-Y. Chen, T.-F. M. Chang, T. Hashimoto, H. Kurosu, and M. Sone. 2019. "Ni–P and TiO_2 Codeposition on Silk Textile via Supercritical CO_2 Promoted Electroless Plating for Flexible and Wearable Photocatalytic Devices." *Electrochimica Acta* 294: 68–75. doi:10.1016/J.ELECTACTA.2018.10.076.

Chuankrerkkul, N., Y. Boonyongmaneerat, K. Saengkiettiyut, P. Rattanawaleedirojn, and S. Saenapitak. 2013. "Injection Moulding of Tungsten Carbide-Nickel Powders Prepared by Electroless Deposition." *Key Engineering Materials* 545: 148–152. doi:10.4028/www.scientific.net/KEM.545.148.

Czagány, M., and P. Baumli. 2019. "Effect of Surfactants on the Behavior of the Ni–P Bath and on the Formation of Electroless Ni-P-TiC Composite Coatings." *Surface and Coatings Technology* 361: 42–49. doi:10.1016/j.surfcoat.2019.01.046.

Czagány, M., P. Baumli, and G. Kaptay. 2017. "The Influence of the Phosphorous Content and Heat Treatment on the Nano-Micro-Structure, Thickness and Micro-Hardness of Electroless Ni–P Coatings on Steel." *Applied Surface Science* 423: 160–169. doi:10.1016/j.apsusc.2017.06.168.

Dai, H., H. Li, and F. Wang. 2006. "Electroless Ni-P Coating Preparation of Conductive Mica Powder by a Modified Activation Process." *Applied Surface Science* 253: 2474–2480. doi:10.1016/j.apsusc.2006.05.010.

De Hazan, Y., F. Knies, D. Burnat, T. Graule, Y. Yamada-Pittini, C. Aneziris, and M. Kraak. 2012. "Homogeneous Functional Ni–P/Ceramic Nanocomposite Coatings via Stable Dispersions in Electroless Nickel Electrolytes." *Journal of Colloid and Interface Science* 365 (1): 163–171.

De Hazan, Y., T. Reuter, D. Werner, R. Clasen, and T. Graule. 2008a. "Interactions and Dispersion Stability of Aluminum Oxide Colloidal Particles in Electroless Nickel Solutions in the Presence of Comb Polyelectrolytes." *Journal of Colloid and Interface Science* 323 (2): 293–300. doi:10.1016/j.jcis.2008.03.036.

De Hazan, Y., D. Werner, M. Z'graggen, M. Groteklaes, and T. Graule. 2008b. "Homogeneous Ni-P/Al_2O_3 Nanocomposite Coatings from Stable Dispersions in Electroless Nickel Baths." *Journal of Colloid and Interface Science* 328 (1): 103–109. doi:10.1016/j.jcis.2008.08.033.

Ding, X., W. Wang, A. Zhang, L. Zhang, and D. Yu. 2019. "Efficient Visible Light Degradation of Dyes in Wastewater by Nickel–phosphorus Plating–titanium Dioxide Complex Electroless Plating Fabric." *Journal of Materials Research* 34 (6): 999–1010. doi:10.1557/jmr.2019.16.

Dong, D., X. H. Chen, W. T. Xiao, G. B. Yang, and P. Y. Zhang. 2009. "Preparation and Properties of Electroless Ni–P–SiO_2 Composite Coatings." *Applied Surface Science* 255 (2009): 7051–7055. doi:10.1016/j.apsusc.2009.03.039.

Du, S., Z. Li, Z. He, H. Ding, X. Wang, and Y. Zhang. 2018. "Effect of Temperature on the Friction and Wear Behavior of Electroless Ni–P–MoS_2–CaF_2 Self-Lubricating Composite Coatings." *Tribology International*. doi:10.1016/j.triboint.2018.07.026.

Ebrahimian-Hosseinabadi, M., K. Azari-Dorcheh, and S. M. Vaghefi. 2006. "Wear Behavior of Electroless Ni–P-B4C Composite Coatings." *Wear* 260: 123–127. doi:10.1016/j.wear.2005.01.020.

Ekmekci, D., and F. Bülbül. 2015. "Preparation and Characterization of Electroless Ni–B/Nano-SiO_2, Al_2O_3, TiO_2 and CuO Composite Coatings." *Bulletin of Materials Science* 38 (3): 761–768. doi:10.1007/s12034-015-0912-1.

Elansezhian, R., B. Ramamoorthy, and P. K. Nair. 2009. "The Influence of SDS and CTAB Surfactants on the Surface Morphology and Surface Topography of Electroless Ni–P Deposits." *Journal of Materials Processing Technology* 209 (1): 233–240. doi:10.1016/j.jmatprotec.2008.01.057.

Elansezhian, R., B. Ramamoorthy, and P. K. Nair. 2008. "Surface & Coatings Technology Effect of Surfactants on the Mechanical Properties of Electroless (Ni–P) Coating." *Surface & Coatings Technology* 203 (5–7): 709–712. doi:10.1016/j.surfcoat.2008.08.021.

Erfanmanesh, M., R. Shoja-Razavi, H. Abdollah-Pour, and H. Mohammadian-Semnani. 2018. "Influence of Using Electroless Ni–P Coated WC-Co Powder on Laser Cladding of Stainless Steel." *Surface and Coatings Technology* 348: 41–54. doi:10.1016/j.surfcoat.2018.05.016.

Etcheverry, B. 2006. "Adhérence, Mécanique et Tribologie Des Revêtements Composites NiP–Talc Multifonctionnels à Empreinte Écologique Réduite." *Interface* 158: 1–158.

Farzaneh, A., M. Ehteshamzadeh, M. Can, O. Mermer, and S. Okur. 2016. "Effects of SiC Particles Size on Electrochemical Properties of Electroless Ni–P–SiC Nanocomposite Coatings." *Protection of Metals and Physical Chemistry of Surfaces* 52 (4): 632–636. doi:10.1134/S2070205116040109.

Farzaneh, A., M. Mohammadi, M. Ehteshamzadeh, and F. Mohammadi. 2013. "Electrochemical and Structural Properties of Electroless Ni–P–SiC Nanocomposite Coatings." *Applied Surface Science* 276: 697–704. doi:10.1016/j.apsusc.2013.03.156.

Fayyad, E. M., A. M. Abdullah, M. K. Hassan, A. M. Mohamed, C. Wang, G. Jarjoura, Z. Farhat. 2018. "Synthesis, Characterization, and Application of Novel Ni–P-Carbon Nitride Nanocomposites." *Coatings* 8 (1): 37. doi:10.3390/coatings8010037.

Fayyad, E. M., A. M. Abdullah, A. M. A. Mohamed, G. Jarjoura, Z. Farhat, and M. K. Hassan. 2019. "Effect of Electroless Bath Composition on the Mechanical, Chemical, and Electrochemical Properties of New NiP–C_3N_4 Nanocomposite Coatings." *Surface and Coatings Technology* 362: 239–251. doi:10.1016/J.SURFCOAT.2019.01.087.

Feldstein, N., T. Lancsek, D. Lindsay, and L. Salerno. 1983. "Electroless Composite Plating." *Metal Finishing.* doi:10.1016/B978-0-12-802685-4/00003-0.

Franco, M., W. Sha, G. Aldic, S. Malinov, and H. Çimenoğlu. 2016. "Author's Accepted Manuscript." *Tribology International.* doi:10.1016/j.triboint.2016.01.047.

Gadhari, P., and P. Sahoo. 2016. "Effect Of Tio_2 Particles on Micro-Hardness, Corrosion, Wear and Friction of Ni–P–Tio_2 Composite Coatings at Different Annealing Temperatures." *Surface Review and Letters* 23(1): 1550082. doi:10.1142/S0218625X15500821.

Gawad, S. A., A. M. Baraka, M. S. Morsi, and M. A. Eltoum. 2013. "Development of Electroless Ni-P-Al_2O_3 and Ni-P-TiO_2 Composite Coatings from Alkaline Hypophosphite Gluconate Baths and Their Properties." *International Journal of Electrochemical Science* 8 (2): 1722–1734.

Ge, J. P., R. X. Che, and X. Z. Wang. 1998. "Structure & Properties of Electroless Ni–P-B4C Composite Coatings." *Plating and Surface Finishing* 85 (10): 69–73.

Georgiza, E., V. Gouda, and P. Vassiliou. 2017. "Production and Properties of Composite Electroless Ni-B-SiC Coatings." *Surface and Coatings Technology* 325: 46–51. doi:10.1016/j.surfcoat.2017.06.019.

Ger, M.-D. 2004. "Electrochemical Deposition of Nickel/SiC Composites in the Presence of Surfactants." *Materials Chemistry and Physics* 87 (1): 67–74. doi:10.1016/J.MATCHEMPHYS.2004.04.022.

Grosjean, A., M. Rezrazi, P. Bercot, and M. Tachez. 1998. "Adaptation of a Mathematical Model to the Incorporation of Silicon Carbide Particles in an Electroless Nickel Deposit." *Metal Finishing* 96: 14–17. doi:10.1016/S0026-0576(97)86615-9.

Grosjean, A., M. Rezrazi, and P. Berçot. 2000. "Some Morphological Characteristics of the Incorporation of Silicon Carbide (SiC) Particles into Electroless Nickel Deposits." *Surface and Coatings Technology* 130 (2–3): 252–256. doi:10.1016/S0257-8972(00)00714-3.

Guglielmi, N. 1972. "Kinetics of the Deposition of Inert Particles from Electrolytic Baths." *Journal of The Electrochemical Society* 119 (8): 1009. doi:10.1149/1.2404383.

Gui, C., C. Yao, J. Huang, and G. Yang. 2018. "Effect of Ultrasonic Vibration on Activation-Electroless Nickel Plating on Nylon 12 Powders." *Journal of Materials Science: Materials in Electronics* 29 (7): 5561–5565. doi:10.1007/s10854-018-8524-3.

Guo, L., L. R. Xiao, X. J. Zhao, Y. F. Song, Z. Y. Cai, H. J. Wang, and C. B. Liu. 2017. "Preparation of WC/Co Composite Powders by Electroless Plating." *Ceramics International* 43 (5): 4076–4082. doi:10.1016/j.ceramint.2016.11.220.

Guo, Z. Q., H. R. Geng, and S. S. Feng. 2010. "Study on Mechanical Properties and Physical Properties of Copper Based Electrical Contact Materials Reinforced by CNT." *Advanced Materials Research* 139–141: 67–71. doi:10.4028/www.scientific.net/AMR.139-141.67.

Guo, Z., H. Liu, Z. Wang, and M. Wang. 1995. "Process and Properties of Electroless Plating RE-Ni-B-SiC Composite Coatings." *Acta Metallurgica Sinica(English Letters)(China)* 8 (2): 118–122.

Hamid, Z. A., S. A. El Badry, and A. A. Aal. 2007a. "Electroless Deposition and Characterization of Ni–P–WC Composite Alloys." *Surface and Coatings Technology* 201: 5948–5953. doi:10.1016/j.surfcoat.2006.11.001.

Hamid, Z. A., S. A. El Badry, and A. A. Aal. 2007b. "Electroless Deposition and Characterization of Ni–P–WC Composite Alloys." *Surface and Coatings Technology* 201 (12): 5948–5953. doi:10.1016/j.surfcoat.2006.11.001.

Han, Y., M. Ma, G. Li, J. Yu, G. Liu, Q. Jing, X. Chen, Q. Wang, and R. Liu. 2008. "(Ni–P)/Graphite Composite Film Plated on Bulk Metallic Glass." *Materials Letters* 62 (10–11): 1707–1710. doi:10.1016/J.MATLET.2007.09.067.

Hashemi, S. H., and R. Shoja-Razavi. 2016a. "Optics & Laser Technology Laser Surface Heat Treatment of Electroless Ni–P–SiC Coating on Al356 Alloy." *Optics and Laser Technology* 85: 1–6. doi:10.1016/j.optlastec.2016.05.018.

Hashemi, S. H., and R. Shoja-Razavi. 2016b. "Laser Surface Heat Treatment of Electroless Ni–P–SiC Coating on Al356 Alloy." *Optics & Laser Technology* 85: 1–6. doi:10.1016/j.optlastec.2016.05.018.

Heydari, M. S., H. R. Baharvandi, and S. R. Allahkaram. 2018. "Electroless Nickel-Boron Coating on B4C-Nano TiB_2 Composite Powders." *International Journal of Refractory Metals and Hard Materials* 76: 58–71. doi:10.1016/j.ijrmhm.2018.05.012.

Hou, J. Y., and H. J. Gao. 2016. "Influence of Magnetic Field on the Mass Transfer of Ni–P-Diamond Electroless Composite Plating and Deposit Properties." *Transactions of the IMF* 94 (6): 305–312. doi:10.1080/00202967.2016.1209293.

Hou, J. Y., S. R. Wang, and Z. W. Zhou. 2013. "The Effect of Ni–P Alloy Pre-Plating on the Performance of Ni-P/Ni-P-PTFE Composite Coatings." *Key Engineering Materials* 561: 537–541. doi:10.4028/www.scientific.net/KEM.561.537.

Hsu, C.-I., K.-H. Hou, M.-D. Ger, and G.-L. Wang. 2015. "The Effect of Incorporated Self-Lubricated BN(h) Particles on the Tribological Properties of Ni–P/BN(h) Composite Coatings." *Applied Surface Science* 357: 1727–1735. doi:10.1016/j.apsusc.2015.09.207.

Hu, J., L. Fang, X.-L. Liao, and L.-T. Shi. 2017. "Influences of Different Reinforcement Particles on Performances of Electroless Composites." *Surface Engineering* 33 (5): 362–368. doi:10.1080/02670844.2016.1230975.

Hu, R., Y. Su, Y. Liu, H. Liu, Y. Chen, C. Cao, and H. Ni. 2018. "Deposition Process and Properties of Electroless Ni–P–Al_2O_3 Composite Coatings on Magnesium Alloy." *Nanoscale Research Letters* 13 (1): 198. doi:10.1186/s11671-018-2608-0.

Huang, Y. S., X. T. Zeng, X. F. Hu, and F. M. Liu. 2005. "Heat Treatment Effects on EN-PTFE-SiC Composite Coatings." *Surface and Coatings Technology* 198 (1–3): 173–177. doi:10.1016/j.surfcoat.2004.10.045.

Huang, Y. S. 2011. "Nickel-Diamond Compound Electroless Plating on Cast Aluminum Alloys." *Advanced Materials Research* 189–193: 265–268. doi:10.4028/www.scientific.net/AMR.189-193.265.

Huang, Z. H., Y. J. Zhou, and T. T. Nguyen. 2019. "Study of Nickel Matrix Composite Coatings Deposited from Electroless Plating Bath Loaded with TiB_2, ZrB_2 and TiC Particles for Improved Wear and Corrosion Resistance." *Surface and Coatings Technology* 364: 323–329. doi:10.1016/j.surfcoat.2019.01.060.

Hung, C. C., C. C. Lin, and H. C. Shih. 2008. "Tribological Studies of Electroless Nickel/Diamond Composite Coatings on Steels." *Diamond and Related Materials* 17 (4–5): 853–859. doi:10.1016/j.diamond.2007.12.055.

Islam, M., M. R. Azhar, Y. Khalid, R. Khan, H. S. Abdo, M. A. Dar, O. R. Oloyede, and T. D. Burleigh. 2015. "Electroless Ni–P/SiC Nanocomposite Coatings With Small Amounts of SiC Nanoparticles for Superior Corrosion Resistance and Hardness." *Journal of Materials Engineering and Performance* 24 (12): 4835–4843. doi:10.1007/s11665-015-1801-x.

Izzard, M., and J. K. Dennis. 1987. "Deposition and Properties of Electroless Nickel/Graphite Coatings." *Transactions of the IMF* 65 (1): 85–89. doi:10.1080/00202967.1987.11870778.

Jafari, M., M. H. Enayati, M. Salehi, S. M. Nahvi, and C. G. Park. 2013a. "Microstructural and Mechanical Characterizations of a Novel HVOF-Sprayed WC-Co Coating Deposited from Electroless Ni–P Coated WC-12Co Powders." *Materials Science and Engineering: A* 578: 46–53. doi:10.1016/j.msea.2013.04.064.

Jafari, M., M. H. Enayati, M. Salehi, S. M. Nahvi, and C. G. Park. 2013b. "Improvement in Tribological Properties of HVOF Sprayed WC–Co Coatings Using Electroless Ni–P Coated Feedstock Powders." *Surface and Coatings Technology* 235: 310–317. doi:10.1016/j.surfcoat.2013.07.059.

Jafari, M., M. H. Enayati, M. Salehi, S. M. Nahvi, and C. G. Park. 2014. "Microstructural Evolution of Nanosized Tungsten Carbide during Heatup Stage of Sintering of Electroless Nickel-Coated Nanostructured WC–Co Powder." *Ceramics International* 40 (7): 11031–11039. doi:10.1016/j.ceramint.2014.03.118.

Jang, J. W., J. H. Kim, J. H. Hwang, T. Y. Lim, J. H. Lee, and M. J. Lee. 2014. "Powder Morphology of Ni/YSZ Core-Shell According to the Plating Conditions of Using the Electroless Plating Method." *Korean Journal of Metals and Materials* 52 (5): 373–377. doi:10.3365/KJMM.2014.52.5.373.

Jappes, J. W., B. Ramamoorthy, and P. K. Nair. 2009. "Novel Approaches on the Study of Wear Performance of Electroless Ni–P/Diamond Composite Deposits." *Journal of Materials Processing Technology* 209 (2): 1004–1010. doi:10.1016/j.jmatprotec.2008.03.040.

Jiang, J., H. Chen, L. Zhu, W. Qian, S. Han, H. Lin, and H. Wu. 2016. "Effect of Heat Treatment on Structures and Mechanical Properties of Electroless Ni–P–GO Composite Coatings." *RSC Advances* 6 (110): 109001–109008. doi:10.1039/C6RA22330C.

Jiang, J., H. Chen, L. Zhu, Y. Sun, W. Qian, H. Lin, and S. Han. 2019. "Microstructure and Electrochemical Properties of Ni–B/Go Ultrasonic-Assisted Composite Coatings." *Surface Review and Letters*: 1950080. doi:10.1142/S0218625X1950080X.

Jiang, J., H. Lu, L. Zhang, and N. Xu. 2007. "Preparation of Monodisperse Ni/PS Spheres and Hollow Nickel Spheres by Ultrasonic Electroless Plating." *Surface and Coatings Technology* 201 (16–17): 7174–7179. doi:10.1016/J.SURFCOAT.2007.01.031.

Jiang, N., C. Q. Fu, and Z. Wang. 2013. "Research on the Microstructure and Tribological Properties of the Electroless Ni–P-PTFE Composite Repairing Coating on Gears." *Applied Mechanics and Materials* 327: 136–139. doi:10.4028/www.scientific.net/AMM.327.136.

Kaczmar, J.W., K. Pietrzak, and W. Wlosinski, K. Kanayo, M. Oluwatosin, A. A. Awe, Shahjahan and Anna Jesil Siham Said Rabeea Al-Siyabi, Rajesh Purohit Dinesh Kumar Koli, Geeta Agnihotri, H.R. Ezatpour and M. Torabi Parizi H. Beygi S.A. Sajjadi, et al. 2017. "Metallic and Oxide Electrodeposition." *Journal of Materials Processing Technology* 30 (1): 367–373. doi:10.5772/55684.

Karthikeyan, S., and B. Ramamoorthy. 2014. "Effect of Reducing Agent and Nano Al_2O_3 particles on the Properties of Electroless Ni–P Coating." *Applied Surface Science* 307 (2014): 654–660. doi:10.1016/j.apsusc.2014.04.092.

Kaya, B., T. Gulmez, M. Demirkol, and S.-I. Ao. 2009. "Study on the Electroless Ni–B Nano-Composite Coatings." *AIP Conference Proceedings*, 62–73. AIP. doi:10.1063/1.3146199.

Kim, S., J. W. Kim, and J. H. Kim. 2017. "Enhancement of Corrosion Resistance in Carbon Steels Using Nickel-Phosphorous/Titanium Dioxide Nanocomposite Coatings under High-Temperature Flowing Water." *Journal of Alloys and Compounds* 698: 267–275. doi:10.1016/J.JALLCOM.2016.12.027.

Kumari, S., A. Panigrahi, S. K. Singh, and S. K. Pradhan. 2018. "Corrosion-Resistant Hydrophobic Nanostructured Ni-Reduced Graphene Oxide Composite Coating with Improved Mechanical Properties." *Journal of Materials Engineering and Performance* 27 (11): 5889–5897. doi:10.1007/s11665-018-3706-y.

Kundu, S., S. K. Das, and P. Sahoo. 2016. "Tribological Behaviour of Electroless Ni–P Deposits Under Elevated Temperature." *Silicon*. doi:10.1007/s12633-016-9450-8.

Kuo, S.-L., Y.-C. Chen, M.-D. Ger, and W.-H. Hwu. 2004. "Nano-Particles Dispersion Effect on Ni/Al_2O_3 Composite Coatings." *Materials Chemistry and Physics* 86 (1): 5–10. doi:10.1016/J.MATCHEMPHYS.2003.11.040.

Larson, C., J. R. Smith, and G. J. Armstrong. 2013. "Current Research on Surface Finishing and Coatings for Aerospace Bodies and Structures—A Review." *Transactions of the IMF* 91 (3): 120–132. doi:10.1179/0020296713Z.000000000102.

Lee, C. K. 2012. "Electroless Ni-Cu-P/Nano-Graphite Composite Coatings for Bipolar Plates of Proton Exchange Membrane Fuel Cells." *Journal of Power Sources* 220: 130–137. doi:10.1016/j.jpowsour.2012.07.022.

Lee, C. K., C. S. Chang, A. H. Tan, C. Y. Yang, and S. L. Lee. 2015a. "Preparation of Electroless Nickel-Phosphorous-TiO_2 Composite Coating for Improvement of Wear and Stress Corrosion Cracking Resistance of AA7075 in 3.5% NaCl." *Key Engineering Materials* 656–657: 74–79. doi:10.4028/www.scientific.net/KEM.656-657.74.

Lee, C. K., C. L. Teng, A. H. Tan, C. Y. Yang, and S. L. Lee. 2015b. "Electroless Ni-P/Diamond/Graphene Composite Coatings and Characterization of Their Wear and Corrosion Resistance in Sodium Chloride Solution." *Key Engineering Materials* 656–657: 51–56. doi:10.4028/www.scientific.net/KEM.656-657.51.

León-Patiño, C. A., J. García-Guerra, and E. A. Aguilar-Reyes. 2019. "Tribological Characterization of Heat-Treated Ni–P and Ni–P–Al_2O_3 Composite Coatings by Reciprocating Sliding Tests." *Wear* 426–427: 330–340. doi:10.1016/j.wear.2019.02.015.

León, O. A, M. H Staia, and H. E Hintermann. 2003. "High Temperature Wear of an Electroless Ni–P–BN (h) Composite Coating." *Surface and Coatings Technology* 163–164: 578–584. doi:10.1016/S0257-8972(02)00663-1.

León, O. A. 2003. "High Temperature Wear of an Electroless Ni–P–BN (h) Composite Coating." *Surface and Coatings Technology* 164: 578–584.

León, O. A., M. H. Staia, and H. E. Hintermann. 1998. "Deposition of Ni–P–BN(h) Composite Autocatalytic Coatings." *Surface and Coatings Technology* 108–109: 461–465.

León, O. A., M. H. Staia, and H. E. Hintermann. 1998. "Deposition of Ni–P–BN(h) Composite Autocatalytic Coatings." *Surface and Coatings Technology* 108–109: 461–465. doi:10.1016/S0257-8972(98)00595-7.

Li, Z. H., X. Q. Wang, M. Wang, F. F. Wang, and H. L. Ge. 2006a. "Preparation and Tribological Properties of the Carbon Nanotubes-Ni-P Composite Coating." *Tribology International* 39 (9): 953–957. doi:10.1016/j.triboint.2005.10.001.

Li, Z. H., X. Q. Wang, M. Wang, F. F. Wang, and H. L. Ge. 2006b. "Preparation and Tribological Properties of the Carbon Nanotubes–Ni–P Composite Coating." *Tribology International* 39 (9): 953–957. doi:10.1016/J.TRIBOINT.2005.10.001.

Li, Z., J. Wang, J. Lu, and J. Meng. 2013. "Tribological Characteristics of Electroless Ni–P–MoS_2 Composite Coatings at Elevated Temperatures." *Applied Surface Science* 264: 516–521. doi:10.1016/j.apsusc.2012.10.055.

Liew, K. W., S. Y. Chia, C. K. Kok, and K.O. Low. 2013. "Evaluation on Tribological Design Coatings of Al_2O_3, Ni–P-PTFE and MoS_2 on Aluminium Alloy 7075 under Oil Lubrication." *Materials & Design* 48: 77–84. doi:10.1016/j.matdes.2012.08.010.

Lin, K.-J., H.-M. Wu, Y.-H. Yu, C.-Y. Ho, M.-H. Wei, F.-H. Lu, and W. J. Tseng. 2013. "Preparation of PMMA-Ni Core-Shell Composite Particles by Electroless Plating on Polyelectrolyte-Modified PMMA Beads." *Applied Surface Science* 282: 741–745. doi:10.1016/j.apsusc.2013.04.175.

Liu, C., and Q. Zhao. 2011. "Influence of Surface-Energy Components of Ni–P–TiO_2–PTFE Nanocomposite Coatings on Bacterial Adhesion." *Langmuir* 27 (15): 9512–9519. doi:10.1021/la200910f.

Liu, D., Y. Yan, K. Lee, and J. Yu. 2009. "Effect of Surfactant on the Alumina Dispersion and Corrosion Behavior of Electroless Ni–P–Al_2 O_3 Composite Coatings." *Materials and Corrosion* 60 (9): 690–694. doi:10.1002/maco.200805170.

Liu, P., and Y. Zhu. 2015. "Interaction Between Fine Diamond Particles in Electroless Nickel Solutions." *Journal of Dispersion Science and Technology* 36 (8): 1170–1177. doi:10.1080/01932691.2014.960525.

Liu, P., Y. Zhu, G. Zhong, X. Zhao, S. Wang, and S. Yang. 2019. "Influence of Inorganic Coating over Diamond Particles on Interaction Force and Dispersibility in Electroless Solution." *Powder Technology* 342: 899–906. doi:10.1016/j.powtec.2018.10.059.

Liu, S., X. Bian, J. Liu, C. Yang, X. Zhao, J. Fan, K. Zhang et al. 2015. "Structure and Properties of Ni–P–graphite (C_g)–TiO_2 Composite Coating." *Surface Engineering* 31 (6): 420–426. doi:10.1179/1743294414Y.0000000445.

Liu, S., J. Feng, X. Bian, J. Liu, H. Xu, and Y. An. 2017. "A Controlled Red Phosphorus@ Ni–P Core@shell Nanostructure as an Ultralong Cycle-Life and Superior High-Rate Anode for Sodium-Ion Batteries." *Energy & Environmental Science* 10 (5): 1222–1233. doi:10.1039/C7EE00102A.

Liu, Y. Y., J. Yu, H. Huang, B. H. Xu, X. L. Liu, Y. Gao, and X. L. Dong. 2007a. "Synthesis and Tribological Behavior of Electroless Ni-P-WC Nanocomposite Coatings." *Surface and Coatings Technology* 201: 7246–7251. doi:10.1016/j.surfcoat.2007.01.035.

Liu, Y. Y., J. Yu, H. Huang, B. H. Xu, X. L. Liu, Y. Gao, and X. L. Dong. 2007b. "Synthesis and Tribological Behavior of Electroless Ni–P–WC Nanocomposite Coatings." *Surface and Coatings Technology* 201 (16–17): 7246–7251. doi:10.1016/j.surfcoat.2007.01.035.

Liu, Y.-M., Y. Sung, Y.-C. Chen, C.-T. Lin, Y.-H. Chou, and M.-D. Ger. 2007c. "A Method to Fabricate Field Emitters Using Electroless Codeposited Composite of MWNTs and Nickel." *Electrochemical and Solid-State Letters* 10 (9): J101. doi:10.1149/1.2749416.

Liu, Z., L. Jian, and W. Tao. 2006. "Preparation of WC-Co Composite Powder by Electroless Plating and Its Application in Laser Cladding." *Materials Letters* 60 (16): 1956–1959. doi:10.1016/j.matlet.2005.12.073.

Lukschandel, J. 1978. "Diamond-Containing Electroless Nickel Coatings." *Transactions of the IMF* 56 (3): 118–120.

Luo, H., M. Leitch, Y. Behnamian, Y. Ma, H. Zeng, and J.-L. Luo. 2015. "Development of Electroless Ni–P/Nano-WC Composite Coatings and Investigation on Its Properties." *Surface and Coatings Technology* 277: 99–106. doi:10.1016/j.surfcoat.2015.07.011.

Ma, C., F. Wu, Y. Ning, F. Xia, and Y. Liu. 2014. "Effect of Heat Treatment on Structures and Corrosion Characteristics of Electroless Ni-P-SiC Nanocomposite Coatings." *Ceramics International* 40: 9279–9284. doi:10.1016/j.ceramint.2014.01.150.

Mafi, I. R., and C. Dehghanian. 2011a. "Comparison of the Coating Properties and Corrosion Rates in Electroless Ni–P/PTFE Composites Prepared by Different Types of Surfactants." *Applied Surface Science* 257 (20): 8653–8658. doi:10.1016/j.apsusc.2011.05.043.

Mafi, I. R., and C. Dehghanian. 2011b. "Studying the Effects of the Addition of TiN Nanoparticles to Ni–P Electroless Coatings." *Applied Surface Science* 258 (5): 1876–1880. doi:10.1016/j.apsusc.2011.10.095.

Mallory G. O., and J. B. Hajdu 1991. *Electroless Plating: Fundamentals and Applications.* American Electroplaters and Surface Finishing Society, Orlando, FL.

Matik, U. 2016. "Surface & Coatings Technology Structural and Wear Properties of Heat-Treated Electroless Ni–P Alloy and Ni–P–Si_3 N_4 Composite Coatings on Iron Based PM Compacts" 302: 528–534. doi:10.1016/j.surfcoat.2016.06.054.

Mazaheri, H., and S. R. Allahkaram. 2012. "Deposition, Characterization and Electrochemical Evaluation of Ni–P–nano Diamond Composite Coatings." *Applied Surface Science* 258 (10): 4574–4580. doi:10.1016/j.apsusc.2012.01.031.

Meng, Z.-Q., X.-B. Li, Y.-J. Xiong, and J. Zhan. 2012. "Preparation and Tribological Performances of Ni–P-Multi-Walled Carbon Nanotubes Composite Coatings." *Transactions of Nonferrous Metals Society of China* 22 (11): 2719–2725. doi:10.1016/S1003-6326(11)61523-9.

Metzger, W., and T. Florian. 1976. "The Deposition of Dispersion Hardened Coatings by Means of Electroless Nickel." *Transactions of the IMF* 54 (1): 174–177. doi:10.1080/00202967.1976.11870394.

Modaressi, A., H. Sifaoui, B. Grzesiak, R. Solimando, U. Domanska, and M. Rogalski. 2007. "CTAB Aggregation in Aqueous Solutions of Ammonium Based Ionic Liquids; Conductimetric Studies." *Colloids and Surfaces A* 296 (2007): 104–108. doi:10.1016/j.colsurfa.2006.09.031.

Mohammadi, M., and M. Ghorbani. 2011. "Wear and Corrosion Properties of Electroless Nickel Composite Coatings with PTFE and/or MoS_2 Particles." *Journal of Coatings Technology and Research* 8 (4): 527–533. doi:10.1007/s11998-011-9329-y.

Mohtar, M. M., Z. M. Ripin, and Z. A. Ahmad. 2013a. "Effect of Heat Treatment on Ni–P–Cg(Graphite)–SiC Composite Coated Cast AlSi Alloy." *Advanced Materials Research* 795: 545–549. doi:10.4028/www.scientific.net/AMR.795.545.

Mohtar, M. M., Z. M. Ripin, and Z. A. Ahmad. 2013b. "Electroless Ni–P–Cg(Graphite)–SiC Composite Coating on Cast AlSi Alloy." *Advanced Materials Research* 795: 540–544. doi:10.4028/www.scientific.net/AMR.795.540.

Monir Vaghefi, S.M., A. Saatchi, M. Ebrahimian-Hoseinabadi, S. M. M. Vaghefi, and A. Saatchi. 2003. "Deposition and Properties of Electroless Ni–P–B_4C Composite Coatings." *Surface & Coatings Technology* 168 (2–3): 259–262. doi:10.1016/S0257-8972(02)00926-X.

Mukhopadhyay, A., T. K. Barman, and P. Sahoo. 2017. "Effects of Heat Treatment on Tribological Behavior of Electroless Ni–B Coating at Elevated Temperatures." *Surface Review and Letters* 24 (Supp 1): 1–22. doi:10.1142/S0218625X18500142.

Nad, E. E., and M. Ehteshamzadeh. 2014. "Effects of TiO_2 Particles Size and Heat Treatment on Friction Coefficient and Corrosion Performance of Electroless Ni–P/TiO_2 Composite Coatings." *Surface Engineering and Applied Electrochemistry* 50 (1): 50–56. doi:10.3103/S1068375514010116.

Necula, B. S., I. Apachitei, L. E. Fratila-apachitei, C. Teodosiu, and J. Duszczyk. 2007. "Stability of Nano/Microsized Particles in Deionized Water and Electroless Nickel Solutions." *Journal of Colloid and Interface Science* 314 (1): 514–522. doi:10.1016/j.jcis.2007.05.073.

Niksefat, V., and M. Ghorbani. 2015. "Mechanical and Electrochemical Properties of Ultrasonic-Assisted Electroless Deposition of Ni–B–TiO_2 Composite Coatings." *Journal of Alloys and Compounds* 633: 127–136. doi:10.1016/j.jallcom.2015.01.250.

Novák, M., D. Vojtěch, M. Zelinková, and T. Vítů. 2009. "Influence of Heat Treatment on Tribological Properties of Ni-P Electroless Coatings." *Proceedings of Metal* 21 (5): 1–6.

Novakovic, J., and P. Vassiliou. 2009. "Vacuum Thermal Treated Electroless NiP–TiO_2 Composite Coatings." *Electrochimica Acta* 54 (9): 2499–2503.

Nwosu, N. O., A. M. Davidson, and N. W. Shearer. 2013. "Effect of SDS on the Dispersion Stability of YSZ Particles in an Electroless Nickel Solution." *Chemical Engineering Communications* 200 (12): 1623–1634. doi:10.1080/00986445.2012.757549.

Odekerken, J. M. 1968. Process for Coating an Object with a Bright Nickel/Chromium Coating. US3644183A, issued February 1968.

Ogihara, H., A. Hara, K. Miyamoto, N. K. Shrestha, T. Kaneda, S. Ito, and T. Saji. 2010. "Synthesis of Super Hard Ni–B/Diamond Composite Coatings by Wet Processes." *Chemical Communication* 46 (3): 442–444. doi:10.1039/B914242H.

Omidvar, H., M. Sajjadnejad, G. Stremsdoerfer, Y. Meas, and A. Mozafari. 2016a. "Composite NiB–Graphite and NiB–PTFE Surface Coatings Deposited by the Dynamic Chemical Plating Technique." *Materials and Manufacturing Processes* 31 (1): 24–30. doi:10.1080/10426914.2015.1004691.

Omidvar, H., M. Sajjadnejad, G. Stremsdoerfer, Y. Meas, and A. Mozafari. 2016b. "Manufacturing Ternary Alloy NiBP-PTFE Composite Coatings by Dynamic Chemical Plating Process." *Materials and Manufacturing Processes* 31 (1): 31–36. doi:10.1080/10426914.2014.994753.

Omidvar, H., M. Sajjadnejad, G. Stremsdoerfer, Y. Meas, and A. Mozafari. 2015. "Characterization of NiBP-Graphite Composite Coatings Deposited by Dynamic Chemical Plating." *Anti-Corrosion Methods and Materials* 62 (2): 116–122. doi:10.1108/ACMM-11-2013-1320.

Palaniappa, M., G. V. Babu, and K. Balasubramanian. 2007. "Electroless Nickel–phosphorus Plating on Graphite Powder." *Materials Science and Engineering: A* 471 (1–2): 165–168. doi:10.1016/j.msea.2007.03.004.

Pancrecious, J. K., J. P. Deepa, V. Jayan, U. S. Bill, T. P. D. Rajan, and B. C. Pai. 2018. "Nanoceria Induced Grain Refinement in Electroless Ni–B–CeO_2 Composite Coating for Enhanced Wear and Corrosion Resistance of Aluminium Alloy." *Surface and Coatings Technology* 356: 29–37. doi:10.1016/j.surfcoat.2018.09.046.

Park, H.-W., J.-W. Jang, Y.-J. Lee, J.-H. Kim, D.-W. Jeon, J.-H. Lee, H.-J. Hwang, and M.-J. Lee. 2017. "Catalysts Characteristics of Ni/YSZ Core-Shell According to Plating Conditions Using Electroless Plating." *Metals and Materials International* 23 (6): 1227–1233. doi:10.1007/s12540-017-6401-x.

Parker, K. 1972. "Recent Advances in Electroless Nickel Deposits." *8th Interfinish Conference*, Basel, Switzerland, 85–90.

Parkinson, R. 1997. *Properties and Applications of Electroless Nickel.* Nickel Development Institute and Metal finishing Industry review, Toronto, ON, p. 1–37.

Peterson, M. B., S. F. Murray, and J. J. Florek. 1959. "Consideration of Lubricants for Temperatures above 1000 F." *ASLE Transactions* 2 (2): 225–234. doi:10.1080/05698195908972374.

Petrova, M., M. Georgieva, V. Chakarova, and E. Dobreva. 2016. "Electroless Deposition of Composite Nickel-Phosphorous Coatings with Diamond Dispersoid." *Archives of Metallurgy and Materials* 61 (2): 493–498. doi:10.1515/amm-2016-0086.

Petrova, M., Z. Noncheva, and E. Dobreva. 2011. "Electroless Deposition of Diamond Powder Dispersed Nickel-Phosphorus Coatings on Steel Substrate." *Transactions of the Institute of Metal Finishing* 89 (2): 89–94. doi:10.1179/174591911X12971865404438.

Qui, H., M. Vélez, H. Quiñones, R. A. Di Giampaolo, J. Lira, and I. C. Grigorescu. 1999. "Electroless Ni-B Coated WC and VC Powders as Precursors for Liquid Phase Sintering." *International Journal of Refractory Metals and Hard Materials* 17: 99–102. doi:10.1016/S0263-4368(98)00035-3.

Rabizadeh, T., and S. R. Allahkaram. 2011. "Corrosion Resistance Enhancement of Ni–P Electroless Coatings by Incorporation of Nano-SiO_2 Particles." *Materials and Design* 32 (1): 133–138. doi:10.1016/j.matdes.2010.06.021.

Radhika, N., R. Ramprasad, and S. Nivethan. 2018. "Experimental Investigation on Adhesive Wear Behavior of Al–Si_6Cu/Ni Coated SiC Composite Under Unlubricated Condition." *Transactions of the Indian Institute of Metals* 71 (5): 1073–1082. doi:10.1007/s12666-017-1242-3.

Ramalho, A., and J. C. Miranda. 2005. "Friction and Wear of Electroless NiP and NiP+PTFE Coatings." *Wear* 259 (7–12): 828–834. doi:10.1016/j.wear.2005.02.052.

Ranganatha, S., and T. V. Venkatesha. 2017. "Fabrication and Anticorrosion Performance of Ni–P–BN Nanocomposite Coatings on Mild Steel." *Surface Engineering and Applied Electrochemistry* 53 (5): 449–455. doi:10.3103/S106837551705009X.

Ranganatha, S., T. V. Venkatesha, and K. Vathsala. 2012. "Electroless Ni–W–P Coating and Its Nano-WS 2 Composite: Preparation and Properties." *Industrial and Engineering Chemistry Research* 51 (23): 7932–7940. doi:10.1021/ie300104w.

Ranganatha, S., T. V. Venkatesha, and K. Vathsala. 2010. "Development of Electroless Ni–Zn–P/Nano-TiO_2 Composite Coatings and Their Properties." *Applied Surface Science* 256 (24): 7377–7383. doi:10.1016/J.APSUSC.2010.05.076.

Ranganatha, S., and T. V. Venkatesha. 2012. "Studies on the Preparation and Properties of Electroless Ni–W–P Alloy Coatings and Its Nano-MoS_2 Composite." *Physica Scripta* 85 (3): 035601. doi:10.1088/0031-8949/85/03/035601.

Rattanawaleedirojn, P., K. Saengkiettiyut, Y. Boonyongmaneerat, N. Chuankrerkkul, and S. Saenapitak. 2014. "Effects of Complexing Agent Concentration and Bath PH on Electroless Nickel Deposition for Tungsten Carbide Powders." *Advanced Materials Research* 970: 240–243. doi:10.4028/www.scientific.net/AMR.970.240.

Rattanawaleedirojn, P., K. Saengkiettiyut, Y. Boonyongmaneerat, S. Sangsuk, N. Promphet, and N. Rodthongkum. 2016. "TiO_2 Sol-Embedded in Electroless Ni–P Coating: A Novel Approach for an Ultra-Sensitive Sorbitol Sensor." *RSC Advances* 6 (73): 69261–69269. doi:10.1039/C6RA05090E.

Reddy, V. V. N., B. Ramamoorthy, and P. K. Nair. 2000. "A Study on the Wear Resistance of Electroless Ni–P/Diamond Composite Coatings." *Wear* 239 (1): 111–116. doi:10.1016/S0043-1648(00)00330-6.

Rezagholizadeh, M., M. Ghaderi, and A. Heidary. 2015. "Electroless Ni–P/Ni-BB4C Duplex Composite Coatings for Improving the Corrosion and Tribological Behavior of Ck45 Steel." *Protection of Metals and Physical Chemistry of Surfaces* 51 (2): 234–239.

Rossi, S., F. Chini, G. Straffelini, P. L Bonora, R. Moschini, and A. Stampali. 2003. "Corrosion Protection Properties of Electroless Nickel/PTFE, Phosphate/MoS_2 and Bronze/PTFE Coatings Applied to Improve the Wear Resistance of Carbon Steel." *Surface and Coatings Technology* 173 (2–3): 235–242. doi:10.1016/S0257-8972(03)00662-5.

Sadhir, M. H., M. Saranya, M. Aravind, A. Srinivasan, A. Siddharthan, and N. Rajendran. 2014. "Comparison of in Situ and Ex Situ Reduced Graphene Oxide Reinforced Electroless Nickel Phosphorus Nanocomposite Coating." *Applied Surface Science* 320: 171–176. doi:10.1016/j.apsusc.2014.09.001.

Sahoo, P., and S. K. Das. 2011. "Tribology of Electroless Nickel Coatings—A Review." *Materials and Design* 32 (4): 1760–1775. doi:10.1016/j.matdes.2010.11.013.

Sarret, M., C. Müller, and A. Amell. 2006. "Electroless NiP Micro- and Nano-Composite Coatings." *Surface and Coatings Technology* 201 (1–2): 389–395. doi:10.1016/j.surfcoat.2005.11.127.

Sathiyamoorthy, P. 2017. "Surface Engineering, Highlights." *Material Science and Engineering* www.barc.gov.in/publications/eb/golden/material/toc/chapter5/5.pdf.

Schaller, E. J., P. R. Sperry, and L. J. Calbo. 1992. *Handbook of Coating Additives*, Vol. 2, pp. 105–163. Dekker, New York.

Seifzadeh, D., and A. R. Hollagh. 2014. "Corrosion Resistance Enhancement of AZ91D Magnesium Alloy by Electroless Ni–Co–P Coating and Ni–Co–P–SiO_2 Nanocomposite." *Journal of Materials Engineering and Performance* 23 (11): 4109–4121. doi:10.1007/s11665-014-1210-6.

Shaffer, S. J., and M. J. Rogers. 2007. "Tribological Performance of Various Coatings in Unlubricated Sliding for Use in Small Arms Action Components—A Case Study." *Wear* 263 (7–12): 1281–1290. doi:10.1016/j.wear.2007.01.115.

Sharma, A., and A. K. Singh. 2013. "Electroless Ni–P-PTFE-Al_2O_3 Dispersion Nanocomposite Coating for Corrosion and Wear Resistance." *Journal of Materials Engineering and Performance* 23 (Ref 24): 142–151. doi:10.1007/s11665-013-0710-0.

Sharma, S. B., R. C. Agarwala, V. Agarwala, and S. Ray. 2002. "Dry Sliding Wear and Friction Behavior of Ni–P–Zro–Al_2O_3 Composite Electroless Coatings on Aluminum." *Materials and Manufacturing Processes* 17 (5): 637–649. doi:10.1081/AMP-120016088.

Sheela, G., and M. Pushpavanam. 2002. "Diamond Dispersed Electroless Nickel Coating." *Metal Finishing* 100: 45–47.

Sherkat, A. A., A. Shafei, and S. M. Monirvaghefi. 2011. "Fabrication and Synthesis of Electroless Nickel-Phosphorus Coatings Reinforced by Incorporated Chromium Carbide-Nickel-Chrome Particles." *Revue de Métallurgie* 108 (2): 95–99. doi:10.1051/metal/2011044.

Shi, L. T., J. Hu, L. Fang, F. Wu, X. L. Liao, and F. M. Meng. 2016. "Effects of Cobalt Content on Mechanical and Corrosion Properties of Electroless Ni–Co–P/TiN Nanocomposite Coatings." *Materials and Corrosion* 67 (10): 1034–1041. doi:10.1002/maco.201608844.

Shibli, S. M. A., and V. S. Dilimon. 2008. "Development of TiO_2-Supported Nano-RuO_2-Incorporated Catalytic Nickel Coating for Hydrogen Evolution Reaction." *International Journal of Hydrogen Energy* 33 (4): 1104–1111. doi:10.1016/j.ijhydene.2007.12.038.

Shibli, S. M. A., B. Jabeera, and R. I. Anupama. 2006a. "Development of ZnO Incorporated Composite Ni–ZnO–P Alloy Coating." *Surface and Coatings Technology* 200 (12–13): 3903–3906. doi:10.1016/J.SURFCOAT.2004.10.017.

Shibli, S. M. A., B. Jabeera, and R. I. Anupama. 2006b. "Incorporation of Nano Zinc Oxide for Improvement of Electroless Nickel Plating." *Applied Surface Science* 253 (3): 1644–1648. doi:10.1016/J.APSUSC.2006.02.063.

Shibli, S. M. A., and J. N. Sebeelamol. 2013. "Development of Fe_2O_3–TiO_2 Mixed Oxide Incorporated Ni–P Coating for Electrocatalytic Hydrogen Evolution Reaction." *International Journal of Hydrogen Energy* 38 (5): 2271–2282. doi:10.1016/j.ijhydene.2012.12.009.

Shibli, S. M. A., and V. S. Dilimon. 2007. "Effect of Phosphorous Content and TiO_2-Reinforcement on Ni–P Electroless Plates for Hydrogen Evolution Reaction." *International Journal of Hydrogen Energy* 32 (2007): 1694–1700. doi:10.1016/j.ijhydene.2006.11.037.

Shu, X., Y. Wang, C. Liu, and W. Gao. 2015. "Microstructure and Properties of Ni–B–TiO_2 Nano-Composite Coatings Fabricated by Electroless Plating." *Materials Technology* 30 (sup 1): A41–A45. doi:10.1179/1753555714Y.0000000190.

Sineva, A. V., A. M. Parfenova, and A. A. Fedorova. 2007. "Adsorption of Micelle Forming and Non-Micelle Forming Surfactants on the Adsorbents of Different Nature." *Colloids and Surfaces A* 306 (2007): 68–74. doi:10.1016/j.colsurfa.2007.04.061.

Soleimani, R., F. Mahboubi, S. Y. Arman, M. Kazemi, and A. Maniee. 2015a. "Development of Mathematical Model to Evaluate Microstructure and Corrosion Behavior of Electroless Ni–P/Nano-SiC Coating Deposited on 6061 Aluminum Alloy." *Journal of Industrial and Engineering Chemistry* 23: 328–337. doi:10.1016/j.jiec.2014.09.002.

Soleimani, R., F. Mahboubi, M. Kazemi, and S. Y. Arman. 2015b. "Corrosion and Tribological Behaviour of Electroless Ni–P/Nano-SiC Composite Coating on Aluminium 6061." *Surface Engineering* 31 (9): 714–721. doi:10.1179/1743294415Y.0000000012.

Song, D., J. Zhou, W. Jiang, X. Zhang, Y. Yan, and F. Li. 2009. "A Novel Activation for Electroless Plating on Preparing Ni/PS Microspheres." *Materials Letters* 63: 282–284. doi:10.1016/j.matlet.2008.10.011.

Srinivasan, K. N., and S. John. 2005. "Studies on Electroless Nickel–PTFE Composite Coatings." *Surface Engineering* 21 (2): 156–160. doi:10.1179/174329405X40902.

Staia, M. H., A. Conzoño, M. R. Cruz, A. Roman, J. Lesage, D. Chicot, and G. Mesmacque. 2002. "Wear Behaviour of Silicon Carbide/Electroless Nickel Composite Coatings at High Temperature." *Surface Engineering* 18 (4): 265–269. doi:10.1179/026708401225005359.

Su, W. 2013. "Microhardness of Electroless Composite Coating of Ni–P with SiC Nano-Particles." *Advanced Materials Research* 662: 223–226. doi:10.4028/www.scientific.net/AMR.662.223.

Sudagar, J., J. S. Lian, Q. Jiang, Z. H. Jiang, G. Y. Li, and R. Elansezhian. 2012. "The Performance of Surfactant on the Surface Characteristics of Electroless Nickel Coating on Magnesium Alloy." *Progress in Organic Coatings* 74 (4): 788–793. doi:10.1016/j.porgcoat.2011.10.022.

Sudagar, J., J. Lian, and W. Sha. 2013. "Electroless Nickel, Alloy, Composite and Nano Coatings—A Critical Review." *Journal of Alloys and Compounds* 571 (31): 183–204. doi:10.1016/j.jallcom.2013.03.107.

Sun, W.-C., J.-M. Xu, Y. Wang, F. Guo, and Z.-W. Jia. 2017. "Effect of Cerium Oxide on Morphologies and Electrochemical Properties of Ni–W–P Coating on AZ91D Magnesium." *Journal of Materials Engineering and Performance* 26 (12): 5753–5759. doi:10.1007/s11665-017-3038-3.

Szczygieł, B., and A. Turkiewicz. 2009. "The Rate of Electroless Deposition of a Four-Component Ni–W–P–ZrO_2 Composite Coating from a Glycine Bath." *Applied Surface Science* 255 (20): 8414–8418. doi:10.1016/j.apsusc.2009.05.145.

Szczygieł, B., and A. Turkiewicz. 2008. "The Effect of Suspension Bath Composition on the Composition, Topography and Structure of Electrolessly Deposited Composite Four-Component Ni–W–P–ZrO_2 Coatings." *Applied Surface Science* 254 (22): 7410–7416. doi:10.1016/J.APSUSC.2008.05.342.

Tamilarasan, T. R., U. Sanjith, R. Rajendran, G. Rajagopal, and J. Sudagar. 2018. "Effect of Reduced Graphene Oxide Reinforcement on the Wear Characteristics of Electroless Ni-P Coatings." *Journal of Materials Engineering and Performance* 27: 3044–3053. doi:10.1007/s11665-018-3246-5.

Tamilarasan, T. R., R. Rajendran, M. Siva shankar, U. Sanjith, G. Rajagopal, and J. Sudagar. 2016. "Wear and Scratch Behaviour of Electroless Ni–P-Nano-TiO_2: Effect of Surfactants." *Wear* 346–347: 148–157.

Tamilarasan, T. R., R. Rajendran, G. Rajagopal, and J. Sudagar. 2015. "Effect of Surfactants on the Coating Properties and Corrosion Behaviour of Ni–P–nano-TiO_2 Coatings." *Surface and Coatings Technology* 276: 320–326. doi:10.1016/J.SURFCOAT.2015.07.008.

Tamilarasan, T. R., U. Sanjith, M. Siva Shankar, and G. Rajagopal. 2017. "Effect of Reduced Graphene Oxide (RGO) on Corrosion and Erosion-Corrosion Behaviour of Electroless Ni-P Coatings." *Wear* 390–391: 385–391. doi:10.1016/j.wear.2017.09.004.

Tian, M. M., W. C. Sun, Q. Shi, Y. Wang, and Q. H. Yang. 2014. "Influence of CNTs Concentration on the Microstructure and Electrochemical Properties of Ni-P-MWNTs Composite Coating." *Materials Science Forum* 809–810: 610–614. doi:10.4028/www.scientific.net/MSF.809-810.610.

Ulicny, J. C., and A. M. Mance. 2004. "Evaluation of Electroless Nickel Surface Treatment for Iron Powder Used in MR Fluids." *Materials Science and Engineering: A* 369 (1–2): 309–313. doi:10.1016/j.msea.2003.11.039.

Uysal, M., R. Karslioğlu, A. Alp, and H. Akbulut. 2011. "Nanostructured Core–shell Ni Deposition on SiC Particles by Alkaline Electroless Coating." *Applied Surface Science* 257 (24): 10601–10606. doi:10.1016/j.apsusc.2011.07.057.

Uysal, M. 2019. "Electroless Codeposition of Ni–P Composite Coatings: Effects of Graphene and TiO_2 on the Morphology, Corrosion, and Tribological Properties." *Metallurgical and Materials Transactions A* 50 (5): 2331–2341. doi:10.1007/s11661-019-05161-9.

Uysal, M., H. Akbulut, M. Tokur, H. Algül, and T. Çetinkaya. 2016. "Structural and Sliding Wear Properties of Ag/Graphene/WC Hybrid Nanocomposites Produced by Electroless Co-Deposition." *Journal of Alloys and Compounds* 654: 185–195. doi:10.1016/j.jallcom.2015.08.264.

Uysal, M., R. Karslioğlu, A. Alp, and H. Akbulut. 2013. "The Preparation of Core–shell Al_2O_3/Ni Composite Powders by Electroless Plating." *Ceramics International* 39 (5): 5485–5493. doi:10.1016/j.ceramint.2012.12.060.

Vélez, M., H. Quiñones, A. R. Di Giampaolo, J. Lira, and I. C. Grigorescu. 1999. "Electroless Ni–B Coated WC and VC Powders as Precursors for Liquid Phase Sintering." *International Journal of Refractory Metals and Hard Materials* 17 (1–3): 99–102. doi:10.1016/S0263-4368(98)00035-3.

Vitry, V., and L. Bonin. 2017. "Increase of Boron Content in Electroless Nickel-Boron Coating by Modification of Plating Conditions." *Surface and Coatings Technology* 311: 164–171. doi:10.1016/j.surfcoat.2017.01.009.

Vojtěch, D. 2009. "Properties of Hard Ni–P–Al_2O_3 and Ni–P–SiC Coatings on Al-Based Casting Alloys." *Materials and Manufacturing Processes* 24 (7–8): 754–757. doi:10.1080/10426910902809784.Wang, J. Z., S. W. Jiang, S. Song, and Z. Q. Wang. 2019. "Addition of Molybdenum Disulfide Solid Lubricant to WC-12Ni Thermal Spray Cemented Carbide Powders through Electroless Ni–MoS_2 Co-Deposition." *Journal of Alloys and Compounds* 786: 594–606. doi:10.1016/j.jallcom.2019.01.338.

Wang, J., F.-L. Zhang, T. Zhang, W.-G. Liu, W.-X. Li, and Y.-M. Zhou. 2018. "Preparation of Ni–P-Diamond Coatings with Dry Friction Characteristics and Abrasive Wear Resistance." *International Journal of Refractory Metals and Hard Materials* 70: 32–38. doi:10.1016/j.ijrmhm.2017.09.012.

Wang, S., X. Huang, M. Gong, and W. Huang. 2015. "Microstructure and Mechanical Properties of Ni–P–Si_3N_4 nanowire Electroless Composite Coatings." *Applied Surface Science* 357 (2015): 328–332. doi:10.1016/j.apsusc.2015.09.011.

Wu, H., F. Liu, W. Gong, F. Ye, L. Hao, J. Jiang, and S. Han. 2015a. "Preparation of Ni–P–GO Composite Coatings and Its Mechanical Properties." *Surface and Coatings Technology* 272: 25–32.

Wu, X., J. Mao, Z. Zhang, and Y. Che. 2015b. "Improving the Properties of 211Z Al Alloy by Enhanced Electroless Ni–P–TiO_2 Nanocomposite Coatings with TiO_2 Sol." *Surface and Coatings Technology* 270: 170–174. doi:10.1016/J.SURFCOAT.2015.03.006.

Wu, Y. T., L. Lei, B. Shen, and W. B. Hu. 2006a. "Investigation in Electroless Ni–P–Cg(Graphite)–SiC Composite Coating." *Surface and Coatings Technology* 201 (1–2): 441–445.

Wu, Y. T., L. Lei, B. Shen, and W. B. Hu. 2006b. "Investigation in Electroless Ni-P-Cg(Graphite)-SiC Composite Coating." *Surface and Coatings Technology* 201: 441–445.

Wu, Y., H. Liu, B. Shen, L. Liu, and W. Hu. 2006c. "The Friction and Wear of Electroless Ni–P Matrix with PTFE and/or SiC Particles Composite." *Tribology International* 39 (6): 553–559. doi:10.1016/j.triboint.2005.04.032.

Wu, Y., B. Shen, L. Liu, and W. Hu. 2006d. "The Tribological Behaviour of Electroless Ni-P-Gr–SiC Composite." *Wear* 261: 201–207. doi:10.1016/j.wear.2005.09.008.

Xiang, Y., J. Zhang, and C. Jin. 2001. "Study of Electroless Ni–P-Nanometer Diamond Composite Coatings." *Plating and Surface Finishing* 2: 64–68.

Xu, H., Z. Yang, M.-K. Li, Y.-L. Shi, Y. Huang, and H.-L. Li. 2005. "Synthesis and Properties of Electroless Ni–P–Nanometer Diamond Composite Coatings." *Surface and Coatings Technology* 191 (2–3): 161–165. doi:10.1016/j.surfcoat.2004.03.045.

Xu, S., Y. C. Chan, X. Zhu, H. Lu, and C. Bailey. 2014. "Effective Method to Disperse and Incorporate Carbon Nanotubes in Electroless Ni–P Deposits." *2014 IEEE 64th Electronic Components and Technology Conference (ECTC)*, 1342–1347. IEEE. doi:10.1109/ECTC.2014.6897466.

Xue, F., L. Zhu, J. Wang, and Z. Tu. 2015. "Catalytic Role of Surface Pre-Treatment of Noble-Metal-like Tungsten Carbide Powder on Electroless Deposition of Nickel." *Surface and Coatings Technology* 265: 32–37. doi:10.1016/j.surfcoat.2015.01.066.

Yang, Y., W. Chen, and C. Zhou. 2011. "Fabrication and Characterization of Electroless Ni–P–ZrO_2 Nano-Composite Coatings." *Applied Nano Science* 1 (2011): 19–26. doi:10.1007/s13204-011-0003-6.

Yang, Z., H. Xu, M.-K. Li, Y.-L. Shi, Y. Huang, and H.-L. Li. 2004. "Preparation and Properties of Ni/P/Single-Walled Carbon Nanotubes Composite Coatings by Means of Electroless Plating." *Thin Solid Films* 466 (1–2): 86–91. doi:10.1016/j.tsf.2004.02.016.

Yang, Z., H. Xu, Y. L. Shi, M. K. Li, Y. Huang, and H. L. Li. 2005. "The Fabrication and Corrosion Behavior of Electroless Ni–P–Carbon Nanotube Composite Coatings." *Materials Research Bulletin* 40 (6): 1001–1009. doi:10.1016/j.materresbull.2005.02.015.

Yazdani, S., R. Tima, and F. Mahboubi. 2018. "Investigation of Wear Behavior of As-Plated and Plasma-Nitrided Ni–B–CNT Electroless Having Different CNTs Concentration." *Applied Surface Science* 457: 942–955. doi:10.1016/j.apsusc.2018.07.020.

Yu, L., W. Huang, and X. Zhao. 2011. "Preparation and Characterization of Ni–P–NanoTiN Electroless Composite Coatings." *Journal of Alloys and Compounds* 509 (10): 4154–4159. doi:10.1016/j.jallcom.2011.01.025.

Yu, Q., T. Zhou, Y. Jiang, X. Yan, Z. An, X. Wang, D. Zhang, and T. Ono. 2018. "Preparation of Graphene-Enhanced Nickel-Phosphorus Composite Films by Ultrasonic-Assisted Electroless Plating." *Applied Surface Science* 435: 617–625. doi:10.1016/j.apsusc.2017.11.169.

Yucheng, W., R. Rong, W. Fengtao, Y. Zaoshi, W. Tugen, and H. Xiaoye. 2008. "Preparation and Characterization of Ni–Cu–P/CNTs Quaternary Electroless Composite Coating." *Materials Research Bulletin* 43 (12): 3425–3432. doi:10.1016/j.materresbull.2008.02.019.

Zangeneh-Madar*, K, and A. Jafari. 2012. "Characterisation of Electroless Nickel Plated Titanium Powder." *Surface Engineering* 28 (6): 393–399. doi:10.1179/1743294411Y.0000000067.

Zarebidaki, A., and S. R. Allahkaram. 2012. "Effect of Heat Treatment on the Properties of Electroless Ni–P–carbon Nanotube Composite Coatings." *Micro & Nano Letters* 7 (1): 90. doi:10.1049/mnl.2011.0482.

Zhang, X. M., X. G. Wang, and J. T. Liu. 2016. "Effect of β-SiC Particle Concentration on the Properties of the Duplex Ni–P/Ni–P–β-SiC Coating." *Materials Science Forum* 852: 1070–1074. doi:10.4028/www.scientific.net/MSF.852.1070.

Zhao, Q. 2004. "Effect of Surface Free Energy of Graded Ni–P–PTFE Coatings on Bacterial Adhesion." *Surface and Coatings Technology* 185 (2–3): 199–204. doi:10.1016/j.surfcoat.2003.12.009.

Zhao, Q., and Y. Liu. 2005. "Electroless Ni–Cu–P–PTFE Composite Coatings and Their Anticorrosion Properties." *Surface and Coatings Technology* 200: 2510–2514. doi:10.1016/j.surfcoat.2004.06.011.

Zhao, Q., Y. Liu, and E. W. Abel. 2004. "Effect of Cu Content in Electroless Ni–Cu–P–PTFE Composite Coatings on Their Anti-Corrosion Properties." *Materials Chemistry and Physics* 87 (2–3): 332–335. doi:10.1016/j.matchemphys.2004.05.028.

Zhou, H.-M., Y. Jia, and J. Li. 2016. "Corrosion and Wear Resistance Behaviors of Electroless Ni–Cu–P–TiN Composite Coating." *Rare Metals*. doi:10.1007/s12598-015-0663-6.

Zhou, S., L. Wang, and S. Shen. 2011. "Fabrication of Ni–P/Palygorskite Core–shell Linear Powder via Electroless Deposition." *Applied Surface Science* 257 (23): 10211–10217. doi:10.1016/j.apsusc.2011.07.023.

Zhu, C., and Y. Zhu. 2013. "Dispersion of Micro Diamond Particles in Electroless Nickel Solution." *Journal of Wuhan University of Technology-Materials Science Edition* 28 (1): 57–61. doi:10.1007/s11595-013-0640-6.

Zhu, X., H. Dong, and K. Lu. 2008. "Coating Different Thickness Nickel-Boron Nanolayers onto Boron Carbide Particles." *Surface and Coatings Technology* 202 (13): 2927–2934. doi:10.1016/j.surfcoat.2007.10.021.

Zielińska, K., A. Stankiewicz, and I. Szczygieł. 2012. "Electroless Deposition of Ni–P–nano-ZrO_2 Composite Coatings in the Presence of Various Types of Surfactants." *Journal of Colloid and Interface Science* 377 (1): 362–367.

Zou, G., M. Cao, H. Lin, H. Jin, Y. Kang, and Y. Chen. 2006a. "Nickel Layer Deposition on SiC Nanoparticles by Simple Electroless Plating and Its Dielectric Behaviors." *Powder Technology* 168: 84–88. doi:10.1016/j.powtec.2006.07.002.

Zou, G.-Z., M.-S. Cao, L. Zhang, J. G. Li, H. Xu, and Y.-J. Chen. 2006b. "A Nanoscale Core-Shell of β-SiCP–Ni Prepared by Electroless Plating at Lower Temperature." *Surface and Coatings Technology* 201 (1–2): 108–112. doi:10.1016/j.surfcoat.2005.11.026.

Zou, T. Z., J. P. Tu, S. C. Zhang, L. M. Chen, Q. Wang, L. L. Zhang, and D. N. He. 2006c. "Friction and Wear Properties of Electroless Ni–P- (IF-MoS_2) Composite Coatings in Humid Air and Vacuum." *Materials Science and Engineering: A* 426 (1–2): 162–168. doi:10.1016/j.msea.2006.03.068.

Zuleta, A. A., O. A. Galvis, J. G. Castaño, F. Echeverría, F. J. Bolivar, M. P. Hierro, and F. J. Pérez-Trujillo. 2009. "Preparation and Characterization of Electroless Ni–P–Fe_3O_4 Composite Coatings and Evaluation of Its High Temperature Oxidation Behaviour." *Surface and Coatings Technology* 203 (23): 3569–3578. doi:10.1016/J.SURFCOAT.2009.05.025.

9 Applications of Electroless Nickel and Practical Aspects

Luiza Bonin and Véronique Vitry

CONTENTS

9.1 INTRODUCTION TO ELECTROLESS NICKEL APPLICATIONS

Electroless nickel coatings are products engineered specifically for their protective and functional properties. However, while protection is their underlying goal, they can also be aesthetic. Most industrial coatings are used for corrosion and wear control of steel and nonferrous alloys, but they lie under a complex market umbrella and work in numerous areas.

In addition, the continuous creation of new alloys and posttreatment coatings continue to enhance material surface characteristics to higher levels of performance. As it was presented in the previous chapters of this book, formulations impart definable finishes, roughness, hardness, and porosity. These coatings provide barrier protection towards aggressive substances and materials. The adhesion of the coatings to the substrate is also a critical point for enhanced performance during application.

Electroless deposition has been extensively studied for many industrial applications. Low-cost materials such as aluminum and cast iron can be upgraded by applying 2–15 μm of nickel coating, which increases the surface hardness and corrosion resistance of these base metals and extends their useful life.

This chapter is divided into three parts: nickel-boron (NiB) traditional and modern applications, nickel-phosphorus (NiP) traditional and modern applications, and a more practical section with some recommendations for the application of electroless nickel coatings.

9.2 NiB TRADITIONAL AND MODERN APPLICATIONS

Electroless NiB coatings are not popular as electroless NiP coatings, due to the higher cost of that alloy. However, the properties of electroless NiB coatings are quite impressive and in some aspects superior to NiP coatings. The hardness of NiB coatings is higher than NiP wear resistance that is better than hard chromium coatings (Shakoor, Kahraman, Gao, & Wang, 2016). In addition, NiB coatings have their typical cauliflower-like surface morphology, which has the ability to retain lubricant and improve the wear properties. Furthermore, structural changes upon heat treatment process enable them to improve further their hardness and wear characteristics. Due to all the characteristics and properties well described in this book, electroless NiB coating is currently used in many industries such as electronics, petroleum, petrochemical, firearms, chemical, plastics injection, aerospace, and vehicle industries. Recently, the catalytic properties of NiB coatings have been studied, generating a new range of industries to be explored.

9.2.1 Automobile Industry

The applications of NiB coatings in the vehicle industry are commonly related to the high wear properties of this material, and with the possibility of coating polymers, the vehicle-use generator pulley is one of the examples.

It is frequent that the vehicle-use generator is provided with a metal pulley that is belt-driven by the engine (Figure 9.1d). However, in view of the reduction of manufacturing cost and vehicle weight, a resin pulley can be used instead of the metal one. However, the pulley is still required to have a high mechanical strength and a high wear resistance. To meet these requirements, glass-reinforcing resin pulley covered with NiB coating can be used (Yuji Ito, 2017).

9.2.2 Machinery

In the case of machinery industry, here again, the attractiveness of NiB coatings relies on their advantageous mechanical and wear properties.

The use of NiB coatings on the inner wall of air cylinder sleeve was described ("Method for forming nickel-boron coating on the inner wall of air cylinder sleeve and air cylinder sleeve comprising nickel-boron coating," 2014) (Figure 9.1h). The application of electroless NiB coatings in this case is particularly interesting due to the high hardness of this coating in the as-plated state (800 HV_{100}). Consequently,

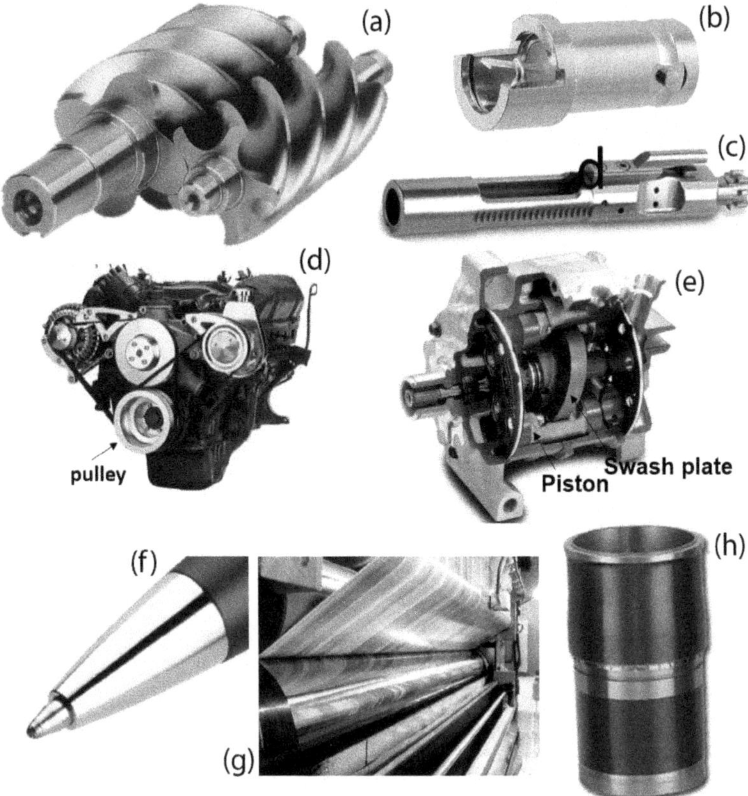

FIGURE 9.1 Industrial patented applications of electroless NiB: (a) rotors of the screw type, (b) electrical contact, (c) firearms, (d) metal pulley, (e) sliding member for compressor, (f) ballpoint pen, (g) electrostatographic printing, and (h) air cylinder sleeve.

a high-temperature preparation process can be avoided for parts with high dimensional precision requirements and the degree of deformation of the parts is reduced.

Another example in the machinery industry is the treatment of the surfaces of rotors of the screw-type rotary machine (Enzymes, 1973). In general, a screw-type rotary fluid machine has a male rotor and a female rotor, which are accommodated in a casing and both rotate in meshing engagement with each other (Figure 9.1a). In this case, NiB coatings are used due to the coating uniformity, low friction, and wear resistance.

Rolling bearing manufacturers are also using NiB coatings. In this case, the coatings are used to enhance the rigidity of a rolling bearing cage, generating excellent durability even in high-speed high-load environment. The rolling bearing cage has, on its outside diameter face, an electroless NiB film (Kirschbaum, H. 2017).

In the case of aluminum sliding member for compressor (Figure 9.1e), a ternary coating is used (Sugiura, 2003). An electroless nickel plating layer is formed on the surface of the base body that contains phosphorous and boron. In this case, the NiB coating is used due to its outstanding hardness, excellent sliding characteristics, and high degree of resistance to peeling and corrosion.

9.2.3 Electronics Industry

In the electronics industry, the use of electroless nickel coatings aims to minimize the use of gold or other noble metal. In addition, electrical contact materials (Figure 9.1b) have critical tolerances and must be of especially high reliability and consistency. The use of electroless NiB connectors is described as over-plating. An electroless nickel barrier layer is plated between the conductive substrate layer, usually copper, and the noble or precious metal layer, usually gold; this layer provides electrical contacts on conductors that exhibit superior contact resistance properties and exceptional corrosion resistance with reduced gold use (Baudrand, 1983).

In addition, the electronics industry has been using electroless NiB plating to form a conductive and protective film for protecting the surface of semiconductor and nonconducting substrates (Ezawa et al., 2000).

9.2.4 Miscellaneous Applications

A large range of blades can be coated with NiB coating: blades for cutting, slicing, sawing, or slitting various materials (McComas, 2009). The advantages, again, are the high hardness and wear resistance, but principally the property of maintaining the hardness at high temperatures.

Usually, printing rolls and press beds are plated with 2–5 μm of electroless nickel. These components are usually made of steel or cast iron that needs to be protected from the corrosive inks and continuous wear. In the case of corona generating device of electrostatographic printing machines (Figure 9.1g), the control screen is coated with a continuous layer of electroless NiB (Gross, 1993).

In everyday life, ballpoint pen tips (Figure 9.1f) of special pens are coated with NiB. The objective in this case is to improve wear resistance and lubrication (Tsuruo, 1990).

Finally, NiB coatings have also been used in firearms (Figure 9.1c) to enhance wear resistance and lubrication properties (McComas, 2004). The columnar structure with nodules at the surface presents impressive results for wear and lubrication at high temperatures. Standard firearms with wet lubricants need to be cleaned and greased on a regular basis after firing. If not regularly greased, the firearm might jam or misfire. Phosphate-reduced electroless coatings have also been used on metal substrates for firearms. However, the best results are reached with boron-reduced coatings.

9.2.5 NiB Amorphous Nanoparticles Catalyst for Selective Hydrogenation

Recently, environment-friendly hydrogen generation by $NaBH_4$ hydrolysis using improved method has attracted much attention (Liang, Li, Li, Zhao, & Xia, 2017). The borohydride fuel cell is basically composed of an anode, a cathode, and a cation exchange membrane (CEM). On the anode, borohydride will be oxidized to generate electrons, water, and borate. On the cathode, hydrogen peroxide combines with protons and electrons to form water, and CEM can conduct cationic Na^+ from the anode to the cathode (Yin, Wang, Duan, Liu, & Wang, 2019). In recent years, amorphous alloy catalysts have received research attention. The special features of amorphous alloys, long-range disordered and short-range ordered structure, show unique properties: especially good catalytic activity, because the dislocations and the presence of more unsaturated atoms in the amorphous structure increase borohydride adsorption.

Electroless nickel with the introduction of boron represents a catalyst with reduced cost (Singh & Das, 2017). Therefore, many efforts have recently been devoted to NiB amorphous materials for fuel cells. In addition, the size control, distribution, and morphology of an amorphous alloy are of vital importance in improving their catalytic performance (Feng et al., 2018).

Yin et al. (2019) have produced amorphous NiB and NiB-Cu catalyst particles. The catalytic properties were studied using alkali solutions. The electro-oxidation of sodium borohydride suggested that the NiB-Cu catalyst had a higher anodic current density than the NiB catalyst.

Liang et al. (2017) have developed a novel catalyst $NiB/NiFe_2O_4$, which was utilized in hydrogen production from $NaBH_4$ hydrolysis. A hydrogen generation rate as high as 299.88 mL min^{-1} g^{-1} for $NiB/NiFe_2O_4$ catalyst can be achieved using 5 wt% $NaBH_4$ solution at 298 K.

With the addition of PEG800, Feng et al. (2018) were able to produce NiB particles as small as 4 nm, which are free of agglomeration. Upon the generation of a catalytic system (good stability in catalytic hydrogenation), the authors reported superior catalytic behaviour in comparison to all unsupported NiB amorphous alloy.

In addition to the use of NiB as catalyst of $NaBH_4$, this catalyst has also been studied in other hydrogenation-dehydrogenation functions, especially, in selective hydrogenation reactions, such as the hydrogenation of furfural, and benzene (Chen et al., 2016).

Chen et al. (2016) showed that NiB particles can act as a catalyst on the isomerization of n-hexane. A unique structure of B-Ni-H was discovered in catalyst, which could provide good hydrogenation-dehydrogenation sites. Meanwhile, the catalyst showed good stability and the isomerization activity remained stable for at least 40 h.

The hydrogenation of phenol and its derivatives was also studied in the presence of NiB catalyst. PVP-NiB catalyst for this reaction was demonstrated by the selective hydrogenation of other phenol derivatives, which showed that the PVP-NiB catalyst was selective for the formation of cyclohexanol. The hydrogenation of phenol is of commercial and environmental significance because the cyclohexanone and cyclohexanol produced are the intermediates for caprolactam and adipic acid used to manufacture nylon 6 and nylon 66, respectively.

Sulfolene to sulfolane hydrogenation in the presence of NiB catalyst was reported by Wang et al. (2004). Hydrogenation of sulfolene to sulfolane is an important industrial process, typically catalyzed by Raney Ni, because sulfolane is widely used as an excellent solvent to dissolve many kinds of organic compounds and polymers. Their studies revealed that the unsupported NiB alloys showed high catalytic activity during the hydrogenation of sulfolene. Furthermore, the NiB/TiO_2 was an excellent catalyst in the hydrogenation of sulfolene with high industrial interest.

A high-active supported nickel catalyst of NiB/SiO_2 was reported by Chiang et al. (2007). The NiB/SiO_2 catalysts had an ultrafine and amorphous structure, which were much more active than NiB and Ni/SiO_2. The optimal catalyst of NiB/SiO_2 was used to hydrogenate citral to citronellal and citronellol, which was approximately 14 times as active as NiB, but less selective than NiB. A high yield of citronellal/citronellol of approximately 98% over 5% NiB/SiO_2 was obtained.

PVP-stabilized NiB catalysts were prepared by Liaw et al. (2005). The PVP-NiB catalysts were examined for their catalysis on the hydrogenation of furfural, crotonaldehyde, and citral. On catalysis, the PVP-NiB catalyst was significantly more active than NiB for hydrogenating furfural to furfuryl alcohol and crotonaldehyde to butyraldehyde. A good yield of citronellal, approximately 90%, could be obtained by reducing citral in cyclohexane over the PVP-NiB catalyst.

The effect of nanoparticles of NiB on hydrogen desorption properties of MgH_2 was investigated by Liu et al. (2012). MgH_2-10 wt% NiB mixture started to release hydrogen at 180°C, whereas it had to heat up to 300°C to release hydrogen for the pure MgH_2. In addition, a hydrogen desorption capacity of 6.0 wt% was reached within 10 min at 300°C for the MgH_2-10 wt% NiB mixture; in contrast, just 2.0 wt% hydrogen was desorbed for pure MgH_2 after 120 min only under the same conditions.

9.3 NiP TRADITIONAL AND MODERN APPLICATIONS

There are many applications for electroless NiP plating in industry, because it provides high corrosion resistance along with good mechanical properties. That is why this coating method is widely used, for example, for hydraulic cylinders, automotive carburetors, valves, petrochemical tubes, and so on (Colaruotolo & Tramontana, 1990; Duffek et al., 1990).

Electroless deposition has been extensively studied for many industrial applications. The most investigated systems by far include nickel, copper, silver, gold, and their alloys. There are many applications for electroless nickel plating in industry, because it provides some very accurate features compared to other surface finishing and plating methods. That is why this coating method is widely used, for example, for hydraulic cylinders, automotive carburetors, valves, electronic conductors, transistor and diode package bases, and so on.

9.3.1 Chemical and Plastics Industry

Electroless NiP is largely used in the chemical process industry. Principally, due to the high corrosion resistance of this coatings, applications of electroless composite in chemical and plastic industries involve coatings for autoclaves, filters, pipe works, heat exchangers, valves (Figure 9.2b), pump components, and so on. When used in conjunction with NiP coatings, innovative and cheap substrate materials have found a place, as alternatives to some high cost or not reliable traditional materials. Concerning applications in the chemical process industry, electroless nickel competes with fiberglass reinforced plastics, mild steel, S316 stainless steel, N02200, glass-lined steel, and Teflon-lined steel (Parkinson, 1997).

For example, the use of stainless steel costs roughly twice that of NiP. The applications of 1 μm thick NiP on mild steel butterfly and ball valves can double the life of

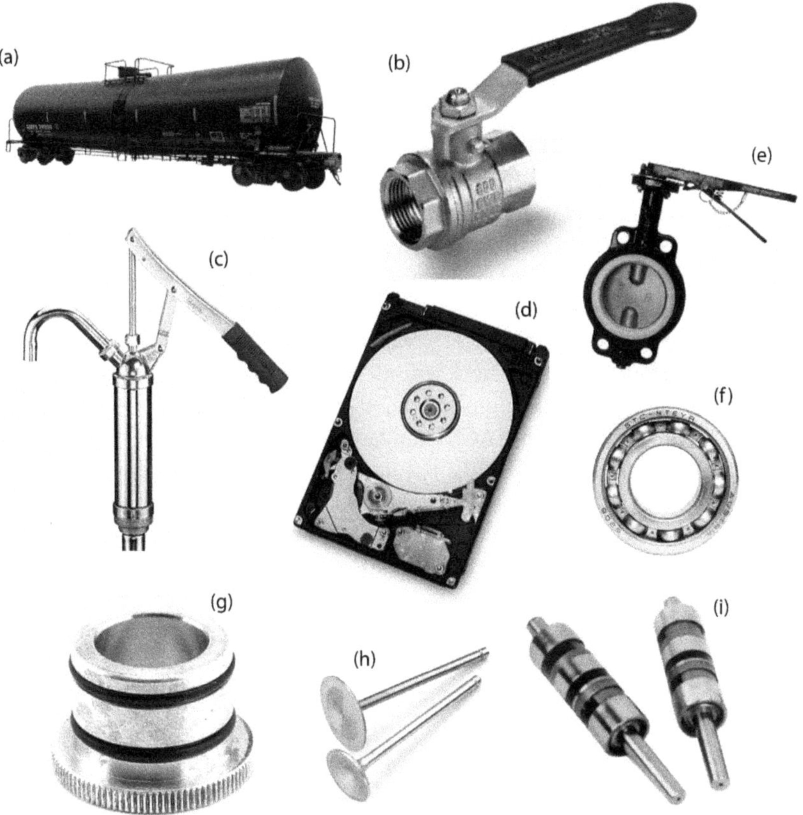

FIGURE 9.2 Miscellaneous applications of electroless NiP: (a) tank car for hydroxyl transport, (b) brass NiP-coated ball valve, (c) pump barrel, (d) rigid memory disc, (e) butterfly valves coated with NiP-PTFE, (f) bearing, (g) carburettor restrictor, (h) engine valves, (i) rotor bearing from textile machine coated with NiP-diamond.

such valves in corrosive environments. The use of 3 μm coating of high-phosphorus electroless nickel increased the service life of the pumps up to 30 times.

In addition to the attractive cost, chemical process industry needs to maintain product purity, preventing corrosion/erosion problems that may affect the environment and service life of equipment.

Electroless nickel has been used extensively in the chloralkali industry, in which the two main products are chlorine and sodium hydroxide. In one example, six diaphragms in a compressor required repairs annually and the total life cycle of the diaphragms was approximately 8 years. Replacement cost was approximately $200,000 but coating with high-phosphorus electroless nickel every 4 years cost only $15,000 (Parkinson, 1997). Other components on rail cars used for transporting sodium hydroxide (Figure 9.2a) have provided excellent service after coating with electroless nickel.

9.3.2 Oil and Gas Industry

Oil and gas production units have extensively used electroless nickel coatings. The coatings are widely used to protect downhole applications such as tubulars, sucker rods, pump barrels/plungers (Figure 9.2c), electric submersible pump (ESP) components, progressive cavity rotary pump components, completion tools, packer assemblies, and components of hydraulic fracturing systems (Kay & Chen, 2006).

The NiP's success in the oil and gas industry can easily be understood; it (high phosphorus) has excellent corrosion resistance when exposed to carbon dioxide (CO_2) and hydrogen sulfide (H_2S)-containing brines. It has also been observed that high-phosphorus electroless nickel deposits develop a passive layer in sour service oil and gas environments containing HS, which effectively inhibits corrosion of the coating. Thus, high-phosphorus electroless nickel deposits have been used successfully in many oil and gas applications involving corrosion caused by CO_2 and H_2S in brines (e.g., as recommended in NACE Standard MR0176-2006).

NiP-polytetrafluoroethylene (PTFE) has been recently found to provide nonstick, non-galling, high dry-lubricity, low friction, good wear, and corrosion-resistant surfaces. Although these composite coatings also provide wear-resistance benefits, they are considered in a separate category based on the unique characteristics they embody: dry lubrication, improved release properties, and repellence of contaminants, such as water and oil. They are applied to pumps and valves butterfly valves (Figure 9.2e) in the oil and gas industry (Mallory & Hajdu, 1991). It cuts the leak rate and permits safe operation of the valve for cryogenic applications (Feldstein, 2019).

9.3.3 Aerospace and Automobile Industry

Aerospace industries have extensively used electroless nickel over the years. Aluminum alloys are mostly used in airplanes because of their low density. Electroless nickel deposition complements aluminum, adding wear resistance, hardness, corrosion protection, and solderability. Only thin layers of 3–10 μm are applied in aircraft engines, turbine, or compressor blades exposed to corrosive environments, due to the weight control. The layer of nickel generated ensures the necessary protection. In addition, NiP easily provides a uniform coating over the various shapes of waveguides. Aluminum radar waveguides are plated with 2 μm of electroless nickel to protect them against aggressive conditions present in marine environments (Parkinson, 1997).

Furthermore, the automotive industry has also been attracted by the high properties and low cost of NiP coatings (Baudrand & Durkin, 1998). The performance of steel and aluminum components has been enhanced by the use of NiP over the last 50 years. The typical applications are: on heat sinks, engine bearings, pistons, gear assemblies, carburetor parts, shock absorbers, and exhaust system components.

The use of new fuel mixtures in the automotive industry (such as the use of alcohol in Brazil) creates corrosion problems in the fuel systems based in zinc-die-cast-components. That is the reason why the automotive industry has been using electroless nickel on fuel systems (Parkinson, 1997). Nickel plating is also used when the parts have a tight tolerance such as fuel injectors, due to the uniformity of the plating.

FIGURE 9.3 NiP-coated connection responsible by brace the ABS covers in light trucks.

Figure 9.3 shows the NiP-coated connection responsible for bracing the ABS covers in light trucks. NiP itself is an ideal coating for this application that has intensive corrosion requirements due to the external use and tribological requirements due to repetitive connection process.

In recent times, composite NiB coatings have attracted the attention of the automobile industry. A large range of composites have been used successfully in many applications, involving various types of wear, of which electroless NiP with PTFE and electroless NiP with fluorinated carbon (CFX) are the most used composites. Both provide an excellent nonstick, low friction, dry lubricant surface. The advantage of NiP-PTFE is its cost; however, NiP-CFX has the advantage of superior temperature resistance. These composites have been used on carburetor and clutch parts, engine valves (Figure 9.2h), gears, and bearings.

NiP-SiC has also found a place in the automobile industry, particularly due to the ability of electroless coatings to coat plastics. With approximately 50 μm thick, NiP-SiC increases the life of plastics and rubber by 15 times. It can be used in automobile components such as reinforced plastic front-end pieces (Balaraju et al., 2003).

9.3.4 Energy Industry

The use of composite NiP coatings for the wind energy industry has been recently suggested by Surface Technology Inc. (Feldstein, 2019). Composite NiP coatings can be advantageous in different aspects in the wind industry. First, the parts require less wear, lower friction, and heat transfer. Second, the use of a reliable coating facilitates the use of new substrate materials, such as titanium, aluminum, low-cost steel alloys, ceramics, and plastics. These coatings allow high productivity due to greater speeds and less wear and can replace environmentally problematic coatings, such as electroplated chromium.

Coatings designed for increased wear resistance have proven to be the most widely used composite coatings in the wind industry to date. Different particles have been tested, such as diamond, silicon carbide, aluminum oxide, tungsten carbide, and boron carbide. However, the unsurpassed hardness and heat transfer of diamond have made this composite the most common.

The use of under-layer coating of high phosphorus is often used to prevent corrosion of different parts as wind-energy equipment are normally exposed to corrosive environments. This provides a barrier layer for corrosion and the outward functional layer will still be the composite coating for better wear performance.

9.3.5 Miscellaneous Applications

Electroless nickel coatings can be applied to nonconductors and workpieces with complicated shapes that present a large range of applications.

Usually, printing rolls and press beds are plated with 2–5 μm of electroless NiP. Steel or cast iron is generally the material of these components, needing to be protected from the corrosive inks and continuous wear.

In the machinery construction, components usually coated with electroless nickel are lathe beds, shafts, levers, calipers, and fasteners.

Wear in the textile industry is another example of problem solved by electroless nickel. The textile industry has hundreds of miles of thread passing through, around, and over components (Figure 9.2h). Steel thread feeds and guides are plated with 5μm electroless nickel to wear protection.

Finally, electroless NiP is an alloy currently used in the precision optics industry as a polishable surface layer on difficult-to-finish mirror substrate materials, such as beryllium. The applications with high performance of these reflective components can be found not only in commercial laboratory instruments but also in large space telescopes and cryogenically cooled optical sensor systems (Hibbard, 1997).

9.4 RECOMMENDATIONS FOR APPLICATION OF ELECTROLESS NICKEL COATINGS

If you are reading this book and impressed with all the impressive properties and applications of the electroless nickel coatings, some practical recommendations might be useful prior to starting your own electroless deposits:

- The applications of electroless coatings are seemingly endless; thus, first, we must understand the reason for using them. The primary focus for applying coatings is to preserve the item that lies beneath it in some way.
- The choice of the correct composition for the properties you are looking for is important. In the case of NiB coating, two different reducing agents (sodium borohydride and dimethylamine borane) can be used. It is noteworthy to mention that they are prepared at different pH and temperature and generated coating with variation in the B content, thus completely impacting the properties. In addition, stabilizers are normally co-deposited with NiB coatings that also influence the properties. In the case of NiP, the same reducing agent is used in all the cases; however, the different concentrations

generate three distinct coatings: low-phosphorous, mid-phosphorous, and high-phosphorous. Whenever corrosion is the main criterion, the high-phosphorus coating is the best candidate, and if the main criterion is hardness and wear resistance, NiB reduced by sodium borohydride is the best candidate.

- Upon choosing the most adequate composition for your coating, care should be taken regarding all bath parameters, namely agitation, bath load, and temperature. For instance, Vitry & Bonin (2017) demonstrated that the increase of 1.5°C bath temperature in a NiB bath changes the B content in the coating and consequently its properties.
- Furthermore, prior to plating, the substrate surface needs to be completely clean to achieve the desired roughness, as electroless Ni coatings will follow the morphology of the substrates. If the surface is too smooth, it is not a major concern; it is however recommended to sandblast the substrate with fine glass beads. This procedure not only helps to remove any contamination from the surface of the substrate, but also results in a better adhesion due to the increase in contact area between coating and substrate.
- Considering the hanging of samples, different techniques can be used for this purpose: individual hangers, hooks, baskets, or hanger bars, as shown in Figure 9.4.
- As-plated coatings already present impressive properties. However, if you need to improve the properties with a heat treatment, specific temperature and time should be chosen for each degree of crystallization desired. Longer treatment time or too high temperature might induce some inverse effects due to the grain growth.
- Electroless nickel can also be used to protect welded parts, as can be seen in Figure 9.5. Electroless Ni act as a barrier in the affected zones that are more

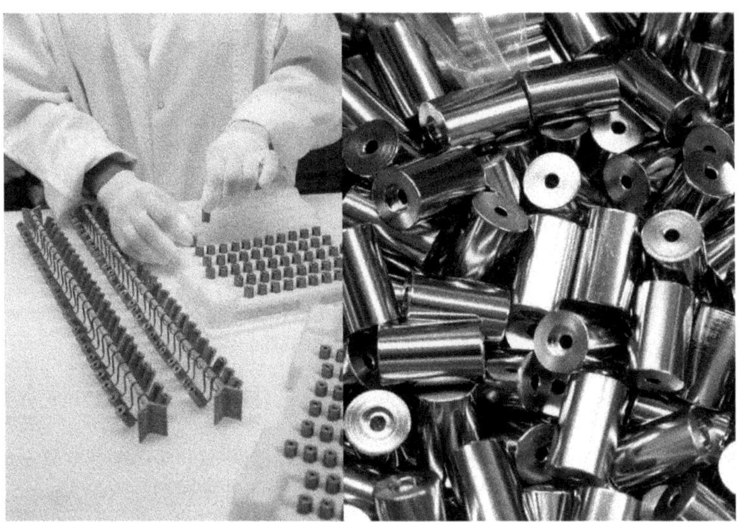

FIGURE 9.4 Hanger bars used for uniform plating.

FIGURE 9.5 Welded parts protected by electroless nickel.

susceptible to corrosion and wear. The application of electroless nickel is indicated in that case due to its ability of coating complicated shapes and small holes.

- On the other hand, electroless nickel does not perform well in highly acidic environments. For that reason, application in areas where hydrolysis of chemical compounds produce acids is not recommended.
- Finally, the importance of the adhesion of the deposit to the substrate and the continuity of the coating without porosity cannot be overlooked: surface defaults will definitely difficult the improvement of the envisaged properties.

REFERENCES

Balaraju, J. N., Sankara Narayanan, T. S. N., & Seshadri, S. K. (2003). Electroless Ni-P composite coatings. *Journal of Applied Electrochemistry*, *33*(9), 807–816. doi:10.1023/A:1025572410205.

Baudrand, D., & Durkin, B. (1998). Automotive applications of electroless nickel. *Metal Finishing*, *96*(May), 20–23. doi:10.1016/S0026-0576(98)80080-9.

Baudrand, D. W. (1983). *Electrical contact materials*. EP 0084 937A1. European Patent Office.

Chen, J., Cai, T., Jing, X., Zhu, L., Zhou, Y., Xiang, Y., & Xia, D. (2016). Surface chemistry and catalytic performance of amorphous NiB/Hβ catalyst for n-hexane isomerization. *Applied Surface Science*, *390*, 157–166. doi:10.1016/j.apsusc.2016.08.030.

Chiang, S. J., Yang, C. H., Chen, Y. Z., & Liaw, B. J. (2007). High-active nickel catalyst of NiB/SiO_2 for citral hydrogenation at low temperature. *Applied Catalysis A: General*, *326*(2), 180–188. doi:10.1016/j.apcata.2007.04.019.

Colaruotolo, J., & Tramontana, D. (1990). Engineering applications of electroless nickel. *Electroless Plating*, 207–226. Retrieved from www.knowel.com.

Duffek, E. F., Baudrand, D. W., & Donaldson, J. G. (1990). Electroless nickel applications in electronics. *Electroless Plating: Fundamentals and Applications*, 229–259.

Enzymes, P. (1973). Method of treating surfaces of rotors of the screw type rotary machine. Japan.

Ezawa, H. et al. (2000). Electroless Ni-B plating liquid, electronic device and method for manufacturing the same. EP1211334A2 European Patent Office, Germany.

Feng, W., Ma, Y., Niu, L., Zhang, H., & Bai, G. (2018). Confined preparation of ultrafine NiB amorphous alloys for hydrogenation. *Catalysis Communications*, *109*(November 2017), 20–23. doi:10.1016/j.catcom.2018.02.015.

Feldstein, M. D. (2019). Composite EN coatings for the wind energy industry varieties. Products Finishing. www.pfonline.com.

Gross, R. A. (1993). Corona Generating Device. US005257073 A. United States Patent.

Hibbard, D. L. (1997). Advanced materials for optics and precision structures: A critical review. *Optical Science, Engineering and Instrumentation*. doi:10.1117/12.279806.

Inc., M. E. (n.d.). WO2017213866A1. US005257073 A. United States Patent.

Kay, M. A., & Chen, Z.-Y. (2006). US 2006/0222585 A1. United States. doi:10.1037/t24245-000.

Kirschbaum, H., Sontgerath, K. Schafer, S. (2017). Use of water soluble lantathanide coupounds as stabilizer in electrolyres for electroless metal deposition. WO2017213866A1. United States Patent.

Liang, Z., Li, Q., Li, F., Zhao, S., & Xia, X. (2017). Hydrogen generation from hydrolysis of $NaBH_4$ based on high stable $NiB/NiFe_2O_4$ catalyst. *International Journal of Hydrogen Energy*, *42*(7), 3971–3980. doi:10.1016/j.ijhydene.2016.10.115.

Liaw, B. J., Chiang, S. J., Tsai, C. H., & Chen, Y. Z. (2005). Preparation and catalysis of polymer-stabilized NiB catalysts on hydrogenation of carbonyl and olefinic groups. *Applied Catalysis A: General*, *284*(1–2), 239–246. doi:10.1016/j.apcata.2005.02.002.

Liu, G., Qiu, F., Li, J., Wang, Y., Li, L., Yan, C., ... Yuan, H. (2012). NiB nanoparticles: A new nickel-based catalyst for hydrogen storage properties of MgH_2. *International Journal of Hydrogen Energy*, *37*(22), 17111–17117. doi:10.1016/j.ijhydene.2012.07.106.

Mallory, G. O., & Hajdu, J. B. (1991). *Electroless Plating: Fundamentals and Applications*. American Electroplaters and Surface Finishing Society, Orlando, FL.

McComas, E. (2004). Nodular nickel boron coating field. US20040111947A1. United States Patent.

McComas, E. (2009). Blade coated with a nickel boron metal coating. US2009151525A1. United States Patent.

Parkinson, R. (1997). Properties and applications of electroless nickel. *Nickel Development Institute Publication*, 33. Retrieved from http://www.nickelinstitute.org/Technical Literature/Technical Series/PropertiesandApplicationsofElectrolessNickel_10081_.aspx.

Shakoor, R. A., Kahraman, R., Gao, W., & Wang, Y. (2016). Synthesis, characterization and applications of electroless Ni-B coatings-A review. *International Journal of Electrochemical Science*, *11*, 2486–2512.

Singh, P. K., & Das, T. (2017). Generation of hydrogen from $NaBH_4$ solution using metal-boride (CoB, FeB, NiB) catalysts. *International Journal of Hydrogen Energy*, *42*(49), 29360–29369. doi:10.1016/j.ijhydene.2017.10.030.

Sugiura, M. (2003). Sliding member for compressor. US2003096134A1. United States.

Tsuruo, N. (1990). Ball point pen tip. JPH02175195A Japan.

Vitry, V., & Bonin, L. (2017). Increase of boron content in electroless nickel-boron coating by modi fi cation of plating conditions. *Surface & Coatings Technology*, *311*, 164–171. doi:10.1016/j.surfcoat.2017.01.009.

Wang, L., Li, W., Zhang, M., & Tao, K. (2004). The interactions between the NiB amorphous alloy and TiO_2 support in the NiB/TiO_2 amorphous catalysts. *Applied Catalysis A: General*, *259*(2), 185–190. doi:10.1016/j.apcata.2003.09.037.

Wang D. (2014). Method for forming nickel-boron coating on inner wall of air cylinder sleeve and air cylindersleeve comprising nickel-boron coating. CN104152876A. China.

Yin, X., Wang, Q., Duan, D., Liu, S., & Wang, Y. (2019). Amorphous NiB alloy decorated by Cu as the anode catalyst for a direct borohydride fuel cell. *International Journal of Hydrogen Energy*, *44*(21), 10971–10981. doi:10.1016/j.ijhydene.2019.02.150.

Yuji Ito, H. (2017). VEHICLE-USE GENERATOR (75)(12) Patent Application Publication (10) Pub. No.: US 2017/0139001 A1. Japan.

Index